Matrix Algebra and Its Applications to Statistics and Econometrics

Matrix Algebra and Its Applications to Statistics and Econometrics

C. Radhakrishna Rao
Pennsylvania State University, USA

M. Bhaskara Rao
North Dakota State University, USA

World Scientific

NEW JERSEY • LONDON • SINGAPORE • BEIJING • SHANGHAI • HONG KONG • TAIPEI • CHENNAI

Published by

World Scientific Publishing Co. Pte. Ltd.

5 Toh Tuck Link, Singapore 596224

USA office: 27 Warren Street, Suite 401-402, Hackensack, NJ 07601

UK office: 57 Shelton Street, Covent Garden, London WC2H 9HE

Library of Congress Cataloging-in-Publication Data

Rao, C. Radhakrishna (Calyampudi Radhakrishna), 1920–

Matrix algebra and its applications to statistics and econometrics / C. Radhakrishna Rao and M. Bhaskara Rao.

p. cm.

Includes bibliographical references and index.

ISBN-13 978-981-02-3268-9 -- ISBN-10 981-02-3268-3 (alk. paper)

1. Matrices. 2. Statistics. 3. Econometrics. I. Bhaskara Rao, M.

QA188.R36 1998

512.9'434--dc21 98-5596

CIP

British Library Cataloguing-in-Publication Data

A catalogue record for this book is available from the British Library.

First published 1998

Reprinted 2001, 2004, 2007

Printed in Singapore by B & JO Enterprise

To our wives

BHARGAVI (Mrs. C.R. Rao)

JAYASRI (Mrs. M.B. Rao)

PREFACE

Matrix algebra and matrix computations have become essential prerequisites for study and research in many branches of science and technology. It is also of interest to know that statistical applications motivated new lines of research in matrix algebra, some examples of which are generalized inverse of matrices, matrix approximations, generalizations of Chebychev and Kantorovich type inequalities, stochastic matrices, generalized projectors, Petrie matrices and limits of eigenvalues of random matrices. The impact of linear algebra on statistics and econometrics has been so substantial, in fact, that a number of books devoted entirely to matrix algebra oriented towards applications in these two subjects are now available. It has also become a common practice to devote one chapter or a large appendix on matrix calculus in books on mathematical statistics and econometrics.

Although there is a large number of books devoted to matrix algebra and matrix computations, most of them are somewhat specialized in character. Some of them deal with purely mathematical aspects and do not give any applications. Others discuss applications using limited matrix theory. We have attempted to bridge the gap between the two types. We provide a rigorous treatment of matrix theory and discuss a variety of applications especially in statistics and econometrics. The book is aimed at different categories of readers: graduate students in mathematics who wish to study matrix calculus and get acquainted with applications in other disciplines, graduate students in statistics, psychology, economics and engineering who wish to concentrate on applications, and to research workers who wish to know the current developments in matrix theory for possible applications in other areas.

This book provides a self-contained, updated and unified treatment of the theory and applications of matrix methods in statistics and econometrics. All the standard results and the current developments, such as the generalized inverse of matrices, matrix approximations, matrix differential calculus and matrix decompositions, are brought together to produce a most comprehensive treatise to serve both as a text in graduate courses and a reference volume for research students and consultants.

It has a large number of examples from different applied areas and numerous results as complements to illustrate the ubiquity of matrix algebra in scientific and technological investigations.

It has 16 chapters with the following contents. Chapter 1 introduces the concept of vector spaces in a very general setup. All the mathematical ideas involved are explained and numerous examples are given. Of special interest is the construction of orthogonal latin squares using concepts of vector spaces. Chapter 2 specializes to unitary and Euclidean spaces, which are vector spaces in which distances and angles between vectors are defined. They play a special role in applications.

Chapter 3 discusses linear transformations and matrices. The notion of a transformation from one vector space to another is introduced and the operational role of matrices for this purpose is explained. Thus matrices are introduced in a natural way and the relationship between transformations and matrices is emphasized throughout the rest of the book. Chapters 4, 5, 6 and 7 cover all aspects of matrix calculus. Special mention may be made of theorems on rank of matrices, factorization of matrices, eigenvalues and eigenvectors, matrix derivatives and projection operators. Chapter 8 is devoted to generalized inverse of matrices, a new area in matrix algebra which has been found to be a valuable tool in developing a unified theory of linear models in statistics and econometrics. Chapters 9, 10 and 11 discuss special topics in matrix theory which are useful in solving optimization problems. Of special interest are inequalities on singular values of matrices and norms of matrices which have applications in almost all areas of science and technology. Chapters 12 and 13 are devoted to the use of matrix methods in the estimation of parameters in univariate and multivariate linear models.

Concepts of quadratic subspaces and new strategies of solving linear equations are introduced to provide a unified theory and computational techniques for the estimation of parameters. Some modern developments in regression theory such as total least squares, estimation of parameters in mixed linear models and minimum norm quadratic estimation are discussed in detail using matrix methods. Chapter 14 deals with inequalities which are useful in solving problems in statistics and econometrics. Chapter 15 is devoted to non-negative matrices and Perron-Frobenius theorem which are essential for the study of and research in econometrics, game theory, decision theory and genetics. Some miscellaneous results not covered in the main themes of previous chapters are put together in Chapter 16.

It is a pleasure to thank Marina Tempelman for her patience in typing numerous revisions of the book.

March 1998

C.R. Rao
M.B. Rao

NOTATION

The following symbols are used throughout the text to indicate certain elements and the operations based on them.

Scalars

$\mathbf{R}$	real numbers
$\mathbf{C}$	complex numbers
$\mathbf{F}$	general field of elements
$x = x_1 + ix_2$	a complex number
$\bar{x} = x_1 - ix_2$	conjugate of x
$\|x\| = (x_1^2 + x_2^2)^{1/2}$	modulus of x

General

$\{a_n\}$	a sequence of elements
$\mathbf{A}, \mathbf{B}, \ldots$	sets of elements
$\mathbf{A} \subset \mathbf{B}$	set $\mathbf{A}$ is contained in set $\mathbf{B}$
$x \in \mathbf{A}$	x is an element of set $\mathbf{A}$
$\mathbf{A} + \mathbf{B}$	$\{x_1 + x_2 : x_1 \in \mathbf{A},\ x_2 \in \mathbf{B}\}$
$\mathbf{A} \cup \mathbf{B}$	$\{x : x \in \mathbf{A}$ and/or $\mathbf{B}\}$
$\mathbf{A} \cap \mathbf{B}$	$\{x : x \in \mathbf{A}$ and $x \in \mathbf{B}\}$

Vector Spaces

$(\mathbf{V}, \mathbf{F})$	vector space over field $\mathbf{F}$
$\dim \mathbf{V}$	dimension of $\mathbf{V}$
$a_1, a_2, \ldots$	vectors in $\mathbf{V}$
$Sp(a_1, \ldots, a_k)$	the set $\{\alpha_1 a_1 + \ldots + \alpha_k a_k : \alpha_1, \ldots, \alpha_k \in \mathbf{C}\}$
$\mathbf{F}^n$	n dimensional coordinate (Euclidean) space
$\mathbf{R}^n$	same as $\mathbf{F}^n$ with $\mathbf{F} = \mathbf{R}$
$\mathbf{C}^n$	same as $\mathbf{F}^n$ with $\mathbf{F} = \mathbf{C}$

$\oplus$	direct sum, $\{x+y : x \in \mathbf{V},\ y \in \mathbf{W};\ \mathbf{V} \cap \mathbf{W} = 0\}$
$< \cdot, \cdot >$	inner product
$(\cdot, \cdot)$	semi inner product

Transformations

$T : \mathbf{V} \to \mathbf{W}$	transformation from space $\mathbf{V}$ to space $\mathbf{W}$
$R(T)$	the range of T, i.e., the set $\{Tx : x \in \mathbf{V}\}$
$K(T)$	the kernel of T, i.e., the set $\{Tx = 0 : x \in \mathbf{V}\}$
$\nu(T)$	nullity (dimension of $K(T)$)

Matrices

$A, B, C, \ldots$	general matrices or linear transformations
$\underset{m \times n}{A}$	$m \times n$ order matrix
$\mathbf{M}_{m,n}$	the class of matrices with m rows and n columns
$\mathbf{M}_{m,n}(\cdot)$	$m \times n$ order matrices with specified property $(\cdot)$
$\mathbf{M}_n$	the class of matrices with n rows and n columns
$A = [a_{ij}]$	a_{ij} is the (i, j) the entry of A (i-th row and j-th column)
$A \in \mathbf{M}_{m,n}$	A is a matrix with m rows and n columns
$Sp(A)$	the vector space spanned by the column vectors of A, also indicated by $R(A)$ considering A as transformation
$\bar{A} = [\bar{a}_{ij}]$	$\bar{a}_{ij}$ is complex conjugate of a_{ij}
A'	obtained from A interchanging rows and columns, i.e., if $A = (a_{ij})$ then $A' = (a_{ji})$
$A^* = \bar{A}'$	Conjugate transpose or transpose of $\bar{A}$ defined above
$A^* = A$	Hermitian or self adjoint
$A^*A = AA^* = I$	unitary
$A^*A = AA^*$	normal
$A^{\#}$	adjoint ($A \in \mathbf{M}_{m,n}$, $< Ax, z >_m = < x, A^{\#}z >_n$)
A^{-1}	inverse of $A \in \mathbf{M}_n$ such that $AA^{-1} = A^{-1}A = I$
A^-	generalized or g-inverse of $A \in \mathbf{M}_{m,n}$, $(AA^-A = A)$
A^+	Moore-Penrose inverse
A^-_{LMN}	Rao-Yanai ($\mathbf{LM}N$) inverse
I_n	identity matrix of order n with all diagonal elements as unities and the rest as zeros
I	identity matrix when the order is implicit

0	zero scalar, vector or matrix
$\rho(A)$	rank of matrix A
$\rho_s(A)$	spectral radius of A
vec A	vector of order mn formed by writing the columns of $A \in \mathbf{M}_{m,n}$ one below the other
$(a_1\|\cdots\|a_n)$	matrix partitioned by column vectors $a_1, \ldots, a_n$
$[A_1\|A_2]$	matrix partitioned by two matrices A_1 and A_2
tr A	trace A, the sum of diagonal elements of $A \in \mathbf{M}_n$
$\|A\|$ or det A	determinant of A
$A \cdot B$	Hadamard-Schur product
$A \otimes B$	Kronecker product
$A \odot B$	Khatri-Rao product
$A \circ B$	matrix with $< b_i, a_j >$ as the (j,i)-th entry, where $A = (a_1\|\ldots\|a_m),\ B = (b_1\|\ldots\|b_n)$
$\\|x\\|$	norm of vector x
$\\|x\\|_{se}$	semi-norm of vector x
$\\|A\\|$	norm or matrix norm of A
$\\|A\\|_F$	Frobenius norm of $A = ([\mathrm{tr}(A^*A)]^{1/2}$
$\\|A\\|_{in}$	induced matrix norm: max $\\|Ax\\|$ for $\\|x\\| = 1$
$\\|A\\|_s$	spectral norm of A
$\\|A\\|_{ui}$	unitarily invariant norm, $\\|U^*AV\\| = \\|A\\|$ for all unitary U and V, $A \in \mathbf{M}_{m,n}$
$\\|A\\|_{wui}$	weakly unitarily invariant norm, $\\|U^*AU\\| = \\|A\\|$ for all unitary U, $A \in \mathbf{M}_n$
$\\|A\\|_{MNi}$	M, N invariant norm
$m(A)$	matrix obtained from $A = (a_{ij})$ by replacing a_{ij} by $\|a_{ij}\|$, the modulus of the number $a_{ij} \in \mathbf{C}$
pd	positive definite matrix ($x^*Ax > 0$ for $x \neq 0$)
nnd	non-negative definite matrix ($x^*Ax \geq 0$)
s.v.d.	singular value decomposition
$B \leq_L A$	or simply $B \leq A$ to indicate (Löwner partial order) $A - B$ is nnd

$x \geq_e y$	$x_i \geq y_i,\ i = 1, \ldots, n$, where $x' = (x_1, \ldots, x_n)$ and $y' = (y_1, \ldots, y_n)$
$B \leq_e A$	entry wise inequality $b_{ij} \leq a_{ij},\ A = (a_{ij}),\ B = (b_{ij})$
$B \geq_e A$	entry wise inequality $b_{ij} \geq a_{ij}$
$A \geq_e 0$	non-negative matrix (all elements are non-negative)
$A >_e 0$	positive matrix (all elements are positive)
$y \ll x$	vector x majorizes vector y
$y \ll_w x$	vector x weakly majorizes vector y
$y \ll_s x$	vector x soft majorizes vector y
$\{\lambda_i(A)\}$	eigenvalues of $A \in \mathbf{M}_n,\ [\lambda_1(A) \geq \ldots \geq \lambda_n(A)]$
$\{\sigma_i(A)\}$	singular values of $A \in \mathbf{M}_{m,n},\ [\sigma_1(A) \geq \ldots \geq \sigma_r(A)]$, $r = \min\{m, n\}$

CONTENTS

CHAPTER 1

VECTOR SPACES

The use of matrix theory is now widespread in both physical and social sciences. The theory of vector spaces and transformations (of which matrices are a special case) have not, however, found a prominent place, although they are more fundamental and offer a better understanding of applied problems. The concept of a vector space is essential in the discussion of topics such as the theory of games, economic behavior, prediction in time series, and the modern treatment of univariate and multivariate statistical methods.

1.1. Rings and Fields

Before defining a vector space, we briefly recall the concepts of groups, rings and fields. Consider a set G of elements with one binary operation defined on them. We call this operation multiplication. If α and β are two elements of G, the binary operation gives an element of G denoted by $\alpha\beta$. The set G is called a group if the following hold:

(g_1) $\alpha(\beta\gamma) = (\alpha\beta)\gamma$ for every α, β and γ in G (associative law).

(g_2) The equations $\alpha y = \beta$ and $y\alpha = \beta$ have unique solutions for y for all α and β in G.

From these axioms, the following propositions (**P**) follow. (We use the symbol **P** for any property, proposition or theorem. The first two digits after **P** denote the section number.)

P 1.1.1 There exists a unique element, which we denote by 1 (the unit element of G), such that

$$\alpha 1 = \alpha \text{ and } 1\alpha = \alpha \text{ for every } \alpha \text{ in } G.$$

P 1.1.2 For every α in G, there exists a unique element, which we denote by α^{-1} (multiplicative inverse of α, or simply, the inverse of α)

such that

$$\alpha\alpha^{-1} = \alpha^{-1}\alpha = 1.$$

A group G is said to be commutative if $\alpha\beta = \beta\alpha$ for every α and β in G. If the group is commutative, it is customary to call the binary operation as addition and use the addition symbol $+$ for the binary operation on the elements of G. The unit element of G is called the zero element of G and is denoted by the symbol 0. The inverse of any element α in G is denoted by $-\alpha$. A commutative group is also called an abelian group.

A simple example of an abelian group is the set of all real numbers with the binary operation being the usual addition of real numbers. Another example of an abelian group is the set $G = (0, \infty)$, the set of all positive numbers, with the binary operation being the usual multiplication of real numbers. We will present more examples later.

A subgroup of a group G is any subset H of G with the properties that $\alpha\beta \in H$ whenever $\alpha, \beta \in H$. A subgroup is a group in its own right under the binary operation of G restricted to H. If H is a subgroup of a group G and $x \in G$, then $xH = \{xy : y \in H\}$ is called a left coset of H. If $x \in H$, then $xH = H$. If x_1H and x_2H are two left cosets, then either $x_1H = x_2H$ or $x_1H \cap x_2H = \emptyset$. A right coset Hx is also defined analogously. A subgroup H of a group G is said to be invariant if $xH = Hx$ for all $x \in G$. Let H be an invariant subgroup of a group G. Let $G|H$ be the collection of all distinct cosets of H. One can introduce multiplication between elements of $G|H$. If H_1 and H_2 are two cosets, define $H_1H_2 = \{\alpha\beta : \alpha \in H_1 \text{ and } \beta \in H_2\}$. Under this binary operation, $G|H$ is a group. Its unit element is H. The group $G|H$ is called the quotient group of G modulo H. It can also be shown that the union of all cosets is G. More concretely, the cosets of H form a partition of G.

There is a nice connection between finite groups and latin squares. Let us give a formal definition of a latin square.

DEFINITION 1.1.3. Let T be a set of n elements. A latin square $L = (t_{ij})$ of order n based on T is a square grid of n^2 elements $t_{ij}, 1 \leq i \leq n, 1 \leq j \leq n$ arranged in n rows and n columns such that

(1) $t_{ij} \in T$ for every i and j,
(2) each element of T appears exactly once in each row,
(3) each element of T appears exactly once in each column.

In a statistical context, T is usually the set of treatments which we wish to compare for their effects over a certain population of experimental units. We select n^2 experimental units arranged in n rows and n columns. The next crucial step is the allocation of treatments to experimental units. The latin square arrangement of treatments is one way of allocating the treatments to experimental units. This arrangement will enable us to compare the effects of any pair of treatments, rows, and columns.

Latin squares are quite common in parlor games. One of the problems is to arrange the kings (K), queens (Q), jacks (J) and aces (A) of a pack of cards in the form of a 4×4 grid so that each row and each column contains one from each rank and each suit. If we denote spades by S, hearts by H, diamonds by D and clubs by C, the following is one such arrangement.

$$\begin{matrix} SA & DK & HQ & CJ \\ CQ & HJ & DA & SK \\ DJ & SQ & CK & HA \\ HK & CA & SJ & DQ \end{matrix}$$

The above arrangement is a superimposition of two latin squares. The suits and ranks each form a latin square of order 4. We now spell out the connection between finite groups and latin squares.

P 1.1.4 Let G be any group with finitely many elements. Then the table of the group operation on the elements of G constitutes a latin square of order n on G.

PROOF. Assume that G has n elements. Let $G = \{\alpha_1, \alpha_2, \ldots, \alpha_n\}$. Assume, without loss of generality, that the group is commutative with the group operation denoted by $+$. Let us consider a square grid of size $n \times n$, where the rows and columns are each indexed by $\alpha_1, \alpha_2, \ldots, \alpha_n$ and the entry located in the i-th row and j-th column is given by $\alpha_i + \alpha_j$. This is precisely the table of the group operation. We claim that no two elements in each row are identical. Suppose not. If $\alpha_i + \alpha_j = \alpha_i + \alpha_k$ for some $1 \leq i, j, k \leq n$ and $j \neq k$, then $\alpha_j = \alpha_k$. This is a contradiction. Similarly, one can show that no two elements in each column are identical.

It is not difficult to construct latin squares on any n symbols. But it is nice to know that the group table of any finite group gives a latin square. However, it is not true that every latin square arises from a group table. We will talk more about latin squares when we discuss fields later.

We now turn our attention to rings. Let $\mathbf{K}$ be a set equipped with two binary operations, which we call as addition and multiplication. The set $\mathbf{K}$ is said to be a ring if the following hold:

(1) With respect to addition, $\mathbf{K}$ is an abelian group.
(2) With respect to multiplication, the associative law holds, i.e.,

$$\alpha(\beta\gamma) = (\alpha\beta)\gamma \quad \text{for every } \alpha, \beta \text{ and } \gamma \text{ in } \mathbf{K}.$$

(3) The multiplication is distributive with respect to addition, i.e.,

$$\alpha(\beta + \gamma) = \alpha\beta + \alpha\gamma \quad \text{for every } \alpha, \beta \text{ and } \gamma \text{ in } \mathbf{K}.$$

If the multiplication operation in the ring $\mathbf{K}$ is commutative, then $\mathbf{K}$ is called a commutative ring. As a simple example, let $K = \{0, 1, 2, 3, 4, 5, 6\}$. The addition and multiplication on $\mathbf{K}$ are the usual addition and multiplication of real numbers but modulo 7. Then K is a commutative ring. Let $\mathbf{Z}$ be the set of all integers with the usual operations of addition and multiplication. Then $\mathbf{Z}$ is a commutative ring.

Finally, we come to the definition of a field. Let $\mathbf{F}$ be a set with the operations of addition and multiplication (two binary operations) satisfying the following:

(1) With respect to the addition, $\mathbf{F}$ is an abelian group.
(2) With respect to the multiplication, $\mathbf{F} - \{0\}$ is an abelian group.
(3) Multiplication is distributive with respect to addition, i.e.,

$$\alpha(\beta + \gamma) = \alpha\beta + \alpha\gamma \quad \text{for every } \alpha, \beta \text{ and } \gamma \text{ in } \mathbf{F}.$$

The members of a field $\mathbf{F}$ are called scalars. Let $\mathbf{Q}$ be the set of all rational numbers, $\mathbf{R}$ the set of all real numbers, and $\mathbf{C}$ the set of all complex numbers. The sets $\mathbf{Q}, \mathbf{R}$ and $\mathbf{C}$ are standard examples of a field. The reader may verify the following from the properties of a field.

P 1.1.5 If $\alpha + \beta = \alpha + \gamma$ for α, β and γ in $\mathbf{F}$, then $\beta = \gamma$.

P 1.1.6 $(-1)\alpha = -\alpha$ for any α in $\mathbf{F}$.

P 1.1.7 $0\,\alpha = 0$ for any α in $\mathbf{F}$.

P 1.1.8 If $\alpha \neq 0$ and β are any two scalars, then there exists a unique scalar x such that $\alpha x = \beta$. In fact, $x = \alpha^{-1}\beta$, which we may also write as β/α.

P 1.1.9 If $\alpha\beta = 0$ for some α and β in $\mathbf{F}$, then at least one of α and β is zero.

Another way of characterizing a field is that it is a commutative ring in which there is a unit element with respect to multiplication and any non-zero element has a multiplicative inverse. In the commutative ring $\mathbf{K} = \{0, 1, 2, 3\}$ with addition and multiplication modulo 4, there are elements α and β none of which is zero and yet $\alpha\beta = 0$. In a field, $\alpha\beta = 0$ implies that at least one of α and β is zero.

EXAMPLE 1.1.10. Let p be any positive integer. Let $\mathbf{F} = \{0, 1, 2, \dots, p-1\}$. Define addition in $\mathbf{F}$ by $\alpha + \beta = \alpha + \beta$ (modulo p) for α and β in $\mathbf{F}$. Define multiplication in $\mathbf{F}$ by $\alpha\beta = \alpha\beta$ (modulo p) for α and β in $\mathbf{F}$. More precisely, define addition and multiplication in $\mathbf{F}$ by

$$\begin{aligned}
\alpha + \beta &= \alpha + \beta && \text{if } \alpha + \beta \leq p - 1, \\
&= \alpha + \beta - p && \text{if } \alpha + \beta > p - 1; \\
\alpha\beta &= \alpha\beta && \text{if } \alpha\beta \leq p - 1, \\
&= \gamma && \text{if } \alpha\beta = rp + \gamma \text{ for some integers} \\
& && r \geq 1 \text{ and } 0 \leq \gamma \leq p - 1.
\end{aligned}$$

If p is a prime number, then $\mathbf{F}$ is a field.

EXAMPLE 1.1.11. Let $\mathbf{F} = \{0, 1, \alpha, \beta\}$, addition and multiplication on the elements of $\mathbf{F}$ be as in the following tables.

Addition table

	0	1	α	β
0	0	1	α	β
1	1	0	β	α
α	α	β	0	1
β	β	α	1	0

Multiplication table

	0	1	α	β
0	0	0	0	0
1	0	1	α	β
α	0	α	β	1
β	0	β	1	α

The binary operations so defined above on $\mathbf{F}$ make $\mathbf{F}$ a field. Finite fields, i.e., fields consisting of a finite number of elements are called Galois fields. One of the remarkable results of a Galois field is that the number of elements in any Galois field is p^m for some prime number p and positive integer m. Example 1.1.10 is a description of the Galois field, $GF(p)$, where p is a prime number. Example 1.1.11 is a description of the Galois field, $GF(2^2)$. As one can see, the description of $GF(p)$ with p being a prime number is easy to provide. But when it comes to describing $GF(p^m)$ with p being prime and $m \geq 2$, additional work is needed. Some methods for construction of such fields are developed in papers by Bose, Chowla, and Rao (1944, 1945a, 1945b). See also Mann (1949) for the use of $GF(p^m)$ in the construction of designs.

Construction of orthogonal latin squares and magic squares are two of the benefits that accrue from a study of finite fields. Let us start with some definitions.

DEFINITION 1.1.12. Let L_1 and L_2 be two latin squares each on a set of n symbols. They are said to be orthogonal if we superimpose one latin square upon the other, every ordered pair of symbols occurs exactly once in the composite square.

The following are two latin squares, one on the set $S_1 = \{S, H, D, C\}$ and the other on the set $S_2 = \{K, Q, J, A\}$.

$$L_1 : \begin{matrix} S & D & H & C \\ C & H & D & S \\ D & S & C & H \\ H & C & S & D \end{matrix} \qquad L_2 : \begin{matrix} A & K & Q & J \\ Q & J & A & K \\ J & Q & K & A \\ K & A & J & Q \end{matrix}$$

The latin squares L_1 and L_2 are orthogonal. Way back in 1779, Leonard Euler posed the following famous problem. There are 36 officers of six different ranks with six officers from each rank. They also come from six different regiments with each regiment contributing six officers. Euler conjectured that it is impossible to arrange these officers in a 6×6 grid so that each row and each column contains one officer from each regiment and one from each rank. In terms of the notation introduced above, can one build a latin square L_1 on the set of regiments and a latin square L_2 on the set of ranks such that L_1 and L_2 are orthogonal? By an exhaustive enumeration, it has been found that Euler was right. But if $n > 6$, one can always find a pair of orthogonal latin squares as shown by Bose, Shrikhande and Parker (1960). In the example presented after Definition 1.1.3, the suits are the regiments, the kings, queens, jacks and aces are ranks, and $n = 4$.

The problem of finding pairs of orthogonal latin squares has some statistical relevance. Suppose we want to compare the effect of some m dose levels of a drug, Drug A say, in combination with some m levels of another drug, Drug B say. Suppose we have m^2 experimental units classified according to two attributes C and D each at m levels. The attribute C, for example, might refer to m different age groups of experimental units and the attribute D might refer to m different social groups. The basic problem is how to assign the $n = m^2$ drug combinations to the experimental units in such a way that the drug combinations and the cross-classified experimental units constitute a pair of orthogonal latin squares. If such an arrangement is possible, it is called a graeco-latin square.

As an illustration, consider the following example. Suppose Drug A is to be applied at two levels: High (A_1) and Low (A_2), and Drug B at two levels: High (B_1) and Low (B_2). The four drug combinations constitute the first set S_1 of symbols, i.e.,

$$S_1 = \{A_1B_1,\ A_1B_2,\ A_2B_1,\ A_2B_2\},$$

for which a latin square L_1 is sought with $n = 4$. Suppose the attribute C has two age groups: $C_1(\leq 40$ years old) and $C_2(> 40$ years old), and D has two groups: D_1 (White) and D_2 (Black). The second latin square L_2 is to be built on the set

$$S_2 = \{C_1D_1,\ C_1D_2,\ C_2D_1,\ C_2D_2\}.$$

Choosing L_1 and L_2 to be orthogonal confers a distinct statistical advantage. Comparisons can be made between the levels of each drug and attribute.

The concept of orthogonality between a pair of latin squares can be extended to any finite number of latin squares.

DEFINITION 1.1.13. Let $L_1, L_2, \ldots, L_m$ be a set of latin squares each of order n. The set is said to be mutually orthogonal if L_i and L_j are orthogonal for every $i \neq j$.

The construction of a set of mutually orthogonal latin squares is of statistical importance. Galois fields provide some help in this connection. Let $GF(s)$ be a Galois field of order s. Using the Galois field, one can construct a set of $s-1$ mutually orthogonal latin squares. Let $GF(s) = \{\alpha_0, \alpha_1, \dots, \alpha_{s-1}\}$ with the understanding that $\alpha_0 = 0$.

P 1.1.14 Let L_r be the square grid in which the entry in the i-th row and j-th column is given by

$$\alpha_{ij}(r) = \alpha_r \alpha_i + \alpha_j, \ 0 \leq i, j \leq s-1,$$

for $1 \leq r \leq s-1$. Then $L_1, L_2, \dots, L_{s-1}$ is a set of mutually orthogonal latin squares.

PROOF. First, we show that each L_r is a latin square. We claim that any two entries in any row are distinct. Consider the i-th row, and p-th and q-th elements in it with $p \neq q$. Look at

$$\alpha_{ip}(r) - \alpha_{iq}(r) = \alpha_p - \alpha_q \neq 0.$$

Consequently, no two entries in any row are identical. Consider now the j-th column, and p-th and q-th entries in it with $p \neq q$. Look at

$$\alpha_{pj}(r) - \alpha_{qj}(r) = (\alpha_r \alpha_p + \alpha_j) - (\alpha_r \alpha_q + \alpha_j) = \alpha_r(\alpha_p - \alpha_q) \neq 0$$

in view of the fact that $r \geq 1$ and $\alpha_r \neq 0$. Now, we need to show that L_r and L_s are orthogonal for any $r \neq s$ and $r, s = 1, 2, \dots, s-1$. Superimpose L_r upon L_s. Suppose $(\alpha_{ij}(r), \alpha_{ij}(s)) = (\alpha_{pq}(r), \alpha_{pq}(s))$ for some $0 \leq i, j \leq s-1$ and $0 \leq p, q \leq s-1$. Then $\alpha_r\alpha_i + \alpha_j = \alpha_r\alpha_p + \alpha_q$ and $\alpha_s\alpha_i + \alpha_j = \alpha_s\alpha_p + \alpha_q$. By subtracting, we obtain

$$(\alpha_r - \alpha_s)\alpha_i = (\alpha_r - \alpha_s)\alpha_p,$$

or, equivalently,

$$(\alpha_r - \alpha_s)(\alpha_i - \alpha_p) = 0.$$

Since $r \neq s$, we have $\alpha_i - \alpha_p = 0$, or $i = p$. We see immediately that $j = q$. This shows that L_r and L_s are orthogonal. This completes the proof.

Pairs of orthogonal latin squares are useful in drawing up schedules for competitions between teams. Suppose Teams A and B each consisting of 4 players want to organize chess matches between members of the teams. The following are to be fulfilled.

(1) Every member of Team A plays every member of Team B.
(2) All the sixteen matches should be scheduled over a span of four days with four matches per day.
(3) Each player plays only one match on any day.
(4) On every day, each team plays an equal number of games with white and black pieces.
(5) Each player plays an equal number of games with white and black pieces.

Drawing a 16-match schedule spread over four days fulfilling Conditions 1, 2, and 3 is not difficult. One could use a latin square on the set of days the games are to be played. The tricky part is to have the schedule fulfilling Conditions 4 and 5.

A pair of orthogonal latin squares can be used to draw up a schedule of matches. Let Di stand for Day i, $i = 1, 2, 3, 4$. Let L_1 and L_2 be the pair of orthogonal latin squares on the sets

$$S_1 = \{D1, D2, D3, D4\}$$

and

$$S_2 = \{1, 2, 3, 4\},$$

respectively, given by

$$L_1 : \begin{array}{cccc} D1 & D2 & D3 & D4 \\ D4 & D3 & D2 & D1 \\ D2 & D1 & D4 & D3 \\ D3 & D4 & D1 & D2 \end{array}$$

$$L_2 : \begin{array}{cccc} 1 & 2 & 3 & 4 \\ 3 & 4 & 1 & 2 \\ 4 & 3 & 2 & 1 \\ 2 & 1 & 4 & 3 \end{array}$$

Replace even numbers in L_2 by white (W), odd numbers by black (B) and then superimpose the latin squares. The resultant composition is given by

Team A/Team B	1	2	3	4
1	$(D1, B)$	$(D2, W)$	$(D3, B)$	$(D3, W)$
2	$(D4, B)$	$(D3, W)$	$(D2, B)$	$(D1, W)$
3	$(D2, W)$	$(D1, B)$	$(D4, W)$	$(D3, B)$
4	$(D3, W)$	$(D4, B)$	$(D1, W)$	$(D2, B)$

The schedule of matches can be drawn up using the composite square.

Day	Team A players		Team B players	Color of pieces by team A players
$D1$	1	vs	1	B
	2	vs	4	W
	3	vs	2	B
	4	vs	3	W
$D2$	1	vs	2	W
	2	vs	3	B
	3	vs	1	W
	4	vs	4	B
$D3$	1	vs	3	B
	2	vs	2	W
	3	vs	4	B
	4	vs	1	W
$D4$	1	vs	4	W
	2	vs	1	B
	3	vs	3	W
	4	vs	2	B

This schedule fulfills all the five requirements 1 to 5 stipulated above.

A pair of orthogonal latin squares can also be used to build magic squares. Let us define formally what a magic square is.

DEFINITION 1.1.15. A magic square of order n is an $n \times n$ square grid consisting of the numbers $1, 2, \ldots, n^2$ such that the entries in each row, each column and each of the two main diagonals sum up to the same number.

We can determine what each row in a magic square sums up to. The sum of all integers from 1 to n^2 is $n^2(n^2+1)/2$. Then each row in the magic square sums up to $(1/n)n^2(n^2+1)/2 = n(n^2+1)/2$. The following are magic squares of orders 3 and 4:

2	9	4
7	5	3
6	1	8

16	3	2	13
5	10	11	8
9	6	7	12
4	15	14	1

(from an engraving of Albrecht Dürer entitled "Melancholia" (1514)).

Many methods are available for the construction of magic squares. What we intend to do here is to show how a pair of orthogonal latin squares can be put to use to pull out a magic square. Let $L_1 = (\ell_{ij}^1)$ and $L_2 = (\ell_{ij}^2)$ be two orthogonal latin squares on the set $\{0, 1, 2, \ldots, n-1\}$. Let $M = (m_{ij})$ be an $n \times n$ square grid in which the entry in the i-th row and j-th column is given by

$$m_{ij} = n\ell_{ij}^1 + \ell_{ij}^2,$$

for $i, j = 1, 2, \ldots, n$. What can we say about the numbers m_{ij}'s? Since L_1 and L_2 are orthogonal, every ordered pair $(i, j), i, j, = 0, 1, 2, \ldots, n-1$ occurs exactly once when we superimpose L_1 upon L_2. Consequently, each of the numbers $0, 1, 2, \ldots, n^2-1$ will appear somewhere in the square grid M. We are almost there. Define a new grid $M' = (m'_{ij})$ of order $n \times n$ with $m'_{ij} = m_{ij} + 1$. Now each of the numbers $1, 2, \ldots, n^2$ appears somewhere in the grid M'.

P 1.1.16 In the grid M', each row and each column sums up to the same number.

PROOF. Since L_1 and L_2 are latin squares, for any $i = 1, 2, \ldots, n$,

$$\begin{aligned}\sum_{j=1}^{n} m_{ij} &= \text{Sum of all entries in the } i\text{-th row of } M \\ &= n\sum_{j=1}^{n} \ell_{ij}^1 + \sum_{j=1}^{n} \ell_{ij}^2 \\ &= n(0+1+\ldots+n-1) + (0+1+\ldots+n-1) \\ &= n(n-1)n/2 + (n-1)n/2 = n(n^2-1)/2,\end{aligned}$$

which is independent of i. In a similar vein, one can show that each column of M sums up to the same number $n(n^2-1)/2$. Thus M' has the desired properties stipulated above.

The grid M' we have obtained above is not quite a magic square. The diagonals of M' may not sum up to the same number. We need to select the latin squares L_1 and L_2 carefully.

P 1.1.17 Let L_1 and L_2 be two orthogonal latin squares of order n each on the same set $\{0, 1, 2, \ldots, n-1\}$. Suppose that each of the two main diagonals of each of the latin squares L_1 and L_2 add up to the same number $(n^2-1)n/2$. Then the grid M' constructed above is a magic square.

PROOF. It is not hard to show that each of the diagonals of M sums up to $n(n^2-1)/2$. We now have M' truly a magic square.

EXAMPLE 1.1.18. In the following L_1 and L_2 are two latin squares of order 5 on the set $\{0, 1, 2, 3, 4\}$. These latin squares satisfy all the conditions stipulated in Proposition 1.1.17. We follow the procedure outlined above.

$$L_1: \begin{matrix} 0 & 1 & 2 & 3 & 4 \\ 2 & 3 & 4 & 0 & 1 \\ 4 & 0 & 1 & 2 & 3 \\ 1 & 2 & 3 & 4 & 0 \\ 3 & 4 & 0 & 1 & 2 \end{matrix} \qquad L_2: \begin{matrix} 0 & 1 & 2 & 3 & 4 \\ 3 & 4 & 0 & 1 & 2 \\ 1 & 2 & 3 & 4 & 0 \\ 4 & 0 & 1 & 2 & 3 \\ 2 & 3 & 4 & 0 & 1 \end{matrix}$$

$$M: \begin{matrix} 0 & 6 & 12 & 18 & 24 \\ 13 & 19 & 21 & 1 & 7 \\ 21 & 2 & 8 & 14 & 15 \\ 9 & 10 & 16 & 22 & 3 \\ 17 & 23 & 4 & 5 & 11 \end{matrix} \qquad M': \begin{matrix} 1 & 7 & 13 & 19 & 25 \\ 14 & 20 & 22 & 2 & 8 \\ 22 & 3 & 9 & 15 & 16 \\ 10 & 11 & 17 & 23 & 4 \\ 18 & 24 & 5 & 6 & 12 \end{matrix}$$

Note that M' is a Magic square of order 5.

We will wind up this section by talking about sub-fields. A subset $\mathbf{F}_1$ of a field $\mathbf{F}$ is said to be a sub-field of $\mathbf{F}$ if $\mathbf{F}_1$ is a field in its own right under the same operations of addition and multiplication of $\mathbf{F}$ restricted to $\mathbf{F}_1$. For example, the set $\mathbf{Q}$ of all rational numbers is a subfield of the set $\mathbf{R}$ of all real numbers. A field $\mathbf{F}$ is said to be algebraically closed if every polynomial equation with the coefficients belonging to $\mathbf{F}$ has at least one root belonging to the field. For example, the set $\mathbf{C}$ of all complex numbers is algebraically closed, whereas the set $\mathbf{R}$ of all real numbers is not algebraically closed.

Complements

As has been pointed out in **P 1.1.4**, the multiplication table of a finite group provides a latin square. We do not need the full force of a group to generate a latin square. A weaker structure would do. Let G be a finite set with a binary operation. The set G is said to be a quasigroup if each of the equations, $\alpha y = \beta$ and $y\alpha = \beta$ has a unique solution in y for every α, β in G.

1.1.1 Show that the multiplication table of a quasigroup with n elements is a latin square of order n.

1.1.2 Show that every latin square of order n gives rise to a quasigroup. (If we look at the definition of a group, it is clear that if the binary operation of a quasigroup G is associative, then G is a group.)

1.1.3 Let $G = \{0, 1, 2\}$ be a set with the following multiplication table.

	0	1	2
0	1	2	0
1	0	1	2
2	2	0	1

Show that G is a quasigroup but not a group.

1.1.4 Let n be an integer ≥ 2 and $G = \{0, 1, 2, \ldots, n-1\}$. Define a binary operation $*$ on G by

$$\alpha * \beta = a\alpha + b\beta + c \text{ (modulo } n)$$

for all α and β in G, where a and b are prime to n. Show that G is a quasigroup.

1.1.5 If $L_1, L_2, \ldots, L_m$ is a set of mutually orthogonal latin squares of order n, show that $m \leq n-1$.

Let $X = \{1, 2, \ldots, n\}$, say, be a finite set. Let G be the collection of all subsets of X. Define a binary operation on G by

$$\alpha\beta = \alpha\Delta\beta, \quad \alpha, \beta \in G,$$

where Δ is the set-theoretic operation of symmetric difference, i.e.,

$$\alpha\Delta\beta = (\alpha - \beta) \cup (\beta - \alpha),$$

where

$$\alpha - \beta = \{x \in X : x \in \alpha,\ x \notin \beta\}.$$

1.1.6 How many elements are there in G?
1.1.7 Show that G is a group.
1.1.8 Set out the multiplication table of the group G when $n = 3$.
1.1.9 Let $\mathbf{F} = \{a + b\sqrt{2};\ a, b \text{ rational}\}$. The addition and multiplication of elements in $\mathbf{F}$ are defined in the usual way. Show that $\mathbf{F}$ is a field.
1.1.10 Show that the set of all integers under the usual operations of addition and multiplication of numbers is not a field.

1.2. Mappings

In the subsequent discussion of vector spaces and matrices, we will be considering transformations or mappings from one set to another. We give some basic ideas for later reference.

Let S and T be two sets. A map, a mapping, or a function f from S to T is a rule which associates to each element of S a unique element of T. If s is any element of S, its associate in T is denoted by $f(s)$. The set S is called the domain of f and the set of all associates in T of elements of S is called the range of f. The range is denoted by $f(S)$. The map f is usually denoted by $f : S \to T$.

Consider a map $f : S \to T$. The map f is said to be surjective or onto if $f(S) = T$, i.e., given any $t \in T$, there exists $s \in S$ such that $f(s) = t$. The map f is said to be injective or one-to-one if any two distinct elements of S have distinct associates in T, i.e., $s_1, s_2 \in S$ and $f(s_1) = f(s_2)$ imply that $s_1 = s_2$. The map f is said to be bijective if f

is one-to-one and onto or surjective and injective. If f is bijective, one can define the inverse map which we denote by $f^{-1} : T \to S$; for t in T, $f^{-1}(t) = s$, where s is such that $f(s) = t$. The map f^{-1} is called the inverse of f.

DEFINITION 1.2.1. Let f be a mapping from a group $\mathbf{G}_1$ to a group $\mathbf{G}_2$. Then f is said to be a homomorphism if

$$f(\alpha\beta) = f(\alpha)f(\beta) \quad \text{for every } \alpha \text{ and } \beta \text{ in } \mathbf{G}_1.$$

If f is bijective, f is said to be an isomorphism and $\mathbf{G}_1$ and $\mathbf{G}_2$ isomorphic.

Suppose $\mathbf{G}$ is a group and H an invariant subgroup of $\mathbf{G}$, i.e., $xH = Hx$ for all $x \in G$. Let $\mathbf{G}|H$ be the quotient group of $\mathbf{G}$ modulo H, i.e., $\mathbf{G}|H$ is the collection of all distinct cosets of H. [Note that a coset of H is the set $\{xy : y \in H\}$ as defined in Section 1.1]. There is a natural map π from $\mathbf{G}$ to $\mathbf{G}|H$. For every α in $\mathbf{G}$, define $\pi(\alpha) =$ the coset of H to which α belongs. The map π is surjective and a homomorphism from $\mathbf{G}$ onto $\mathbf{G}|H$. This map π is called the projection of $\mathbf{G}$ onto the quotient group $\mathbf{G}|H$.

DEFINITION 1.2.2. Let f be a mapping from a field $\mathbf{F}_1$ to a field $\mathbf{F}_2$. Then f is said to be a homomorphism if

$$f(\alpha + \beta) = f(\alpha) + f(\beta),$$
$$f(\alpha\beta) = f(\alpha)f(\beta)$$

for every α and β in $\mathbf{F}_1$. If f is bijective, then f is called an isomorphism and the fields $\mathbf{F}_1$ and $\mathbf{F}_2$ are called isomorphic.

Complements

1.2.1 Let S and T be two finite sets consisting of the same number of elements. Let $f : S \to T$ be a map. If f is surjective, show that f is bijective.

1.2.2 Let $S = \{1, 2, 3, 4\}$ and G be the collection of all bijective maps from S to S. For any two maps f and g in G, define the composite map $f \circ g$ by $(f \circ g)(x) = f(g(x)), x \in S$. Show that under the binary operation of composition of maps, G is a group. Let H be the collection of all maps f in G such that $f(1) = 1$. Show that H is a subgroup but not invariant. Identify all distinct left cosets of H. Is this a group under the usual multiplication of cosets?

1.3. Vector Spaces

The concept of a vector space is central in any discussion of multivariate methods. A set of elements (called vectors) is said to be a vector space or a linear space over a field of scalars $\mathbf{F}$ if the following axioms are satisfied. (We denote the set of elements by $\mathbf{V}(\mathbf{F})$ to indicate its dependence on the underlying field $\mathbf{F}$ of scalars. Sometimes, we denote the vector space simply by $\mathbf{V}$ if the underlying field of scalars is unambiguously clear. We denote the elements of the set $\mathbf{V}(\mathbf{F})$ by Roman letters and the elements of $\mathbf{F}$ by Greek letters.)

(1) To every pair of vectors x and y, there corresponds a vector $x+y$ in such a way that under the binary operation $+$, $\mathbf{V}(\mathbf{F})$ is an abelian group.
(2) To every vector x and a scalar α, there corresponds a vector αx, called the scalar product of α and x, in such a way that $\alpha_1(\alpha_2 x) = (\alpha_1\alpha_2)x$ for every α_1, α_2 in $\mathbf{F}$ and x in $\mathbf{V}(\mathbf{F})$, and $1x = x$ for every x in $\mathbf{V}(\mathbf{F})$, where 1 is the unit element of $\mathbf{F}$.
(3) The distributive laws hold for vectors as well as scalars, i.e., $\alpha(x+y) = \alpha x + \alpha y$ for every α in $\mathbf{F}$ and x, y in $\mathbf{V}(\mathbf{F})$, and $(\alpha_1 + \alpha_2)x = \alpha_1 x + \alpha_2 x$ for every α_1, α_2 in $\mathbf{F}$ and x in $\mathbf{V}(\mathbf{F})$.

We now give some examples. The first example plays an important role in many applications.

EXAMPLE 1.3.1. Let $\mathbf{F}$ be a field of scalars and $k \geq 1$ an integer. Consider the following collection of ordered tuples:

$$\mathbf{F}^k = \{(\alpha_1, \alpha_2, \dots, \alpha_k) : \alpha_i \in \mathbf{F}, i = 1, 2, \dots, k\}.$$

Define addition and scalar multiplication in $\mathbf{F}^k$ by

$$(\alpha_1, \alpha_2, \dots, \alpha_k) + (\beta_1, \beta_2, \dots, \beta_k) = (\alpha_1 + \beta_1, \alpha_2 + \beta_2, \dots, \alpha_k + \beta_k)$$

for every $(\alpha_1, \alpha_2, \dots, \alpha_k)$ and $(\beta_1, \beta_2, \dots, \beta_k)$ in $\mathbf{F}^k$;

$$\delta(\alpha_1, \alpha_2, \dots, \alpha_k) = (\delta\alpha_1, \delta\alpha_2, \dots, \delta\alpha_k)$$

for every δ in $\mathbf{F}$ and $(\alpha_1, \alpha_2, \dots, \alpha_k)$ in $\mathbf{F}^k$. It can be verified that $\mathbf{F}^k$ is a vector space over the field $\mathbf{F}$ with $(0, 0, \dots, 0)$ as the zero-vector. We call $\mathbf{F}^k$ a k-dimensional coordinate space. Strictly speaking, we should write the vector space $\mathbf{F}^k$ as $\mathbf{F}^k(\mathbf{F})$. We will omit the symbol in the parentheses, which will not cause any confusion.

Special cases of $\mathbf{F}^k$ are $\mathbf{R}^k$ and $\mathbf{C}^k$, i.e., when $\mathbf{F}$ is the field $\mathbf{R}$ of real numbers and $\mathbf{C}$ of complex numbers, respectively. They are also called real and complex arithmetic spaces.

EXAMPLE 1.3.2. Let $n \geq 1$. The collection of all polynomials of degree less than n with coefficients from a field $\mathbf{F}$ with the usual addition and scalar multiplication of polynomials is a vector space. Symbolically, we denote this collection by

$$\mathbf{P}_n(\mathbf{F})(t) = \{\alpha_0 + \alpha_1 t + \alpha_2 t^2 + \ldots + \alpha_{n-1} t^{n-1} : \alpha_i \in \mathbf{F}, \\ i = 0, 1, 2, \ldots, n-1\},$$

which is a vector space over the field $\mathbf{F}$. The entity $\alpha_0 + \alpha_1 t + \alpha_2 t^2 + \ldots + \alpha_{n-1} t^{n-1}$ is called a polynomial in t with coefficients from the field $\mathbf{F}$.

EXAMPLE 1.3.3. Let $\mathbf{V}$ be the collection of all real valued functions of a real variable which are differentiable. If we take $\mathbf{F} = \mathbf{R}$, and define sum of two functions in $\mathbf{V}$ and scalar multiplication in the usual way, then $\mathbf{V}$ is a vector space over the field $\mathbf{R}$ of real numbers.

EXAMPLE 1.3.4. Let $\mathbf{V} = \{(\alpha, \beta) : \alpha > 0 \text{ and } \beta > 0\}$. Define vector addition and scalar multiplication in $\mathbf{V}$ as follows.

(1) $(\alpha_1, \beta_1) + (\alpha_2, \beta_2) = (\alpha_1\alpha_2, \beta_1\beta_2)$ for every (α_1, β_1) and (α_2, β_2) in $\mathbf{V}$.

(2) $\delta(\alpha, \beta) = (\alpha^\delta, \beta^\delta)$ for every δ in $\mathbf{R}$ and (α, β) in $\mathbf{V}$.

Then $\mathbf{V}$ is a vector space over the field $\mathbf{R}$ of real numbers.

EXAMPLE 1.3.5. Let p be an odd integer. Let $\mathbf{V} = \{(\alpha, \beta) : \alpha$ and β real$\}$. Define vector addition and scalar multiplication in $\mathbf{V}$ as below:

(1) $(\alpha_1, \beta_1) + (\alpha_2, \beta_2) = ((\alpha_1^p + \alpha_2^p)^{1/p}, (\beta_1^p + \beta_2^p)^{1/p})$ for every (α_1, β_1) and (α_2, β_2) in $\mathbf{V}$.

(2) $\delta(\alpha, \beta) = (\delta^{1/p}\alpha, \delta^{1/p}\beta)$ for every δ in $\mathbf{R}$ and (α, β) in $\mathbf{V}$.

Then $\mathbf{V}$ is a vector space over the field $\mathbf{R}$ of real numbers. This statement is not correct if p is an even integer.

EXAMPLE 1.3.6. Let $\mathbf{F} = \{0, 1, 2\}$. With addition and multiplication modulo 3, $\mathbf{F}$ is a field. See Example 1.1.10. Observe that the vector space $\mathbf{F}^k$ has only 3^k elements, while $\mathbf{R}^k$ has an uncountable number of elements.

The notion of isomorphic vector spaces will be introduced now. Let $\mathbf{V}_1$ and $\mathbf{V}_2$ be two vector spaces over the same field $\mathbf{F}$ of scalars. The spaces $\mathbf{V}_1$ and $\mathbf{V}_2$ are said to be isomorphic to each other if there exists a bijection $h : \mathbf{V}_1 \to \mathbf{V}_2$ such that

$$h(x + y) = h(x) + h(y) \quad \text{for all } x, y \in \mathbf{V}_1,$$

$$h(\alpha x) = \alpha h(x) \quad \text{for all } \alpha \in \mathbf{F} \text{ and } x \in \mathbf{V}_1.$$

Complements

1.3.1 Examine which of the following are vector spaces over the field $\mathbf{C}$ of complex numbers. Explain why or why not?

(1) $\mathbf{V} = \{(\alpha, \beta);\ \alpha \in \mathbf{R},\ \beta \in \mathbf{C}\}$.
Addition:

$$(\alpha_1, \beta_1) + (\alpha_2, \beta_2) = (\alpha_1 + \alpha_2, \beta_1 + \beta_2);\ (\alpha_1, \beta_1), (\alpha_2, \beta_2) \in \mathbf{V}.$$

Scalar multiplication:

$$\delta(\alpha, \beta) = (\delta\alpha, \delta\beta),\ \delta \in \mathbf{C}, (\alpha, \beta) \in \mathbf{V}.$$

(2) $\mathbf{V} = \{(\alpha, \beta) : \alpha + \beta = 0, \alpha, \beta \in \mathbf{C}\}$.
Addition:

$$(\alpha_1, \beta_1) + (\alpha_2, \beta_2) = (\alpha_1 + \alpha_2, \beta_1 + \beta_2);\ (\alpha_1, \beta_1), (\alpha_2, \beta_2) \in \mathbf{V}.$$

Scalar multiplication:

$$\delta(\alpha, \beta) = (\delta\alpha, \delta\beta),\ \delta \in \mathbf{C},\ (\alpha, \beta) \in \mathbf{V}.$$

1.3.2 Let $\mathbf{V}_1 = (0, \infty)$ and $\mathbf{F} = \mathbf{R}$. The addition in $\mathbf{V}_1$ is the usual operation of multiplication of real numbers. The scalar multiplication is defined by

$$\alpha x = x^\alpha,\ \alpha \in \mathbf{R},\ x \in \mathbf{V}_1.$$

Show that $\mathbf{V}_1$ is a vector space over $\mathbf{R}$. Identify the zero vector of $\mathbf{V}_1$.

1.3.3 Show that $\mathbf{V}_1$ of Complement 1.3.2 and the vector space $\mathbf{V}_2 = \mathbf{R}$ over the field $\mathbf{R}$ of real numbers are isomorphic. Exhibit an explicit isomorphism between $\mathbf{V}_1$ and $\mathbf{V}_2$.

1.3.4 Let $\mathbf{V}(\mathbf{F})$ be a vector space over a field $\mathbf{F}$. Let, for any fixed positive integer n,

$$\mathbf{V}^n(\mathbf{F}) = \{(x_1, x_2, \ldots, x_n) : x_i \in \mathbf{V}(\mathbf{F}) \text{ for each } i\}.$$

Define addition in $\mathbf{V}^n(\mathbf{F})$ by

$$(x_1, x_2, \ldots, x_n) + (y_1, y_2, \ldots, y_n) = (x_1 + y_1, x_2 + y_2, \ldots, x_n + y_n)$$

for $(x_1, x_2, \ldots, x_n), (y_1, y_2, \ldots, y_n) \in \mathbf{V}^n(\mathbf{F})$. Define scalar multiplication in $\mathbf{V}^n(\mathbf{F})$ by

$$\alpha(x_1, x_2, \ldots, x_n) = (\alpha x_1, \alpha x_2, \ldots, \alpha x_n), \ \alpha \in \mathbf{F} \quad \text{and} \quad (x_1, x_2, \ldots, x_n) \in \mathbf{V}^n(\mathbf{F}).$$

Show that $\mathbf{V}^n(\mathbf{F})$ is a vector space over the field $\mathbf{F}$.

1.4. Linear Independence and Basis of a Vector Space

Through out this section, we assume that we have a vector space $\mathbf{V}$ over a field $\mathbf{F}$ of scalars. The notions of linear independence, linear dependence and basis form the core in the development of vector spaces.

DEFINITION 1.4.1. A finite set $x_1, x_2, \ldots, x_k$ of vectors is said to be linearly dependent if there exist scalars $\alpha_1, \alpha_2, \ldots, \alpha_k$, not all zeros, such that $\alpha_1 x_1 + \alpha_2 x_2 + \ldots + \alpha_k x_k = 0$. Otherwise, it is said to be linearly independent.

P 1.4.2 The set consisting of only one vector, which is the zero vector 0, is linearly dependent.

P 1.4.3 The set consisting of only one vector, which is a non-zero vector, is linearly independent.

P 1.4.4 Any set of vectors containing the zero vector is linearly dependent.

P 1.4.5 A set $x_1, x_2, \ldots, x_k$ of non-zero vectors is linearly dependent if and only if there exists $2 \leq i \leq k$ such that

$$x_i = \beta_1 x_1 + \beta_2 x_2 + \beta_{i-1} x_{i-1}$$

for some scalars $\beta_1, \beta_2, \ldots, \beta_{i-1}$, i.e., there is a member in the set which can be expressed as a linear combination of its predecessors.

PROOF. Let $i \in \{1, 2, \ldots, k\}$ be the smallest integer such that the set of vectors $x_1, x_2, \ldots, x_i$ is linearly dependent. Obviously, $2 \leq i \leq k$. There exist scalars $\alpha_1, \alpha_2, \ldots, \alpha_i$, not all zero, such that $\alpha_1 x_1 + \alpha_2 x_2 + \ldots + \alpha_i x_i = 0$. By the very choice of $i, \alpha_i \neq 0$. Thus we can write

$$x_i = -(\alpha_1/\alpha_i)x_1 - (\alpha_2/\alpha_i)x_2 - \ldots - (\alpha_{i-1}/\alpha_i)x_{i-1}.$$

P 1.4.6 Let A and B be two finite sets of vectors such that $A \subset B$. If A is linearly dependent, so is B. If B is linearly independent, so is A.

DEFINITION 1.4.7. Let B be any subset (finite or infinite) of $\mathbf{V}$. The set B is said to be linearly independent if every finite subset of B is linearly independent.

DEFINITION 1.4.8. (Basis of a vector space) A linearly independent set B of vectors is said to be a (Hamel) basis of $\mathbf{V}$ if every vector of $\mathbf{V}$ is a linear combination of the vectors in B. The vector space $\mathbf{V}$ is said to be finite dimensional if there exists a Hamel basis B consisting of finitely many vectors.

It is not clear at the outset whether a vector space possesses a basis. Using Zorn's lemma, one can demonstrate the existence of a maximal linearly independent system of vectors in any vector space. (A discussion of this particular feature is beyond the scope of the book.) Any maximal set is indeed a basis of the vector space.

From now on, we will be concerned with finite dimensional vector spaces only. Occasionally, infinite dimensional vector spaces will be presented as examples to highlight some special features of finite dimensional vector spaces. The following results play an important role.

P 1.4.9 If $x_1, x_2, \ldots, x_k$ and $y_1, y_2, \ldots, y_s$ are two sets of bases for the vector space $\mathbf{V}$, then $k = s$.

PROOF. Suppose $k \neq s$. Let $s > k$. It is obvious that the set $y_1, x_1, x_2, \ldots, x_k$ is linearly dependent. By **P 1.4.5**, there is a vector x_i which is a linear combination of its predecessors in the above set. Consequently, every vector in **V** is a linear combination of the vectors $y_1, x_1, x_2, \ldots, x_{i-1}, x_{i+1}, \ldots, x_k$. Observe now that the set $y_2, y_1, x_1, x_2, \ldots, x_{i-1}, x_{i+1}, \ldots, x_k$ is linearly dependent. Again by **P 1.4.5**, there exists a $j \in \{1, 2, \ldots, i-1, i+1, \ldots, k\}$ such that x_j is a linear combination of its predecessors. (Why?) Assume, without loss of generality, that $i < j$. It is clear that every vector in **V** is a linear combination of the vectors $y_2, y_1, x_1, x_2, \ldots, x_{i-1}, x_{i+1}, \ldots, x_{j-1}, x_{j+1}, \ldots, x_k$. Continuing this process, we will eventually obtain the set $y_k, y_{k-1}, \ldots, y_2, y_1$ such that every vector in **V** is a linear combination of members of this set. This is a contradiction to the assumption that $s > k$. Even if we assume that $s < k$, we end up with a contradiction. Hence $s = k$.

In finite dimensional vector spaces, one can now introduce the notion of the dimension of a vector space. It is precisely the cardinality of any Hamel basis of the vector space. We use the symbol dim(**V**) for the dimension of the vector space **V**.

P 1.4.10 Any given set $x_1, x_2, \ldots, x_r$ of linearly independent vectors can be enlarged to a basis of **V**.

PROOF. Let $y_1, y_2, \ldots, y_k$ be a basis of **V**, and consider the set $x_1, y_1, y_2, \ldots, y_k$ of vectors, which is linearly dependent. Using the same method as enunciated in the proof of **P 1.4.9**, we drop one of y_i's and then add one of x_j's until we get a set $x_r, x_{r-1}, \ldots, x_1, y_{(1)}, y_{(2)}, \ldots, y_{(k-r)}$, which is a basis for **V**, where $y_{(i)}$'s are selections from $y_1, y_2, \ldots, y_k$. This completes the proof.

P 1.4.11 Every vector x in **V** has a unique representation in terms of any given basis of **V**.

PROOF. Let $x_1, x_2, \ldots, x_k$ be a basis for **V**. Let

$$x = \alpha_1 x_1 + \alpha_2 x_2 + \ldots + \alpha_k x_k$$

and, also

$$x = \beta_1 x_1 + \beta_2 x_2 + \ldots + \beta_k x_k,$$

for some scalars α_i's and β_j's. Then

$$0 = (\alpha_1 - \beta_1)x_1 + (\alpha_2 - \beta_2)x_2 + \ldots + (\alpha_k - \beta_k)x_k,$$

from which it follows that $\alpha_i - \beta_i = 0$ for every i, in view of the fact that the basis is linearly independent.

In view of the unique representation presented above, one can define a map from $\mathbf{V}$ to $\mathbf{F}^k$. Let $x_1, x_2, \ldots, x_k$ be a basis for $\mathbf{V}$. Let $x \in \mathbf{V}$, and $x = \alpha_1 x_1 + \alpha_2 x_2 + \ldots + \alpha_k x_k$ be the unique representation of x in terms of the vectors of the basis. The ordered tuple $(\alpha_1, \alpha_2, \ldots, \alpha_k)$ is called the set of coordinates of x with respect to the given basis. Define

$$\varphi(x) = (\alpha_1, \alpha_2, \ldots, \alpha_k) \in \mathbf{F}^k,$$

which one can verify to be a bijective map from $\mathbf{V}$ to $\mathbf{F}^k$. Further, $\varphi(\cdot)$ is a homomorphism from the vector space $\mathbf{V}$ to the vector space $\mathbf{F}^k$. Consequently, the vector spaces $\mathbf{V}$ and $\mathbf{F}^k$ are isomorphic. We record this fact as a special property below.

P 1.4.12 Any vector space $\mathbf{V}(\mathbf{F})$ of dimension k is isomorphic to the vector space $\mathbf{F}^k$.

The above result also implies that any two vector spaces over the same field of scalars and of the same dimension are isomorphic to each other.

It is time to take stock of the complete meaning and significance of **P 1.4.12**. If a vector space $\mathbf{V}$ over a field $\mathbf{F}$ is isomorphic to the vector space $\mathbf{F}^k$ for some $k \geq 1$, why bother to study vector spaces in the generality they are introduced? The vector space $\mathbf{F}^k$ is simple to visualize and one could restrict oneself to the vector space $\mathbf{F}^k$ in subsequent dealings. There are two main reasons against pursuing such a seemingly simple trajectory. The isomorphism that is built between the vector spaces $\mathbf{V}(\mathbf{F})$ and $\mathbf{F}^k$ is based on a given basis of the vector space $\mathbf{V}(\mathbf{F})$. In the process of transformation, the intrinsic structural beauty of the space $\mathbf{V}(\mathbf{F})$ is usually lost in its metamorphism. For the second reason, suppose we establish a certain property of the vector space $\mathbf{F}^k$. If we would like to examine how this property comports itself in the space $\mathbf{V}(\mathbf{F})$, we could use any one of the isomorphisms operational between $\mathbf{V}(\mathbf{F})$ and $\mathbf{F}^k$, and translate this property into the space $\mathbf{V}(\mathbf{F})$. The isomorphism used is heavily laced with the underlying basis and an understanding of the property devoid of its external trappings provided by the isomorphism would then become a herculean task.

As a case in point, take $\mathbf{F} = \mathbf{R}$ and $\mathbf{V} = \mathbf{P}_k$, the set of all polynomials with real coefficients of degree $< k$. The vector space $\mathbf{P}_k$ is isomorphic to $\mathbf{R}^k$. Linear functionals on vector spaces are introduced in Section 1.7. One could introduce a linear functional f on $\mathbf{P}_k$ as follows. Let μ be a measure on the Borel σ-field of $[a, b]$, a non-degenerate interval. For $x \in \mathbf{P}_k$, let

$$f(x) = \int_a^b x(t)\mu(dt).$$

One can verify that

$$f(x + y) = f(x) + f(y), \; x, y \in \mathbf{P}_k$$

and

$$f(\alpha x) = \alpha f(x), \; \alpha \in \mathbf{R}, \, x \in \mathbf{P}_k.$$

Two distinct measures μ_1 and μ_2 on $[a, b]$ might produce the same linear functional. For example, if

$$\int_a^b t^m \mu_1(dt) = \int_a^b t^m \mu_2(dt)$$

for $m = 0, 1, 2, \ldots, k-1$, then

$$f_1(x) = \int_a^b x(t)\mu_1(dt) = \int_a^b x(t)\mu_2(dt) = f_2(x)$$

for all $x \in \mathbf{P}_k$. A discussion of features such as this in $\mathbf{P}_k$ is not possible in $\mathbf{R}^k$. The vector space $\mathbf{P}_k$ has a number facets allied to it which would be lost if we were to work only with $\mathbf{R}^k$ using some isomorphism between $\mathbf{P}_k$ and $\mathbf{R}^k$. We will work with vector spaces as they come and ignore **P 1.4.12.**

Complements.

1.4.1 Let $\mathbf{V} = \mathbf{C}$, the set of all complex numbers. Then $\mathbf{V}$ is a vector space over the field $\mathbf{C}$ of complex numbers with the usual addition and multiplication of complex numbers. What is the dimension of the vector space $\mathbf{V}$?

1.4.2 Let $\mathbf{V} = \mathbf{C}$, the set of all complex numbers. Then $\mathbf{V}$ is a vector space over the field $\mathbf{R}$ of real numbers with the usual addition of complex numbers. The scalar multiplication in **V** is the usual multiplication of a complex number by a real number. What is the dimension of **V**? How does this example differ from the one in Complement 1.4.1?

1.4.3 Let $\mathbf{V} = \mathbf{R}$, the set of all real numbers. Then **V** is a vector space over the field **Q** of rational numbers. The addition in **V** is the usual addition of real numbers. The scalar multiplication in **V** is the multiplication of a real number by a rational number. What is the dimension of **V**?

1.4.4 Let **R** be the vector space over the field **Q** of rational numbers. See Complement 1.4.3. Show that $\sqrt{2}$ and $\sqrt{3}$ are linearly independent.

1.4.5 Determine the dimension of the vector space introduced in Example 1.3.4. Identify a basis of this vector space.

1.4.6 Let $\mathbf{F} = \{0, 1, 2, 3, 4\}$ be the field in which addition and multiplication are carried out in the usual way but modulo 5. How many points are there in the vector space $\mathbf{F}^3$?

1.5. Subspaces

In any set with a mathematical structure on it, subsets which exhibit all the features of the original mathematical structure deserve special scrutiny. A study of such subsets aids a good understanding of the mathematical structure itself.

DEFINITION 1.5.1. A subset **S** of a vector space **V** is said to be a subspace of **V** if $\alpha x + \beta y \in \mathbf{S}$ whenever $x, y \in \mathbf{S}$ and $\alpha, \beta \in \mathbf{F}$.

P 1.5.2 A subspace **S** of a vector space **V** is a vector space over the same field **F** of scalars under the same definition of addition of vectors and scalar multiplication operational in **V**. Further, $\dim(\mathbf{S}) \leq \dim(\mathbf{V})$.

PROOF. It is clear that **S** is a vector space in its own right. In order to show that $\dim(\mathbf{S}) \leq \dim(\mathbf{V})$, it suffices to show that the vector space **S** admits a basis. For then, any basis of **S** is a linearly independent set in **V** which can be extended to a basis of **V**. It is known that every vector space admits a basis.

If **S** consists of only the zero-vector, then **S** is a zero-dimensional subspace of **V**. If every vector in **S** is of the form αx for some fixed non-zero vector x and for some α in **F**, then **S** is a one-dimensional subspace of **V**. If every vector in **S** is of the form $\alpha x_1 + \beta x_2$ for some fixed set of linearly independent vectors x_1 and x_2 and for some α and β in **F**, then **S** is a two-dimensional subspace of **V**. The schematic way we have described above is the way one generally obtains subspaces of various dimensions. The sets $\{0\}$ and **V** are extreme examples of subspaces of **V**.

P 1.5.3 The intersection of any family of subspaces of **V** is a subspace of **V**.

P 1.5.4 Given an r-dimensional subspace **S** of **V**, we can find a basis $x_1, x_2, \ldots, x_r, x_{r+1}, x_{r+2}, \ldots, x_k$ of **V** such that $x_1, x_2, \ldots, x_r$ is a basis of **S**.

The result of **P 1.5.4** can also be restated as follows: given a basis $x_1, x_2, \ldots, x_r$ of **S**, it can be completed to a basis of **V**.

The subspaces spanned by a finite set of vectors need special attention. If $x_1, x_2, \ldots, x_r$ is a finite collection of vectors from a vector space **V(F)**, then the set

$$\{\alpha_1 x_1 + \alpha_2 x_2 + \ldots + \alpha_r x_r : \alpha_1, \alpha_2, \ldots, \alpha_r \in \mathbf{F}\}$$

is a subspace of **V(F)**. This subspace is called the span of $x_1, x_2, \ldots, x_r$ and is denoted by $Sp(x_1, x_2, \ldots, x_r)$. Of course, any subspace of **V(F)** arises this way. The concept of spanning plays a crucial role in the following properties.

P 1.5.5 Given a subspace **S** of **V**, we can find a subspace $\mathbf{S}^c$ of **V** such that $\mathbf{S} \cap \mathbf{S}^c = \{0\}$, $\dim(\mathbf{S}) + \dim(\mathbf{S}^c) = \dim(\mathbf{V})$, and

$$\mathbf{V} = \mathbf{S} \oplus \mathbf{S}^c = \{x + y : x \in \mathbf{S},\ y \in \mathbf{S}^c\}.$$

Further, any vector x in **V** has a unique decomposition $x = x_1 + x_2$ with $x_1 \in \mathbf{S}$ and $x_2 \in \mathbf{S}^c$.

PROOF. Let $x_1, x_2, \ldots, x_r, x_{r+1}, \ldots, x_k$ constitute a basis for the vector space **V** such that $x_1, x_2, \ldots, x_r$ is a basis for **S**. Let $\mathbf{S}^c$ be the subspace of **V** spanned by $x_{r+1}, x_{r+2}, \ldots, x_k$. The subspace $\mathbf{S}^c$ meets all the properties mentioned above.

We have introduced a special symbol $\oplus$ above. The mathematical operation $\mathbf{S} \oplus \mathbf{S}^c$ is read as the direct sum of the subspaces **S** and $\mathbf{S}^c$.

The above result states that the vector space $\mathbf{V}$ is the direct sum of two disjoint subspaces of $\mathbf{V}$. We use the phrase that the subspaces $\mathbf{S}$ and $\mathbf{S}^c$ are disjoint even though they have the zero vector in common! We would like to emphasize that the subspace $\mathbf{S}^c$ is not unique. Suppose $\mathbf{V} = \mathbf{R}^2$ and $\mathbf{S} = \{(x, 0) : x \in \mathbf{R}\}$. One can take $\mathbf{S}^c = \{(x, x) : x \in \mathbf{R}\}$ or $\mathbf{S}^c = \{(x, 2x) : x \in \mathbf{R}\}$. We will introduce a special phraseology to describe the subspace $\mathbf{S}^c$: $\mathbf{S}^c$ is a complement of $\mathbf{S}$. More formally, two subspaces $\mathbf{S}_1$ and $\mathbf{S}_2$ are complement to each other if $\mathbf{S}_1 \cap \mathbf{S}_2 = \{0\}$ and $\{x + y : x \in \mathbf{S}_1, \, y \in \mathbf{S}_2\} = \mathbf{V}$.

P 1.5.6 Let $K = \{x_1, x_2, \ldots, x_r\}$ be a subset of the vector space $\mathbf{V}$ and $Sp(K)$ be the vector space spanned by the vectors in K, i.e., $Sp(K)$ is the space of all linear combinations of the vectors $x_1, x_2, \ldots, x_r$. Then

$$Sp(K) = \cap_\nu \mathbf{S}_\nu,$$

where the intersection is taken over all subspaces $\mathbf{S}_\nu$ of $\mathbf{V}$ containing K.

Let $\mathbf{S}_1$ and $\mathbf{S}_2$ be two subspaces of a vector space $\mathbf{V}$. Let

$$\mathbf{S}_1 + \mathbf{S}_2 = \{x + y : x \in \mathbf{S}_1, \, y \in \mathbf{S}_2\}.$$

The operation + defined between subspaces of $\mathbf{V}$ is analogous to the operation of direct sum $\oplus$. We reserve the symbol $\oplus$ for subspaces $\mathbf{S}_1$ and $\mathbf{S}_2$ which are disjoint, i.e., $\mathbf{S}_1 \cap \mathbf{S}_2 = \{0\}$. The following results give some properties of the operation + defined for subspaces.

P 1.5.7 Let $\mathbf{S}_1$ and $\mathbf{S}_2$ be two subspaces of a vector space $\mathbf{V}$. Let $\mathbf{S}$ be the smallest subspace of $\mathbf{V}$ containing both $\mathbf{S}_1$ and $\mathbf{S}_2$. Then

(1) $\mathbf{S} = \mathbf{S}_1 + \mathbf{S}_2$,
(2) $\dim(\mathbf{S}) = \dim(\mathbf{S}_1) + \dim(\mathbf{S}_2) - \dim(\mathbf{S}_1 \cap \mathbf{S}_2)$.

PROOF. It is clear that $\mathbf{S}_1 + \mathbf{S}_2 \subseteq \mathbf{S}$. Note that $\mathbf{S}_1 + \mathbf{S}_2$ is a subspace of $\mathbf{V}$ containing both $\mathbf{S}_1$ and $\mathbf{S}_2$. Consequently, $\mathbf{S} \subseteq \mathbf{S}_1 + \mathbf{S}_2$. This establishes (1). To prove (2), let $x_1, x_2, \ldots, x_r$ be a basis for $\mathbf{S}_1 \cap \mathbf{S}_2$, where $r = \dim(\mathbf{S}_1 \cap \mathbf{S}_2)$. Let $x_1, x_2, \ldots, x_r, x_{r+1}, x_{r+2}, \ldots, x_m$ be the completion of the basis of $\mathbf{S}_1 \cap \mathbf{S}_2$ to $\mathbf{S}_1$, where $\dim(\mathbf{S}_1) = m$. Refer to **P 1.5.4**. Let $x_1, x_2, \ldots, x_r, y_{r+1}, y_{r+2}, \ldots, y_n$ be the completion of the basis of $\mathbf{S}_1 \cap \mathbf{S}_2$ to $\mathbf{S}_2$, where $\dim(\mathbf{S}_2) = n$. It now follows that a basis of $\mathbf{S}_1+\mathbf{S}_2$ is given by $x_1, x_2, \ldots, x_r, x_{r+1}, x_{r+2}, \ldots, x_m, y_{r+1}, y_{r+2}, \ldots, y_n$. (Why?) Consequently,

$$\begin{aligned}\dim(\mathbf{S}_1+\mathbf{S}_2) &= r+(m-r)+(n-r)\\ &= m+n-r\\ &= \dim(\mathbf{S}_1)+\dim(\mathbf{S}_2)-\dim(\mathbf{S}_1\cap\mathbf{S}_2).\end{aligned}$$

P 1.5.8 Let $\mathbf{S}_1$ and $\mathbf{S}_2$ be two subspaces of $\mathbf{V}$. Then the following statements are equivalent.

(1) Every vector x in $\mathbf{V}$ has a unique representation x_1+x_2 with $x_1 \in \mathbf{S}_1$ and $x_2 \in \mathbf{S}_2$.
(2) $\mathbf{S}_1 \cap \mathbf{S}_2 = \{0\}$.
(3) $\dim(\mathbf{S}_1)+\dim(\mathbf{S}_2) = \dim(\mathbf{V})$.

Complements.

1.5.1 Let x, y and z be three vectors in a vector space $\mathbf{V}$ satisfying $x+y+z=0$. Show that the subspaces of $\mathbf{V}$ spanned by x and y and by x and z are identical.

1.5.2 Show that the subspace $\mathbf{S} = \{0\}$ of a vector space $\mathbf{V}$ has a unique complement.

1.5.3 Consider the vector space $\mathbf{R}^3$. The vectors $(1,0,0), (0,1,0)$ generate a subspace of $\mathbf{R}^3$, say $\mathbf{S}$. Show that $\mathbf{S}\{(0,0,1)\}$ and $\mathbf{S}\{(1,1,1)\}$ are two possible complementary one-dimensional subspaces of $\mathbf{S}$. Show that, in general, the choice of a complementary subspace $\mathbf{S}^c$ of $\mathbf{S} \subset \mathbf{V}$ is not unique.

1.5.4 Let $\mathbf{S}_1$ and $\mathbf{S}_2$ be the subspaces of the vector space $\mathbf{R}^3$ spanned by $\{(1,0,0),(0,0,1)\}$ and $\{(0,1,1),(1,2,3)\}$, respectively. Find a basis for each of the subspaces $\mathbf{S}_1 \cap \mathbf{S}_2$ and $\mathbf{S}_1+\mathbf{S}_2$.

1.5.5 Let $\mathbf{F} = \{0,1,2\}$ with addition and multiplication defined modulo 3. Let $\mathbf{S}$ be the subspace of $\mathbf{F}^3$ spanned by $(0,1,2)$ and $(1,1,2)$. Identify a complement of $\mathbf{S}$.

1.5.6 Let $\mathbf{F} = \{0,1,2\}$ with addition and multiplication modulo 3. Make a complete list of all subspaces of the vector space $\mathbf{F}^3$. Count how many subspaces are there for each of the dimensions 1,2, and 3.

1.5.7 Show that the dimension of the subspace of $\mathbf{R}^6$ spanned by the following row vectors is 4.

$$\begin{matrix}
1 & 1 & 0 & 1 & 0 & 0 \\
1 & 1 & 0 & 0 & 1 & 0 \\
1 & 1 & 0 & 0 & 0 & 1 \\
1 & 0 & 1 & 1 & 0 & 0 \\
1 & 0 & 1 & 0 & 1 & 0 \\
1 & 0 & 1 & 0 & 0 & 1
\end{matrix}$$

1.5.8 Consider pq row vectors each consisting of $p+q+1$ entries arranged in q blocks of p rows each in the following way. The last p columns in each block have the same structure with ones in the diagonal and zeros elsewhere.

$$\text{Block 1} \quad \left\{\begin{matrix}
1 & 1 & 0 & \dots & 0 & 1 & 0 & \dots & 0 \\
1 & 1 & 0 & \dots & 0 & 0 & 1 & \dots & 0 \\
\vdots & \vdots & \vdots & \dots & \vdots & \vdots & \vdots & \dots & \vdots \\
1 & 1 & 0 & \dots & 0 & 0 & 0 & \dots & 1
\end{matrix}\right.$$

$$\text{Block 2} \quad \left\{\begin{matrix}
1 & 0 & 1 & \dots & 0 & 1 & 0 & \dots & 0 \\
1 & 0 & 1 & \dots & 0 & 0 & 1 & \dots & 0 \\
\vdots & \vdots & \vdots & \dots & \vdots & \vdots & \vdots & \dots & \vdots \\
1 & 0 & 1 & \dots & 0 & 0 & 0 & \dots & 1
\end{matrix}\right.$$

$$\dots \qquad \dots \qquad \dots \qquad \dots \qquad \dots$$

$$\text{Block } q \quad \left\{\begin{matrix}
1 & 0 & 0 & \dots & 1 & 1 & 0 & \dots & 0 \\
1 & 0 & 0 & \dots & 1 & 0 & 1 & \dots & 0 \\
\vdots & \vdots & \vdots & \dots & \vdots & \vdots & \vdots & \dots & \vdots \\
1 & 0 & 0 & \dots & 1 & 0 & 0 & \dots & 1
\end{matrix}\right.$$

Show that the subspace of $\mathbf{R}^{p+q+1}$ spanned by the row vectors is of dimension $p+q-1$.

1.5.9 If pq numbers a_{ij}, $i = 1, 2, \dots, p$; $j = 1, 2, \dots, q$ are such that the tetra difference

$$a_{ij} + a_{rs} - a_{is} - a_{rj} = 0$$

for all i, j, r, and s, show that

$$a_{ij} = a_i + b_j$$

for all i and j for some suitably chosen numbers $a_1, a_2, \ldots, a_p$ and $b_1, b_2, \ldots, b_q$.

(Complements 1.5.7-1.5.9 are applied in the analysis of variance of two-way-classified data in statistics.)

1.6. Linear Equations

Let $x_1, x_2, \ldots, x_m$ be fixed vectors in any vector space $\mathbf{V}(\mathbf{F})$. Consider the following homogeneous linear equation,

$$\beta_1 x_1 + \beta_2 x_2 + \ldots + \beta_m x_m = 0, \tag{1.6.1}$$

with β_i's in $\mathbf{F}$. The word "homogeneous" refers to the vector 0 that appears on the right hand side of the equality (1.6.1). If we have a non-zero vector, the equation is called non-homogeneous. The basic goal in this section is to determine β_i's satisfying equation (1.6.1). Let $b = (\beta_1, \beta_2, \ldots, \beta_m)$ be a generic symbol which is a solution of (1.6.1). The entity b can be regarded as a vector in the vector space $\mathbf{F}^m$. Let $\mathbf{S}$ be the collection of all such vectors b satisfying equation (1.6.1). We will establish some properties of the set $\mathbf{S}$.

Some comments are in order before we spell out the properties of $\mathbf{S}$. The vector $(0, 0, \ldots, 0)$ is always a member of $\mathbf{S}$. The equation (1.6.1) is intimately related to the notion of linear dependence or independence of the vectors $x_1, x_2, \ldots, x_m$ in $V(\mathbf{F})$. If $x_1, x_2, \ldots, x_m$ are linearly independent, $\beta_1 = 0, \beta_2 = 0, \ldots, \beta_m = 0$ is the only solution of (1.6.1). The set $\mathbf{S}$ has only one vector. If $x_1, x_2, \ldots, x_m$ are linearly dependent, the set $\mathbf{S}$ has more than one vector of $\mathbf{F}^m$. The objective is to explore the nature of the set $\mathbf{S}$. Another point of inquiry is why one is confined to only one equation in (1.6.1). The case of more than one equation can be handled in an analogous manner. Suppose $x_1, x_2, \ldots, x_m$ and $y_1, y_2, \ldots, y_m$ are two sets of vectors in $\mathbf{V}(\mathbf{F})$. Suppose we are interested in solving the equations

$$\beta_1 x_1 + \beta_2 x_2 + \ldots + \beta_m x_m = 0$$

$$\beta_1 y_1 + \beta_2 y_2 + \ldots + \beta_m y_m = 0$$

in unknown $\beta_1, \beta_2, \ldots, \beta_m$ in **F**. These two equations can be rewritten as a single equation

$$\beta_1(x_1, y_1) + \beta_2(x_2, y_2) + \ldots + \beta_m(x_m, y_m) = (0, 0),$$

with $(x_1, y_1), (x_2, y_2), \ldots, (x_m, y_m) \in \mathbf{V}^2(\mathbf{F})$. The treatment can now proceed in exactly the same way as for the equation (1.6.1).

P 1.6.1 **S** is a subspace of $\mathbf{F}^m$.

P 1.6.2 Let $\mathbf{V}_1$ be the vector subspace of **V** spanned by $x_1, x_2, \ldots, x_m$. Then $\dim(\mathbf{S}) = m - \dim(\mathbf{V}_1)$.

PROOF. If each $x_i = 0$, then it is obvious that $\mathbf{S} = \mathbf{F}^m, \dim(\mathbf{S}) = m$, and $\dim(\mathbf{V}_1) = 0$. Consequently, $\dim(\mathbf{S}) = m - \dim(\mathbf{V}_1)$. Assume that there exists at least one $x_i \neq 0$. Assume, without loss of generality, that $x_1, x_2, \ldots, x_r$ are linearly independent and each of $x_{r+1}, x_{r+2}, \ldots, x_m$ is a linear combination of $x_1, x_2, \ldots, x_r$. This implies that $\dim(\mathbf{V}_1) = r$. Accordingly, we can write

$$x_j = \beta_{j,1}x_1 + \beta_{j,2}x_2 + \ldots + \beta_{j,r}x_r \tag{1.6.2}$$

for each $j = r+1, r+2, \ldots, m$ and for some $\beta_{j,s}$'s in **F**. Then the vectors,

$$\begin{aligned}
b_1 &= (\beta_{r+1,1}, \beta_{r+1,2}, \ldots, \beta_{r+1,r}, -1, \ 0, \ldots, \ 0),\\
b_2 &= (\beta_{r+2,1}, \beta_{r+2,2}, \ldots, \beta_{r+2,r}, \ 0, -1, \ldots, \ 0),\\
&\ldots \qquad \ldots \qquad \ldots \qquad \ldots \qquad \ldots\\
b_{m-r} &= (\beta_{m,1}, \ \beta_{m,2}, \ldots, \ \beta_{m,r}, \ 0, \ 0, \ldots, -1),
\end{aligned} \tag{1.6.3}$$

are all linearly independent (why?) and satisfy equation (1.6.1). If we can show that the collection of vectors in (1.6.3) spans all solutions, then it follows that they form a basis for the vector space **S**, and consequently,

$$\dim(\mathbf{S}) = m - r = m - \dim(\mathbf{V}_1).$$

If $b = (\beta_1, \beta_2, \ldots, \beta_m)$ is any solution of (1.6.1), one can verify that

$$b = -\beta_{r+1}b_1 - \beta_{r+2}b_2 - \ldots - \beta_m b_{m-r},$$

i.e., b is a linear combination of $b_1, b_2, \ldots, b_{m-r}$. Use the fact that $x_1, x_2, \ldots, x_r$ are linearly independent and equation (1.6.2). This completes the proof.

A companion to the linear homogeneous equation (1.6.1) is the so-called non-homogeneous equation,

$$\beta_1 x_1 + \beta_2 x_2 + \ldots + \beta_m x_m = y,$$

for some known vector $y \neq 0$. Note that while a homogeneous equation (1.6.1) always has a solution, namely, the null vector in $\mathbf{F}^m$, a non-homogeneous equation may not have a solution. Such an equation is said to be inconsistent. For example, let $x_1 = (1,1,1), x_2 = (1,0,1)$ and $x_3 = (2,1,2)$ be three vectors in the vector space $\mathbf{R}^3(\mathbf{R})$. Then the non-homogeneous equation,

$$\beta_1 x_1 + \beta_2 x_2 + \beta_3 x_3 = (1,0,2),$$

has no solution.

P 1.6.3 The non-homogeneous equation,

$$\beta_1 x_1 + \beta_2 x_2 + \ldots + \beta_m x_m = y, y \neq 0, \tag{1.6.4}$$

admits a solution if and only if y is dependent on $x_1, x_2, \ldots, x_m$.

The property mentioned above is a reformulation of the notion of dependence of vectors. We now identify the set of solutions (1.6.4) if it admits at least one solution. If (1.6.4) admits a solution, we will use the phrase that (1.6.4) is consistent.

P 1.6.4 Assume that equation (1.6.4) has a solution. Let $b_0 = (\beta_1, \beta_2, \ldots, \beta_m)$ be any particular solution of (1.6.4). Let $\mathbf{S}_1$ be the set of all solutions of (1.6.4). Then

$$\mathbf{S}_1 = \{b_0 + b : b \in \mathbf{S}\}, \tag{1.6.5}$$

where $\mathbf{S}$ is the set of all solutions of the homogeneous equation (1.6.1).

PROOF. It is clear that for any $b \in \mathbf{S}$, $b_0 + b$ is a solution of (1.6.4). Conversely, if c is a solution of (1.6.4), we can write $c = b_0 + (c - b_0)$. Note that $c - b_0 \in \mathbf{S}$.

Note that the consistent non-homogeneous equation (1.6.4) admits a unique solution if and only if the subspace **S** contains only one vector, namely, the zero vector. Equivalent conditions are that $\dim(\mathbf{S}) = 0 = m - \dim(\mathbf{V}_1)$ or $x_1, x_2, \ldots, x_m$ are linearly independent.

A special and important case of the linear equation (1.6.4) arises when $x_1, x_2, \ldots, x_m$ belong to the vector space $\mathbf{V}(\mathbf{F}) = \mathbf{F}^k$, for some $k \geq 1$. If we write $x_i = (x_{1i}, x_{2i}, \ldots, x_{ki})$ for $i = 1, 2, \ldots, m$, with each $x_{ji} \in \mathbf{F}$, and $y = (y_1, y_2, \ldots, y_k)$ with each $y_i \in \mathbf{F}$, then the linear equation (1.6.4) can be rewritten in the form,

$$\begin{aligned} x_{11}\beta_1 + x_{12}\beta_2 + \ldots + x_{1m}\beta_m &= y_1, \\ x_{21}\beta_1 + x_{22}\beta_2 + \ldots + x_{2m}\beta_m &= y_2, \\ \ldots \qquad \ldots \qquad \ldots \qquad &. \\ x_{k1}\beta_1 + x_{k2}\beta_2 + \ldots + x_{km}\beta_m &= y_k, \end{aligned} \tag{1.6.6}$$

which is a system of k simultaneous linear equations in m unknowns $\beta_1, \beta_2, \ldots, \beta_m$. Associated with the system (1.6.6), we introduce the following vectors:

$$\begin{aligned} u_i &= (x_{i1}, x_{i2}, \ldots, x_{im}), \ i = 1, 2, \ldots, k, \\ v_i &= (x_{i1}, x_{i2}, \ldots, x_{im}, y_i), \ i = 1, 2, \ldots, k. \end{aligned}$$

For reasons that will be clear when we take up the subject of matrices, we call $x_1, x_2, \ldots, x_m$ and y as column vectors, and $u_1, u_2, \ldots, u_k, v_1, v_2, \ldots, v_k$ as row vectors. The following results have special bearing on the system (1.6.6) of equations.

P 1.6.5 The maximal number, g, of linearly independent column vectors among $x_1, x_2, \ldots, x_m$ is the same as the maximal number, s, of linearly independent row vectors among $u_1, u_2, \ldots, u_k$.

PROOF. The vector y has no bearing on the property enunciated above. Assume that each $y_i = 0$. If we arrange mk elements from **F** in the form of a rectangular grid consisting of k rows and m columns, each row can be viewed as a vector in the vector space $\mathbf{F}^m$ and each column can be viewed as a vector in the vector space $\mathbf{F}^k$. The property under discussion is concerned about the maximal number of linearly independent rows and of independent columns. We proceed with the proof as follows. The case that every $u_i = 0$ can be handled easily. Assume that there is at least one vector $u_i \neq 0$. Assume, without loss of generality, that $u_1, u_2, \ldots, u_s$ are linearly independent and each u_j, for $j = s+1, s+2, \ldots, k$, is a linear combination of $u_1, u_2, \ldots, u_s$. Consider the subsystem of equations (1.6.6) with y_i's taken as zeros consisting of the first s equations

$$x_{i1}\beta_1 + x_{i2}\beta_2 + \ldots + x_{im}\beta_m = 0, \; i = 1, \ldots, s. \tag{1.6.7}$$

Let $\mathbf{S}$ be the collection of all solutions of (1.6.6) and $\mathbf{S}^*$ that of (1.6.7). It is clear that $\mathbf{S} = \mathbf{S}^*$. Let $\mathbf{V}_1$ be the vector space spanned by $x_1, x_2, \ldots, x_m$. Let $\dim(\mathbf{V}_1) = g$. By **P 1.6.2**, $\dim(\mathbf{S}) = m - \dim(\mathbf{V}_1) = m - g$. The reduced system of equations (1.6.7) can be rewritten in the format of (1.6.1) as

$$\beta_1 x_1^* + \beta_2 x_2^* + \ldots + \beta_m x_m^* = 0,$$

with $x_1^*, x_2^*, \ldots, x_m^*$, now, in $\mathbf{F}^s$. Let $\mathbf{V}_1^*$ be the subspace of $\mathbf{F}^s$ spanned by $x_1^*, x_2^*, \ldots, x_m^*$. (Observe that the components of each x_i^* are precisely the first s components of x_i.) Consequently, $\dim(\mathbf{V}_1^*) \leq \dim(\mathbf{F}^s) = s$. By **P 1.6.2**, $\dim(\mathbf{S}^*) = m - \dim(\mathbf{V}_1^*) \geq m - s$, which implies that $m - g \geq m - s$, or, $g \leq s$. By interchanging the roles of rows and columns, we would obtain the inequality $s \leq g$. Hence $s = g$.

The above result can be paraphrased from an abstract point of view. Let the components $x_1, x_2, \ldots, x_m$ be arranged in the form of a rectangular grid consisting of k rows and m columns so that the entries in the i-th column are precisely the entries of x_i. We have labelled the rows of the rectangular grid by $u_1, u_2, \ldots, u_k$. The above result establishes that the maximal number of linearly independent vectors among $x_1, x_2, \ldots, x_m$ is precisely the maximal number of linearly independent vectors among $u_1, u_2, \ldots, u_k$. We can stretch this analogy a little further. The type of relationship that exists between $x_1, x_2, \ldots, x_m$ and $u_1, u_2, \ldots, u_k$ is precisely the same that exists between $x_1, x_2, \ldots, x_m, y$ and $v_1, v_2, \ldots, v_k$. Consequently, the maximal number of linearly independent vectors among $x_1, x_2, \ldots, x_m, y$ is the same as the maximal number of linearly independent vectors among $v_1, v_2, \ldots, v_k$. This provides a useful characterization of consistency of a system of nonhomogeneous linear equations.

P 1.6.6 A necessary and sufficient condition that the non-homogeneous system (1.6.6) of equations has a solution is that the maximal number, g, of linearly independent vectors among $u_1, u_2, \ldots, u_k$ is the same as the maximal number, h, of linearly independent vectors among the augmented vectors $v_1, v_2, \ldots, v_k$.

PROOF. By **P 1.6.3**, equations (1.6.6) admit a solution if and only if the maximal number of linearly independent vectors among $x_1, x_2, \ldots, x_m$ is the same as the maximal number of linearly independent vectors among $x_1, x_2, \ldots, x_m, y$. But the maximal number of linearly independent vectors among $x_1, x_2, \ldots, x_m, y$ is the same as the maximal number of linearly independent vectors among $v_1, v_2, \ldots, v_k$. Consequently, a solution exists for (1.6.6) if and only if $g = s = h$.

The systems of equations described in (1.6.6) arises in many areas of scientific pursuit. One of the pressing needs is to devise a criterion whose verification guarantees a solution to the system. One might argue that **P 1.6.5** and **P 1.6.6** do provide criteria for the consistency of the system. But these criteria are hard to verify. The following proposition provides a necessary and sufficient condition for the consistency of the system (1.6.6). At a first glance, the condition may look very artificial. But time and again, this is the condition that becomes easily verifiable to check on the consistency of the system (1.6.6).

P 1.6.7 The system (1.6.6) of non-homogeneous linear equations admits a solution if and only if

$$\varepsilon_1 y_1 + \varepsilon_2 y_2 + \ldots + \varepsilon_k y_k = 0$$

whenever

$$\varepsilon_1 u_1 + \varepsilon_2 u_2 + \ldots + \varepsilon_k u_k = 0 \tag{1.6.8}$$

for any $\varepsilon_1, \varepsilon_2, \ldots, \varepsilon_k$ in $\mathbf{F}$.

PROOF. Suppose the system (1.6.6) admits a solution. Suppose $\varepsilon_1 u_1 + \varepsilon_2 u_2 + \ldots + \varepsilon_k u_k = 0$ for some $\varepsilon_1, \varepsilon_2, \ldots, \varepsilon_k$ in $\mathbf{F}$. Multiply the i-th equation of (1.6.6) by ε_i and then sum over i. It now follows that $\varepsilon_1 y_1 + \varepsilon_2 y_2 + \ldots + \varepsilon_k y_k = 0$. Conversely, view

$$\varepsilon_1 u_1 + \varepsilon_2 u_2 + \ldots + \varepsilon_k u_k = 0$$

as a system of homogeneous linear equations in k unknowns $\varepsilon_1, \varepsilon_2, \ldots, \varepsilon_k$. Consider the system of homogeneous linear equations

$$\varepsilon_1 v_1 + \varepsilon_2 v_2 + \ldots + \varepsilon_k v_k = 0$$

in k unknowns $\varepsilon_1, \varepsilon_2, \ldots, \varepsilon_k$. By (1.6.8), these two systems of equations have the same set of solutions. The dimensions of the spaces of solutions are $k - s$ and $k - h$, respectively. Thus we have $k - s = k - h$, or $s = h$. By **P 1.6.6**, the system (1.6.6) has a solution.

Complements

1.6.1 Let $\mathbf{Q}$ be the field of rational numbers. Consider the system of equations.

$$2\beta_1 + \beta_3 - \beta_4 = 0$$
$$\beta_2 - 2\beta_3 - 3\beta_4 = 0$$

in unknown $\beta_1, \beta_2, \beta_3, \beta_4 \in \mathbf{Q}$. Determine the dimension of the solution subspace $\mathbf{S}$ of $\mathbf{Q}^4$. Show that

$$2\beta_1 + \beta_3 - \beta_4 = y_1$$
$$\beta_2 - 2\beta_3 - 3\beta_4 = y_2$$

admit a solution for every y_1 and y_2 in $\mathbf{Q}$.

1.6.2 Consider the system (1.6.6) of equations with $y_1 = y_2 = \ldots = y_k = 0$. Show that the system has a non-trivial solution if $k < m$.

1.7. Dual Space

One way to understand the intricate structure of a vector space is to pursue the linear functionals defined on the vector space. The duality that reigns between the vector space and its space of linear functionals aids and reveals what lies inside a vector space.

DEFINITION 1.7.1. A function f defined on a vector space $\mathbf{V}(\mathbf{F})$ taking values in $\mathbf{F}$ is said to be a linear functional if

$$f(\alpha_1 x_1 + \alpha_2 x_2) = \alpha_1 f(x_1) + \alpha_2 f(x_2)$$

for every x_1, x_2 in $\mathbf{V}(\mathbf{F})$ and α_1, α_2 in $\mathbf{F}$.

One can view the field $\mathbf{F}$ as a vector space over the field $\mathbf{F}$ itself. Under this scenario, a linear functional is simply a homomorphism from the vector space $\mathbf{V}(\mathbf{F})$ to the vector space $\mathbf{F}(\mathbf{F})$.

EXAMPLE 1.7.2. Consider the vector space $\mathbf{R}^n$. Let $\alpha_1, \alpha_2, \ldots, \alpha_n$ be fixed real numbers. For $x = (\xi_1, \xi_2, \ldots, \xi_n) \in \mathbf{R}^n$. let

$$f(x) = \alpha_1\xi_1 + \alpha_2\xi_2 + \ldots + \alpha_n\xi_n.$$

The map f is a linear functional. If $\alpha_i = 1$ and $\alpha_j = 0$ for $j \neq i$ for some fixed $1 \leq i \leq n$, then the map f is called the i-th co-ordinate functional.

EXAMPLE 1.7.3. Let $\mathbf{P}_n$ be the collection of all polynomials $x(\cdot)$ of degree $< n$ with coefficients in the field $\mathbf{C}$ of complex numbers. We have seen that $\mathbf{P}_n$ is a vector space over the field $\mathbf{C}$. Let $\alpha(\cdot)$ be any complex-valued integrable function defined on a finite interval $[a, b]$. Then for $x(\cdot)$ in $\mathbf{P}_n$, let

$$f(x) = \int_a^b \alpha(t)x(t)\,dt.$$

Then f is a linear functional on $\mathbf{P}_n$.

It is time to introduce the notion of a dual space. Later, we will also determine the structure of a linear functional on a finite dimensional vector space.

DEFINITION 1.7.4. Let $\mathbf{V}(\mathbf{F})$ be any vector space and $\mathbf{V}'$ the space of all linear functionals defined on $\mathbf{V}(\mathbf{F})$. Let us denote by 0 the linear functional which assigns the value zero of $\mathbf{F}$ for every element in $\mathbf{V}(\mathbf{F})$. The set $\mathbf{V}'$ is called the dual space of $\mathbf{V}(\mathbf{F})$.

We will now equip the space $\mathbf{V}'$ with a structure so that it becomes a vector space over the field $\mathbf{F}$. Let $f_1, f_2 \in \mathbf{V}'$ and $\alpha_1, \alpha_2 \in \mathbf{F}$. Then the function f defined by

$$f(x) = \alpha_1 f_1(x) + \alpha_2 f_2(x), \ x \in \mathbf{V}(\mathbf{F})$$

is clearly a linear functional on $\mathbf{V}(\mathbf{F})$. We denote the functional f by $\alpha_1 f_1 + \alpha_2 f_2$. This basic operation includes, in its wake, the binary operation of addition and scalar multiplication on $\mathbf{V}'$ by the elements of the field $\mathbf{F}$. Under these operations of addition and scalar multiplication, $\mathbf{V}'$ becomes a vector space over the field $\mathbf{F}$.

P 1.7.5 Let $x_1, x_2, \ldots, x_k$ be a basis of a finite dimensional vector space $\mathbf{V}(\mathbf{F})$. Let $\alpha_1, \alpha_2, \ldots, \alpha_k$ be a given set of scalars from $\mathbf{F}$. Then there exists one and only one linear functional f on $\mathbf{V}(\mathbf{F})$ such that

$$f(x_i) = \alpha_i, \ i = 1, 2, \ldots, k.$$

PROOF. Any vector x in $\mathbf{V}(\mathbf{F})$ has a unique representation $x = \xi_1 x_1 + \xi_2 x_2 + \ldots + \xi_k x_k$ for some scalars $\xi_1, \xi_2, \ldots, \xi_k$ in $\mathbf{F}$. If f is any linear functional on $\mathbf{V}(\mathbf{F})$, then

$$f(x) = \xi_1 f(x_1) + \xi_2 f(x_2) + \ldots + \xi_k f(x_k),$$

which means that the value $f(x)$ is uniquely determined by the values of f at $x_1, x_2, \ldots, x_k$. The function f defined by

$$f(x) = \xi_1 \alpha_1 + \xi_2 \alpha_2 + \ldots + \xi_k \alpha_k$$

for $x = \xi_1 x_1 + \xi_2 x_2 + \ldots + \xi_k x_k \in \mathbf{V}(\mathbf{F})$ is clearly a linear functional satisfying $f(x_i) = \alpha_i$ for each i. Thus the existence and uniqueness follow.

P 1.7.6 Let $x_1, x_2, \ldots, x_k$ be a basis of a finite dimensional vector space $\mathbf{V}$. Then there exists a unique set $f_1, f_2, \ldots, f_k$ of linear functionals in $\mathbf{V}'$ such that

$$f_i(x_j) = \begin{cases} 1, & \text{if } \ i = j, \\ 0, & \text{if } \ i \neq j, \end{cases} \tag{1.7.1}$$

and these functionals form a basis for the vector space $\mathbf{V}'$. Consequently, $\dim(\mathbf{V}) = \dim(\mathbf{V}')$.

PROOF. From **P 1.7.5**, the existence of k linear functionals satisfying (1.7.1) is established. We need to demonstrate that these linear functionals are linearly independent and form a basis for the vector space $\mathbf{V}'$. Let f be any linear functional in $\mathbf{V}'$. Let $f(x_i) = \alpha_i$, $i = 1, 2, \ldots, k$. Note that $f = \alpha_1 f_1 + \alpha_2 f_2 + \ldots + \alpha_k f_k$. The linear functionals $f_1, f_2, \ldots, f_k$ do indeed span the vector space $\mathbf{V}'$. As for their linear independence, suppose $\beta_1 f_1 + \beta_2 f_2 + \ldots + \beta_k f_k = 0$ for some scalars $\beta_1, \beta_2, \ldots, \beta_k$ in $\mathbf{F}$. Observe that $0 = (\beta_1 f_1 + \beta_2 f_2 + \ldots + \beta_k f_k)(x_i) = \beta_i$

for each $i = 1, 2, \ldots, k$. Hence linear independence of these functionals follows. The result that the dimensions of the vector space $\mathbf{V}$ and its dual space are identical is obvious now.

The basis $f_1, f_2, \ldots, f_k$ so arrived above is called the dual basis of $x_1, x_2, \ldots, x_k$. Now we are ready to prove the separation theorem.

P 1.7.7 Let u and v be two distinct vectors in a vector space $\mathbf{V}$. Then there exists a linear functional f in $\mathbf{V}'$ such that $f(u) \neq f(v)$. Equivalently, for any non-zero vector x in $\mathbf{V}$, there exists a linear functional f in $\mathbf{V}'$ such that $f(x) \neq 0$.

PROOF. Let $x_1, x_2, \ldots, x_k$ be a basis of $\mathbf{V}$ and $f_1, f_2, \ldots, f_k$ its dual basis. Write $x = \xi_1 x_1 + \xi_2 x_2 + \ldots + \xi_k x_k$ for some scalars $\xi_1, \xi_2, \ldots, \xi_k$ in $\mathbf{F}$. If x is non-zero, there exists $1 \leq i \leq k$ such that ξ_i is non-zero. Note that $f_i(x) = \xi_i \neq 0$. The first statement of **P 1.7.7** follows if we take $x = u - v$.

Since $\mathbf{V}'$ is a vector space, we can define its dual vector space $\mathbf{V}''$ as the space of all linear functionals defined on $\mathbf{V}'$. From **P 1.7.6**, we have $\dim(\mathbf{V}) = \dim(\mathbf{V}') = \dim(\mathbf{V}'')$. Consequently, all these three vector spaces are isomorphic. But there is a natural isomorphic map from $\mathbf{V}$ to $\mathbf{V}''$, which we would like to identify explicitly.

P 1.7.8 For every linear functional z_0 in $\mathbf{V}''$, there exists a unique x_0 in $\mathbf{V}$ such that

$$z_0(f) = f(x_0) \quad \text{for every } f \text{ in } \mathbf{V}'.$$

The correspondence $z_0 \Leftrightarrow x_0$ is an isomorphism between $\mathbf{V}''$ and $\mathbf{V}$.

PROOF. Let $f_1, f_2, \ldots, f_k$ be a basis of $\mathbf{V}'$. Given x_0 in $\mathbf{V}$, there exists a unique z_0 in $\mathbf{V}''$ such that

$$z_0(f_1) = f_1(x_0), z_0(f_2) = f_2(x_0), \ldots, z_0(f_k) = f_k(x_0).$$

We refer to **P 1.7.5**. Consequently, $z_0(f) = f(x_0)$ for all f in $\mathbf{V}'$. If x_1 and x_2 are two distinct vectors in $\mathbf{V}$, then the corresponding vectors z_1 and z_2 in $\mathbf{V}''$ must be distinct. If not, $(z_1 - z_2)(f) = 0 = f(x_1) - f(x_2) = f(x_1 - x_2)$ for all f in $\mathbf{V}'$. But this is impossible in view of **P 1.7.7**. Thus we observe that the correspondence $x_0 \Leftrightarrow z_0$

enunciated above is an injection. It is also clear that this association is a homomorphism. The isomorphism of this map now follows from the fact that $\dim(\mathbf{V}) = \dim(\mathbf{V}'')$.

Now that we have dual spaces in circulation, we can introduce the notion of annihilator of any subset (subspace or not) of a vector space.

DEFINITION 1.7.9. The annihilator of a subset $\mathbf{S}$ of a vector space $\mathbf{V}$ is the set $\mathbf{S}^a$ of linear functionals given by

$$\mathbf{S}^a = \{f \in \mathbf{V}' : f(x) = 0 \quad \text{for every } x \text{ in } \mathbf{S}\}.$$

It is clear that the annihilator $\mathbf{S}^a$ is a subspace of the vector space $\mathbf{V}'$ regardless of whether $\mathbf{S}$ is a subspace or not. If $\mathbf{S}$ contains only the null vector, then $\mathbf{S}^a = \mathbf{V}'$. If $\mathbf{S} = \mathbf{V}$, then $\mathbf{S}^a = \{0\}$. If $\mathbf{S}$ contains a non-zero vector, then $\mathbf{S}^a \neq \mathbf{V}'$ in view of **P 1.7.7**.

P 1.7.10 If $\mathbf{S}$ is a subspace of a vector space $\mathbf{V}$, then $\dim(\mathbf{S}^a) = \dim \mathbf{V} - \dim(\mathbf{S})$.

PROOF. Let $x_1, x_2, \dots, x_r$ be a basis of the subspace $\mathbf{S}$ which can be extended to a full basis $x_1, x_2, \dots, x_r, x_{r+1}, x_{r+2}, \dots, x_k$ of $\mathbf{V}$. Let $f_1, f_2, \dots, f_k$ be the dual basis of $\mathbf{V}'$. Let $f \in \mathbf{S}^a$. We can write $f = \alpha_1 f_1 + \alpha_2 f_2 + \dots + \alpha_k f_k$ for some scalars $\alpha_1, \alpha_2, \dots, \alpha_k$ in $\mathbf{F}$. Observe that for each $1 \leq i \leq r, 0 = f(x_i) = \alpha_i$. Consequently, f is a linear combination of $f_{r+1}, f_{r+2}, \dots, f_k$ only, i.e., $f = \alpha_{r+1} f_{r+1} + \alpha_{r+2} f_{r+2} + \dots + \alpha_k f_k$. This implies that $\mathbf{S}^a$ is a subspace of the span $Sp(f_{r+1}, f_{r+2}, \dots, f_k)$ of $f_{r+1}, f_{r+2}, \dots, f_k$. By the very construction of the dual basis, $f_i(x_j) = 0$ for every $1 \leq j \leq r$ and $r + 1 \leq i \leq k$. Consequently, each f_i, $r + 1 \leq i \leq k$, belongs to $\mathbf{S}^a$. Thus we observe that $Sp(f_{r+1}, f_{r+2}, \dots, f_k) \subseteq \mathbf{S}^a$. We have now identified precisely what $\mathbf{S}^a$ is, i.e., $\mathbf{S}^a = Sp(f_{r+1}, f_{r+2}, \dots, f_k)$. From this it follows that $\dim(\mathbf{S}^a) = k - r = \dim(\mathbf{V}) - \dim(\mathbf{S})$.

The operation of annihilation can be extended. We start with a subspace $\mathbf{S}$ of a vector space $\mathbf{V}$, and arrive at its annihilator $\mathbf{S}^a$ which is a subspace of $\mathbf{V}'$. Now we can look at the annihilator $\mathbf{S}^{aa}$ of the subspace $\mathbf{S}^a$. Of course, $\mathbf{S}^{aa}$ would be a subspace of $\mathbf{V}''$. This chain could go on forever.

P 1.7.11 If $\mathbf{S}$ is a subspace of a vector space $\mathbf{V}$, then $\mathbf{S}^{aa}$ is isomorphic to $\mathbf{S}$.

PROOF. Consider the bijection as identified between $\mathbf{V}$ and $\mathbf{V}''$ in **P 1.7.8**. For every x_0 in $\mathbf{V}$ there exists a unique z_0 in $\mathbf{V}''$ such that $z_0(f) = f(x_0)$ for every f in $\mathbf{V}'$. If $z_0 \in \mathbf{S}^{aa}$, then $z_0(f) = f(x_0) = 0$ for every f in $\mathbf{S}^a$. Since $\mathbf{S}$ is a subspace, this implies that $x_0 \in \mathbf{S}$. In a similar vein, one can show that if $x_0 \in \mathbf{S}$, then $z_0 \in \mathbf{S}^{aa}$. The isomorphism that has been developed between $\mathbf{V}$ and $\mathbf{V}''$ in **P 1.7.8** when restricted to the subspace $\mathbf{S}$ is an isomorphism between $\mathbf{S}$ and $\mathbf{S}^{aa}$.

P 1.7.12 If $\mathbf{S}_1$ and $\mathbf{S}_2$ are subspaces of a vector space $\mathbf{V}$, then

$$(\mathbf{S}_1 \cap \mathbf{S}_2)^a = \mathbf{S}_1^a + \mathbf{S}_2^a$$

$$(\mathbf{S}_1 + \mathbf{S}_2)^a = \mathbf{S}_1^a \cap \mathbf{S}_2^a.$$

These identities follow from the definition of annihilator.

Complements

1.7.1 If f is a non-zero linear functional from a vector space $\mathbf{V}(\mathbf{F})$ to a field $\mathbf{F}$, show that f is a surjection (onto map).
1.7.2 If f_1 and f_2 are two linear functionals on a vector space $\mathbf{V}(\mathbf{F})$ satisfying $f_1(x) = 0$ whenever $f_2(x) = 0$ for x in $\mathbf{V}(\mathbf{F})$, show that $f_1 = \alpha f_2$ for some α in $\mathbf{F}$.
1.7.3 Let $\mathbf{F}$ be a field and $\mathbf{P}_n(t)$ the vector space of all polynomials of degree less than n with coefficients from the field $\mathbf{F}$. For any $x = \sum_{i=0}^{n-1} d_i t^i$ in $\mathbf{P}_n$ with d_i's in $\mathbf{F}$, define

$$f(x) = \sum_{i=0}^{n-1} d_i \beta_i$$

for any fixed choice $\beta_0, \beta_1, \beta_2, \ldots, \beta_{n-1}$ of scalars from $\mathbf{F}$. Show that f is a linear functional. Show that any linear functional on $\mathbf{P}_n(t)$ arises this way.
1.7.4 Let $\mathbf{F} = \{0, 1, 2\}$ with addition and multiplication modulo 3. Spell out all the linear functionals explicitly on $\mathbf{F}^3$.
1.7.5 The vectors (1,1,1,1), (1,1,-1,-1), (1,-1,1,-1), and (1,-1,-1,1) form a basis of the vector space $\mathbf{R}^4$. Let f_1, f_2, f_3, and f_4 be the dual basis. Evaluate each of these linear functionals at $x = (1, 2, 3, 4)$.

1.7.6 Let f be a linear functional on a vector space $\mathbf{V}(\mathbf{F})$ and

$$\mathbf{S} = \{x \in \mathbf{V}(\mathbf{F}) : f(x) = 0\}.$$

Show that $\mathbf{S}$ is a subspace of $\mathbf{V}(\mathbf{F})$. Comment on the possible values of the dimension of $\mathbf{S}$.

1.7.7 If $\mathbf{S}$ is any subset of a vector space $\mathbf{V}$, show that $\mathbf{S}^{aa}$ is isomorphic with the subspace spanned by $\mathbf{S}$.

1.8.8 If $\mathbf{S}_1$ and $\mathbf{S}_2$ are two subsets of a vector space $\mathbf{V}$ such that $\mathbf{S}_1 \subset \mathbf{S}_2$, show that $\mathbf{S}_2^a \subset \mathbf{S}_1^a$.

1.8. Quotient Space

There are many ways of generating new vector spaces from a given vector space. Subspaces are one lot. Dual spaces are another. In this section, we will introduce quotient spaces.

DEFINITION 1.8.1. Let $\mathbf{S}$ be a subspace of a vector space $\mathbf{V}$. Let x be an element of $\mathbf{V}$. Then $\mathbf{S}_x = x + \mathbf{S}$ is said to be a coset of $\mathbf{S}$.

We have seen what cosets are in the context of groups. The idea is exactly the same. The group structure under focus here is the addition of vectors of a vector space. We define the following operations on the cosets of $\mathbf{S}$.

Addition: For x, y in $\mathbf{V}$, let

$$\mathbf{S}_x + \mathbf{S}_y = \{u + v : u \in \mathbf{S}_x,\ v \in \mathbf{S}_y\}.$$

Scalar multiplication: For α in $\mathbf{F}$ and x in $\mathbf{V}$, let

$$\alpha\mathbf{S}_x = \begin{cases} \{\alpha u : u \in \mathbf{S}_x\} & \text{if } \alpha \neq 0, \\ \mathbf{S} & \text{if } \alpha = 0. \end{cases}$$

The operation of addition defined above is nothing new. We have introduced this operation in the context of complementary subspaces. The following properties of these operations can be verified easily.

(1) $\mathbf{S}_x + \mathbf{S}_y = \mathbf{S}_{x+y}$ for any x, y in $\mathbf{V}$. This means that $\mathbf{S}_x + \mathbf{S}_y$ is also a coset of $\mathbf{S}$.

(2) $\mathbf{S}_x + \mathbf{S} = \mathbf{S}_x$ for all x in $\mathbf{V}$.

(3) $\mathbf{S}_x + \mathbf{S}_{-x} = \mathbf{S}$ for all x in $\mathbf{V}$.

In addition to the above properties, the operation of addition satisfies commutative and associative laws. The set of all distinct cosets of $\mathbf{S}$ thus becomes a commutative group. The zero element of the group is $\mathbf{S}$. The negative of $\mathbf{S}_x$ is $\mathbf{S}_{-x}$. The scalar multiplication introduced above on the cosets satisfies all the rules of a vector space. Consequently, the set of all cosets form a vector space which is called the quotient space associated with the subspace $\mathbf{S}$ and is denoted by $\mathbf{V}/\mathbf{S}$. The following result identifies what the quotient space is like.

P 1.8.2 The quotient space is isomorphic to every complement of the subspace $\mathbf{S}$ of $\mathbf{V}$.

PROOF. Let $\mathbf{S}^c$ be a complement of $\mathbf{S}$. Let $f : \mathbf{S}^c \to \mathbf{V}/\mathbf{S}$ be defined by $f(x) = \mathbf{S}_x, x \in \mathbf{S}^c$. We show that f is an isomorphism. Let x_1 and x_2 be two distinct points of $\mathbf{S}^c$. Then $\mathbf{S}_{x_1} \neq \mathbf{S}_{x_2}$. If not, for any given z_1 in $\mathbf{S}$ there exists z_2 in $\mathbf{S}$ such that $x_1 + z_1 = x_2 + z_2 = x$, say. What this means is that x has two distinct decompositions, which is not possible. Consequently, f is an injection. Let $\mathbf{K}$ be any coset of $\mathbf{S}$. Then $\mathbf{K} = x + \mathbf{S}$ for some x in $\mathbf{V}$. Since x admits a unique decomposition, we can write $x = x_0 + x_1$ with $x_0 \in \mathbf{S}$ and $x_1 \in \mathbf{S}^c$. Consequently, $\mathbf{K} = x + \mathbf{S} = x_0 + x_1 + \mathbf{S} = x_1 + (x_0 + \mathbf{S}) = x_1 + \mathbf{S}$. Thus $\mathbf{K}$ is of the form $x_1 + \mathbf{S}$ for some x_1 in $\mathbf{S}^c$. This shows that f is a surjection. It can be verified that f is a linear map. Hence f is an isomorphism.

P 1.8.3 For any subspace $\mathbf{S}$ of a vector space $\mathbf{V}, \dim(\mathbf{V}/\mathbf{S}) = \dim(\mathbf{V}) - \dim(\mathbf{S})$.

This result follows from the fact that $\mathbf{V}/\mathbf{S}$ is isomorphic to a complement of $\mathbf{S}$. If one is searching for a complement of $\mathbf{S}, \mathbf{V}/\mathbf{S}$ is a natural candidate!

Complements

1.8.1 Let $\mathbf{F} = \{0, 1, 2\}$ with addition and multiplication modulo 3. Let $\mathbf{S}$ be the subspace of $\mathbf{F}^3$ spanned by (1,0,0) and (1,0,2). Construct the quotient space $\mathbf{F}^3/S$. Exhibit a complement of the subspace $\mathbf{S}$ different from the quotient space $\mathbf{F}^3/\mathbf{S}$.

1.9. Projective Geometry

It is time to enjoy the fruits of labor expended so far. We will present some rudiments of projective geometry just sufficient for our needs.

Some applications of projective geometry include the construction of orthogonal latin squares and balanced incomplete block designs.

DEFINITION 1.9.1. Let S be a set of elements and $\mathbf{S}$ a collection of subsets of S. The pair $(S, \mathbf{S})$ is said to be a projective geometry if

(1) given any two distinct elements in S there is a unique set in $\mathbf{S}$ containing these two points, and
(2) any two distinct sets in $\mathbf{S}$ have only one member of S in common.

In the picturesque language of geometry, the members of S are called *points* and the sets in $\mathbf{S}$ *lines*. The condition (1) translates into the dictum that there is one and only one line passing through any two given distinct points. The condition (2) aligns to the statement that any two distinct lines meet at one and only one point. If the set S is finite, the associated geometry is called a finite projective geometry. In this section, we show that such geometries can be constructed from vector spaces.

Consider a three-dimensional vector space $\mathbf{V}(\mathbf{F})$ over a finite field $\mathbf{F}$ consisting of s elements, say. If x_1, x_2, and x_3 are linearly independent vectors in $\mathbf{V}(\mathbf{F})$, we can identify $\mathbf{V}(\mathbf{F})$ as

$$\mathbf{V}(\mathbf{F}) = \{\alpha_1 x_1 + \alpha_2 x_2 + \alpha_3 x_3 : \alpha_1, \alpha_2, \alpha_3 \in \mathbf{F}\}.$$

Since each α_i can be chosen in different ways, the number of distinct vectors in $\mathbf{V}(\mathbf{F})$ is s^3. We now build a finite projective geometry out of $\mathbf{V}(\mathbf{F})$. Let S be the collection of all one-dimensional subspaces (points) of $\mathbf{V}(\mathbf{F})$. Consider any two-dimensional subspace of $\mathbf{V}(\mathbf{F})$. Any such subspace can be written as the union of all its one-dimensional subspaces. Instead of taking the union, we identify the two-dimensional subspace by the set of all its one-dimensional subspaces. With this understanding, let $\mathbf{S}$ be the collection of all two-dimensional subspaces (lines) of $\mathbf{V}(\mathbf{F})$. We provide examples in the later part of the section. The important point that emerges out of the discussion carried out so far is that the pair $(S, \mathbf{S})$ is a projective geometry.

P 1.9.2 The pair $(S, \mathbf{S})$ is a finite projective geometry. Moreover:

(1) The number of points, i.e., the cardinality of S is $s^2 + s + 1$.
(2) The number of lines, i.e., the cardinality of $\mathbf{S}$ is $s^2 + s + 1$.
(3) The number of points on each line is $s + 1$.

PROOF. A one-dimensional subspace of $\mathbf{V}(\mathbf{F})$ is spanned by a non-zero vector of $\mathbf{V}(\mathbf{F})$. For each non-zero vector x in $\mathbf{V}(\mathbf{F})$, let $M(x)$ be the one-dimensional subspace spanned by x. There are $s^3 - 1$ non-zero vectors in $\mathbf{V}(\mathbf{F})$. But the one-dimensional subspaces spanned by each of these vectors are not necessarily distinct in view of the fact that $M(x) = M(\alpha x)$ for each non-zero α in $\mathbf{F}$ and non-zero x in $\mathbf{V}(\mathbf{F})$. There are $s - 1$ vectors giving rise to the same one-dimensional subspace. Consequently, the total number of one-dimensional subspaces is $(s^3 - 1)/(s - 1) = s^2 + s + 1$. Thus the cardinality of the set S is $s^2 + s + 1$. This proves (1). Any two-dimensional subspace of $\mathbf{V}(\mathbf{F})$ is spanned by two linearly independent vectors of $\mathbf{V}(\mathbf{F})$. For any two linearly independent vectors x_1, x_2 of $\mathbf{V}(\mathbf{F})$, let $M(x_1, x_2)$ be the two-dimensional subspace of $\mathbf{V}(\mathbf{F})$ spanned by x_1, x_2. The total number of pairs of linearly independent vectors is $(s^3 - 1)(s^3 - s)/2$. (Why?) The total number of non-zero vectors in any two-dimensional subspace is $s^2 - 1$. The subspace $M(x_1, x_2)$ can also be spanned by any two linearly independent vectors in $M(x_1, x_2)$. The total number of pairs of linearly independent vectors in $M(x_1, x_2)$ is $(s^2 - 1)(s^2 - s)/2$. (Why?) Consequently, the total number of different two-dimensional subspaces is $[(s^3-1)(s^3-s)/2]/[(s^2-1)(s^2-s)/2] = s^2+s+1$. This proves (2). Using the argument similar to the one used in establishing (1), the total number of distinct one-dimensional subspaces of a two-dimensional subspace is $(s^2 - 1)/(s - 1) = s + 1$. This proves (3). It remains to be shown that $(S, \mathbf{S})$ is a projective geometry. Let $M(x_1)$ and $M(x_2)$ be two distinct one-dimensional spaces of $\mathbf{V}(\mathbf{F})$, i.e., distinct points of S. Let $M(x_3, x_4)$ be any two-dimensional subspace containing both $M(x_1)$ and $M(x_2)$. Since $x_1, x_2 \in M(x_3, x_4)$, it follows that $M(x_1, x_2) = M(x_3, x_4)$. Consequently, there is one and only one line containing any two points. Consider now two different two-dimensional subspaces $M(x_1, x_2)$ and $M(x_3, x_4)$ of $\mathbf{V}(\mathbf{F})$. The vectors x_1, x_2, x_3 and x_4 are linearly dependent as our vector space is only three dimensional. There exist scalars α_1, α_2, not both zero, and α_3, α_4, not both zero, in $\mathbf{F}$ such that

$$\alpha_1 x_1 + \alpha_2 x_2 = \alpha_3 x_3 + \alpha_4 x_4 = y, \quad \text{say.}$$

(Why?) Obviously, y is non-zero. Clearly, the point y belongs to both the lines $M(x_1, x_2)$ and $M(x_3, x_4)$. This means that there is at least one point common to any two distinct lines. Suppose $M(y_1)$ and $M(y_2)$

are two distinct one-dimensional subspaces common to both $M(x_1, x_2)$ and $M(x_3, x_4)$. Then it follows that

$$M(y_1, y_2) = M(x_1, x_2) = M(x_3, x_4).$$

This is a contradiction. This shows that any two distinct lines intersect at one and only one point. The proof is complete.

The projective geometry described above is denoted by $PG(2, s)$. The relevance of the number s in the notation is clear. The number two is precisely the dimension of the vector space $\mathbf{F}^3$ less one. The number s could not be any integer. Since the cardinality of the Galois field $\mathbf{F}$ is s, $s = p^m$ for some prime number p and integer $m \geq 1$. See Section 1.1.

A concrete construction of the projective geometry $PG(2, s)$ is not hard. For $PG(2, s)$, what we need is a Galois field $\mathbf{F}$ with s elements, which we have described in Example 1.1.10 in a special case. Let the s elements of $\mathbf{F}$ be denoted by $\alpha_0 = 0, \alpha_1 = 1, \alpha_2, \ldots, \alpha_{s-1}$. The underlying three-dimensional vector space $\mathbf{V}(\mathbf{F})$ over the field $\mathbf{F}$ can be taken to be $\mathbf{F}^3$. We now identify explicitly one- and two-dimensional subspaces of $\mathbf{F}^3$.

(a) *one-dimensional subspaces*

Any one-dimensional subspace of $\mathbf{F}^3$ is one of the following types:

(1) Span $\{(1, \alpha_i, \alpha_j)\}$, $i, j = 0, 1, 2, \ldots, s-1$.

(2) Span $\{(0, 1, \alpha_i)\}$, $i = 0, 1, 2, \ldots, s-1$.

(3) Span $\{(0, 0, 1)\}$.

One way to see this is to observe, first, that the vectors $(1, \alpha_i, \alpha_j), (0, 1, \alpha_k)$ and $(0, 0, 1)$ are always (Why?) linearly independent for any $i, j, k = 0, 1, 2, \ldots, s-1$. Secondly, the totality of all one-dimensional subspaces listed above is exactly $s^2 + s + 1$.

In our projective geometry $PG(2, s)$, each of these one-dimensional subspaces constitute points of the set S. For ease of identification, we codify these one-dimensional subspaces.

One-dimensional subspace	*Code assigned*
$\text{Span}\{(1, \alpha_i, \alpha_j)\}$	$s^2 + is + j$, $i, j = 0, 1, 2, \ldots, s-1$
$\text{Span}\{(0, 1, \alpha_i)\}$	$s + i$, $i = 0, 1, 2, \ldots, s-1$
$\text{Span}\{(0, 0, 1)\}$	1

The integer codes for the points of S are all different, although they may not be successive numbers.

We now try to identify the two-dimensional subspaces of $\mathbf{F}^3$.

Two and associated one dimensional subspaces

Type	*Two-dimensional subspaces*	*Constituent one-dimensional subspaces*
1	$\{(0,0,1),(0,1,0)\}$	$\{(0,1,0)\}$; $\{(0,\alpha_j,1)\}$, $j=0,1,2,\dots,s-1$
2	$\{(0,0,1),(1,\alpha_i,0)\}$ $i=0,1,2,\dots,s-1$	$\{(0,0,1\}$; $\{(1,\alpha_i,\alpha_j)\}$, $j=0,1,2,\dots,s-1$
3	$\{(0,1,0),(1,0,\alpha_i)\}$ $i=0,1,2,\dots,s-1$	$\{(0,1,0\}$; $\{(1,\alpha_j,\alpha_i)\}$, $j=0,1,2,\dots,s-1$
4	$\{(0,1,\alpha_i),(1,\alpha_k,0)\}$ $i=1,2,3\dots,s-1$, $k=0,1,2,\dots,s-1$	$\{(0,1,\alpha_i\}$; $\{(1,\alpha_k+\alpha_j,\alpha_i\alpha_j)\}$, $j=0,1,2,\dots,s-1$

(For typographical convenience, the qualifying phrase "span" is omitted for the subspaces indicated.) The above table needs some justification. When we wrote down all one-dimensional subspaces of $\mathbf{F}^3$ in a systematic fashion, the following vectors in $\mathbf{F}^3$ arranged in three distinct groups played a pivotal role.

Group 1 $(0,0,1)$

Group 2 $(0,1,0)\quad (0,1,1),\quad (0,1,\alpha_2);\quad \dots;\ (0,1,\alpha_{s-1})$

Group 3
$$\begin{array}{lllll}
(1,0,0) & (1,0,1), & (1,0,\alpha_2); & \dots; & (1,0,\alpha_{s-1}) \\
(1,1,0) & (1,1,1), & (1,1,\alpha_2); & \dots; & (1,1,\alpha_{s-1}) \\
(1,\alpha_2,0) & (1,\alpha_2,1), & (1,\alpha_2,\alpha_2); & \dots; & (1,\alpha_2,\alpha_{s-1}) \\
\dots & \dots & \dots & \dots & \dots \\
(1,\alpha_{s-1},0) & (1,\alpha_{s-1},1), & (1,\alpha_{s-1},\alpha_2); & \dots; & (1,\alpha_{s-1},\alpha_{s-1})
\end{array}$$

Take any two vectors from anywhere in the above pool. Their span will give a two-dimensional subspace of $\mathbf{F}^3$. But there will be duplications. We need to select carefully pairs of vectors from the above so as to avoid duplications. Let us begin with the vector from Group 1 and find partners for this vector to generate a two-dimensional subspace. To start with, take the vector $(0,1,0)$ from Group 2 as a partner. Their span would give us a two-dimensional subspace of $\mathbf{F}^3$. The one-dimensional subspaces of Span $\{(0,0,1),(0,1,0)\}$ can be identified in terms of the notation of the one-dimensional subspaces we have employed earlier. The one-dimensional subspaces are given by Span $\{(0,1,0)\}$, Span $\{\alpha_j(0,1,0)+(0,0,1)\}, j=0,1,2,\ldots,s-1$. These one-dimensional subspaces can be rewritten in succinct from as: Span $\{(0,1,0)\}$, Span $\{(0,\alpha_j,1)\}, j=0,1,2,\ldots,s-1$, which are $s+1$ in number, as expected. This particular two-dimensional subspace is categorized as of Type 1. Let us consider the span of $(0,0,1)$ and any one of the remaining vectors in Group 2. We will not get anything new. Let us now consider the span of the vector in Group 1 and any vector from Group 3. Consider, in particular, the span of $(0,0,1)$ and any vector in the first column of Group 3. Any such two-dimensional subspace is categorized as of Type 2. You might ask why. Before we answer this question, observe that there are s such two-dimensional subspaces and they are all distinct. If we consider the span of $(0,0,1)$ with any vector in any of the remaining columns of vectors in Group 3, it would coincide with one of the two-dimensional subspaces we have stored under Type 2. The operation of finding mates for the vector $(0,0,1)$ ends here. Let us work with the vector $(0,1,0)$ from Group 2. The vector space spanned by $(0,1,0)$ and any one of the remaining vectors in Group 2 would coincide with the one we have already got under Type 1. Consider the vector space spanned by $(0,1,0)$ and any one of the vectors in the first row of Group 3. Each of these two-dimensional subspaces is categorized as of Type 3. There are s many of these two-dimensional spaces. These are all distinct among themselves and are also distinct from what we have got under Types 1 and 2. Further, the vector space spanned by $(0,1,0)$ and any one the vectors from the remaining rows of Group 3 would coincide with one of those we have already got. This completes the search of mates for the vector $(0,1,0)$. Consider any vector $(0,1,\alpha_i), i=1,2,\ldots,s-1$ from Group 2. The vector space spanned by $(0,1,\alpha_i)$ and any one of the vectors from Group 3 is categorized as of Type 4. All these two-dimensional spaces are distinct and are also distinct from what we have already got. There are $s(s-1)$ vector subspaces in Type 4. So far, we have got $s(s-1)+s+s+1=s^2+s+1$ distinct two-dimensional subspaces. We have no more! The identification of the one-dimensional subspaces in any two-dimensional subspace listed above is similar to the one we have explained for the two-dimensional subspace listed under Type 1. It would be quite instructive to use the integer codes for the one-dimensional subspaces listed under each two-dimensional subspace delineated above.

EXAMPLE 1.9.3. Let us look at a very specific example. Take $s = 3$. The Galois field $\mathbf{F}$ can be identified as $\{0, 1, 2\}$ with the usual operations of addition and multiplication modulo 3. The total number of one-dimensional subspaces (points) in $PG(2, 3)$ is 13 and the total number of two-dimensional subspaces (lines) is 13. They are identified explicitly in the accompanying table along with the integer codes of the one-dimensional subspaces.

The numbers involved in the integer code are:1,3,4,5,9,10,11,12,13,14, 15,16 and 17. For an inkling of what is to come, suppose we want to compare the performance of 13 treatments on some experimental units. Suppose that the treatments are numbered as 1,3,4,5,9,10,11,12,13,14, 15,16 and 17. In the above description of the projective geometry $PG(2, 3)$, each line can be viewed as a block and each point as a treatment. Corresponding to each line which has four points (integer codes) on it, we create a block with four experimental units and assign to them the treatments associated with the integer codes. We thus have 13 blocks corresponding to 13 lines, with four treatments in each block. We then have what is called a balanced incomplete block design! Each treatment is replicated exactly four times and every pair of treatments appears exactly in one block. We will now describe formally what a balanced incomplete block design is and explain how projective geometry provides such a design.

One of the primary goals in design of experiments is to compare the effect of some ν treatments over a certain population of experimental units. The experimental units can be arranged in blocks in such a way that the units in each block are homogeneous in all perceivable aspects. Ideally, we would like to have blocks each containing ν experimental units, i.e., of size ν, so that each treatment can be tried once in each block. In practice, blocks of size k with $k < \nu$ may only be available. Suppose we have b blocks each of size k. A design is simply an allocation of treatments to experimental units in the blocks. One of the basic problems is to allocate the treatments to units in a judicious manner so that we can compare the performance of any two treatments statistically with the same precision. A design is said to be a balanced incomplete block design if the following conditions are met:

(1) Each treatment appears in r blocks, i.e., each treatment is replicated r times.
(2) Every pair of treatments appears together in λ blocks.

Such a design is denoted by BIBD with parameters b, k, ν, r and λ. The basic question is how to construct such a design. Projective geometry is a vehicle to realize our goal. If the number of treatments $\nu = s^2 + s + 1$ for some s of the form p^m, where p is a prime number and m is a positive integer, and the number of blocks available is $b = s^2 + s + 1$ each of size $s + 1$, then the projective geometry $PG(2, s)$ will deliver the goods. Identify the points of the projective geometry with treatments and the lines with blocks. We have a balanced incomplete block design with parameters $b = s^2 + s + 1,\ k = s + 1,\ \nu = s^2 + s + 1,\ r = s + 1$ and $\lambda = 1$. The example presented in the accompanying Table is a BIBD with $b = 13,\ k = 4,\ \nu = 13,\ r = 4$ and $\lambda = 1$.

After having gone through the gamut outlined above, one gets the uncomfortable feeling that the technique of projective geometries in the construction of balanced incomplete block designs is of limited scope. In practice, the number of blocks available, the number of treatments to be tested and block size may not conform to the above specifications. (At this juncture, we would like to point out that a BIBD may not be available for any given b, k and ν.) In order to have more flexibility, we need to extend the ideas germane to the projective geometry $PG(2, s)$. Start with a Galois field $\mathbf{F}$ consisting of s elements. Consider the vector space $\mathbf{F}^{m+1}$. The projective geometry $PG(m, s)$ consists of the set S of all one-dimensional subspaces of the vector space $\mathbf{F}^{m+1}$ and the set $\mathbf{S}$ of all k-dimensional subspaces of $\mathbf{F}^{m+1}$ for some $k \geq 2$. The elements of S are called points and elements of $\mathbf{S}$ are called k-planes. One could treat each point as a treatment and each k-plane as a block. This is a more general way of developing a BIBD. We will not pursue in detail the general construction. We only want to provide a rudimentary introduction to what vector spaces and their ilk can solve a variety of statistical problems.

For a discussion of finite projective geometries of dimensions more than two, the reader is referred to Rao (1945c, 1946a) and to the references to other papers given there.

TABLE: BIBD DESIGN

Type	*Two-dimensional subspaces*	*One-dimensional subspaces*	*Integer codes*
1	$\{(0,0,1),(0,1,0)\}$	$\{(0,0,1)\}:\{(0,1,0)\}$ $\{(0,1,1)\}:\{(0,1,2)\};$	$1,3,4,5$
2	$\{(0,0,1),(1,0,0)\}$	$\{(0,0,1)\}:\{(1,0,0)\}$ $\{(1,0,1)\}:\{(1,0,2)\};$	$1,9,10,11$
	$\{(0,0,1),(1,1,0)\}$	$\{(0,0,1)\}:\{(1,1,0)\}$ $\{(1,1,1)\}:\{(1,1,2)\};$	$1,12,13,14$
	$\{(0,0,1),(1,2,0)\}$	$\{(0,0,1)\}:\{(1,2,0)\}$ $\{(1,2,1)\}:\{(1,2,2)\};$	$1,15,16,17$
3	$\{(0,1,0),(1,0,0)\}$	$\{(0,1,0)\}:\{(1,0,0)\}$ $\{(1,1,0)\}:\{(1,2,0)\};$	$3,9,12,15$
	$\{(0,1,0),(1,0,1)\}$	$\{(0,1,0)\}:\{(1,0,1)\}$ $\{(1,1,1)\}:\{(1,2,1)\};$	$3,10,13,16$
	$\{(0,1,0),(1,0,2)\}$	$\{(0,1,0)\}:\{(1,0,2)\}$ $\{(1,1,2)\}:\{(1,2,2)\};$	$3,11,14,17$
4	$\{(0,1,1),(1,0,0)\}$	$\{(0,1,1)\}:\{(1,0,0)\}$ $\{(1,1,1)\}:\{(1,2,2)\};$	$4,9,13,17$
	$\{(0,1,1),(1,1,0)\}$	$\{(0,1,1)\}:\{(1,0,2)\}$ $\{(1,1,0)\}:\{(1,2,1)\};$	$4,11,12,16$
	$\{(0,1,1),(1,2,0)\}$	$\{(0,1,1)\}:\{(1,0,1)\}$ $\{(1,1,2)\}:\{(1,2,0)\};$	$4,10,14,15$
	$\{(0,1,2),(1,0,0)\}$	$\{(0,1,2)\}:\{(1,0,0)\}$ $\{(1,1,2)\}:\{(1,2,1)\};$	$5,9,14,16$
	$\{(0,1,2),(1,1,0)\}$	$\{(0,1,2)\}:\{(1,0,1)\}$ $\{(1,1,0)\}:\{(1,2,2)\};$	$5,10,12,17$
	$\{(0,1,2),(1,2,0)\}$	$\{(0,1,2)\}:\{(1,0,2)\}$ $\{(1,1,1)\}:\{(1,2,0)\}.$	$5,11,13,15$

Note: Some references to material covered in this Chapter, where further details can be obtained, are Bose, Shrikhande and Parker (1960), Halmos (1958), Raghava Rao (1971) and Rao (1947, 1949).

CHAPTER 2

UNITARY AND EUCLIDEAN SPACES

So far we have studied the relationship between the elements of a vector space through the notion of independence. It would be useful to consider other concepts such as distance and angle between vectors as in the case of two and three dimensional Euclidean spaces. It appears that these concepts can easily be extended to vector spaces over the field of complex or real numbers by defining a function called the inner product of two vectors.

2.1. Inner Product

However abstract a vector space may be, when it comes to practicality, we would like to relate the vectors either with real numbers or complex numbers. One useful way of relating vectors with real numbers is to associate a norm, which is a non-negative real number, with every vector. We will see more of this later. Another way is to relate pairs of vectors with complex numbers leading to the notion of inner product between vectors. We will present rudiments of these ideas now.

DEFINITION 2.1.1. Let $\mathbf{V}$ be a vector space over a field $\mathbf{F}$, where $\mathbf{F}$ is either the field $\mathbf{C}$ of complex numbers or $\mathbf{R}$ of real numbers. A map $< \cdot, \cdot >$ from $\mathbf{V} \times \mathbf{V}$ to $\mathbf{F}$ is called an *inner product* if the following properties hold for all x, y, z in $\mathbf{V}$ and α, β in $\mathbf{F}$.

$$(1)\quad < x, y > \quad = \begin{cases} \overline{< y, x >}, & \text{if } \mathbf{F} = \mathbf{C}, \text{ (anti-symmetry)} \\ < y, x >, & \text{if } \mathbf{F} = \mathbf{R}. \text{ (symmetry)} \end{cases}$$

$$(2)\quad < x, x > \quad > 0, \text{ if } x \neq 0, \text{ (positivity)}$$
$$= 0, \text{ if } x = 0.$$

$$(3)\quad < \alpha x + \beta y, z > = \alpha < x, z > + \beta < y, z > . \text{ (linearity in the first argument)}$$

The bar that appears in (1) above is the operation of conjugation on complex numbers. A vector space furnished with an inner product is called an inner product space. It is customary to call such a space unitary when $\mathbf{F} = \mathbf{C}$, and Euclidean when $\mathbf{F} = \mathbf{R}$.

We have the following proposition as a consequence of the conditions (1), (2) and (3) of Definition 2.1.1. In the sequel, most of the results are couched with reference to the field $\mathbf{C}$ of complex numbers. Only minor modifications are needed when the underlying field is $\mathbf{R}$.

P 2.1.2 For any x, y, z in $\mathbf{V}(\mathbf{C})$ and α, β in $\mathbf{C}$, the following hold.

(a) $< x, \alpha y + \beta z > = \bar{\alpha} < x, y > + \bar{\beta} < x, z >$.
(b) $< x, 0 > = < 0, x > = 0$.
(c) $< \alpha x, \beta y > = \alpha < x, \beta y > = \alpha \bar{\beta} < x, y >$.

Some examples of inner product spaces are provided below.

EXAMPLE 2.1.3. Consider the vector space $\mathbf{R}^k$ for some $k \geq 1$. For any two vectors $x = (\alpha_1, \alpha_2, \ldots, \alpha_k)$ and $y = (\beta_1, \beta_2, \ldots, \beta_k)$ in $\mathbf{R}^k$, define

$$< x, y > = \alpha_1 \beta_1 + \alpha_2 \beta_2 + \ldots + \alpha_k \beta_k,$$

which can be shown to be an inner product on the vector space $\mathbf{R}^k$. This is the standard inner product of the space $\mathbf{R}^k$.

EXAMPLE 2.1.4. Consider the vector space $\mathbf{C}^k$ for some $k \geq 1$. Let $\delta_1, \delta_2, \ldots, \delta_k$ be fixed positive numbers. For any two vectors $x = (\alpha_1, \alpha_2, \ldots, \alpha_k)$ and $y = (\beta_1, \beta_2, \ldots, \beta_k)$ in $\mathbf{C}^k$, define

$$< x, y > = \delta_1 \alpha_1 \bar{\beta}_1 + \delta_2 \alpha_2 \bar{\beta}_2 + \ldots + \delta_k \alpha_k \bar{\beta}_k,$$

which can be shown to be an inner product on $\mathbf{C}^k$. If $\delta_1 = \delta_2 = \ldots = \delta_k = 1$, the resultant inner product is the so-called standard inner product on the space $\mathbf{C}^k$.

One might wonder about the significance of the way the inner product is defined above on $\mathbf{C}^k$. If one defines

$$< x, y > = \delta_1 \alpha_1 \beta_1 + \delta_2 \alpha_2 \beta_2 + \ldots + \delta_k \alpha_k \beta_k,$$

then one of the conditions (which one?) of Definition 2.1.1 is violated.

EXAMPLE 2.1.5. Let $\mathbf{P}_n$ be the space of all polynomials of degree less than n with coefficients from the field $\mathbf{C}$ of complex numbers. For any two polynomials $x(\cdot)$ and $y(\cdot)$ in $\mathbf{P}_n$, define

$$< x, y > = \int_0^1 x(t)\overline{y(t)}\, dt,$$

which can be shown to be an inner product on $\mathbf{P}_n$.

EXAMPLE 2.1.6. Let $x_1, x_2, \ldots, x_k$ be a basis of a vector space $\mathbf{V}(\mathbf{C})$. For any two vectors x and y in $\mathbf{V}$, we will have unique representations

$$x = \alpha_1 x_1 + \alpha_2 x_2 + \ldots + \alpha_k x_k \quad \text{and} \quad y = \beta_1 x_1 + \beta_2 x_2 + \ldots + \beta_k x_k$$

in terms of the basis vectors. Let $\delta_1, \delta_2, \ldots, \delta_k$ be some fixed positive numbers. Define

$$< x, y > = \delta_1 \alpha_1 \bar{\beta}_1 + \delta_2 \alpha_2 \bar{\beta}_2 + \ldots + \delta_k \alpha_k \bar{\beta}_k,$$

which can be shown to be an inner product on the vector space $\mathbf{V}$.

Note that an inner product on a vector space can be defined in many ways. The choice of a particular inner product depends on its usefulness in solving a given problem. We will see instances of several inner products in subsequent sections and chapters. Every inner product gives rise to what is known as a norm.

DEFINITION 2.1.7. (Norm) Let $< \cdot, \cdot >$ be an inner product on a vector space $\mathbf{V}$. The positive square root of $< x, x >$ for any x in $\mathbf{V}$ is called the norm of x and is denoted by $\|x\|$.

There is a more general definition of a norm on a vector space. The norm we have introduced above arises from an inner product. The more general version of a norm will be considered in a later chapter. As a consequence of Definitions 2.1.1 and 2.1.7, the following inequality follows.

P 2.1.8 (Cauchy-Schwartz Inequality). Let $(\mathbf{V}, < \cdot, \cdot >)$ be an inner product space with the associated norm $\| \cdot \|$ in $\mathbf{V}$. Then for any two vectors x and y in $\mathbf{V}$, the inequality

$$| < x, y > | \leq \|x\| \, \|y\|, \tag{2.1.1}$$

holds. Moreover, equality holds in the above if and only if $\gamma x + \delta y = 0$ for some γ and δ, not both zero, in $\mathbf{C}$.

PROOF. Let $\beta = < x, x >$ and $\alpha = - < y, x >$. Observe that $\bar{\alpha} = - < x, y >$. We are required to establish that $|\alpha|^2 = \alpha\bar{\alpha} = | < x, y > |^2 \leq \beta < y, y >$. By the definition of an inner product,

$$\begin{aligned} 0 \leq < \alpha x + \beta y, \alpha x + \beta y > &= \alpha < x, \alpha x + \beta y > + \beta < y, \alpha x + \beta y > \\ &= \alpha\bar{\alpha} < x, x > + \alpha\bar{\beta} < x, y > \\ &\quad + \beta\bar{\alpha} < y, x > + \beta\bar{\beta} < y, y > \\ &= |\alpha|^2\beta - |\alpha|^2\beta - |\alpha|^2\beta + \beta^2 < y, y > \\ &= -|\alpha|^2\beta + \beta^2 < y, y >, \end{aligned} \tag{2.1.2}$$

from which (2.1.1) follows. If $\gamma x + \delta y = 0$ for some γ and δ, not both zero, in $\mathbf{C}$, it is clear that equality holds in (2.1.1). On the other hand, if equality holds in (2.1.1), equality must hold in (2.1.2) throughout. This implies that $\alpha x + \beta y = 0$. If $x = 0$, take γ to be any non-zero scalar and $\delta = 0$. If $x \neq 0$, take $\gamma = \alpha$ and $\delta = \beta$. This completes the proof.

P 2.1.9 For any two vectors x and y in a vector space $\mathbf{V}$ equipped with an inner product $< \cdot, \cdot >$, the following inequality holds.

$$< x, y > + < y, x > \leq 2\|x\| \, \|y\|.$$

PROOF. Observe that $2| < x, y > | \geq < x, y > + < y, x >$. The result now follows from the Cauchy-Schwartz inequality.

We now establish some properties of a norm. For any two vectors x and y, $\|x - y\|$ can be regarded as the distance between x and y.

P 2.1.10 Let x and y be any two vectors in an inner product space $\mathbf{V}$ with inner product $< \cdot, \cdot >$ and norm $\| \cdot \|$. Then the following hold.

(1) $\|x + y\| \leq \|x\| + \|y\|$.
(2) $\|x - y\| + \|y\| \geq \|x\|$ (triangle inequality of distance).
(3) $\|x + y\|^2 + \|x - y\|^2 = 2\|x\|^2 + 2\|y\|^2$ (parallelogram law).
(4) $\|x + y\|^2 = \|x\|^2 + \|y\|^2$ if $< x, y >= 0$ (Pythagorous theorem).

PROOF. By the definition of the norm,

$$\begin{aligned}\|x+y\|^2 &= \|x\|^2+\|y\|^2+<x,y>+<y,x>\\ &\le \|x\|^2+\|y\|^2+2\|x\|\ \|y\| \quad \text{(by Cauchy-Schwartz inequality)}\\ &= (\|x\|+\|y\|)^2,\end{aligned}$$

from which (1) follows. In (1), if we replace x by y and y by $x-y$, we obtain (2). The relations expostulated in (3) and (4) can be established in an analogous fashion.

We now formally define the distance and angle between any two vectors in any inner product space.

DEFINITION 2.1.11. Let x and y be any two vectors in a vector space $\mathbf{V}$ equipped with an inner product $<\cdot,\cdot>$ and the associated norm $\|\cdot\|$. The distance $\delta(x,y)$ between x and y is defined by $\delta(x,y)=\|x-y\|$.

P 2.1.12 The distance function $\delta(\cdot,\cdot)$ defined above has the following properties.

(1) $\delta(x,y)=\delta(y,x)$ for any x and y in $\mathbf{V}$.
(2) $\delta(x,y)\ge 0$ for any x and y in $\mathbf{V}$,
$=0$ if and only if $x=y$.
(3) $\delta(x,y)\le\delta(x,z)+\delta(y,z)$ for any x,y and z in $\mathbf{V}$ (triangle inequality).

PROOF. The properties (1) and (2) follow from the very definition of the distance function. If we replace x by $x-y$ and y by $x-z$ in **P 2.1.10** (2), we would obtain the triangle inequality (3) above.

DEFINITION 2.1.13. Let $\mathbf{V}$ be a Euclidean space equipped with an inner product $<\cdot,\cdot>$ and the associated norm $\|\cdot\|$. For any two non-zero vectors x and y in $\mathbf{V}$, the angle θ between x and y is defined by

$$\cos\theta = <x,y>/[\|x\|\ \|y\|].$$

Observe that, in view of the Cauchy-Schwartz inequality, $\cos\theta$ always lies in the interval $[-1,1]$. This definition does not make sense in unitary spaces because $<x,y>$ could be a complex number.

The notion of angle between two non-zero vectors of a Euclidean vector space is consonant with the usual perception of angle in vogue in the two-dimensional Euclidean space. Let $x=(x_1,x_2)$ and $y=(y_1,y_2)$

be two non-zero vectors in the first quadrant of the two-dimensional Euclidean space $\mathbf{R}^2$. Let L_1 be the line joining the vectors $0 = (0, 0)$ and $x = (x_1, x_2)$ and L_2, the line joining $0 = (0, 0)$ and $y = (y_1, y_2)$. Let θ_1 and θ_2 be the angles the lines L_1 and L_2 make with the *x-axis*, respectively. Then the angle θ between the lines L_1 and L_2 at the origin is given by $\theta = \theta_1 - \theta_2$. Further,

$$\cos\theta = \cos(\theta_1 - \theta_2) = \cos\theta_1\cos\theta_2 + \sin\theta_1\sin\theta_2$$
$$= \frac{x_1}{\|x\|}\cdot\frac{y_1}{\|y\|} + \frac{x_2}{\|x\|}\cdot\frac{y_2}{\|y\|} = \frac{< x, y >}{\|x\|\,\|y\|}.$$

Complements

2.1.1 Let $\mathbf{V}$ be a real inner product space and α, β be two positive real numbers. Show that the angle between two non-zero vectors x and y of $\mathbf{V}$ is the same as the angle between the vectors αx and βy.
2.1.2 Compute the angle between the vectors $x = (3, -1, 1, 0)$ and $y = (2, 1, -1, 1)$ in $\mathbf{R}^4$ with respect to the standard inner product of the space $\mathbf{R}^4$.
2.1.3 Let α, β, γ and δ be four complex numbers. For $x = (\xi_1, \xi_2)$ and $y = (\eta_1, \eta_2)$ in $\mathbf{C}^2$, define $< x, y > = \alpha\xi_1\bar{\eta}_1 + \beta\xi_2\bar{\eta}_1 + \gamma\xi_1\bar{\eta}_2 + \delta\xi_2\bar{\eta}_2$. Under what conditions on α, β, γ, and δ, is $< \cdot, \cdot >$ an inner product on $\mathbf{C}^2$?
2.1.4 Suppose $\|x + y\|^2 = \|x\|^2 + \|y\|^2$ for some two vectors x and y in an unitary space $\mathbf{V}$. Is it true that $< x, y > = 0$? What happens when $\mathbf{V}$ is a Euclidean space?

2.2. Orthogonality

Let us, for a moment, look at two points $x = (x_1, x_2)$ and $y = (y_1, y_2)$ in the two-dimensional Euclidean space $\mathbf{R}^2$. Draw a line joining the points $0 = (0, 0)$ and x and another line joining 0 and y. We would like to enquire under what circumstances the angle between these lines is 90°. Equivalently, we ask under what conditions the triangle $\Delta 0xy$ formed by the points 0, x, and y is a right-angled triangle with the angle $\angle x0y$ at the origin $= 90^o$. It turns out that the condition is $< x, y > = x_1y_1 + x_2y_2 = 0$. (Draw a picture.)

This is the idea that we would like to pursue in inner product spaces.

DEFINITION 2.2.1. Two vectors x and y in an inner product space **V** are said to be orthogonal if the inner product between x and y is zero, i.e., $< x, y > = 0$.

In the case of a Euclidean space, orthogonality of x and y implies that the angle between x and y is 90°. Trivially, if $x = 0$, then x is orthogonal to every vector in **V**. The notion of orthogonality can be extended to any finite set of vectors.

DEFINITION 2.2.2. A collection, $x_1, x_2, \ldots, x_r$, of vectors in an inner product space **V** is said to be orthonormal if

$$< x_i, x_j >= \begin{cases} 0 & \text{if } \ i \neq j, \\ 1 & \text{if } \ i = j. \end{cases}$$

If a vector x is such that $< x, x > = \|x\|^2 = 1$, then x is said to be of unit length. If we drop the condition that each vector above be of unit length, then the vectors are said to be an orthogonal set of vectors.

P 2.2.3 Let $x_1, x_2, \ldots, x_r$ be an orthogonal set of non-zero vectors in an inner product space **V**. Then $x_1, x_2, \ldots, x_r$ are necessarily linearly independent.

PROOF. Suppose $y = \alpha_1 x_1 + \alpha_2 x_2 + \ldots + \alpha_r x_r = 0$. Then for each $1 \leq i \leq r$, $0 = < y, x_i >= \alpha_i < x_i, x_i >= 0$. Since x_i is a non-zero vector, $\alpha_i = 0$. This shows that the orthogonal set under discussion is linearly independent.

One of the most useful techniques in the area of orthogonality is the well-known Gram-Schmidt orthogonalization process. The process transforms a given bunch of vectors in an inner product space into an orthogonal set.

P 2.2.4 (Gram-Schmidt Orthogonalization Process). Given a linearly independent set $x_1, x_2, \ldots, x_r$ of vectors in an inner product space, it is possible to construct an orthonormal set $z_1, z_2, \ldots, z_r$ of vectors such that $Sp(x_1, \ldots, x_i) = Sp(z_1, \ldots, z_i)$ for every $i = 1, \ldots, r$. [Note the definition: $Sp(a_1, \ldots, a_k) = \{\alpha_1 a_1 + \ldots + \alpha_k a_k : \alpha_1, \ldots, \alpha_k \in \mathbf{C}\}$.]

PROOF. Define vectors $y_1, y_2, \ldots, y_r$ in the following way:

$y_1 = x_1$,

$y_2 = x_2 - \alpha_{2,1} y_1$,

$$y_3 = x_3 - \alpha_{3,2}y_2 - \alpha_{3,1}y_1,$$
$$\dots \quad \dots \quad \dots$$
$$y_r = x_r - \alpha_{r,r-1}y_{r-1} - \alpha_{r,r-2}y_{r-2} - \dots - \alpha_{r,1}y_1,$$

for some scalars $\alpha_{i,j}$'s. We will choose $\alpha_{i,j}$'s carefully so that the new vectors $y_1, y_2, \dots, y_r$ form an orthogonal set of non-zero vectors. The determination of $\alpha_{i,j}$'s is done sequentially. Choose $\alpha_{2,1}$ so that y_1 and y_2 are orthogonal. Setting

$$0 = < y_2, y_1 > = < x_2, x_1 > - \alpha_{2,1} < y_1, y_1 >,$$

we obtain $\alpha_{2,1} = < x_2, x_1 > / < y_1, y_1 >$. (Note that $< y_1, y_1 > > 0$.) Thus y_2 is determined. Further, the vector y_2 is non-zero since x_1 and x_2 are linearly independent. Choose $\alpha_{3,2}$ and $\alpha_{3,1}$ so that y_1, y_2 and y_3 are pairwise orthogonal. Set

$$0 = < y_3, y_2 > = < x_3, y_2 > - \alpha_{3,2} < y_2, y_2 >,$$
$$0 = < y_3, y_1 > = < x_3, y_1 > - \alpha_{3,1} < y_1, y_1 > .$$

From these two equations, we can determine $\alpha_{3,1}$ and $\alpha_{3,2}$ which would meet our requirements. Thus y_3 is determined. Note that the vector y_3 is a linear combination of the vectors x_1, x_2, and x_3. Consequently, $y_3 \neq 0$. (Why?) Continuing this process down the line successively, we will obtain a set $y_1, y_2, \dots, y_r$ of orthogonal non-zero vectors. The computation of the coefficients $\alpha_{i,j}$'s is very simple. For the desired orthonormal set, set $z_i = y_i/\|y_i\|$, $i = 1, 2 \dots, r$. From the above construction, it is clear that

(1) each y_i is a linear combination of $x_1, x_2, \dots, x_i$, and
(2) each x_i is a linear combination of $y_1, y_2, \dots, y_i$,

from which we have

$$Sp(x_1, x_2, \dots, x_i) = Sp(y_1, y_2, \dots, y_i) = Sp(z_1, z_2, \dots, z_i).$$

EXAMPLE 2.2.5. Let $\mathbf{P}_4$ be the vector space of all real polynomials of degree less than 4. The polynomials $1, x, x^2, x^3$ form a linearly independent set of vectors in $\mathbf{P}_4$. For $p(\cdot)$ and $q(\cdot)$ in $\mathbf{P}_4$, let

$$< p(\cdot), q(\cdot) >= \int_{-1}^{+1} p(x)\, q(x)\, dx.$$

Observe that $<\cdot,\cdot>$ is an inner product on $\mathbf{P}_4$. The vectors $1, x, x^2, x^3$ are not orthogonal under the above inner product. We can invoke the Gram-Schmidt orthogonalization process on these vectors to obtain an orthonormal set. The process gives

$$p_1(x) = 1, \; p_2(x) = x,$$
$$p_3(x) = x^2 - 1/3, \; p_4(x) = x^3 - (3/5)x.$$

This process can be continued forever. The sequence $p_1, p_2, \ldots$ of polynomials so obtained is the well-known Legendre orthogonal polynomials. We can obtain an orthonormal set by dividing each polynomial by its norm.

We can create other sequences of polynomials from $1, x, x^2, \ldots$ by defining inner products of the type

$$< p_i(x), p_j(x) > = \int p_i p_j f(x) dx$$

choosing a suitable function $f(x)$.

We know that every vector space has a basis. If the vector space comes with an inner product, it is natural to enquire whether it has a basis consisting of orthonormal vectors, i.e., an orthonormal basis. The Gram-Schmidt orthogonalization process provides the sought-after basis. We will record this in the form of a proposition.

P 2.2.6 Every inner product space has an orthonormal basis.

If we have a basis $x_1, x_2, \ldots, x_k$ for an inner product space $\mathbf{V}$, we can write every vector x in $\mathbf{V}$ as a linear combination of $x_1, x_2, \ldots, x_k$,

$$x = \alpha_1 x_1 + \alpha_2 x_2 + \ldots + \alpha_k x_k,$$

for some $\alpha_1, \alpha_2, \ldots, \alpha_k$ in $\mathbf{C}$. Determining these coefficients α_i's is a hard problem. If $x_1, x_2, \ldots, x_k$ happen to be orthonormal, then these coefficients can be calculated in a simple way. More precisely, we have $\alpha_i = < x, x_i >$ for each i and

$$x = < x, x_1 > x_1 + < x, x_2 > x_2 + \ldots + < x, x_k > x_k.$$

This is not hard to see. There are other advantages that accrue if we happen to have an orthonormal basis. The inner product between any two vectors x and y in $\mathbf{V}$ can be computed in a straightforward manner. More precisely,

$$\begin{aligned} < x, y > = < x, x_1 >< x_1, y > &+ < x, x_2 >< x_2, y > \\ &+ \ldots + < x, x_k >< x_k, y > . \end{aligned} \quad (2.2.1)$$

The above is the well-known Parseval Identity. Once we know the coefficients α_i's in the representation of x in terms of $x_1, x_2, \ldots, x_k$ and the coefficients β_i's in the representation of y in terms of $x_1, x_2, \ldots, x_k$, we can immediately jot down the inner product of x and y, courtesy Parseval identity, as

$$< x, y > = \alpha_1 \bar{\beta}_1 + \alpha_2 \bar{\beta}_2 + \ldots + \alpha_k \beta_k.$$

One consequence of (2.2.1) is that the norm of x can be written down explicitly in terms of these coefficients. More precisely,

$$\|x\|^2 = | < x, x_1 > |^2 + | < x, x_2 > |^2 + \ldots + | < x, x_k > |^2. \quad (2.2.2)$$

In this connection, it is worth bringing into focus the Bessel Inequality. The statement reads as follows: if $x_1, x_2, \ldots, x_r$ is a set of orthonormal vectors in an inner product space, then for any vector x in the vector space, the following inequality holds

$$\|x\|^2 \geq \sum_{i=1}^{r} | < x, x_i > |^2.$$

The identity (2.2.1) is not hard to establish. What is interesting is that the characteristic property (2.2.1) of an orthonormal basis characterizes the orthonormal property. We record this phenomenon in the following proposition.

P 2.2.7 Let $\mathbf{V}$ be an inner product space of dimension k. Let $x_1, x_2, \ldots, x_k$ be some k vectors in $\mathbf{V}$ having the property that for any two vectors x and y in $\mathbf{V}$,

$$\begin{aligned} < x, y > = < x, x_1 >< x_1, y > &+ < x, x_2 >< x_2, y > \\ &+ \ldots + < x, x_k >< x_k, y > . \end{aligned} \quad (2.2.3)$$

Then $x_1, x_2, \ldots, x_k$ is an orthonormal basis for **V**.

PROOF. Let us see what we can get out of the property (2.2.3) enunciated above. By plugging $x = x_1$ and $y = x_1$ in (2.2.3), we observe that

$$\|x_1\|^2 = \|x_1\|^4 + |< x_1, x_2 >|^2 + \ldots + |< x_1, x_k >|^2. \qquad (2.2.4)$$

Since all the terms involved are non-negative, the only way that the above equality could hold is that $\|x_1\| \leq 1$. As a matter of fact, we have $\|x_i\| \leq 1$ for every i. Let $u_1, u_2, \ldots, u_k$ be an orthonormal basis of the vector space **V**. Each x_i has a unique representation in terms of the given orthonormal basis. By (2.2.2),

$$\|x_i\|^2 = |< x_i, u_1 >|^2 + |< x_i, u_2 >|^2 + \ldots + |< x_i, u_k >|^2$$

for each i. By plugging $x = y = u_i$ in (2.2.3), we observe that

$$1 = |< x_1, u_i >|^2 + |< x_2, u_i >|^2 + \ldots + |< x_k, u_i >|^2, \qquad (2.2.5)$$

for each i. Summing (2.2.5) over i, we obtain

$$\begin{aligned} k &= \sum_{i=1}^{k} \sum_{j=1}^{k} |< x_j, u_i >|^2 \\ &= \sum_{j=1}^{k} \sum_{i=1}^{k} |< x_j, u_i >|^2 = \sum_{j=1}^{k} \|x_j\|^2. \end{aligned} \qquad (2.2.6)$$

We have seen earlier that each $\|x_i\| \leq 1$. This can coexist with (2.2.6) only when each $\|x_i\| = 1$. In that case, if we look at (2.2.4) and related identities, it follows that $< x_i, x_j >= 0$ for all $i \neq j$. This completes the proof.

If we look at the proof, one wonders whether the assumption that the vector space has dimension k can be dropped at all. It is not feasible. Try to prove the above proposition by dropping the assumption on the dimension!

In Chapter 1, we talked about complements of subspaces of a vector space. We have also seen that the complement need not be unique. If we have, additionally, an inner product on the vector space, the whole idea of seeking a complement for the given subspace has to be reexamined under the newer circumstances. This is what we propose to do now.

P 2.2.8 Let $\mathbf{V}$ be a vector space equipped with an inner product $<\cdot,\cdot>$ and $\mathbf{S}$ a subspace of $\mathbf{V}$. Then there exists a subspace $\mathbf{S}^\perp$ of $\mathbf{V}$ with the following properties.

(1) $<x, y> = 0$ whenever $x \in \mathbf{S}$ and $y \in \mathbf{S}^\perp$.
(2) $\mathbf{S} \cap \mathbf{S}^\perp = \{0\}$ and $\mathbf{V} = \mathbf{S} \oplus \mathbf{S}^\perp$.
(3) $\dim(\mathbf{S}) + \dim(\mathbf{S}^\perp) = \dim(\mathbf{V})$.

PROOF. Let $x_1, x_2, \ldots, x_r$ be a basis of the subspace $\mathbf{S}$, and extend it to a full basis $x_1, x_2, \ldots, x_r, x_{r+1}, \ldots, x_k$ of $\mathbf{V}$. Let $z_1, z_2, \ldots, z_r, z_{r+1}, \ldots, z_k$ be the orthonormal basis of $\mathbf{V}$ obtained by the Gram-Schmidt orthogonalization process carried out on the x_i's. We now have a natural candidate to fit the bill. Let $\mathbf{S}^\perp$ be the vector subspace spanned by $z_{r+1}, z_{r+2}, \ldots, z_k$. Trivially, (1) follows. To prove (2), note that every vector x in $\mathbf{V}$ has a unique representation,

$$\begin{aligned} x &= (\alpha_1 z_1 + \alpha_2 z_2 + \ldots + \alpha_r z_r) + (\alpha_{r+1} z_{r+1} + \ldots + \alpha_k z_k) \\ &= y_1 + y_2, \quad \text{say,} \end{aligned}$$

for some scalars α_i's in $\mathbf{C}$, where $y_1 = \alpha_1 z_1 + \alpha_2 z_2 + \ldots + \alpha_r z_r$ and $y_2 = x - y_1$. It is clear that $y_1 \in \mathbf{S}$ and $y_2 \in \mathbf{S}^\perp$. By the very construction, we have $\mathbf{S} \cap \mathbf{S}^\perp = \{0\}$ and $\dim(\mathbf{S}) + \dim(\mathbf{S}^\perp) = \dim(\mathbf{V})$.

We have talked about complements of a subspace in Chapter 1. The subspace $\mathbf{S}^\perp$ is a complement of the subspace $\mathbf{S}$ after all. But the subspace $\mathbf{S}^\perp$ is special. It has an additional property (1) listed above. In order to distinguish it from the plethora of complements available, let us call the subspace $\mathbf{S}^\perp$ an orthogonal complement of the subspace $\mathbf{S}$. When we say that $\mathbf{S}^\perp$ is an orthogonal complement, we sound as though it is not unique. There could be other subspaces exhibiting the properties (1), (2) and (3) listed above. The proof given above is not much of help to settle the question of uniqueness. The subspace $\mathbf{S}^\perp$ is indeed unique and can be characterized in the following way.

P 2.2.9 Let $\mathbf{S}$ be a subspace of an inner product space $\mathbf{V}$. Then any subspace $\mathbf{S}^\perp$ having the properties (1) and (2) of **P 2.2.8** can be characterized as

$$\mathbf{S}^\perp = \{x \in \mathbf{V} : < x, y > = 0 \quad \text{for every} \quad y \in \mathbf{S}\}.$$

PROOF. Let $\mathbf{S}^*$ be a subspace of $\mathbf{V}$ having the properties (1) and (2) of **P 2.2.8**. Let $\mathbf{S}^{**} = \{x \in \mathbf{V} :< x, y >= 0 \text{ for every } y \in \mathbf{S}\}$. We will show that $\mathbf{S}^* = \mathbf{S}^{**}$. As in the proof of **P 2.2.8**, let $z_1, z_2, \ldots, z_r, z_{r+1}, \ldots, z_k$ be an orthonormal basis for $\mathbf{V}$ such that $z_1, z_2, \ldots, z_r$ is a basis for the subspace $\mathbf{S}$. Then for $x \in \mathbf{S}$,

$$\begin{aligned} x &= < x, z_1 > z_1 + < x, z_2 > z_2 + \ldots + < x, z_r > z_r \\ &\quad + < x, z_{r+1} > z_{r+1} + \ldots + < x, z_k > z_k, \end{aligned}$$

with $< x, z_1 > z_1 + < x, z_2 > z_2 + \ldots + < x, z_r > z_r \in \mathbf{S}$. If $x \in \mathbf{S}^*$, then, by (2), $< x, z_i > = 0$ for every $i = 1, 2, \ldots, r$. Consequently, $x \in \mathbf{S}^{**}$. Conversely, if $x \in \mathbf{S}^{**}, < x, z_i > = 0$ for every $i = 1, 2, \ldots, r$, in particular. Hence $x \in \mathbf{S}^*$. This completes the proof.

To stretch matters beyond what was outlined in **P 2.2.8**, one could talk about the orthogonal complement $(\mathbf{S}^\perp)^\perp$ of the subspace $\mathbf{S}^\perp$. If we look at the conditions, especially (1), of **P 2.2.8** for the orthogonal complement $\mathbf{S}^\perp$ of a given subspace $\mathbf{S}$ should meet, we perceive some symmetry in the way the condition (1) is arraigned. **P 2.2.9** provides a strong motivation for the following result.

P 2.2.10 For any subspace $\mathbf{S}$ of an inner product space $\mathbf{V}$, the relation, $(\mathbf{S}^\perp)^\perp = \mathbf{S}$, holds true.

PROOF. Does this really require a proof? Well, let us try one. By **P 2.2.8** (1) and (2), it follows that $(\mathbf{S}^\perp)^\perp \subset \mathbf{S}$. Since $\dim(\mathbf{S}) + \dim(\mathbf{S}^\perp) = \dim((\mathbf{S}^\perp)^\perp) + \dim(\mathbf{S}^\perp) = \dim(\mathbf{V})$, we have $\dim((\mathbf{S}^\perp)^\perp) = \dim(\mathbf{S})$. This together with $(\mathbf{S}^\perp)^\perp \subset \mathbf{S}$ implies that $(\mathbf{S}^\perp)^\perp = \mathbf{S}$.

In the absence of an inner product on a vector space, we could talk about complements of subspaces of the vector space. One could also talk about the complement of a given complement of a given subspace of a vector space. There is no guarantee that the second generation complement will be identical with the given subspace. Starting with a given subspace, one can keep on taking complements no two of which are alike!

We are now in a position to introduce Orthogonal Projections. These projections have some bearing in some optimization problems. A more general definition of a projection will be provided later. First, we start with a definition.

DEFINITION 2.2.11. (Orthogonal Projection) Let $\mathbf{S}$ be a subspace of an inner product space $\mathbf{V}$ and $\mathbf{S}^{\perp}$ its orthogonal complement. Let x be any vector in $\mathbf{V}$. Then the vector x admits a unique decomposition, $x = y + z$, with $y \in \mathbf{S}$ and $z \in \mathbf{S}^{\perp}$. Define a map $\mathbf{P_S}$ from $\mathbf{V}$ to $\mathbf{S}$ by $\mathbf{P_S}(x) = y$. The map $\mathbf{P_S}$ is called an orthogonal projection from the space $\mathbf{V}$ to the space $\mathbf{S}$.

The orthogonal projection is really a nice map. It is a homomorphism from the vector space onto the vector space $\mathbf{S}$. It is idempotent. These facts are enshrined in the following proposition.

P 2.2.12 Let $\mathbf{P_S}$ be the orthogonal projection from the inner product space $\mathbf{V}$ to its subspace $\mathbf{S}$. Then it has the following properties.

(1) The map $\mathbf{P_S}$ is a linear map from $\mathbf{V}$ onto $\mathbf{S}$.
(2) The map $\mathbf{P_S}$ is idempotent, i.e., $\mathbf{P_S}(\mathbf{P_S}(x)) = \mathbf{P_S}(x)$ for every x in the vector space $\mathbf{V}$.

PROOF. The uniqueness of the decomposition of any vector x as a sum of two vectors y and z, with y in $\mathbf{S}$ and z in $\mathbf{S}^{\perp}$, is the key ingredient for the map $\mathbf{P_S}$ to have such nice properties as (1) and (2). For the record, observe that the map $\mathbf{P_S}$ is an identity map when it is restricted to the subspace $\mathbf{S}$.

In order to define the projection map we do not need an inner product on the vector space. The projection map can always be defined on $\mathbf{V}$ with respect to some fixed complement $\mathbf{S}^c$ of the subspace $\mathbf{S}$ of $\mathbf{V}$. Such a map will have properties (1) and (2) of **P 2.2.12**.

The orthogonal projection expostulated above arises in a certain optimization problem. Given any vector x in an inner product space $\mathbf{V}$ and any subspace $\mathbf{S}$ of $\mathbf{V}$, we would like to compute explicitly the distance between the point x and the subspace $\mathbf{S}$. The notion of distance between any two vectors of an inner product space can be extended to cover vectors and subsets of the vector space. More precisely, if x is a vector in $\mathbf{V}$ and A a subset of $\mathbf{V}$, the distance between x and A can be defined as $\inf_{y \in A} \|x - y\|$. Geometrically speaking, this number is the shortest distance between x and points of A. Generally, this distance is hard to compute and may not be attained. If the subset happens to be a subspace of $\mathbf{V}$, the computation is simple, and in fact, the distance is attained at some point of the subspace.

P 2.2.13 Let x be any vector in an inner product space $\mathbf{V}$ and $\mathbf{S}$ a subspace of **V**. Then the distance $\delta(x, \mathbf{S})$ between x and **S** is given by

$$\delta(\dot{x}, \mathbf{S}) = \inf_{y \in \mathbf{S}} \|x - y\| = \|x - \mathbf{P_S}(x)\|.$$

PROOF. Since $\mathbf{V} = \mathbf{S} \oplus \mathbf{S}^{\perp}$, we can write $x = x_1 + x_2$ with $x_1 \in \mathbf{S}$ and $x_2 \in \mathbf{S}^{\perp}$. Of course, $x_1 = \mathbf{P_S}(x)$. For any vector y in **S**, observe that

$$\|x - y\|^2 = \|x_1 + x_2 - y\|^2 = \|(x_1 - y) + x_2\|^2 = \|x_1 - y\|^2 + \|x_2\|^2.$$

The last equality requires justification. First, the vector $x_1 - y$ belongs to **S**, and of course, $x_2 \in \mathbf{S}^{\perp}$. Consequently, $< x_1 - y, x_2 >= 0$. Pythagorous theorem now justifies the last equality above. See **P 2.1.10** (4). After having split $\|x - y\|^2$ into two parts, we minimize $\|x_1 - y\|^2$ over all $y \in \mathbf{S}$. Since x_1 belongs to **S**, the minimum occurs at $y = x_1 = \mathbf{P_S}(x)$. This completes the proof.

The Pythagorous theorem and the decomposition of an inner product space into two subspaces which are orthogonal to each other are two sides of the same coin. If **S** and $\mathbf{S}^{\perp}$ are complementary subspaces of a inner product space **V**, and $x \in \mathbf{S}$, $y \in \mathbf{S}^{\perp}$, then $\|x + y\|^2 = \|x\|^2 + \|y\|^2 \geq \|y\|^2$. The above inequality can be paraphrased as follows. For any fixed $y \in \mathbf{S}^{\perp}$, the inequality, $\|x + y\| \geq \|y\|$, holds for every $x \in \mathbf{S}$. Does this property characterize membership of y in $\mathbf{S}^{\perp}$? Yes, it does.

P 2.2.14 Let y be any vector in an inner product space **V** and **S** a subspace of **V**. Then $y \in \mathbf{S}^{\perp}$ if and only if

$$\|x + y\| \geq \|y\| \quad \text{for every } x \text{ in } \mathbf{S}. \tag{2.2.7}$$

PROOF. We have already checked the "only if" part of the above statement. To prove the "if part", let $y = y_1 + y_2$ be the orthogonal decomposition of y with $y_1 \in \mathbf{S}$ and $y_2 \in \mathbf{S}^{\perp}$. It suffices to show that $y_1 = 0$. Observe that

$$\|y\|^2 = \|y_1\|^2 + \|y_2\|^2 \geq \|y_2\|^2 = \|(-y_1) + y\|^2 \geq \|y\|^2.$$

A word of explanation is in order. In the above chain of equalities and inequalities, Pythagorous theorem is used as well as (2.2.6). Observe also that $(-y) \in \mathbf{S}$. Thus equality must prevail everywhere in the above chain and hence $y_1 = 0$.

Complements

2.2.1 Show that two vectors x and y are orthogonal if and only if

$$\|\alpha x + \beta y\|^2 = \|\alpha x\|^2 + \|\beta y\|^2$$

for all pairs of scalars α and β. Show that two vectors x and y in a real inner product space are orthogonal if and only if

$$\|x + y\|^2 = \|x\|^2 + \|y\|^2.$$

2.2.2 If x and y are vectors of unit length in a Euclidean space, show that $x + y$ and $x - y$ are orthogonal.

2.2.3 Let $x_1, x_2, \ldots, x_k$ be an orthonormal basis of an inner product space and $y_1 = x_1, y_2 = x_1 + x_2, \ldots, y_k = x_1 + \ldots + x_k$. Apply the Gram-Schmidt orthogonalization process to $y_1, y_2, \ldots, y_k$.

2.2.4 Let $\mathbf{S}_1$ and $\mathbf{S}_2$ be two subspaces of an inner product space. Prove the following.

(1) If $\mathbf{S}_1 \subset \mathbf{S}_2$, then $\mathbf{S}_2^\perp \subset \mathbf{S}_1^\perp$.
(2) $(\mathbf{S}_1 \cap \mathbf{S}_2)^\perp = \mathbf{S}_1^\perp + \mathbf{S}_2^\perp$.
(3) $(\mathbf{S}_1 + \mathbf{S}_2)^\perp = \mathbf{S}_1^\perp \cap \mathbf{S}_2^\perp$.

2.2.5 Let $\mathbf{S}$ be a subspace of $\mathbf{V}$ and consider the set of points $H = \{x_0 + x : x \in \mathbf{S}\}$ for fixed x_0. Find $\min \|y - z\|$ for given y with respect to $z \in H$.

2.3. Linear Equations

In Section 1.6, we have considered a linear equation, homogeneous or non-homogeneous, in the environment of vector spaces involving unknown scalars belonging to the underlying field. A special case of such a linear equation we have considered is the one when the vector space was $\mathbf{F}^k$, for some $k \geq 1$. The linear equation gave rise to a system of linear equations (1.6.6), which, upon close scrutiny, gives the feeling that there is some kind of inner product operation involved. In this section, we will indeed consider inner product spaces and equations involving the underlying inner products. Let $a_1, a_2, \ldots, a_m$ be given vectors in an inner product space $\mathbf{V}$. Let $\alpha_1, \alpha_2, \ldots, \alpha_m$ be given scalars in the underlying field of the vector space. Consider the system of equations,

$$< x, a_i > = \alpha_i, \ i = 1, 2, \ldots, m, \tag{2.3.1}$$

in unknown $x \in \mathbf{V}$.

If $\mathbf{V} = \mathbf{C}^k$ or $\mathbf{R}^k$, and $< \cdot, \cdot >$ is the usual inner product on $\mathbf{V}$, then the above system of equations (2.3.1) identifies with the system (1.6.6). The above system is, in a way, more general than the system (1.6.6). Of course, in (1.6.6), the underlying field $\mathbf{F}$ is quite arbitrary. We now need to explore some methods of solving equation (2.3.1).

P 2.3.1 The system (2.3.1) of equations has a solution (i.e., the equations (2.3.1) are consistent) if and only if

$$\sum_{i=1}^{m} \bar{\beta}_i \alpha_i = 0 \quad \text{whenever} \quad \sum_{i=1}^{m} \beta_i a_i = 0$$
$$\text{for any scalars} \quad \beta_1, \beta_2, \ldots, \beta_m \text{ in } \mathbf{C}. \tag{2.3.2}$$

PROOF. Suppose the system (2.3.1) admits a solution x, say. Suppose for some scalars $\beta_1, \beta_2, \ldots, \beta_m$ in $\mathbf{C}$, $\sum_{i=1}^{m} \beta_i a_i = 0$. Then

$$< x, \sum_{i=1}^{m} \beta_i a_i > = 0 = \sum_{i=1}^{m} \bar{\beta}_i < x, a_i > = \sum_{i=1}^{m} \bar{\beta}_i \alpha_i.$$

For the converse, consider the following system of linear equations

$$\begin{array}{llll} < a_1, a_1 > \gamma_1 + & \cdots & + < a_m, a_1 > \gamma_m = \alpha_1, \\ < a_1, a_2 > \gamma_1 + & \cdots & + < a_m, a_2 > \gamma_m = \alpha_2, \\ \quad \cdot & \cdots & \quad \cdot \\ < a_1, a_m > \gamma_1 + & \cdots & + < a_m, a_m > \gamma_m = \alpha_m \end{array} \tag{2.3.3}$$

in unknown scalars $\gamma_1, \gamma_2, \ldots, \gamma_m$ in $\mathbf{C}$. Our immediate concern is whether the system (2.3.3) admits a solution. We are back into the fold of the system of linear equations (1.6.6). We would like to use

P 1.6.7 which provides a necessary and sufficient condition for the system (1.6.6) to have a solution. Let $u_i = (< a_1, a_i >, < a_2, a_i >, \ldots, < a_m, a_i >)$, $i = 1, 2, \ldots, m$. We need to verify whether the condition,

$$\beta_1\alpha_1 + \beta_2\alpha_2 + \ldots + \beta_m\alpha_m = 0$$

is satisfied whenever

$$\beta_1 u_1 + \beta_2 u_2 + \ldots + \beta_m u_m = 0 \quad \text{for any} \quad \beta_1, \beta_2, \ldots, \beta_m \in \mathbf{C}, \tag{2.3.4}$$

is satisfied to guarantee a solution to the system (2.3.3). If $\beta_1 u_1 + \beta_2 u_2 + \ldots + \beta_m u_m = 0$, then this is equivalent to $\beta_1 < a_i, a_1 > + \beta_2 < a_i, a_2 > + \ldots + \beta_m < a_i, a_m > = 0$ for each $i = 1, 2, \ldots, m$. This, in turn, is equivalent to $< a_i, \sum_{j=1}^{m} \bar{\beta}_j a_j > = 0$ for each $i = 1, 2, \ldots, m$. Suppose $\beta_1 u_1 + \beta_2 u_2 + \ldots + \beta_m u_m = 0$ for some scalars $\beta_1, \beta_2, \ldots, \beta_m$. By what we have discussed above, this is equivalent to $< a_i, \sum_{i=1}^{m} \bar{\beta}_j a_j > = 0$ for each $i = 1, 2, \ldots, m$. This then implies that $\sum_{i=1}^{m} \bar{\beta}_i < a_i, \sum_{j=1}^{m} \bar{\beta}_j a_j > = 0$ $= < \sum_{i=1}^{m} \bar{\beta}_i a_i, \sum_{i=1}^{m} \bar{\beta}_i a_i >$. Consequently, $\sum_{i=1}^{m} \bar{\beta}_i a_i = 0$. By (2.3.2), $\sum_{i=1}^{m} \beta_i \alpha_i = 0$. Thus (2.3.4) is verified. The system (2.3.3) admits a solution. Denote by, with an apology for an abuse of notation, $\gamma_1, \gamma_2, \ldots, \gamma_m$ a solution to the system (2.3.3) of equations. Let

$$x_0 = \gamma_1 a_1 + \gamma_2 a_2 + \ldots + \gamma_m a_m.$$

One can verify that x_0 is a solution of the system (2.3.1) of equations. The verification process merely coincides with the validity of the system (2.3.3) of equations.

As has been commented earlier, there is an uncanny resemblance between the systems (2.3.1) and (1.6.6). In view of **P 1.6.7**, the above result is not surprising.

Suppose the system (2.3.1) of equations is consistent, i.e., the system admits a solution. An immediate concern is the identification of a solution. If we scrutinize the proof of **P 2.3.1** carefully, it will certainly provide an idea of how to obtain a solution to the system. This solution is built upon the solution of the system (2.3.3) of equations operating in the realm of the field of complex numbers. Solving the system (2.3.3) is practical since we are dealing with complex numbers only. There may be more than one solution. We need to determine the structure of the set of all solutions of (2.3.1). The following proposition is concerned about this aspect of the problem.

P 2.3.2 Let $\mathbf{S}_1$ be the collection of all solutions to the system (2.3.1) of equations, assumed to be consistent. Let $\mathbf{S}$ be the collection of all solutions to the system

$$< x, a_i > = \ 0 \quad i = 1, 2, \ldots, m,$$

of equations. Let x_0 be any particular solution of (2.3.1). Then

(1) $\mathbf{S}$ is a subspace of $\mathbf{V}$, and
(2) $\mathbf{S}_1 = x_0 + \mathbf{S} = \{x_0 + y : y \in \mathbf{S}\}$.

The above proposition is modeled on **P 1.6.4**. The same kind of humdrum argument carries through.

Among the solutions available to the system (2.3.1), we would like to pick up that solution x for which $\|x\|$ is minimum. We could label such a solution as a minimum norm solution. The nicest thing about the solution which we have offered in the proof of **P 2.3.1** is that it is indeed a minimum norm solution. Let us solemnize this fact in the following proposition.

P 2.3.3 The unique minimum norm solution of (2.3.1), when it is consistent, is given by

$$x_0 = \gamma_1 a_1 + \gamma_2 a_2 + \ldots + \gamma_m a_m,$$

where $\gamma_1, \gamma_2, \ldots, \gamma_m$ is any solution to the system (2.3.3) of equations. Further, $\|x_0\| = \bar{\gamma}_1 \alpha_1 + \bar{\gamma}_2 \alpha_2 + \ldots + \bar{\gamma}_m \alpha_m$.

PROOF. We have already shown that x_0 is a solution to the system (2.3.1) in **P 2.3.1**. Any general solution to the system (2.3.1) is of the form $x_0 + y$, where y satisfies the conditions that $< y, a_i >= \ 0$ for each $i = 1, 2, \ldots, m$. See **P 2.3.2**. This y is orthogonal to x_0! (Why?) Consequently, by Pythagorous theorem,

$$\|x_0 + y\|^2 = \|x_0\|^2 + \|y\|^2 \geq \|x_0\|^2,$$

which shows that x_0 is a minimum norm solution. Further, in the above the equality is attained only when $y = 0$. This shows that the solution x_0 is unique with respect to the property of minimum norm. As for the norm of x_0, we note that

$$\|x_0\|^2 = < x_0, x_0 > = < x_0, \sum_{i=1}^{m} \gamma_i a_i > = \sum_{i=1}^{m} \bar{\gamma}_i < x_0, a_i > = \sum_{i=1}^{m} \bar{\gamma}_i \alpha_i.$$

This completes the proof.

Complements

2.3.1 Let $\mathbf{V}$ be an inner product space and $x \in \mathbf{V}$. Let $\mathbf{S}$ be a subspace of $\mathbf{V}$.

(1) Show that $\|x - y\|$ is minimized over all $y \in \mathbf{S}$ at any $y = \hat{y} \in \mathbf{S}$ for which $(x - \hat{y})$ is orthogonal to $\mathbf{S}$, i.e., $< x - \hat{y}, y >= 0$ for all $y \in \mathbf{S}$. (This is an alternative formulation of **P 2.2.13**.)

(2) Suppose $\mathbf{S}$ is spanned by the vectors $y_1, y_2, \ldots, y_r$. Show that the problem of determining $\hat{y} \in \mathbf{S}$ such that $x - \hat{y}$ is orthogonal to $\mathbf{S}$ is equivalent to the problem of determining scalars $\beta_1, \beta_2, \ldots, \beta_r$ such that $x - (\beta_1 y_1 + \beta_2 y_2 + \ldots + \beta_r y_r)$ is orthogonal to $\mathbf{S}$ which, in turn, is equivalent to solving the equations

$$\begin{aligned}
< y_1, y_1 > \beta_1 + < y_2, y_1 > \beta_2 + \ldots + < y_r, y_1 > \beta_r &= < x, y_1 >, \\
< y_1, y_2 > \beta_1 + < y_2, y_2 > \beta_2 + \ldots + < y_r, y_2 > \beta_r &= < x, y_2 >, \\
\ldots \qquad \ldots \qquad \ldots & \\
< y_1, y_r > \beta_1 + < y_2, y_r > \beta_2 + \ldots + < y_r, y_r > \beta_r &= < x, y_r >,
\end{aligned}$$

in unknown scalars $\beta_1, \beta_2 \ldots, \beta_r$.

(3) Show that the system of equations is solvable.

(The method outlined in (2) is a practical way of evaluating $\mathbf{P_S}(x)$.)

(4) Let $\mathbf{V} = \mathbf{R}^n$ with its standard inner product. Let $x = (x_1, x_2, \ldots, x_n)$ and $y_i = (y_{i1}, y_{i2}, \ldots, y_{in})$, $i = 1, 2, \ldots, r$. Show that the steps involved in (1), (2), and (3) above lead to the least squares theory of approximating the vector x by a vector from the vector space spanned by $y_1, y_2, \ldots, y_r$.

2.3.2 Let $\mathbf{P}_4$ be the vector space of all polynomials of degree less than 4 with real coefficients. The inner product in $\mathbf{P}_4$ is defined by

$$< y_1(x), y_2(x) > = \int_{-1}^{1} y_1(x) y_2(x)\, dx$$

for $y_1(x), y_2(x) \in \mathbf{P}_4$. Let $\mathbf{S}$ be the vector space spanned by the polynomials $y_1(x) = 1, y_2(x) = x$, and $y_3(x) = x^2$. Determine the best approximation of the polynomial $2x + 3x^2 - 4x^3$ by a polynomial from $\mathbf{S}$.

2.4. Linear Functionals

In Section 1.7, we presented some discussion on linear functionals of vector spaces and dual spaces. Now that we have an additional structure on our vector spaces, namely, inner products, we need to reexamine the concept of linear functional in the new environment. The definitions of a linear functional and that of the dual space remain the same. If we use the inner product available on the underlying vector space, we see that the linear functionals have a nice structural form and we get a clearer understanding of the duality between the vector space and its dual space. First, we deal with linear functionals.

P 2.4.1 (Representation Theorem of a Linear Functional) Let $\mathbf{V}$ be an inner product space and f a linear functional on $\mathbf{V}$. Then there exists a unique vector z in $\mathbf{V}$ such that

$$f(x) \quad = < x, z > \text{ for every } x \text{ in } \mathbf{V}. \qquad (2.4.1)$$

PROOF. If $f(x) = 0$ for every x in $\mathbf{V}$, take $z = 0$. Otherwise, let $\mathbf{S}^{\perp}$ be the orthogonal complement of the subspace $\mathbf{S} = \{x \in \mathbf{V} : f(x) = 0\}$ of $\mathbf{V}$. The subspace $\mathbf{S}^{\perp}$ has at least one non-zero vector. (Why?) Choose any vector u in $\mathbf{S}^{\perp}$ such that $< u, u > = 1$. Set $z = \overline{f(u)}u$. The vector z is the required candidate. Since $u \notin \mathbf{S}$, $f(u) \neq 0$ and hence $z \neq 0$ and $z \in \mathbf{S}^{\perp}$. Let us see whether the vector z does the job. To begin with, we verify

$$f(z) = \overline{f(u)} f(u) \quad = < z, z > .$$

Let $x \in \mathbf{V}$. Set

$$x_1 = x - [f(x)/ < z, z >] z. \qquad (2.4.2)$$

We show that $x = x_1 + x_2$ is the orthogonal decomposition of x with respect to the subspaces $\mathbf{S}$ and $\mathbf{S}^\perp$, where $x_2 = [f(x)/ < z, z >]z$. It is clear that $x_2 \in \mathbf{S}^\perp$. Further, $f(x_1) = 0$. Consequently, $x_1 \in \mathbf{S}$. From (2.4.2),

$$0 = < x_1, z > = < x, z > - [f(x)/ < z, z >]f(z) = < x, z > - f(x).$$

Hence $f(x) = < x, z >$. The uniqueness of the vector z is easy to establish. If z_1 is another vector such that $< x, z_1 > = < x, z >$ for all x in $\mathbf{V}$, then $< x, z_1 - z > = 0$ for all x in $\mathbf{V}$. Hence, we must have $z_1 = z$. This completes the proof.

Thus with every linear functional f on the inner product space $\mathbf{V}$, we have a unique vector z in $\mathbf{V}$ satisfying (2.4.1). This correspondence is an isomorphism between the vector space $\mathbf{V}$ and its dual space $\mathbf{V}'$. This is not hard to establish. We can reap some benefits out of the representation theorem presented above. We recast **P 1.7.6** in our new environment.

P 2.4.2 Let $x_1, x_2, \ldots, x_k$ constitute a basis for an inner product space $\mathbf{V}$. Then we can find another basis $z_1, z_2, \ldots, z_k$ for $\mathbf{V}$ such that

$$< x_i, z_j > = \begin{cases} 1 & \text{if } i = j, \\ 0 & \text{if } i \neq j, \end{cases}$$

for all i and j. Further, for any x in $\mathbf{V}$, we can write

$$x = < x, z_1 > x_1 + < x, z_2 > x_2 + \ldots + < x, z_k > x_k.$$

Also, $z_i = x_i$ for every i if and only if $x_1, x_2, \ldots, x_k$ is an orthonormal basis for $\mathbf{V}$.

PROOF. There are several ways of establishing the veracity of the above proposition. One way is to use the result of **P 1.7.6** to obtain a dual basis of linear functionals $f_1, f_2, \ldots, f_k$ for $\mathbf{V}'$ and then use **P 2.4.1** above to obtain the associated vectors $z_1, z_2, \ldots, z_k$. We would like to describe another way (really?) which is more illuminating. Let x be any vector in $\mathbf{V}$ and $(\xi_1, \xi_2, \ldots, \xi_k)$ its co-ordinates with respect to the basis $x_1, x_2, \ldots, x_k$, i.e.,

$$x = \xi_1 x_1 + \xi_2 x_2 + \ldots, +\xi_k x_k.$$

For each $1 \leq i \leq k$, define $f_i : \mathbf{V} \to \mathbf{F}$ by $f_i(x) = \xi_i$. One can verify that each f_i is a linear functional. By **P 2.4.1**, there exists a unique vector z_i in $\mathbf{V}$ such that $f_i(x) = < x, z_i >$ for every x in $\mathbf{V}$, and for each $1 \leq i \leq k$. One can verify that $z_1, z_2, \ldots, z_k$ constitute a basis for the vector space $\mathbf{V}$. Since the co-ordinates of x_i are $(0, 0, \ldots, 0, 1, 0, \ldots, 0)$ with 1 in the i-th position, we have

$$< x_i, z_j >= \begin{cases} 1 & \text{if } j = i, \\ 0 & \text{if } j \neq i. \end{cases}$$

The other statements of the proposition follow in a simple manner.

As an application of the above ideas, let us consider a statistical prediction problem. Let $(\Omega, \mathcal{A}, P)$ be a probability space and $x_1, x_2, \ldots, x_k$ be square integrable real random variables (defined on the probability space). Then the collection of all random variables of the form $\alpha_1 x_1 + \alpha_2 x_2 + \ldots + \alpha_k x_k$ for all real scalars $\alpha_1, \alpha_2, \ldots, \alpha_k$ is a vector space $\mathbf{V}$ over the field $\mathbf{R}$ of real numbers with the usual operations of addition and scalar multiplication of random variables. We introduce the following inner product on the vector space $\mathbf{V}$. For any x, y in $\mathbf{V}$,

$$< x, y > = E(xy),$$

where E stands for the expectation operator. The above expectation is evaluated with respect to the joint distribution of x and y. In statistical parlance, $E(xy)$ is called the product moment of x and y. Assume that $x_1, x_2, \ldots, x_k$ are linearly independent. What this means, in our context, is that if $\alpha_1 x_1 + \alpha_2 x_2 + \ldots + \alpha_k x_k = 0$ almost surely for some scalars $\alpha_1, \alpha_2, \ldots, \alpha_k$ then each α_i must be equal to zero. This implies that none of the random variables is degenerate almost surely and $\dim(\mathbf{V}) = k$. For applications, it is convenient to adjoin the random variable x_0 which is equal to the constant 1 almost surely to our lot $x_1, x_2, \ldots, x_k$ if it is not already there. Let $\mathbf{V}^*$ be the vector space spanned by $x_0, x_1, \ldots, x_k$. Let p be any positive integer less than k. Let $\mathbf{S}$ be the vector space spanned by $x_0, x_1, \ldots, x_p$. Let y be any random variable in $\mathbf{V}^*$. Now we come to the prediction problem. Suppose we are able to observe $x_1, x_2, \ldots, x_p$. We would like to predict the value of the random variable y. Mathematically, we want to propose a linear predictor $\beta_0 + \beta_1 x_1 + \ldots + \beta_p x_p$ as our prediction of the random variable y. Practically, what this means is that whenever we observe $x_1, x_2, \ldots, x_p$, plug in the observed values into the predictor and then declare that the resultant number is our predicted value of y. Now the question arises as to the choice of the scalars $\beta_0, \beta_1, \ldots, \beta_p$. Of course, we all feel that we must choose the scalars optimally, optimal in some sense. One natural optimality criterion can be developed in the following way. For any choice of the scalars $\beta_0, \beta_1, \ldots, \beta_p$,

$$y - \beta_0 - \beta_1 x_1 - \beta_2 x_2 - \ldots - \beta_p x_p$$

can be regarded as prediction error. We need to minimize the prediction error in some way. One way of doing this is to choose the scalars $\beta_0, \beta_1, \ldots, \beta_p$ in such a way that

$$\|y - \beta_0 - \beta_1 x_1 - \ldots \beta_p x_p\|^2 = E(y - \beta_0 - \beta_1 x_1 - \ldots - \beta_p x_p)^2 \quad (2.4.3)$$

is a minimum.

This kind of scenario arises in a variety of contexts. In Econometrics, x_i could denote the price of a particular stock at time period $i, i = 1, 2, \ldots, k$. After having observed $x_1, x_2, \ldots, x_p$ at p successive time points, we would like to predict the price of the stock at the time point $p + 1$. In such a case, we take $y = x_{p+1}$. In the spatial prediction problem, the objective is to predict a response variable y at a new site given observations $x_1, x_2, \ldots, x_p$ at p existing sites. The spatial prediction problem is known as kriging in geostatistics literature.

Let us get back to the problem of choosing the scalars $\beta_0, \beta_1, \ldots, \beta_p$ in (2.4.3). Observe that for any choice of scalars $\beta_0, \beta_1, \ldots, \beta_p$, the vector $\beta_0 + \beta_1 x_1 + \ldots + \beta_p x_p$ belongs to the subspace **S**. The problem now reduces to finding a vector x in **S** such that $\|y - x\|$ is a minimum. We have already solved this problem. See the result of **P 2.2.13**. The solution is given by $x = \mathbf{P_S}(y)$, the orthogonal projection of y onto the subspace **S**. Let us try to compute explicitly the orthogonal projection of y onto the subspace **S**. Observe that x must be of the form $\beta_0 + \beta_1 x_1 + \ldots + \beta_p x_p$ for some scalars $\beta_0, \beta_1, \ldots, \beta_p$. Write the orthogonal decomposition of y with respect to the subspaces **S** and $\mathbf{S}^\perp$ as

$$y = x + (y - x) = \mathbf{P_S}(y) + (y - x).$$

Observe that $(y - x) \in \mathbf{S}^\perp$ if and only if $< y - x, x_i >= 0$ for every $i = 0, 1, 2, \ldots, p$. But $< y - x, x_i >= 0$ means that $E((y - x)x_i) =$

$0 = E((y - \beta_0 - \beta_1 x_1 - \ldots - \beta_p x_p)x_i)$. Expanding the expectation, we obtain the following equations in $\beta_0, \beta_1, \ldots, \beta_p$:

$$\begin{array}{llllll}
\beta_0 & + \beta_1 E(x_1) & + \beta_2 E(x_2) & + \ldots & + \beta_p E(x_p) & = E(y), \\
\beta_0 E(x_1) & + \beta_1 E(x_1^2) & + \beta_2 E(x_1 x_2) & + \ldots & + \beta_p E(x_1 x_p) & = E(x_1 y), \\
\beta_0 E(x_2) & + \beta_1 E(x_2 x_1) & + \beta_2 E(x_2^2) & + \ldots & + \beta_p E(x_2 x_p) & = E(x_2 y), \\
\ldots & \ldots & \ldots & \ldots & \ldots & \\
\beta_0 E(x_p) & + \beta_1 E(x_p x_1) & + \beta_2 E(x_p x_2) & + \ldots & + \beta_p E(x_p^2) & = E(x_p y).
\end{array}$$

We need to solve these equations in order to build the required predictor. These linear equations can be simplified further. We can eliminate β_0 by using the first equation above, i.e.,

$$\beta_0 = E(y) - \beta_1 E(x_1) - \beta_2 E(x_2) - \ldots - \beta_p E(x_p),$$

from each of the remaining equations. Now we will have p equations in p unknowns $\beta_1, \beta_2, \ldots, \beta_p$:

$$\begin{array}{lllll}
\beta_1 s_{11} & + \beta_2 s_{12} + & \ldots & + \beta_p s_{1p} & = s_{01}, \\
\beta_1 s_{21} & + \beta_2 s_{22} + & \ldots & + \beta_p s_{2p} & = s_{02}, \\
\ldots & & \ldots & & \ldots \\
\beta_1 s_{p1} & + \beta_2 s_{p2} + & \ldots & + \beta_p s_{pp} & = s_{0p},
\end{array}$$

where $s_{ij} = E(x_i x_j) - E(x_i)E(x_j) =$ covariance between x_i and $x_j, 1 \leq i, j \leq p$, and $s_{0i} = E(yx_i) - E(y)E(x_i) =$ covariance between y and $x_i, i = 1, 2, \ldots, p$. The problem of determining the optimal predictor of y reduces to the problem of solving the above p linear equations in p unknowns!

Complements

2.4.1 Let $\mathbf{R}^3$ be the three-dimensional vector space equipped with the standard inner product. Let $f : \mathbf{R}^3 \to \mathbf{R}$ be the linear functional defined by

$$f(x) = f(x_1, x_2, x_3) = 2x_1 + x_2 - x_3$$

for $x = (x_1, x_2, x_3) \in \mathbf{R}^3$. Determine the vector $z \in \mathbf{R}^3$ such that

$$f(x) = < x, z >, \; x \in \mathbf{R}^3.$$

2.4.2 Let **V** be a real vector space with an inner product and T a transformation from **V** to **V**. Define a map $f : \mathbf{V} \rightarrow \mathbf{R}$ by

$$f(x) \ = < Tx, y >, \ x \in \mathbf{V}$$

for some fixed vector $y \in \mathbf{V}$. Show that f is a linear functional on **V**. Determine $z \in \mathbf{V}$ such that

$$f(x) \ = < x, z > \quad \text{for all } x \in \mathbf{V}.$$

2.5. Semi-inner Product

We have seen what inner products are in Section 2.1. We do come across some maps on the product space $\mathbf{V} \times \mathbf{V}$ of a vector space which are almost like an inner product. A semi-inner product, which is the focus of attention in this section, is one such map relaxing one of the conditions of an inner product. In this section, we outline some strategies how to handle semi-inner product spaces. All the definitions and results presented in this section are designed for vector spaces over the field **C** of complex numbers. The modifications should be obvious if the underlying field is that of real numbers.

DEFINITION 2.5.1. Let **V** be a vector space. A complex valued function $(\cdot, \cdot)$ defined over the product space $\mathbf{V} \times \mathbf{V}$ is said to be a semi-inner product if it meets the following conditions:

(1) $(x, y) = \overline{(y, x)}$ for all x and y in **V**.
(2) $(x, x) \geq 0$ for all x in **V**.
(3) $(\alpha_1 x_1 + \alpha_2 x_2, y) = \alpha_1 (x_1, y) + \alpha_2 (x_2, y)$ for all x_1, x_2 and y in **V**, and α_1, α_2 in **C**.

These conditions are the same as those for an inner product except for (2) which admits the possibility of (x, x) vanishing for $x \neq 0$. We use the notation $(\cdot, \cdot)$ for a semi-inner product, and $< \cdot, \cdot >$ for a regular inner product for which $< x, x >= 0$ only when $x = 0$. In the same vein, we define the positive square root of (x, x) as the semi-norm of x and denote it by $\|x\|_{se}$. Note that $\|x\|_{se}$ could be zero for a non-zero vector. The vector space **V** equipped with a semi-inner product is called a semi-inner product space.

Most of the results that are valid for an inner product space are also valid for a semi-inner product space. There are, however, some essential differences. In the following proposition, we highlight some of the salient features of a semi-inner product space.

P 2.5.2 Let $(\cdot,\cdot)$ be a semi-inner product on a vector space $\mathbf{V}$. Then the following are valid.

(1) $(0,0) = 0$.

(2) $(x,y) = 0$ if either $\|x\|_{se} = 0$ or $\|y\|_{se} = 0$

(3) (Cauchy-Schwartz Inequality) $|(x,y)| \leq \|x\|_{se}\|y\|_{se}$ for all x and y in $\mathbf{V}$.

(4) (Triangle Inequality) $\|(x+y)\|_{se} \leq \|x\|_{se} + \|y\|_{se}$ for all x and y in $\mathbf{V}$.

(5) The set $\mathbf{N} = \{x \in \mathbf{V} : \|x\|_{se} = 0\}$ is a subspace of $\mathbf{V}$.

PROOF. Before attempting a proof, it will be instructive to scan some parts of Section 2.1 to get a feeling about where the differences lie. To prove (1), choose $\alpha_1 = \alpha_2 = 0$ and $y = 0$. For (2), one could use the Cauchy-Schwartz inequality stated in (3). But this is not the right thing to do. In the proof of (3), we make use of the fact that the assertion of (2) is true. Suppose $\|y\|_{se} = 0$. Then for any complex number α,

$$\begin{aligned} 0 \leq (x+\alpha y, x+\alpha y) &= \|x\|_{se}^2 + \alpha(y,x) + \bar{\alpha}(x,y) + \alpha\bar{\alpha}(y,y) \\ &= \|x\|_{se}^2 + \alpha(y,x) + \bar{\alpha}\overline{(y,x)}. \end{aligned}$$

If $\alpha = \gamma + i\delta$ and $(x,y) = \xi + i\eta$ for real numbers γ, δ, ξ and η, then

$$\|x\|_{se}^2 + \alpha(y,x) + \bar{\alpha}\overline{(y,x)} = \|x\|_{se}^2 + 2\gamma\xi + 2\delta\eta \geq 0 \tag{2.5.1}$$

for all real numbers γ and δ. Set $\gamma = 0$. Then

$$\|x\|_{se}^2 + 2\delta\eta \geq 0$$

for all real numbers δ. This is possible only if $\eta = 0$. Set $\delta = 0$ in (2.5.1). Then

$$\|x\|_{se}^2 + 2\gamma\xi \geq 0$$

for all real numbers γ. This is possible only if $\xi = 0$. Consequently,

$$(x,y) = \xi + i\eta = 0.$$

To prove (3), we follow the same route as the one outlined in the proof of (2). The statement of (3) is valid when $\|y\|_{se} = 0$ by virtue of (2). Assume that $\|y\|_{se} \neq 0$. For any complex number α, observe that

$$0 \leq (x + \alpha y, x + \alpha y) = \|x\|_{se}^2 + \alpha(y, x) + \bar{\alpha}(x, y) + \bar{\alpha}\alpha\|y\|_{se}^2.$$

Set $\alpha = -(x, y)/\|y\|_{se}^2$. Then

$$0 \leq \|x\|_{se}^2 - |(x, y)|^2/\|y\|_{se}^2 - |(x, y)|^2/\|y\|_{se}^2 + |(x, y)|^2/\|y\|_{se}^2,$$

from which the Cauchy-Schwartz inequality follows. In the above, we have used the fact that $(x, y)(y, x) = |(x, y)|^2$. This proof is essentially the same as the one provided for the inequality in inner product spaces. Observe the role played by the statement (2) in the proof. The proof of (4) hinges on an application of the Cauchy-Schwartz inequality and is analogous to the one provided in the inner product case. Finally, we tackle the statement (5). To show that $\mathbf{N}$ is a subspace, let x and y belong to $\mathbf{N}$, and α and β be complex numbers. We need to show that

$$(\alpha x + \beta y, \alpha x + \beta y) = 0.$$

But $(\alpha x + \beta y, \alpha x + \beta y) = \alpha\bar{\alpha}\|x\|_{se}^2 + \beta\bar{\beta}\|y\|_{se}^2 + \alpha\bar{\beta}(x, y) + \bar{\alpha}\beta(y, x) = 0$, in view of (2). This completes the proof.

We would like to bring into focus certain differences between inner product spaces and semi-inner product spaces. We look at an example.

EXAMPLE 2.5.3. Consider the k-dimensional Euclidean space $\mathbf{R}^k$ for some $k \geq 2$. Let $1 \leq r < k$ be an integer. For $x = (\xi_1, \xi_2, \dots, \xi_k)$ and $y = (\eta_1, \eta_2, \dots, \eta_k)$ in $\mathbf{R}^k$, define

$$(x, y) = \xi_1\eta_1 + \xi_2\eta_2 + \dots + \xi_r\eta_r.$$

The map $(\cdot, \cdot)$ is a semi-inner product on the vector space $\mathbf{R}^k$. The norm of x works out to be

$$\|x\|_{se} = (\xi_1^2 + \xi_2^2 + \dots + \xi_r^2)^{1/2}.$$

The subspace $\mathbf{N} = \{x \in \mathbf{R}^k : \|x\|_{se} = 0\}$ consists of vectors of the form $(0, 0, \dots, \xi_{r+1}, \dots, \xi_k)$. Consequently, $\dim(\mathbf{N}) = k - r$.

The dimensions of orthogonal vector spaces under a semi-inner product may not be additive. To illustrate this point, let $x_1 = (1, 0, 0, \ldots, 0)$ and $x_2 = (0, \ldots, 0, 1, 0, \ldots, 0)$ where 1 in x_2 appears at the $(r+1)$-th position. Let $< \cdot, \cdot >$ be any inner product on $\mathbf{R}^k$ and $(\cdot, \cdot)$ the semi-inner product on $\mathbf{R}^k$ introduced above. Let $\mathbf{S}$ be the vector space spanned by x_1 and x_2. Obviously, the dimension of the subspace $\mathbf{S}$ of $\mathbf{R}^k$ is two. Let $\mathbf{U} = \{x \in \mathbf{R}^k : (x, x_1) = (x, x_2) = 0\}$. Every vector x in $\mathbf{U}$ is of the form $(0, \xi_2, \xi_3, \ldots, \xi_k)$ for some real numbers $\xi_2, \xi_3, \ldots, \xi_k$. Observe that the dimension of the subspace $\mathbf{U}$ of $\mathbf{R}^k$ is $k-1$. The subspace $\mathbf{U}$ can be regarded as the orthogonal complement of the subspace $\mathbf{S}$ with respect to the semi-inner product $(\cdot, \cdot)$. But $\dim(\mathbf{S}) + \dim(\mathbf{U}) \neq k$. Also, $\mathbf{S} \cap \mathbf{U} \neq \{0\}$. On the other hand, if we define $\mathbf{U} = \{x \in \mathbf{R}^k : < x, x_1 > = < x, x_2 > = 0\}$, then the subspace $\mathbf{U}$ is the orthogonal complement of the subspace $\mathbf{S}$ with respect to the inner product $< \cdot, \cdot >$.

There are two ways of manufacturing an inner product from a semi-inner product. Let $\mathbf{V}$ be a semi-inner product space with semi-inner product $(\cdot, \cdot)$. The critical ideas are based on the subspace $\mathbf{N}$ defined above. Take any complement $\mathbf{N}^c$ of $\mathbf{N}$. Look up **P 1.5.5**. The restriction of the semi-inner product $(\cdot, \cdot)$ to the vector space $\mathbf{N}^c$ is an inner product! This is rather easy to see. Another way is to use the idea of a quotient space. Let $\mathbf{W} = \mathbf{V}/\mathbf{N}$ be the quotient space of $\mathbf{V}$ with respect to the subspace $\mathbf{N}$. Look up Section 1.8. The space $\mathbf{W}$ is the collection of all cosets of $\mathbf{N}$ of the form $x + \mathbf{N}, x \in \mathbf{V}$. We define a map $< \cdot, \cdot >$ on the product space $\mathbf{W} \times \mathbf{W}$ by

$$< x + \mathbf{N}, y + \mathbf{N} > = (x, y)$$

for any two distinct cosets of $\mathbf{N}$. It is not hard to show that the map $< \cdot, \cdot >$ is an inner product on the quotient space $\mathbf{W}$. We will record this fact in the form of a proposition for future reference.

P 2.5.4 The map $< \cdot, \cdot >$ defined above on the quotient space $\mathbf{W}$ is an inner product on $\mathbf{W}$.

It is time to define, formally, the orthogonal complement of a subspace $\mathbf{S}$ of a vector space $\mathbf{V}$ equipped with a semi-inner product $(\cdot, \cdot)$. Following the procedure in the case of inner product spaces, define $\mathbf{S}_s^{\perp} = \{x \in \mathbf{V} : (x, y) = 0 \text{ for every } y \in \mathbf{S}\}$ as the orthogonal complement of $\mathbf{S}$ with respect to the semi-inner product $(\cdot, \cdot)$. As we have seen earlier, $\mathbf{S} \cap \mathbf{S}_s^\perp$ may contain non-zero vectors. It is clear that $\mathbf{S}_s^\perp$ is a subspace of $\mathbf{V}$, and then that $\mathbf{S} \cap \mathbf{S}_s^\perp$ is a subspace. The dimension of the space $\mathbf{S} \cap \mathbf{S}_s^\perp$ could be more than one. But one could always decompose any given vector x in $\mathbf{V}$ into a sum $x_0 + x_{00}$ with $x_0 \in \mathbf{S}$ and $x_{00} \in \mathbf{S}_s^\perp$. Let us put this down as a proposition.

P 2.5.5 For any given vector x in a semi-inner product space $\mathbf{V}$, there exists x_0 in $\mathbf{S}$ and x_{00} in $\mathbf{S}_s^\perp$ such that

$$x = x_0 + x_{00}.$$

Further, the subspaces $\mathbf{S}$ and $\mathbf{S}_s^\perp$ together span the space $\mathbf{V}$.

PROOF. Let $x_1, x_2, \ldots, x_r$ be a basis of the vector space $\mathbf{S}$. The basic problem is to determine scalars $\alpha_1, \alpha_2, \ldots, \alpha_r$ such that $x - \alpha_1 x_1 - \alpha_2 x_2 - \ldots - \alpha_r x_r$ belongs to $\mathbf{S}_s^\perp$. If we succeed in this mission, we simply let $x_0 = \alpha_1 x_1 + \alpha_2 x_2 + \ldots + \alpha_r x_r$ and $x_{00} = x - x_0$. We will then have the desired decomposition. The scalars could be found. The condition that $x - \alpha_1 x_1 - \alpha_2 x_2 - \ldots - \alpha_r x_r$ belongs to $\mathbf{S}_s^\perp$ is equivalent to

$$(x - \alpha_1 x_1 - \alpha_2 x_2 - \ldots - \alpha_r x_r, x_i) = 0 \quad \text{for each} \quad i = 1, 2, \ldots, r.$$

These equations can be rewritten in an illuminating way.

$$\begin{array}{llll l} (x_1, x_1)\alpha_1 & + (x_2, x_1)\alpha_2 & + \ldots & + (x_r, x_1)\alpha_r & = (x, x_1), \\ (x_1, x_2)\alpha_1 & + (x_2, x_2)\alpha_2 & + \ldots & + (x_r, x_2)\alpha_r & = (x, x_2), \\ \ldots & \ldots & \ldots & \ldots & \\ (x_1, x_r)\alpha_1 & + (x_2, x_r)\alpha_2 & + \ldots & + (x_r, x_r)\alpha_r & = (x, x_r). \end{array}$$

We have r linear equations in r unknowns $\alpha_1, \alpha_2, \ldots, \alpha_r$. This system of equations is analogous to the one presented in (1.6.6). We can invoke **P 1.6.7** to check whether this system admits a solution. The case $\|x\|_{se} = 0$ is trivial. The decomposition is: $x = 0 + x$. Assume that $\|x\|_{se} \neq 0$. Let $u_i = ((x_1, x_i), (x_2, x_i), \ldots, (x_r, x_i)), i = 1, 2, \ldots, r$. Suppose $\varepsilon_1 u_1 + \varepsilon_2 u_2 + \ldots + \varepsilon_r u_r = 0$ for some scalars $\varepsilon_1, \varepsilon_2, \ldots, \varepsilon_r$. This is equivalent to

$$(x_j, \sum_{i=1}^{r} \bar{\varepsilon}_i x_i) = 0 \quad \text{for each} \quad j = 1, 2, \ldots, r.$$

This implies that $(\sum_{j=1}^{r} \bar{\varepsilon}_j x_j, \sum_{i=1}^{r} \bar{\varepsilon}_i x_i)_s = 0$, from which we have $\|\sum_{i=1}^{r} \bar{\varepsilon}_i x_i\| = 0$. Following the line of thought outlined in **P 1.6.7**, we need only to verify that

$$\varepsilon_1(y, x_1) + \varepsilon_2(y, x_2) + \ldots + \varepsilon_r(y, x_r) = 0.$$

But

$$\varepsilon_1(y, x_1) + \varepsilon_2(y, x_2) + \ldots + \varepsilon_r(y, x_r) = (y, \sum_{i=1}^{r} \bar{\varepsilon}_i x_i) = 0,$$

in view of the fact that $\|\sum_{i=1}^{r} \bar{\varepsilon}_i x_i\| = 0$ and the result of **P 2.5.2** (2). Consequently, the above system of equations is consistent. Finally, the mere fact that the decomposition is possible is good enough to conclude that the spaces $\mathbf{S}$ and $\mathbf{S}_s^{\perp}$ together span $\mathbf{V}$. This completes the proof.

As has been pointed out earlier, the dimensions of $\mathbf{S}$ and $\mathbf{S}_s^{\perp}$ need not add up to the dimension of $\mathbf{V}$. Further, we would like to point out that the decomposition is not unique. This makes it difficult to define the projection of $\mathbf{V}$ onto the subspace $\mathbf{S}$. We need not worry about it. We can get around this difficulty.

Let us consider an optimization problem in the context of a semi-inner product space similar to the one considered in Section 2.2. Let $\mathbf{V}$ be a vector space equipped with a semi-inner product $(\cdot, \cdot)$ and $\mathbf{S}$ a subspace of $\mathbf{V}$. Let x be any vector in $\mathbf{V}$. We raise the question whether the minimum of $\|x - z\|_{se}$ is attained over all $z \in \mathbf{S}$, or in other words, whether there exists a vector $x_0 \in \mathbf{S}$ such that

$$\inf_{z \in \mathbf{S}} \|x - z\|_{se} = \|x - x_0\|_{se}.$$

This has a solution and it is not hard to guess the vector. The experience we have had with inner product spaces should come handy. In fact, the same kind of proof as in the case of regular inner products works. In **P 2.5.5** we showed that any given vector $x \in \mathbf{V}$ admits a decomposition

$$x = x_0 + x_{00}$$

with $x_0 \in \mathbf{S}$ and $x_{00} \in \mathbf{S}_s^{\perp}$. The decomposition is not unique as stated earlier. But any x_0 and x_{00} with the stated inclusion properties will do.

The vector x_0 is indeed the solution to the optimization problem above. Note that for any vector z in $\mathbf{S}$,

$$\begin{aligned}\|x-z\|_{se}^2 &= \|(x-x_0)+(x_0-z)\|_{se}^2 \\ &= ((x-x_0)+(x_0-z),(x-x_0)+(x_0-z)) \\ &= \|(x-x_0)\|_{se}^2+\|(x_0-z)\|_{se}^2+(x-x_0,x_0-z) \\ &\quad +(x_0-z,x-x_0) \\ &= \|(x-x_0)\|_{se}^2+\|(x_0-z)\|_{se}^2 \\ &\geq \|x-x_0\|_{se}^2. \qquad (2.5.2)\end{aligned}$$

This inequality establishes that the vector x_0 is a desired solution to our optimization problem. The semi-inner products that appear above vanish in view of the facts that $x - x_0 \in \mathbf{S}_s^{\perp}$ and $x_0 - z \in \mathbf{S}$. But we must admit that the solution vector x_0 need not to be unique. The solution vector x_0 can be characterized in the following way.

P 2.5.6 Let x be any vector in any semi-inner product space $\mathbf{V}$. Let $\mathbf{S}$ be a subspace of $\mathbf{V}$. The following two statements are equivalent.

(1) There exists a vector x_0 in $\mathbf{S}$ such that

$$\inf_{z\in\mathbf{S}} \|x-z\|_{se} = \|x-x_0\|_{se}.$$

(2) There exists a vector x_0 in $\mathbf{S}$ such that $x - x_0 \in \mathbf{S}_s^{\perp}$.

PROOF. The statement (2) implies the statement (1) from the inequality established in (2.5.2). Suppose (1) is true. Then for any complex number α and for any vector z in $\mathbf{S}$.

$$\begin{aligned}\|x-x_0-\alpha z\|_{se}^2 &= \|x-x_0\|_{se}^2+\alpha\bar{\alpha}\|z\|_{se}^2-\bar{\alpha}(x-x_0,z)-\alpha(z,x-x_0) \\ &\geq \|x-x_0\|_{se}^2,\end{aligned}$$

since $x_0 + \alpha z \in \mathbf{S}$. Since the above inequality is true for any scalar α, it follows that $(x - x_0, z) = 0$. But this equality is true for any z in $\mathbf{S}$. Hence $x - x_0 \in \mathbf{S}_s^{\perp}$.

We consider another problem of minimization. Let $\mathbf{S}$ be a subspace of a semi-inner product space. Let $\mathbf{H}$ be a coset of $\mathbf{S}$. We want to minimize $\|x\|_{se}$ over all x in $\mathbf{H}$. A solution to this problem can be characterized in the following way.

P 2.5.7 Let $\mathbf{H}$ be a coset of the subspace $\mathbf{S}$ of a semi-inner product space $\mathbf{V}$. Suppose that there exists x_0 in $\mathbf{H}$ such that

$$\inf_{x \in \mathbf{H}} \|x\|_{se} = \|x_0\|_{se}.$$

Then $x_0 \in \mathbf{S}_s^{\perp}$.

PROOF. Suppose x_0 is a solution to the minimization problem. Then for any scalar α and any vector z in $\mathbf{S}$,

$$\|x_0 + \alpha z\|_{se}^2 = \|x_0\|_{se}^2 + \alpha\bar{\alpha}\|z\|_{se}^2 + \bar{\alpha}(x_0, z) + \alpha(z, x_0) \geq \|x_0\|_{se}^2.$$

The above inequality follows if we observe that $x_0 + \alpha z \in \mathbf{H}$. Since the above inequality is valid for any scalar α, it follows that $(x_0, z) = 0$. But this equality is valid for any z in $\mathbf{S}$. Hence $x_0 \in \mathbf{S}_s^{\perp}$.

Complements

2.5.1 Let $(\cdot, \cdot)$ be a semi-inner product on a vector space $\mathbf{V}$ and $\mathbf{N} = \{x \in \mathbf{V} : \|x\|_{se} = 0\}$. Show that $\mathbf{N}^{\perp} = \mathbf{V}$.
2.5.2 Let $(\cdot, \cdot)$ be a semi-inner product on a vector space $\mathbf{V}$ and $\mathbf{N} = \{x \in \mathbf{V} : \|x\|_{se} = 0\}$. If $x + \mathbf{N} = y + \mathbf{N}$ for $x, y \in \mathbf{V}$, show that $\|x\|_{se} = \|y\|_{se}$. Show that for the coset $x + \mathbf{N}$ of $\mathbf{N}$,

$$\operatorname*{Inf}_{z \in x+\mathbf{N}} \|z\|_{se} = \|x\|_{se}.$$

2.6. Spectral Theory

The spectral theory of conjugate bilinear functionals, to be introduced shortly, can be regarded as a crowning achievement of the theory of vector spaces. We will see numerous instances of the pivotal role that the spectral theory plays in a variety of problems. Let us begin with a definition.

DEFINITION 2.6.1. Let $\mathbf{V}$ be a vector space over the field $\mathbf{C}$ of complex numbers. A map $K(\cdot, \cdot)$ from $\mathbf{V} \times \mathbf{V}$ into $\mathbf{C}$ is said to be a Hermitian conjugate bilinear functional if it has the following properties.

(1) $K(x, y) = \overline{K(y, x)}$ for all x and y in $\mathbf{V}$. (Hermitian property)
(2) $K(\alpha_1 x_1 + \alpha_2 x_2, y) = \alpha_1 K(x_1, y) + \alpha_2 K(x_2, y)$ for all vectors x_1, x_2, y in $\mathbf{V}$ and scalars α_1 and α_2. (Conjugate bilinearity)

A few words of explanation are needed about the terminology used. If we look at the conditions (1) and (2) carefully, they are part of the ones that define an inner product or a semi-inner product. The only defining condition of a semi-inner product that is missing from the list above is that $K(x,x)$ need not be non-negative. The goal of spectral theory is to express any given Hermitian conjugate bilinear functional as a linear combination of semi-inner products by breaking up the vector space $\mathbf{V}$ into orthogonal subspaces with respect to a specified inner product on $\mathbf{V}$. Another point worth noting is the following. By combining (1) and (2), one can show that

$$K(x, \alpha_1 y_1 + \alpha_2 y_2) = \bar{\alpha}_1 K(x, y_1) + \bar{\alpha}_2 K(x, y_2)$$

for all vectors x, y_1, y_2 in $\mathbf{V}$ and scalars α_1, α_2. The map $K(\cdot,\cdot)$ is not quite bilinear! We see that the phrase "conjugate bilinearity" is very apt to describe the property (2). The final remark is that the Hermitian property gives us immediately that $K(x,x)$ is real for all vectors x.

We now develop a body of results eventually culminating in the spectral theorem for a Hermitian conjugate bilinear functional. We always assume that the vector space $\mathbf{V}$ comes with an inner product $< \cdot, \cdot >$. Whenever we talk about orthogonality it is always with respect to the underlying inner product on the vector space.

P 2.6.2 Let $K(\cdot,\cdot)$ be a Hermitian conjugate bilinear functional and $< \cdot, \cdot >$ an inner product on a vector space. Then the following are valid.

(1) The supremum of $K(x,x)/ < x, x >$ over all non-zero vectors x in $\mathbf{V}$ is attained at some vector x_1 in $\mathbf{V}$.

(2) $K(y, x_1) = 0$ for $y \in (Sp(x_1))^{\perp}$, where x_1 is the vector under focus in (1) and $Sp(x_1)$ is the vector space spanned by x_1.

PROOF. Let $z_1, z_2, \ldots, z_k$ be an orthogonal basis of the vector space $\mathbf{V}$ and $x = \gamma_1 z_1 + \gamma_2 z_2 + \ldots + \gamma_k z_k$ an arbitrary vector in $\mathbf{V}$ in its usual representation in terms of the given basis. Let us compute $K(x,x)$ and $< x, x >$. Note that

$$K(x,x) = \sum_{i=1}^{k} \sum_{i=1}^{k} \gamma_i \bar{\gamma}_j K(z_i, z_j),$$

and

$$< x, x > = \sum_{i=1}^{k} \gamma_i \bar{\gamma}_i.$$

We want to simplify the problem. Observe that for any non-zero scalar α, and non-zero $x \in \mathbf{V}$, $K(x,x)/ < x, x > = K(\alpha x, \alpha x)/ < \alpha x, \alpha x >$. Consequently,

$$\sup_{x \in \mathbf{V}, x \neq 0} K(x,x)/ < x, x > = \sup_{x \in \mathbf{V}, < x, x > = 1} K(x,x).$$

Let $k_{ij} = K(z_i, z_j)$ for all i and j. The maximization problem can now be recast as follows. Maximize the objective function

$$\sum_{i=1}^{k} \sum_{j=1}^{k} \gamma_i \bar{\gamma}_j k_{ij}$$

over all complex numbers $\gamma_1, \gamma_2, \cdots, \gamma_k$ subject to the condition

$$\sum_{i=1}^{k} |\gamma_i|^2 = 1.$$

The set $D = \{(\gamma_1, \gamma_2, \cdots, \gamma_k) \in \mathbf{C}^k : |\gamma_1|^2 + |\gamma_2|^2 + \ldots + |\gamma_k|^2 = 1\}$ is a compact subset of $\mathbf{C}^k$. Further, the objective function is a continuous function on D. By a standard argument in Mathematical Analysis, the supremum of the objective function is attained at some vector $(\gamma_1^*, \gamma_2^*, \cdots, \gamma_k^*)$ of D. Let $x_1 = \gamma_1^* z_1 + \gamma_2^* z_2 + \ldots + \gamma_k^* z_k$. It is clear that the supremum of $K(x,x)/ < x, x >$ over all non-zero x in $\mathbf{V}$ is attained at x_1. This completes the proof of Part (1). For Part (2), let α be any complex number and y any vector in $(Sp(x_1))^{\perp}$. If $y = 0$, (2) is trivially true. Assume that $y \neq 0$. Then for any complex number α, $\alpha x_1 + y$ is non-zero. In view of the optimality of the vector x_1, we have

$$K(\alpha x_1 + y, \alpha x_1 + y)/ < \alpha x_1 + y, \alpha x_1 + y > \leq K(x_1, x_1)/ < x_1, x_1 > .$$

Let us expand the ratio that appears on the left hand side above and then perform cross-multiplication. On writing $\alpha = \alpha_1 + i\alpha_2$ and $K(x_1, y)$

$= \delta_1 + i\delta_2$ for some real numbers $\alpha_1, \alpha_2, \delta_1$, and δ_2, and observing that $< x_1, y > = 0$, we have

$$2 < x_1, x_1 > (\alpha_1\delta_1 - \alpha_2\delta_2) \leq K(x_1, x_1) < y, y > - K(y, y) < x_1, x_1 > .$$

More usefully, we have the following inequality:

$$(\alpha_1\delta_1 - \alpha_2\delta_2) \leq [K(x_1, x_1) < y, y > - K(y, y) < x_1, x_1 >]/[2 < x_1, x_1 >].$$

This inequality is true for all real numbers α_1 and α_2. But the number that appears on the right hand side of the inequality is fixed. Consequently, we must have $\delta_1 = \delta_2 = 0$. Hence $K(x_1, y) = \delta_1 + i\delta_2 = 0$.

Some comments are in order on the above result. The attainment of the supremum as established in Part (1) is purely a topological property. Even though the Hermitian bilinear conjugate functional is not an inner product, it inherits some properties of the inner product. If $x_1 \in Sp(x_1)$ and $y \in (Sp(x_1))^{\perp}$, it is clear that $< x_1, y > = 0$. Part (2) says that the same property holds for $K(\cdot, \cdot)$, i.e., $K(x_1, y) = 0$. But more generally, we would like to know whether $K(x, y) = 0$ whenever $x \in \mathbf{S}$ and $y \in \mathbf{S}^{\perp}$, where $\mathbf{S}$ and $\mathbf{S}^{\perp}$ are a pair of orthogonal complementary subspaces of $\mathbf{V}$. **P 2.6.2** is just a beginning in response to this query. In the following proposition, we do a better job.

P 2.6.3 Let $K(\cdot, \cdot)$ be a Hermitian conjugate bilinear functional on a vector space $\mathbf{V}$ equipped with an inner product $< \cdot, \cdot >$. Then there exists a basis $x_1, x_2, \ldots, x_k$ for the vector space $\mathbf{V}$ such that

$$< x_i, x_j > = K(x_i, x_j) = 0 \quad \text{for all} \quad i \neq j. \tag{2.6.1}$$

PROOF. Choose x_1 as in **P 2.6.2** and apply the result to the vector space $(Sp(x_1))^{\perp}$ with $K(\cdot, \cdot)$ and $< \cdot, \cdot >$ restricted to the subspace $(Sp(x_1))^{\perp}$. There exists a vector $x_2 \in (Sp(x_1))^{\perp}$ such that

(1) $\sup\limits_{x \in (Sp(x_1))^{\perp}, x \neq 0} K(x, x)/ < x, x > = K(x_2, x_2)/ < x_2, x_2 >$,

(2) $K(x_1, x_2) = < x_1, x_2 > = 0$,

(3) $K(u, v) = < u, v > = 0$ whenever $u \in Sp(x_1, x_2)$ and $v \in (Sp(x_1, x_2))^{\perp}$,

where, as usual, $Sp(x_1, x_2)$ is the vector space spanned by the vectors x_1 and x_2. Reflect a little on (3) and see why it holds. Now the focus of attention is the vector space $(Sp(x_1, x_2))^{\perp}$. A repeated application of **P 2.6.2** yields the desired basis.

It is possible that the Hermitian conjugate bilinear functional could be an inner product or a semi-inner product in its own right. What **P 2.6.3** is trying to convey to us is that we can find a common orthonormal basis under both the inner products $K(\cdot,\cdot)$ and $< \cdot,\cdot >$.

Once we have obtained a basis $x_1, x_2, \ldots, x_k$ having the property (2.6.1), it is a simple job to normalize them, i.e., have them satisfy $< x_i, x_i > = 1$ for every i. Assume now that $x_1, x_2, \ldots, x_k$ is an orthonormal basis under the inner product $< \cdot,\cdot >$ satisfying (2.6.1). Let $K(x_i, x_i) = \lambda_i, i = 1, 2, \ldots, k$. Assume, without loss of generality, that $\lambda_1 \geq \lambda_2 \geq \ldots \geq \lambda_k$. The numbers λ_i's are called the eigenvalues of $K(\cdot,\cdot)$ with respect to the inner product $< \cdot,\cdot >$, and the corresponding vectors $x_1, x_2, \ldots, x_k$, the eigenvectors of $K(\cdot,\cdot)$ corresponding to the eigenvalues $\lambda_1, \lambda_2, \ldots, \lambda_k$. There is no reason to believe that all λ_i's to be distinct. Let $\lambda_{(1)}, \lambda_{(2)}, \ldots, \lambda_{(s)}$ be the distinct eigenvalues with multiplicities $r_1, r_2, \ldots, r_s$, respectively. We tabulate the eigenvalues, the corresponding eigenvectors, and the subspaces spanned by the eigenvectors in a systematic fashion.

Eigenvalues	*Corresponding eigenvectors*
$\lambda_1 = \lambda_2 = \ldots = \lambda_{r_1} = \lambda_{(1)}$	$x_1, x_2, \ldots, x_{r_1}$
$\lambda_{r_1+1} = \lambda_{r_1+2} = \ldots = \lambda_{r_1+r_2} = \lambda_{(2)}$	$x_{r_1+1}, x_{r_1+2}, \ldots, x_{r_1+r_2}$
$\ldots \quad \ldots \quad \ldots$	$\ldots \quad \ldots$
$\lambda_{k-r_s+1} = \lambda_{k-r_s+2} = \ldots = \lambda_k = \lambda_{(s)}$	$x_{k-r_s+1}, x_{k-r_s+2}, \ldots, x_k$

The subspace spanned by the i-th set of vectors is denoted by E_i, $i = 1, \ldots, s$.

We want to introduce another phrase. The subspace $\mathbf{E}_i$ is called the eigenspace of $K(\cdot,\cdot)$ corresponding to the eigenvalue $\lambda_{(i)}$. From the way these eigenspaces are constructed, it is clear that the eigenspaces $\mathbf{E}_1, \mathbf{E}_2, \ldots, \mathbf{E}_s$ are mutually orthogonal. What this means is that if $x \in \mathbf{E}_i$ and $y \in \mathbf{E}_j$, then $< x, y > = 0$ for any two distinct i and j. Moreover, the vector space $\mathbf{V}$ can be realized as the direct sum of the subspaces $\mathbf{E}_1, \mathbf{E}_2, \ldots, \mathbf{E}_s$. More precisely, given any vector y in $\mathbf{V}$, we can find y_i in $\mathbf{E}_i$ for each i such that $y = y_1 + y_2 + \ldots + y_s$. This decomposition is unique. Symbolically, we can write

$$\mathbf{V} = \mathbf{E}_1 \oplus \mathbf{E}_2 \oplus \cdots \oplus \mathbf{E}_s.$$

Some more properties of eigenvalues, eigenvectors and eigenspaces are recorded in the following proposition.

P 2.6.4 Let $K(\cdot,\cdot), <\cdot,\cdot>, \lambda_i$'s and $\mathbf{E}_i$'s be as defined above. The following are valid.

(1) $K(x,y) = <x,y> = 0$ for every x in $\mathbf{E}_i$ and y in $\mathbf{E}_j$ for any two distinct i and j.
(2) $K(x,x)/<x,x> = \lambda_{(i)}$ for every x in $\mathbf{E}_i$ and for every i.
(3) If $x, y \in \mathbf{E}_i$ for any i and $<x,y> = 0$, then $K(x,y) = 0$.
(4) If $x, y \in \mathbf{E}_i$ for any i, then $K(x,y) = \lambda_{(i)} <x,y>$.
(5) If $y_{i1}, y_{i2}, \ldots, y_{ir_i}$ is an orthonormal basis for the subspace $\mathbf{E}_i$, $i = 1, 2, \ldots, s$, then the k vectors, $y_{11}, y_{12}, \ldots, y_{1r_1}; y_{21}, y_{22}, \ldots, y_{2r_2}; \ldots; y_{s1}, y_{s2}, \ldots, y_{sr_s}$, constitute an orthonormal basis for the vector space $\mathbf{V}$.

One can establish the above assertion by a repeated application of (2.6.1) and the defining properties of $K(\cdot,\cdot)$ and $<\cdot,\cdot>$. The property (5) above has an interesting connotation. The property (2.6.1) is very critical in understanding the structure of any Hermitian conjugate bilinear functional. Once we obtain the subspaces $\mathbf{E}_i$'s, one could generate a variety of orthonormal bases for the vector space $\mathbf{V}$ satisfying (2.6.1) by piecing together a variety of orthonormal bases for each subspace $\mathbf{E}_i$. If the eigenvalues are all distinct, or equivalently, each subspace $\mathbf{E}_i$ is one-dimensional, we do not have such a kind of freedom. In this case, the normalized vectors $x_1, x_2, \ldots, x_k$ satisfying (2.6.1) are unique. Of course, we need to demonstrate that any orthonormal basis of the vector space $\mathbf{V}$ satisfying (2.6.1) arises in the way Part (5) of the above proposition outlines. Let us put that down as a proposition.

P 2.6.5 Let $K(\cdot,\cdot), <\cdot,\cdot>, x_i$'s and $\mathbf{E}_i$'s be as defined above. Let $z_1, z_2, \ldots, z_k$ be an orthonormal basis having the property (2.6.1). Then each z_i must belong to some subspace $\mathbf{E}_j$. Equivalently, every orthonormal basis of $\mathbf{V}$ satisfying (2.6.1) is generated as outlined in Part (5) of **P 2.6.4**.

PROOF. Since $\mathbf{V} = \mathbf{E}_1 \oplus \mathbf{E}_2 \oplus \ldots \oplus \mathbf{E}_s$, each vector z_i in the given orthonormal basis has a unique decomposition

$$z_i = z_{i1} + z_{i2} + \ldots + z_{is}, \; z_{ij} \in \mathbf{E}_j, \; j = 1, 2, \ldots, s.$$

Let us work with the vector z_1. Since for every $j \neq 1, < z_1, z_j > = 0$, we have

$$< z_{11}, z_{j1} > + < z_{12}, z_{j2} > + \ldots + < z_{1s}, z_{js} > = 0.$$

Since for every $j \neq 1, K(z_1, z_j) = 0$, we have

$$K(z_{11}, z_{j1}) + K(z_{12}, z_{j2}) + \ldots + K(z_{1s}, z_{js}) = 0.$$

This implies, from Part (4) of **P 2.6.4**,

$$\lambda_{(1)} < z_{11}, z_{j1} > + \lambda_{(2)} < z_{12}, z_{j2} > + \ldots + \lambda_{(s)} < z_{1s}, z_{js} > = 0.$$

Consequently,

$$< \lambda_{(1)} z_{11} + \lambda_{(2)} z_{12} + \ldots + \lambda_{(s)} z_{1s}, z_j >= 0.$$

As this is true for every $j \neq 1$, and $z_1, z_2, \ldots, z_s$ is an orthonormal basis, the vector $\lambda_{(1)} z_{11} + \lambda_{(2)} z_{12} + \ldots + \lambda_{(s)} z_{1s}$ must be a multiple of the vector z_1. Since

$$\lambda_{(1)} z_{11} + \lambda_{(2)} z_{12} + \ldots + \lambda_{(s)} z_{1s} = \alpha(z_{11} + z_{12} + \ldots + z_{1s})$$

for some scalar α, we have

$$(\lambda_{(1)} - \alpha) z_{11} + (\lambda_{(2)} - \alpha) z_{12} + \ldots + (\lambda_{(s)} - \alpha) z_{1s} = 0.$$

Now, we claim that $z_{1j} \neq 0$ for exactly one $j \in \{1, 2, \ldots, s\}$. Suppose not. Then there are distinct indices $j_1, j_2, \ldots, j_r \in \{1, 2, \ldots, s\}$ with $r \geq 2$ such that $z_{1j_i} \neq 0$ for $i = 1, 2, \ldots, r$ and $z_{1j} = 0$ for every $j \in \{1, 2, \ldots, s\} - \{j_1, j_2, \ldots, j_r\}$. Since $z_{1j_1}, z_{1j_2}, \ldots, z_{1j_r}$ are linearly independent, we must have $\lambda_{(j_i)} - \alpha = 0$ for every $i = 1, 2, \ldots, r$. But the $\lambda_{(i)}$'s are all distinct. This contradiction establishes the claim. In view of the claim, we have that $z_1 = z_{1j}$. Hence $z_1 \in \mathbf{E}_j$. The same story can be repeated for the other members of the given basis.

One important consequence of the above results is that the eigenvalues $\lambda_{(i)}$'s and the eigenspaces $\mathbf{E}_i$'s are uniquely defined. If we look at the process how the normalized vectors $x_1, x_2, \ldots, x_k$ are chosen satisfying (2.6.1), we had some degree of freedom in the selection of these vectors at every stage of optimization. In the final analysis, it does not matter how the vectors are selected. They lead to the same eigenvalues $\lambda_{(i)}$'s and eigenspaces $\mathbf{E}_i$'s.

We need the following terminology. The rank of semi-inner product $(\cdot,\cdot)$ on $\mathbf{V}$ with respect to an inner product $<\cdot,\cdot>$ on $\mathbf{V}$ is defined to be the dimension of the subspace $\mathbf{N}^{\perp}$, where $\mathbf{N} = \{x \in \mathbf{V} : (x,x) = 0\}$. The orthogonal complement $\mathbf{N}^{\perp}$ is worked out with respect to the inner product $<\cdot,\cdot>$.

Now we are ready to state and prove the spectral theorem for Hermitian conjugate bilinear functionals. The main substance of the spectral theorem is that every Hermitian conjugate bilinear functional is a linear combination of semi-inner products. More precisely, we want to write any given Hermitian conjugate bilinear functional $K(\cdot,\cdot)$ in the following form:

$$K(\cdot,\cdot) = \delta_1(\cdot,\cdot)_1 + \delta_2(\cdot,\cdot)_2 + \ldots + \delta_m(\cdot,\cdot)_m, \tag{2.6.2}$$

having the following features. (In the background, we have an inner product $<\cdot,\cdot>$ on the vector space $\mathbf{V}$.)

(1) The numbers $\delta_1, \delta_2, \ldots, \delta_m$ are strictly decreasing.

(2) The semi-inner products $(\cdot,\cdot)_1, (\cdot,\cdot)_2, \ldots, (\cdot,\cdot)_m$ are all of non-zero ranks.

(3) The subspaces $\mathbf{F}_1, \mathbf{F}_2, \ldots, \mathbf{F}_m$ are pairwise orthogonal, where $\mathbf{F}_i = \mathbf{N}_i^{\perp}$ and $\mathbf{N}_i = \{x \in \mathbf{V} : (x,x)_i = 0\}$. (The orthogonality is with respect to the inner product $<\cdot,\cdot>$.)

(4) For any pair of vectors x and y in $\mathbf{V}$, we have

$$<x,y> = (x,y)_1 + (x,y)_2 + \ldots + (x,y)_m.$$

In abstract terms, when we say we have a spectral form for a Hermitian conjugate bilinear functional with respect to a given inner product, we mean a form of the type (2.6.2) exhibiting all the features (1), (2), (3), and (4) listed above. For such a form, we demonstrate that the vector space $\mathbf{V}$ is the direct sum of the subspaces $\mathbf{F}_1, \mathbf{F}_2, \ldots, \mathbf{F}_m$. Suppose $\mathbf{F}_{m+1}$ is the subspace of $\mathbf{V}$ orthogonal to each of $\mathbf{F}_1, \mathbf{F}_2, \ldots, \mathbf{F}_m$. We show that the subspace $\mathbf{F}_{m+1}$ is zero-dimensional. Observe that the vector space $\mathbf{V}$ is the direct sum of the subspaces $\mathbf{F}_1, \mathbf{F}_2, \ldots, \mathbf{F}_{m+1}$. Any vector x in $\mathbf{V}$ has a unique decomposition

$$x = u_1 + u_2 + \ldots + u_{m+1},$$

with $u_i \in \mathbf{F}_i$. By (4) above,

$$\begin{aligned} < x, x > &= (x,x)_1 + (x,x)_2 + \ldots + (x,x)_m \\ &= (u_1,u_1)_1 + (u_2,u_2)_2 + \ldots + (u_m,u_m)_m \\ &= < u_1,u_1 > + < u_2,u_2 > + \ldots + < u_m,u_m > . \end{aligned}$$

On the other hand,

$$< x,x > = < u_1,u_1 > + < u_2,u_2 > + \ldots + < u_{m+1},u_{m+1} > .$$

Consequently, $< u_{m+1}, u_{m+1} >= 0$. Since x is arbitrary, it follows that the subspace $\mathbf{F}_{m+1}$ is zero-dimensional. In the following result, we identify explicitly the spectral form of a Hermitian conjugate bilinear functional.

P 2.6.6 (Spectral Theorem). Let $K(\cdot,\cdot)$ be a Hermitian conjugate bilinear functional and $< \cdot,\cdot >$ an inner product on a vector space $\mathbf{V}$. Then there exist semi-inner products $(\cdot,\cdot)_i$, $i = 1, 2, \ldots, s$ of non-zero ranks, and distinct real scalars $\lambda_{(i)}, i = 1, 2, \ldots, s$ such that the following hold.

(1) The subspaces $\mathbf{F}_i, i = 1, 2, \ldots, s$ of $\mathbf{V}$ are pairwise orthogonal, where $\mathbf{F}_i = \mathbf{N}_i^{\perp}$ and $\mathbf{N}_i = \{x \in \mathbf{V} : (x,x)_i = 0\}$.

(2) For every $x, y \in \mathbf{V}$,

$$< x,y > = (x,y)_1 + (x,y)_2 + \ldots + (x,y)_s.$$

(3) $K(x,y) = \lambda_{(1)}(x,y)_1 + \lambda_{(2)}(x,y)_2 + \ldots + \lambda_{(s)}(x,y)_s.$

PROOF. The spade work we have carried out so far should come in handy. It is not hard to guess the identity of the scalars $\lambda_{(i)}$'s. We need to identify precisely the subspaces $\mathbf{F}_i$'s before we proceed further. Let the distinct eigenvalues $\lambda_{(i)}$'s and the eigenspaces $\mathbf{E}_i$'s be those as outlined above. We prove that $\mathbf{F}_i = \mathbf{E}_i$ for every i. We define first the semi-inner products. Let $x, y \in \mathbf{V}$. Since the vector space $\mathbf{V}$ is the direct sum of the vector spaces $\mathbf{E}_1, \mathbf{E}_2, \ldots, \mathbf{E}_s$, we can write

$$\begin{aligned} x &= u_1 + u_2 + \ldots + u_s, \\ y &= v_1 + v_2 + \ldots + v_s, \end{aligned}$$

with u_i and v_i in $\mathbf{E}_i$, in a unique way. For each $1 \le i \le s$, define

$$(x, y)_i = < u_i, v_i > .$$

One can check that $(\cdot,\cdot)_i$ is a semi-inner product on the vector space $\mathbf{V}$. Next we show that $(x, x)_i = 0$ for x in $\mathbf{V}$ if and only if $x \in \mathbf{E}_i^{\perp}$. If $(x, x)_i = 0$, then $< u_i, u_i >= 0$ which implies that $u_i = 0$. Thus we observe that

$$x = u_1 + u_2 + \ldots + u_{i-1} + u_{i+1} + \ldots + u_s.$$

Consequently, $x \in \mathbf{E}_i^{\perp}$. (Why?) The converse follows in the same way if we retrace the steps involved above. Thus we identify the null space $\mathbf{N}_i$ of the semi-inner product $(\cdot,\cdot)_i$ as $\mathbf{E}_i^{\perp}$. Hence $\mathbf{N}_i^{\perp} = \mathbf{E}_i$. In view of this identification, (1) follows. By the very definition of the semi-inner products, we have for any $x, y \in \mathbf{V}$,

$$\begin{aligned} < x, y > &= < u_1, v_1 > + < u_2, v_2 > + \ldots + < u_s, v_s > \\ &= (x, y)_1 + (x, y)_2 + \ldots + (x, y)_s. \end{aligned}$$

This establishes (2). Finally, by **P 2.6.4** (4), we have

$$\begin{aligned} K(x, y) &= K(u_1, v_1) + K(u_2, v_2) + \ldots + K(u_s, v_s) \\ &= \lambda_{(1)} < u_1, v_1 > + \lambda_{(2)} < u_2, v_2 > + \ldots + \lambda_{(s)} < u_s, v_s > \\ &= \lambda_{(1)}(x, y)_1 + \lambda_{(2)}(x, y)_2 + \ldots + \lambda_{(s)}(x, y)_s, \end{aligned}$$

from which (3) follows. It is clear that each semi-inner product introduced above is of non-zero rank.

The set $\{\lambda_{(1)}, \lambda_{(2)}, \ldots, \lambda_{(s)}\}$ of eigenvalues is called the spectrum of $K(\cdot,\cdot)$. In the following result, we show that the representation given above is unique.

P 2.6.7 Let $K(\cdot,\cdot)$ be a Hermitian conjugate bilinear functional and $< \cdot,\cdot >$ an inner product on a vector space $\mathbf{V}$. Let $\lambda_{(1)} > \lambda_{(2)} > \ldots > \lambda_{(s)}$ be the eigenvalues and $\mathbf{E}_1, \mathbf{E}_2, \ldots, \mathbf{E}_s$ the corresponding eigenspaces of $K(\cdot,\cdot)$. Suppose

$$K(\cdot,\cdot) = \delta_1(\cdot,\cdot)\tilde{}_1 + \delta_2(\cdot,\cdot)\tilde{}_2 + \ldots + \delta_m(\cdot,\cdot)\tilde{}_m$$

is a spectral form of $K(\cdot,\cdot)$ for some real numbers $\delta_1 > \delta_2 > \ldots > \delta_m$ and semi-inner products $(\cdot,\cdot)\tilde{}_1, (\cdot,\cdot)\tilde{}_2, \ldots, (\cdot,\cdot)\tilde{}_m$ embodying the three features (1), (2), and (3) of the spectral theorem **P 2.6.6** outlined above. Then $m = s$, $\delta_i = \lambda_{(i)}$, and $(\cdot,\cdot)\tilde{}_i = (\cdot,\cdot)_i$ for every i, where the semi-inner product $(\cdot,\cdot)$ is the same as the one defined in **P 2.6.6**.

PROOF. The ideas are essentially contained in the discussion preceding **P 2.6.6**. Let $\mathbf{P}_i = \{x \in \mathbf{V} : (x,x)\tilde{}_i = 0\}$ and $\mathbf{G}_i = \mathbf{P}_i^{\perp}$ for each $1 \le i \le m$. By hypothesis (1), $\mathbf{G}_1, \mathbf{G}_2, \ldots, \mathbf{G}_m$ are pairwise orthogonal. First, we show that the vector space $\mathbf{V}$ is the direct sum of the subspaces $\mathbf{G}_1, \mathbf{G}_2, \ldots, \mathbf{G}_m$. Let $\mathbf{G}_{m+1}$ be a subspace of $\mathbf{V}$ orthogonal to $\mathbf{G}_1, \mathbf{G}_2, \ldots, \mathbf{G}_m$ so that

$$\mathbf{V} = \mathbf{G}_1 \oplus \mathbf{G}_2 \oplus \ldots \oplus \mathbf{G}_m \oplus \mathbf{G}_{m+1}.$$

Let $x \in \mathbf{V}$. We can write

$$x = x_1 + x_2 + \ldots + x_m + x_{m+1}$$

with $x_i \in \mathbf{G}_i$, $1 \le i \le m+1$. By (2),

$$\begin{aligned} <x,x> &= (x,x)\tilde{}_1 + (x,x)\tilde{}_2 + \ldots + (x,x)\tilde{}_m \\ &= (x_1,x_1)\tilde{}_1 + (x_2,x_2)\tilde{}_2 + \ldots + (x_m,x_m)\tilde{}_m \\ &= <x_1,x_1> + <x_2,x_2> + \ldots + <x_m,x_m>. \end{aligned}$$

But

$$\begin{aligned} <x,x> = &<x_1,x_1> + <x_2,x_2> + \ldots \\ &+ <x_m,x_m> + <x_{m+1},x_{m+1}>, \end{aligned}$$

which implies that $<x_{m+1},x_{m+1}> = 0$. Since x is arbitrary, it follows that $\mathbf{G}_{m+1}$ is zero-dimensional. If $y_{i1}, y_{i2}, \ldots, y_{ir_i}$ is an orthonormal basis of $\mathbf{G}_i$, $i = 1, 2, \ldots, m$, then

$$y_{11}, y_{12}, \ldots, y_{1r_1}, y_{21}, y_{22}, \ldots, y_{2r_2}, \ldots, y_{m1}, y_{m2}, \ldots, y_{mr_m}$$

is an orthonormal basis of $\mathbf{V}$. Also

$$<y_{ij}, y_{rt}> = 0 = K(y_{ij}, y_{rt})$$

for every $(i,j) \neq (r,t)$. By **P 2.6.5**, each y_{ij} belongs to some $\mathbf{E}_r$. This immediately leads to the verification of the result.

There is an alternative way of writing down the spectral form of a Hermitian conjugate bilinear functional. In this form, the functional is written explicitly in terms of the underlying inner product.

P 2.6.8 Let $K(\cdot,\cdot)$ be a Hermitian conjugate bilinear functional and $< \cdot,\cdot >$ an inner product on a vector space $\mathbf{V}$. Then there exist orthonormal vectors $x_1, x_2, \ldots, x_r$ in $\mathbf{V}$ and real numbers $\lambda_1 \geq \lambda_2 \geq \ldots \geq \lambda_r$ such that for any pair x and y of vectors in $\mathbf{V}$, we have

$$K(x,y) = \lambda_1 < x, x_1 >< x_1, y > + \lambda_2 < x, x_2 >< x_2, y > + \ldots + \lambda_r < x, x_r >< x_r, y >, \tag{2.6.3}$$

where $r \leq \dim(\mathbf{V})$.

PROOF. Choose orthonormal vectors $x_1, x_2, \ldots, x_k$ in $\mathbf{V}$ satisfying (2.6.1), where $k = \dim(\mathbf{V})$. Let $\lambda_i = K(x_i, x_i)$ for each i. For given vectors x and y in $\mathbf{V}$, write the decompositions of x and y as

$$x = < x, x_1 > x_1 + < x, x_2 > x_2 + \ldots + < x, x_k > x_k$$

and

$$y = < y, x_1 > x_1 + < y, x_2 > x_2 + \ldots + < y, x_k > x_k.$$

Consequently,

$$\begin{aligned} K(x,y) &= K(\sum_{i=1}^{k} < x, x_i > x_i, \sum_{j=1}^{k} < y, x_i > x_i) \\ &= \sum_{i=1}^{k} \lambda_i < x, x_i >< x_i, y > . \end{aligned}$$

In the above representation, we omit those λ_i's which are zero. Thus we have the desired representation (2.6.3). The statement of **P 2.6.8** can be reworded in the following way.

P 2.6.9 Let $K(\cdot,\cdot)$ be a Hermitian conjugate bilinear functional and $< \cdot,\cdot >$ an inner product on a vector space $\mathbf{V}$. Then there exist vectors $x_1, x_2, \ldots, x_k$ in $\mathbf{V}$ and real numbers $\lambda_1, \lambda_2, \ldots, \lambda_k$ such that for any pair x and y of vectors in $\mathbf{V}$, we have

$$K(x,y) = \lambda_1 < x, x_1 >< x_1, y > + \lambda_2 < x, x_2 >< x_2, y > + \dots + \lambda_k < x, x_k >< x_k, y >$$

and

$$< x, y > = < x, x_1 >< x_1, y > + < x, x_2 >< x_2, y > + \dots + < x, x_k >< x_k, y > . \tag{2.6.4}$$

The properties (2.6.1), (2.6.3) and (2.6.4) are all equivalent. The second part of Property (2.6.4) is equivalent to the fact that $x_1, x_2, \dots, x_k$ should constitute an orthonormal basis of the vector space $\mathbf{V}$. See **P 2.2.7**. Consequently, the properties (2.6.3) and (2.6.4) are equivalent. It is clear that (2.6.1) and (2.6.4) are equivalent.

It is time to take stock of what has been accomplished. One crucial point we need to discuss is that what happens to the spectral form when the Hermitian conjugate bilinear functional $K(\cdot,\cdot)$ is itself a semi-inner product or, more restrictively, an inner product. In that case, we will get an additional bonus. If the Hermitian conjugate bilinear functional is a semi-inner product, then all its eigenvalues are non-negative. (Why?) If the Hermitian conjugate bilinear functional is an inner product, then all its eigenvalues are positive. (Again, why?)

The spectral representation of a Hermitian conjugate bilinear functional $K(\cdot,\cdot)$ in (2.6.2) naturally depends on the inner product $< \cdot, \cdot >$ chosen. It is, therefore, of some interest to examine how the representations differ for different choices of the inner product. The following theorems shows that in any representation the number of positive, negative and zero eigenvalues are the same, while the actual eigenvalues and the corresponding eigenvectors may not be the same. We give a representation of $K(\cdot,\cdot)$ in terms of some basic linear functionals which brings into focus the stated facts above.

P 2.6.10 Let $K(\cdot,\cdot)$ be a Hermitian conjugate bilinear functional on a vector space $\mathbf{V}$. Then there exist $p + q\, (\leq \dim(\mathbf{V}))$ linearly independent linear functionals $L_1, L_2, \dots, L_{p+q}$ defined on $\mathbf{V}$ such that for every pair x and y of vectors in $\mathbf{V}$, we have

$$K(x,y) = \sum_{i=1}^{p} L_i(x)\overline{L_i(y)} - \sum_{j=1}^{q} L_{p+j}(x)\overline{L_{p+j}(y)}. \tag{2.6.5}$$

Moreover, the numbers p and q in (2.6.5) are unique for a given $K(\cdot,\cdot)$, while the choice of linear functionals is not unique.

PROOF. Consider the representation of $K(\cdot,\cdot)$ given in (2.6.4). Assume, without loss of generality, that $\lambda_1, \lambda_2, \ldots, \lambda_p$ are positive λ_{p+1}, $\lambda_{p+2}, \ldots, \lambda_{p+q}$ are negative, and $\lambda_{p+q+1}, \lambda_{p+q+2}, \ldots, \lambda_k$ are zeros. Let $\mu_{p+j} = -\lambda_{p+j}, j = 1, 2, \ldots, q$. Then for any pair x and y of vectors in $\mathbf{V}$,

$$\begin{aligned} K(x,y) &= \sum_{i=1}^{p} \lambda_i < x, x_i >< x_i, y > \\ &\qquad - \sum_{j=1}^{q} \mu_{p+j} < x, x_{p+j} >< x_{p+j}, y > \\ &= \sum_{i=1}^{p} < x, (\lambda_i)^{1/2} x_i >< (\lambda_i)^{1/2} x_i, y > \\ &\qquad - \sum_{j=1}^{q} < x, (\mu_{p+j})^{1/2} x_{p+j} >< (\mu_{p+j})^{1/2} x_{p+j}, y > \\ &= \sum_{i=1}^{p} L_i(x)\overline{L_i(y)} - \sum_{j=1}^{q} L_{p+j}(x)\overline{L_{p+j}(y)}, \end{aligned}$$

where $L_i(x) =< x, (\lambda_i)^{1/2} x_i >$, $i = 1, 2, \ldots, p$ and, similarly we have $L_{p+j}(x) = < x, (\mu_{p+j})^{1/2} x_{p+j} >$, $j = 1, 2, \ldots, q$. Observe that for each i, $L_i(\cdot)$ is a linear functional on the vector space $\mathbf{V}$. Since x_i's are orthonormal, the linear functionals $L_i(\cdot), i = 1, 2, \ldots, p+q$ are all linearly independent. Now let us settle the question of uniqueness. Suppose we have two representations of $K(\cdot,\cdot)$ given by

$$K(x,y) = \sum_{i=1}^{p} L_i(x)\overline{L_i(y)} - \sum_{j=1}^{q} L_{p+j}(x)\overline{L_{p+j}(y)}$$

and

$$K(x,y) = \sum_{i=1}^{r} M_i(x)\overline{M_i(y)} - \sum_{j=1}^{s} M_{r+j}(x)\overline{M_{r+j}(y)}$$

for all x and y in $\mathbf{V}$, and for some sets $\{L_i\}$ and $\{M_i\}$ of linearly independent linear functionals on $\mathbf{V}$. Set $x = y$. Then for any x in $\mathbf{V}$,

$$K(x,x) = \sum_{i=1}^{p} L_i(x)\overline{L_i(x)} - \sum_{j=1}^{q} L_{p+j}(x)\overline{L_{p+j}(x)} \tag{2.6.6}$$

$$= \sum_{i=1}^{r} M_i(x)\overline{M_i(x)} - \sum_{j=1}^{s} M_{r+j}(x)\overline{M_{r+j}(x)}. \tag{2.6.7}$$

Suppose $r > p$. Consider the following linear equations,

$$\begin{aligned} L_i(x) &= 0,\ i = 1, 2, \ldots, p, \\ M_{r+j}(x) &= 0,\ j = 1, 2, \ldots, s, \end{aligned} \tag{2.6.8}$$

in x. For any x satisfying (2.6.8), $K(x,x)$ is non-positive as per the representation (2.6.6) and is non-negative as per the representation (2.6.7). This apparent anomaly will not arise if

$$\begin{aligned} M_i(x) &= 0,\ i = 1, 2, \ldots, r, \\ L_{p+j}(x) &= 0,\ j = 1, 2, \ldots, q, \end{aligned} \tag{2.6.9}$$

whenever the vector x satisfies equations (2.6.8). What this means is that each of the linear functionals $M_1, M_2, \ldots, M_r$ is a linear combination of the functionals $L_1, L_2, \ldots, L_p, M_{r+1}, M_{r+2}, \ldots, M_{r+s}$. See Complement 1.7.2. As a consequence, each of the linear functionals $M_1, M_2, \ldots, M_r, M_{r+1}, \ldots, M_{r+s}$ is a linear combination of the functionals $L_1, L_2, \ldots, L_p, M_{r+1}, M_{r+2}, \ldots, M_{r+s}$. This is not possible since $M_1, M_2, \ldots, M_r, M_{r+1}, \ldots, M_{r+s}$ are linearly independent and $p < r$. Thus we must have $r \leq p$. By similar argument, we can show that $p \leq r$. Hence $p = r$. In a similar vein, it follows that $q = s$. This completes the proof.

The numbers $p + q$ and $p - q$ are called the rank and signature of the Hermitian conjugate bilinear functional $K(\cdot,\cdot)$, respectively.

If we recall how we obtained the first eigenvalue λ_1 of the Hermitian conjugate bilinear functional, we identify λ_1 as the largest value of the ratio $K(x,x)/ < x,x >$ as x varies over all non-zero vectors of $\mathbf{V}$. The remaining eigenvalues also do have such optimality properties. In the following theorem, we characterize the intermediate eigenvalues. Before we present the theorem, we would like to reiterate the basic framework under which we operate.

Let $\lambda_1 \geq \lambda_2 \geq \ldots \geq \lambda_k$ be the eigenvalues of the Hermitian conjugate bilinear functional $K(\cdot,\cdot)$ with respect to an inner product $<\cdot,\cdot>$, and $x_1, x_2, \ldots, x_k$ the corresponding eigenvectors which form an orthonormal basis for the vector space $\mathbf{V}$. All these facts have been handed down to us from **P 2.6.2** and the discussion that ensued.

P 2.6.11 **(Minimax Theorems)** Let $\mathbf{M}_s$ be the vector space spanned by the first s eigenvectors of $K(\cdot,\cdot)$ and $\mathbf{M}_{-s}$ the vector space spanned by the last s eigenvectors for each $1 \leq s \leq k$. The following hold:

(1) $\inf\limits_{x\in\mathbf{M}_s, x\neq 0} K(x,x)/<x,x> = \lambda_s$ and the infimum is attained at $x = x_s$.

(2) $\sup\limits_{x\in\mathbf{M}_{-s}, x\neq 0} K(x,x)/<x,x> = \lambda_{k-s+1}$ and the supremum is attained at $x = x_{k-s+1}$.

(3) $\inf\limits_{\mathbf{S}} \sup\limits_{x\in\mathbf{S}, x\neq 0} K(x,x)/<x,x> = \lambda_{k-s+1}$ where the infimum is taken over all subspaces $\mathbf{S}$ of $\mathbf{V}$ with $\dim(\mathbf{S}) \geq s$, and the infimum is attained at $\mathbf{S} = \mathbf{M}_{-s}$.

(4) $\sup\limits_{\mathbf{S}} \inf\limits_{x\in\mathbf{S}, x\neq 0} K(x,x)/<x,x> = \lambda_s$ where the supremum is taken over all subspaces $\mathbf{S}$ of $\mathbf{V}$ with $\dim(\mathbf{S}) \geq s$, and the supremum is attained at $\mathbf{S} = \mathbf{M}_s$.

PROOF. We begin by proving (1). Let x be any non-zero vector in the vector subspace $\mathbf{M}_s$. Write down its representation in terms of the given orthonormal basis of $\mathbf{M}_s$:

$$x = \alpha_1 x_1 + \alpha_2 x_2 + \ldots + \alpha_s x_s,$$

for some scalars $\alpha_1, \alpha_2, \ldots, \alpha_s$. We compute

$$\frac{K(x,x)}{<x,x>} = [\alpha_1\bar{\alpha}_1\lambda_1 + \alpha_2\bar{\alpha}_2\lambda_2 + \ldots + \alpha_s\bar{\alpha}_s\lambda_s]/[\alpha_1\bar{\alpha}_1 + \alpha_2\bar{\alpha}_2 + \ldots + \alpha_s\bar{\alpha}_s].$$

The above computation indicates that we are taking a weighted average of the numbers $\lambda_1 \geq \lambda_2 \geq \ldots \geq \lambda_s$ with the non-negative weights $\alpha_1\bar{\alpha}_1, \alpha_2\bar{\alpha}_2, \ldots, \alpha_s\bar{\alpha}_s$. It is clear that the weighted average is always

$\geq \lambda_s$. It is also clear that when $x = x_s$, the weighted average is exactly equal to λ_s. This proves (1). A similar argument establishes the truth of (2). Let us now prove (3). Let **S** be any subspace with $\dim(\mathbf{S}) \geq s$. Observe that

$$\begin{aligned}
\sup_{x \in \mathbf{S}, x \neq 0} \frac{K(x,x)}{< x,x >} &\geq \sup_{x \in \mathbf{S} \cap \mathbf{M}_{k-s+1}, x \neq 0} K(x,x)/ < x,x > \\
&\geq \inf_{x \in \mathbf{S} \cap \mathbf{M}_{k-s+1}, x \neq 0} K(x,x)/ < x,x > \qquad (2.6.10) \\
&\geq \inf_{x \in \mathbf{M}_{k-s+1}, x \neq 0} K(x,x)/ < x,x > = \lambda_{k-s+1} \text{ by (1).}
\end{aligned}$$

The first inequality above holds as long as $\mathbf{S} \cap \mathbf{M}_{k-s+1} \neq \{0\}$. This is so, in view of the fact that

$$\begin{aligned}
\dim(\mathbf{S} \cap \mathbf{M}_{k-s+1}) &= \dim(\mathbf{S}) + \dim(\mathbf{M}_{k-s+1}) - \dim(\mathbf{S} + \mathbf{M}_{k-s+1}) \\
&\geq (s) + (k - s + 1) - k = 1.
\end{aligned}$$

Taking the infimum of (2.6.10) over all subspaces **S** of dimensions $\geq s$, we obtain

$$\inf_{\mathbf{S}} \sup_{x \in \mathbf{S}, x \neq 0} K(x,x)/ < x,x > \geq \lambda_{k-s+1}.$$

The subspace $\mathbf{S} = \mathbf{M}_{-s}$ is one such subspace with dimension equal to s and for which, by (2),

$$\sup_{x \in \mathbf{S}, x \neq 0} K(x,x)/ < x,x > = \lambda_{k-s+1}.$$

Consequently, the infimum above is attained at the subspace $\mathbf{S} = \mathbf{M}_{-s}$. This completes the proof of (3). In order to prove (4), repeat the above argument used in the proof of (3) with only one modification: in the chain of inequalities, use $\mathbf{S} \cap \mathbf{M}_{-(k-s+1)}$ instead of $\mathbf{S} \cap \mathbf{M}_{k-s+1}$.

There is another line of thinking when it comes to analyzing a Hermitian conjugate functional $K(\cdot,\cdot)$ defined on a vector space **V**. Suppose **Q** is a subspace of **V**. One can restrict the given bilinear functional $K(\cdot,\cdot)$ to the Cartesian product $\mathbf{Q} \times \mathbf{Q}$. The restriction still remains a Hermitian conjugate bilinear functional on the vector space **Q**. Now the question arises as to the type of relationship that prevails between the eigenvalues of $K(\cdot,\cdot)$ as defined on $\mathbf{V} \times \mathbf{V}$ and the eigenvalues of $K(\cdot,\cdot)$ as defined on $\mathbf{Q} \times \mathbf{Q}$. In the following proposition, we establish some inequalities.

P 2.6.12 Let $K(\cdot,\cdot)$ be a Hermitian conjugate bilinear functional on a vector space $\mathbf{V}$ of dimension k. Let $\lambda_1 \geq \lambda_2 \geq \ldots \geq \lambda_k$ be its eigenvalues. Let $\mathbf{Q}$ be a subspace of $\mathbf{V}$ of dimension t. Let $\mu_1 \geq \mu_2 \geq \ldots \geq \mu_t$ be the eigenvalues of $K(\cdot,\cdot)$ restricted to $\mathbf{Q} \times \mathbf{Q}$. Then the following inequalities hold:

(1) $\lambda_s \geq \mu_s$ for $s = 1, 2, \ldots, t$;
(2) $\mu_{t-s+1} \geq \lambda_{k-s+1}$ for $s = 1, 2, \ldots, t$.

PROOF. This result follows by an application of minimax theorems established above. By Part (3) of **P 2.6.11** (Minimax Theorems),

$$\mu_{t-s+1} = \inf_{\mathbf{S} \subset \mathbf{Q}} \sup_{x \in \mathbf{S}, x \neq 0} K(x,x)/ < x, x >,$$

where the infimum is taken over all subspaces $\mathbf{S}$ of the vector space $\mathbf{Q}$ with $\dim(\mathbf{S}) \geq s$. But this infimum,

$$\inf_{\mathbf{S} \subset \mathbf{Q}} \sup_{x \in \mathbf{S}, x \neq 0} \frac{K(x,x)}{< x, x >} \geq \inf_{\mathbf{S} \subset \mathbf{V}} \sup_{x \in \mathbf{S}, x \neq 0} \frac{K(x,x)}{< x, x >} = \lambda_{k-s+1},$$

where the infimum on the right hand side of the above inequality is taken over all subspaces $\mathbf{S}$ of the vector space $\mathbf{V}$ with $\dim(\mathbf{S}) \geq s$. This is true for any $s = 1, 2, \ldots, t$. This proves Part (2) above. Using Part (4) of **P 2.6.11**, one can establish Part (1) above.

The following is a simple consequence of the above result. This result is usually known as the interlace theorem for eigenvalues.

P 2.6.13 (**Interlace Theorem**) Let $K(\cdot,\cdot)$ be a Hermitian conjugate bilinear functional on a vector space $\mathbf{V}$ of dimension k. Let $\lambda_1 \geq \lambda_2 \geq \ldots \geq \lambda_k$ be the eigenvalues of $K(\cdot,\cdot)$. Let $\mathbf{Q}$ be a subspace of $\mathbf{V}$ of dimension $(k-1)$. Let $\mu_1 \geq \mu_2 \geq \ldots \geq \mu_{k-1}$ be the eigenvalues of $K(\cdot,\cdot)$ restricted to the subspace $\mathbf{Q}$. Then

$$\lambda_1 \geq \mu_1 \geq \lambda_2 \geq \mu_2 \geq \ldots \geq \mu_{k-1} \geq \lambda_k.$$

Complements

2.6.1 Let $\mathbf{V} = \mathbf{C}^3$ and consider the following function on the product space $\mathbf{V} \times \mathbf{V}$:

$$\begin{aligned} K(x,y) = {} & \xi_1\eta_1 + (1+i)\xi_1\eta_2 + (1-i)\xi_1\eta_3 \\ & + (1-i)\xi_2\eta_1 + \xi_2\eta_3 + (1+i)\xi_3\eta_1 + \xi_3\eta_2 + \xi_3\eta_3 \end{aligned}$$

for $x = (\xi_1, \xi_2, \xi_3)$ and $y = (\eta_1, \eta_2, \eta_3) \in \mathbf{V}$.

(1) Show that $K(\cdot, \cdot)$ is a Hermitian conjugate bilinear functional on $\mathbf{V}$.

(2) Obtain the spectral form of $K(\cdot, \cdot)$.

2.7. Conjugate Bilinear Functionals and Singular Value Decomposition

It may be puzzling to the reader that the phrase "Conjugate bilinear functional" is cropping up again in a new section. Are we not done with it in Section 2.6? The functionals we are entertaining in this section are defined in a more general framework than hitherto considered. You will see the difference when we introduce the definition.

DEFINITION 2.7.1. Let $\mathbf{V}_1$ and $\mathbf{V}_2$ be two vector spaces both over the field $\mathbf{C}$ of complex numbers. A map $B(\cdot, \cdot)$ from $\mathbf{V}_1 \times \mathbf{V}_2$, the Cartesian product of $\mathbf{V}_1$ and $\mathbf{V}_2$, into $\mathbf{C}$ is said to be a conjugate bilinear functional if

(1) $B(\alpha_1 x_1 + \alpha_2 x_2, y) = \alpha_1 B(x_1, y) + \alpha_2 B(x_2, y)$,

(2) $B(x, \beta_1 y_1 + \beta_2 y_2) = \bar{\beta}_1 B(x, y_1) + \bar{\beta}_2 B(x, y_2)$

hold for every x, x_1, x_2 in $\mathbf{V}_1$; y, y_1, y_2 in $\mathbf{V}_2$; and $\alpha_1, \alpha_2, \beta_1, \beta_2$ in $\mathbf{C}$.

Some remarks are in order. In the above framework, the vector spaces $\mathbf{V}_1$ and $\mathbf{V}_2$ need not be identical. If $\mathbf{V}_1 = \mathbf{V}_2$ and $B(\cdot, \cdot)$ is a Hermitian conjugate bilinear functional on $\mathbf{V}_1$, then $B(\cdot, \cdot)$ is a conjugate bilinear functional on $\mathbf{V}_1 \times \mathbf{V}_1$ in the sense portrayed above. If $\mathbf{V}_1 = \mathbf{V}_2$ and $B(\cdot, \cdot)$ is a conjugate bilinear functional on $\mathbf{V}_1 \times \mathbf{V}_2$, it is not necessary that $B(\cdot, \cdot)$ be Hermitian. Try an example.

In what follows, we choose and fix an inner product $< \cdot, \cdot >_1$ on the vector space $\mathbf{V}_1$ and an inner product $< \cdot, \cdot >_2$ on $\mathbf{V}_2$. We establish what is called the Singular Value Decomposition of the conjugate bilinear functional $B(\cdot, \cdot)$ with respect to the inner products $< \cdot, \cdot >_1$ and $< \cdot, \cdot >_2$. This decomposition is in the same spirit as the spectral form of a Hermitian conjugate bilinear functional on a vector space. In the proof of the singular value decomposition, we make use of the spectral form of a certain Hermitian conjugate bilinear functional on a vector space. Even though the singular value decomposition sounds more general than the spectral form but it is a derivative of the spectral form.

P 2.7.2 **(Singular Value Decomposition)** Let $B(\cdot,\cdot)$, $<\cdot,\cdot>_1$ and $<\cdot,\cdot>_2$ be as defined above. Then there exist positive real numbers $\sigma_1 \geq \sigma_2 \geq \ldots \geq \sigma_r$ with $r \leq \min\{\dim(\mathbf{V}_1), \dim(\mathbf{V}_2)\}$, orthonormal vectors $x_1, x_2, \ldots, x_r$ in $\mathbf{V}_1$ and orthonormal vectors $y_1, y_2, \ldots, y_r$ in $\mathbf{V}_2$ such that for any two vectors x in $\mathbf{V}_1$ and y in $\mathbf{V}_2$,

$$\begin{aligned} B(x,y) &= \sigma_1 <x,x_1>_1<y_1,y>_2 + \sigma_2 <x,x_2>_1<y_2,y>_2 + \ldots \\ &\quad + \sigma_r <x,x_r>_1<y_r,y>_2 . \end{aligned}$$

PROOF. For every fixed y in $\mathbf{V}_2$, $B(\cdot,y)$ is a linear functional on the vector space $\mathbf{V}_1$. Therefore, there exists a unique vector $\eta(y)$ in $\mathbf{V}_1$ such that $B(x,y) = <x,\eta(y)>_1$ for every x in $\mathbf{V}_1$. See **P 2.4.1**. For every fixed vector x in $\mathbf{V}_1$, the map $\overline{B(x,\cdot)}$ is a linear functional on the vector space $\mathbf{V}_2$. Consequently, there exists a unique vector $\xi(x)$ in $\mathbf{V}_2$ such that $\overline{B(x,y)} = <\xi(x),y>_2$ for all y in $\mathbf{V}_2$. We now define a function $K(\cdot,\cdot)$ on the Cartesian product space $\mathbf{V}_1 \times \mathbf{V}_1$ as follows. For any two vectors x and u in $\mathbf{V}_1$, let

$$K(x,u) = <\xi(x),\xi(u)>_2 .$$

Let us put together the maps $\xi(\cdot)$ and $\eta(\cdot)$ in the following way, for any vectors x in $\mathbf{V}_1$ and y in $\mathbf{V}_2$,

$$B(x,y) = <x,\eta(y)>_1 = <\xi(x),y>_2 . \tag{2.7.1}$$

Since the map $K(\cdot,\cdot)$ is defined through an inner product, $K(\cdot,\cdot)$ is a semi-inner product on the vector space $\mathbf{V}_1$! We would like to appeal to the spectral form as exemplified by (2.6.4). Let r be the rank of the spectral form of the semi-inner product $K(\cdot,\cdot)$ with respect to the inner product $<\cdot,\cdot>_1$. Obviously, $r \leq \min\{\dim(\mathbf{V}_1), \dim(\mathbf{V}_2)\}$. It is clear that the eigenvalues of $K(\cdot,\cdot)$ are non-negative. For reasons that will be clear later, let $\sigma_1^2 \geq \sigma_2^2 \ldots \geq \sigma_r^2$ be the positive eigenvalues of $K(\cdot,\cdot)$. As per (2.6.4), there exist orthonormal vectors $x_1, x_2, \ldots, x_r$ in $\mathbf{V}_1$ such that for any pair of vectors x and u in $\mathbf{V}_1$,

$$\begin{aligned} K(x,u) &= \sigma_1^2 <x,x_1><x_1,u>_1 + \sigma_2^2 <x,x_2>_1<x_2,u>_1 + \ldots \\ &\quad + \sigma_r^2 <x,x_r>_1<x_r,u>_1 . \end{aligned} \tag{2.7.2}$$

Further, we note that for any $i, j \in \{1, 2, \ldots, r\}$

$$K(x_i, x_j) = \begin{cases} \sigma_i^2 & \text{if} \quad i = j, \\ 0 & \text{if} \quad i \neq j. \end{cases}$$

Extend the set $x_1, x_2, \ldots, x_r$ of orthonormal vectors to an orthonormal basis $x_1, x_2, \ldots, x_r, x_{r+1}, \ldots, x_k$ in $\mathbf{V}_1$. For the basis, we do have

$$K(x_i, x_j) = \begin{cases} \sigma_i^2 & \text{if } i = j, (i \text{ and } j \in \{1, 2, \ldots, r\}), \\ 0 & \text{if } i = j, (i \text{ and } j \in \{r+1, \ldots, k\}), \\ 0 & \text{if} \quad i \neq j. \end{cases} \tag{2.7.3}$$

Further, for any vector y in $\mathbf{V}_2$, we have the usual expansion

$$\begin{aligned} < x, \eta(y) >_1 = < x, x_1 >_1 < x_1, \eta(y) >_1 \; + \; < x, x_2 >_1 < x_2, \eta(y) >_1 \; + \\ \ldots \; + \; < x, x_k >_1 < x_k, \eta(y) >_1, \end{aligned}$$

which, with the help of (2.7.1), becomes

$$\begin{aligned} < x, \eta(y) >_1 = < x, x_1 >_1 < \xi(x_1), y >_2 \; + \; < x, x_2 >_1 < \xi(x_2), y >_2 \; + \\ \ldots + \; < x, x_k >_1 < \xi(x_k), y >_2 . \end{aligned} \tag{2.7.4}$$

We need to procure an orthonormal set of vectors for the vector space $\mathbf{V}_2$. A natural candidate is:

$$z_i = \xi(x_i), \;\; i = 1, 2, \ldots, k.$$

Note that by (2.7.3), for $i, j \in \{1, 2, \ldots, r\}$,

$$< z_i, z_j >_2 = < \xi(x_i), \xi(x_j) >_2 = \; K(x_i, x_j) = \begin{cases} \sigma_i^2 & \text{if} \quad i = j, \\ 0 & \text{if} \quad i \neq j. \end{cases}$$

Thus $z_1, z_2, \ldots, z_r$ are orthogonal. Let σ_i be the positive square root of σ_i^2 and $y_i = (\sigma_i)^{-1} z_i, i = 1, 2, \ldots, r$. Then $y_1, y_2, \ldots, y_r$ is a set of orthonormal vectors in $\mathbf{V}_2$. Further, $z_i = 0$ if $i = r+1, r+2, \ldots, k$.

Using (2.7.1) and (2.7.4), the puzzle is solved. For any vector x in $\mathbf{V}_1$ and y in $\mathbf{V}_2$, we have

$$\begin{aligned} B(x,y) &= < x, \eta(y) >_1 \\ &= < x, x_1 >_1 < z_1, y >_2 + < x, x_2 >_1 < z_2, y >_2 + \dots \\ &\quad + < x, x_k >_1 < z_k, y >_2 \\ &= \sigma_1 < x, x_1 >_1 < y_1, y >_2 + \sigma_2 < x, x_2 >_1 < y_2, y >_2 + \\ &\quad \dots + \sigma_r < x, x_r >_1 < y_r, y >_2 . \end{aligned}$$

This completes the proof.

In the development of a spectral form for a Hermitian conjugate bilinear functional defined on a vector space, the critical point was to demonstrate the existence of an orthonormal basis for the vector space satisfying (2.6.4). In the context of a conjugate bilinear functional defined on the Cartesian product of two vector spaces, we can exhibit two orthonormal bases one for each of the underlying vector spaces satisfying a property similar to (2.6.4). Such bases are called canonical bases. Let us state formally what was discussed above.

P 2.7.3 **(Canonical Bases of Two Vector Spaces)** Let $(\mathbf{V}_1, < \cdot, \cdot >_1)$ and $(\mathbf{V}_2, < \cdot, \cdot >_2)$ be two inner product spaces with dimensions k and m, say, respectively. Let $B(\cdot, \cdot)$ be a conjugate bilinear functional defined on $\mathbf{V}_1 \times \mathbf{V}_2$. Then there exists an orthonormal basis $x_1, x_2, \dots, x_k$ for the vector space $\mathbf{V}_1$ and an orthonormal basis $y_1, y_2, \dots, y_m$ for $\mathbf{V}_2$ such that

$$\begin{aligned} B(x_i, y_j) &= 0 \quad \text{for all } i \neq j, \\ &\neq 0 \quad \text{for } j = i. \end{aligned} \tag{2.7.5}$$

PROOF. Most of the spade work needed for this result was already done in the proof of **P 2.7.2**. We have already obtained orthonormal vectors $x_1, x_2, \dots, x_r$ in $\mathbf{V}_1$ and orthonormal vectors $y_1, y_2, \dots, y_r$ in $\mathbf{V}_2$ such that (2.7.5) is satisfied for all $i, j \in \{1, 2, \dots, r\}$. Extend $x_1, x_2, \dots, x_r$ to an orthonormal basis $x_1, x_2, \dots, x_k$ of $\mathbf{V}_1$, and extend the same courtesy to the set $y_1, y_2, \dots, y_r$. The property (2.7.5) covers all the basis vectors in view of the singular value decomposition of the bilinear form $B(\cdot, \cdot)$.

It is time we name the numbers $\sigma_1 \geq \sigma_2 \geq \ldots \geq \sigma_r > 0$. These numbers are called the singular values of the conjugate bilinear form $B(\cdot,\cdot)$. Further, the vectors x_i in $\mathbf{V}_1$ and y_i in $\mathbf{V}_2$ are called canonical vectors associated with the singular value $\sigma_i, i = 1, 2, \ldots, r$. In the context of a Hermitian conjugate bilinear functional defined on a vector space, the eigenvalues were obtained as a solution to a certain optimization problem. The singular values also have a similar optimality property.

P 2.7.4 The largest singular value σ_1 of a conjugate bilinear functional $B(\cdot,\cdot)$ defined on the Cartesian product $\mathbf{V}_1 \times \mathbf{V}_2$ of two vector spaces has the following property:

$$\sigma_1 = \sup_{x \in \mathbf{V}_1, x \neq 0, y \in \mathbf{V}_2, y \neq 0} |B(x,y)|/[\|x\|_1 \|y\|_2].$$

Moreover, the supremum above is attained at $x = x_1$ and $y = y_1$, where x_1 and y_1 are a set of canonical vectors associated with the singular value σ_1.

PROOF. This result is a simple consequence of the singular value decomposition of $B(\cdot,\cdot)$. For any vectors x in $\mathbf{V}_1$ and y in $\mathbf{V}_2$, we have

$$\begin{aligned} B(x,y) &= \sigma_1 < x, x_1 >_1 < y_1, y >_2 + \sigma_2 < x, x_2 >_1 < y_2, y >_2 + \ldots \\ &\quad + \sigma_r < x, x_r >_1 < y_r, y >_2 . \end{aligned}$$

Expand each of x and y with respect to their respective orthonormal bases stemming from the singular value decomposition:

$$\begin{aligned} x &= < x, x_1 >_1 x_1 + < x, x_2 >_1 x_2 + \ldots + < x, x_k >_1 x_k \\ &= \alpha_1 x_1 + \alpha_2 x_2 + \ldots + \alpha_k x_k, \quad \text{say}, \\ y &= < y, y_1 >_2 y_1 + < y, y_2 >_2 y_2 + \ldots + < y, y_m >_2 y_m \\ &= \beta_1 y_1 + \beta_2 y_2 + \ldots + \beta_m y_m, \quad \text{say}. \end{aligned}$$

By the Cauchy-Schwartz inequality, it now follows that

$$\begin{aligned} |B(x,y)| &\leq \sigma_1 [\sum_{i=1}^{r} |\alpha_i||\beta_i|] \\ &\leq \sigma_1 (\sum_{i=1}^{r} |\alpha_i|^2)^{1/2} (\sum_{i=1}^{r} |\beta_i|^2)^{1/2} \\ &\leq \sigma_1 (\sum_{i=1}^{k} |\alpha_i|^2)^{1/2} (\sum_{j=1}^{m} |\beta_j|^2)^{1/2} = \sigma_1 \|x\|_1 \|y\|_2. \end{aligned}$$

Consequently, the supremum under discussion is $\leq \sigma_1$. It is clear that $B(x_1, y_1) = \sigma_1$. This completes the proof.

Complements

2.7.1 Let $\mathbf{V}_1 = \mathbf{R}^2$ and $\mathbf{V}_2 = \mathbf{R}^3$. Let B be the functional on $\mathbf{V}_1 \times \mathbf{V}_2$ defined by

$$B(x, y) = \xi_1\eta_1 + \xi_1\eta_3 - \xi_2\eta_1 + \xi_2\eta_2 + \xi_2\eta_3$$

for $x = (\xi_1, \xi_2) \in \mathbf{V}_1$ and $y = (\eta_1, \eta_2, \eta_3) \in \mathbf{V}_2$.

(1) Show that B is a conjugate bilinear functional.

(2) Obtain the singular value decomposition of B.

Note: The material covered in this Chapter is based on Halmos (1958), Rao and Mitra (1968a, 1971a, 1971b) and Rao (1973c).

CHAPTER 3

LINEAR TRANSFORMATIONS AND MATRICES

In Chapters 1 and 2, we have looked at entities called linear functionals. They were maps from a vector space into the associated field of the vector space. The notion of a linear functional can be extended to a more general setting. In this chapter, we study linear transformations from one vector space into another and their representation in matrix form.

3.1. Preliminaries

Let $\mathbf{V}$ and $\mathbf{W}$ be two arbitrary vector spaces over the same field $\mathbf{F}$. A map T from $\mathbf{V}$ to $\mathbf{W}$, written as $T : \mathbf{V} \to \mathbf{W}$, is said to be a linear transformation if

$$T(\alpha x + \beta y) = \alpha T(x) + \beta T(y)$$

for every α and β in $\mathbf{F}$, and x and y in $\mathbf{V}$. There are other names used for the type of map introduced above: a *linear operator*, a *homomorphism*, or a *linear mapping*. In the sequel, any linear transformation is simply called a transformation, and a general transformation will be referred to as a map. The vector space $\mathbf{V}$ is called the domain of the transformation T. If $\mathbf{S}$ is a subspace of $\mathbf{V}$, we can restrict the map T to $\mathbf{S}$, and the restriction is usually denoted by $T|\mathbf{S}$. For the restricted map, the domain is obviously the space $\mathbf{S}$. With $T : \mathbf{V} \to \mathbf{W}$, one can associate two sets

$$\mathbf{R}(T) = \{T(x) \in \mathbf{W} : x \in \mathbf{V}\}, \ \mathbf{K}(T) = \{x \in \mathbf{V} : T(x) = 0\}.$$

The set $\mathbf{R}(T)$ is a subset of the vector space $\mathbf{W}$ and is called the *range* of the transformation T. Further, one can show that $\mathbf{R}(T)$ is a subspace of

$\mathbf{W}$. The set $\mathbf{K}(T)$ is a subset of the space $\mathbf{V}$ and is called the *kernel* of the transformation T. One can show that $\mathbf{K}(T)$ is a subspace of $\mathbf{V}$. To round up the discussion on the range and kernel of a transformation, let us introduce two more notions. The dimension of the subspace $\mathbf{R}(T)$ is called the rank of the transformation T and is usually denoted by $\rho(T)$. The dimension of the subspace $\mathbf{K}(T)$ is called the *nullity* of the transformation T and is usually denoted by $\nu(T)$. If the transformation T is such that $\mathbf{R}(T) = \mathbf{W}$, then the transformation is labeled as onto. Otherwise, it is labeled as an into transformation. T is said to be an isomorphism if T is one-to-one and onto. If T is an isomorphism, we declare that the vector spaces $\mathbf{V}$ and $\mathbf{W}$ are isomorphic. Let us state a few facts surrounding these notions.

P 3.1.1 Let $T : \mathbf{V} \to \mathbf{W}$ and $\mathbf{K}^c$ be a complement of the subspace $\mathbf{K}(T)$ in $\mathbf{V}$. Then the following hold.

(1) The transformation $T|\mathbf{K}^c$, i.e., T restricted to $\mathbf{K}^c$, has the range $\mathbf{R}(T)$, i.e., $\mathbf{R}(T|\mathbf{K}^c) = \mathbf{R}(T)$.
(2) The transformation $T|\mathbf{K}^c : \mathbf{K}^c \to \mathbf{R}(T)$ is one-to-one and onto.
(3) $\dim(\mathbf{K}^c) = \rho(T) = \dim(\mathbf{V}) - \nu(T)$.

PROOF. It is obvious that $\mathbf{R}(T|\mathbf{K}^c) \subset \mathbf{R}(T)$. Let $y \in \mathbf{R}(T)$. There exists a vector x in $\mathbf{V}$ such that $T(x) = y$. Since $\mathbf{K}(T) \oplus \mathbf{K}^c = \mathbf{V}$, we can write $x = x_1 + x_2$ with $x_1 \in \mathbf{K}(T)$ and $x_2 \in \mathbf{K}^c$. As $y = T(x) = T(x_1) + T(x_2) = T(x_2)$ and $x_2 \in \mathbf{K}^c$, we have $y \in \mathbf{R}(T|\mathbf{K}^c)$. This proves (1). For (2), we need to show that the map $T|\mathbf{K}^c$ is one-to-one. Let u and v be two vectors in $\mathbf{K}^c$ such that $T(u) = T(v)$. As $T(u - v) = 0$, we have $u - v \in \mathbf{K}(T)$. The vector $u - v$ also belongs to $\mathbf{K}^c$. Hence $u - v = 0$. This shows that $T|\mathbf{K}^c$ is one-to-one. We have already seen that $\dim(\mathbf{V}) = \dim(\mathbf{K}(T)) + \dim(\mathbf{K}^c)$. (See **P 1.5.8.**) Since $\mathbf{K}^c$ and $\mathbf{R}(T)$ are isomorphic, (3) follows.

The above result can be rephrased as follows. Given any transformation from one vector space $\mathbf{V}$ to another $\mathbf{W}$, we have two subspaces $\mathbf{K}(T)$ and $\mathbf{R}(T)$ with $\mathbf{K}(T)$ being a subspace of $\mathbf{V}$ and $\mathbf{R}(T)$ a subspace of $\mathbf{W}$. The dimensions of these subspaces match the dimension of the vector space $\mathbf{V}$, i.e., $\dim(\mathbf{K}(T)) + \dim(\mathbf{R}(T)) = \dim(\mathbf{V})$. We look at some examples.

EXAMPLE 3.1.2. Let $\mathbf{F}$ be any field. Define a map T from $\mathbf{F}^3$ to $\mathbf{F}^2$ by $T(\xi_1, \xi_2, \xi_3) = (\xi_1, \xi_2)$ for (ξ_1, ξ_2, ξ_3) in $\mathbf{F}^3$. Note that $\mathbf{F}^3$ and $\mathbf{F}^2$ are three-dimensional and two-dimensional vector spaces, respectively, over the same field $\mathbf{F}$. Further, T is a linear transformation.

EXAMPLE 3.1.3. Consider the vector space $\mathbf{P}_n$ of all polynomials of degree less than n with complex numbers as coefficients. Then the map (Differential operator) T defined by,

$$T(\sum_{i=0}^{n-1} \xi_i x^i) = (d/dx)(\sum_{i=0}^{n-1} \xi_i x^i) = (\sum_{i=1}^{n-1} i\xi_i x^{i-1})$$

is a linear transformation from $\mathbf{P}_n$ to $\mathbf{P}_{n-1}$, where $\sum_{i=0}^{n-1} \xi_i x^i$ is a polynomial in x of degree less than n with $\xi_0, \xi_1, \ldots, \xi_{n-1} \in \mathbf{C}$.

EXAMPLE 3.1.4. Consider the vector space $\mathbf{P}_n$ of all polynomials of degree less than n with complex coefficients. Then the map (Integral operator) S defined as

$$S(\sum_{i=0}^{n-1} \xi_i x^i) = (\sum_{i=0}^{n-1}(\xi_i/(i+1)))x^{i+1}$$

is a linear transformation from $\mathbf{P}_n$ to $\mathbf{P}_{n+1}$.

EXAMPLE 3.1.5. Let $\mathbf{V}$ and $\mathbf{W}$ be arbitrary vector spaces over the same field. Let $w_1, w_2, \ldots, w_r$ be any r vectors in $\mathbf{W}$ and $y_1, y_2, \ldots, y_r$ any r linear functionals on $\mathbf{V}$. Then the map T defined by,

$$T(x) = y_1(x)w_1 + y_2(x)w_2 + \ldots + y_r(x)w_r,\ x \in \mathbf{V},$$

is a transformation from $\mathbf{V}$ to $\mathbf{W}$.

EXAMPLE 3.1.6. Let $\mathbf{S}$ be a subspace of a vector space $\mathbf{V}$. Let $\mathbf{S}^c$ be a complement of $\mathbf{S}$ in $\mathbf{V}$. Then any vector x in $\mathbf{V}$ can be written as $x = x_1 + x_2$ with $x_1 \in \mathbf{S}$ and $x_2 \in \mathbf{S}^c$ in a unique way. Define a map T from $\mathbf{V}$ to $\mathbf{S}$ by, $T(x) = x_1$. Then T is a transformation from $\mathbf{V}$ onto $\mathbf{S}$. Such an operator is called a projection onto $\mathbf{S}$ along $\mathbf{S}^c$. We have come across this notion when we were discussing inner product spaces and orthogonal complements of subspaces. We do not have to have an inner product on the vector space in order to have the notion of projection feasible.

Finally, we state some results which connect the notions of range, kernel and isomorphism.

P 3.1.7 Let $T : \mathbf{V} \to \mathbf{W}$. Then the following are valid.

(1) The transformation T is an isomorphism if and only if $\dim(\mathbf{K}(T)) = 0$, i.e., $\mathbf{K}(T)$ contains only the zero vector, and $\mathbf{R}(T) = \mathbf{W}$.
(2) If the vector spaces $\mathbf{V}$ and $\mathbf{W}$ have the same dimension, then $\mathbf{K}(T) = \{0\}$ if and only if $\mathbf{R}(T) = \mathbf{W}$.
(3) If T is an isomorphism from $\mathbf{V}$ to $\mathbf{W}$, and $\mathbf{S}$ is a subspace of $\mathbf{V}$, then $T|\mathbf{S}$ is an isomorphism from $\mathbf{S}$ onto $\mathbf{R}(T|\mathbf{S})$.

PROOF. Part (1) is easy to establish. For part (2), one can use the dimensional identity, $\dim(\mathbf{V}) = \dim(\mathbf{K}(T)) + \dim(\mathbf{R}(T))$. The proof of part (3) is trivial.

Complements

3.1.1 Let $\mathbf{V}$ be the set of all complex numbers regarded as a vector space over the field of real numbers. Let T be the map from $\mathbf{V}$ to $\mathbf{V}$ defined by

$$T(x + iy) = x - iy,$$

the conjugate of the complex number $x + iy \in \mathbf{V}$. Show that T is an isomorphism of $\mathbf{V}$.

3.2. Algebra of Transformations

In this section, we look at the collection of all transformations from one vector space to another. This collection can be endowed with a structure so as to make it a vector space. Of course, the underlying structure of the vector space involved plays a crucial role in passing its features to the collection. In the sequel, assume that all vector spaces are over the same field $\mathbf{F}$.

DEFINITION 3.2.1. Let $T : \mathbf{V} \to \mathbf{W}$ and $S : \mathbf{V} \to \mathbf{W}$. Define a map $T + S$ by,

$$(T + S)(x) = T(x) + S(x), \; x \in \mathbf{V}.$$

For each α in $\mathbf{F}$, define a map αT by,

$$(\alpha T)(x) = \alpha T(x), \; x \in \mathbf{V}.$$

It is clear that the maps $T + S$ and αT are transformations. Thus addition and scalar multiplication of transformations are naturally available on the collection of all transformations. The following proposition clearly spells what these operations mean from a structural point of view.

P 3.2.2 Let $L(\mathbf{V}, \mathbf{W})$ be the collection of all transformations from the vector space $\mathbf{V}$ to the vector space $\mathbf{W}$. With the operations of addition and scalar multiplication of transformations defined above, $L(\mathbf{V}, \mathbf{W})$ is a vector space over the same field $\mathbf{F}$.

One can define one more operation on transformations, namely, composition, subject to some compatibility conditions.

DEFINITION 3.2.3. Let $T : \mathbf{V} \to \mathbf{W}$ and $S : \mathbf{W} \to \mathbf{U}$. Define a map $ST : \mathbf{V} \to \mathbf{U}$ by

$$(ST)(x) = S(T(x)),\ x \in \mathbf{V}.$$

It is clear that the map ST is a transformation. The transformation is, sometimes, called the product of the transformations S and T. The product TS may not be defined. If $\mathbf{V} = \mathbf{W} = \mathbf{U}$, one can always define the product ST as well as TS, and they need not be the same.

After having defined the space $L(\mathbf{V}, \mathbf{W})$, the next item on the agenda is to determine the magnitude of its dimension. The following proposition addresses this problem.

P 3.2.4 If $\dim(\mathbf{V}) = m$ and $\dim(\mathbf{W}) = n$, then $\dim(L(\mathbf{V}, \mathbf{W})) = mn$.

PROOF. Let $x_1, x_2, \ldots, x_m$ be a basis of the vector space $\mathbf{V}$ and $w_1, w_2, \ldots, w_n$ a basis of $\mathbf{W}$. Let T be any transformation from $L(\mathbf{V}, \mathbf{W})$. First, we observe that the value of $T(x)$, for any $x \in \mathbf{V}$, is determined by the set $T(x_1), T(x_2), \ldots, T(x_m)$ of vectors in $\mathbf{W}$. For, we can write

$$x = \alpha_1 x_1 + \alpha_2 x_2 + \ldots + \alpha_m x_m$$

for some $\alpha_1, \alpha_2, \ldots, \alpha_m$ in $\mathbf{F}$ in a unique fashion. It follows that

$$T(x) = \alpha_1 T(x_1) + \alpha_2 T(x_2) + \ldots + \alpha_m T(x_m).$$

Once a basis for the vector space $\mathbf{V}$ is fixed, knowing the vector x is equivalent to knowing its co-ordinates $\alpha_1, \alpha_2, \ldots, \alpha_m$. Once we know the values of the transformation T at $x_1, x_2, \ldots, x_m$, we can immediately write down the value of T at x. This innocuous observation has deep implications. One can build a transformation from **V** to **W** demanding it to take certain values in the vector space **W** at certain vectors in **V**! With this in view, for every fixed $i \in \{1, 2, \ldots, n\}$ and $j \in \{1, 2, \ldots, m\}$, let T_{ij} be a transformation from **V** to **W** satisfying $T_{ij}x_j = \omega_i$, $T_{ij}x_k = 0$ for all $k \neq j$. In other words, we want a transformation T_{ij} such that $T_{ij}x_1 = 0, T_{ij}x_2 = 0, \ldots, T_{ij}x_{j-1} = 0, T_{ij}x_j = w_i, T_{ij}x_{j+1} = 0, \ldots, T_{ij}x_m = 0$. Note that T_{ij} is the only transformation that takes the value 0 at $x_1, x_2, \ldots, x_{j-1}, x_{j+1}, \ldots, x_m$ and the value ω_i at x_j. (Why?) We claim that the set $T_{ij}, 1 \leq i \leq n, 1 \leq j \leq m$ of transformations constitutes a linearly independent set in $L(\mathbf{V}, \mathbf{W})$. Suppose

$$\sum_{i=1}^{n} \sum_{j=1}^{m} \alpha_{ij} T_{ij} = 0$$

for some scalars $\alpha_{ij}, 1 \leq i \leq n, 1 \leq j \leq m$. In particular, for each $1 \leq k \leq m$,

$$0 = (\sum_{i=1}^{n} \sum_{j=1}^{m} \alpha_{ij} T_{ij})(x_k) = \alpha_{1k} w_1 + \alpha_{2k} w_2 + \ldots + \alpha_{nk} w_n.$$

Since $w_1, w_2, \ldots, w_n$ are linearly independent, $\alpha_{ik} = 0$ for every $i = 1, 2, \ldots, n$. Since k is arbitrary, it follows that $\alpha_{ij} = 0$ for every i and j. This establishes the linear independence of the transformations. Next, we show that any transformation T in $L(\mathbf{V}, \mathbf{W})$ is a linear combination of the transformations T_{ij}'s. Let $T(x_i) = y_i$, $1 \leq i \leq m$. Expand each y_i in terms of the basis of **W**, i.e.,

$$y_i = \beta_{i1} w_1 + \beta_{i2} w_2 + \ldots + \beta_{in} w_n$$

for some scalars β_{ij}'s. One simply verifies that

$$T = \sum_{i=1}^{m} \sum_{j=1}^{n} \beta_{ij} T_{ji}.$$

It now follows that $\dim(L(\mathbf{V}, \mathbf{W})) = mn$.

The notion of a linear transformation is not that much different from the notion of a linear functional. By piecing together linear functionals in some suitable manner we can construct a linear transformation from one vector space to another. An inkling of this phenomenon has already been provided in Example 3.1.5. A fuller version of this example is the substance of the following proposition.

P 3.2.5 Let **V** and **W** be two vector spaces. Let $w_1, w_2, \ldots, w_n$ be a basis for **W**. Let T be a map from **V** to **W**. Then T is a transformation if and only if there exist linear functionals $y_1, y_2, \ldots, y_n$ on **V** such that

$$T(x) = y_1(x)w_1 + y_2(x)w_2 + \ldots + y_n(x)w_n$$

for every x in **V**.

PROOF. If $T = y_1w_1 + y_2w_2 + \ldots + y_nw_n$ for some linear functionals $y_1, y_2, \ldots, y_n$ on the vector space **V**, it is clear that T is a transformation. Conversely, let T be a transformation from **V** to **W**. Let $x_1, x_2, \ldots, x_m$ be a basis for the vector space **V**. Let $y_i = T(x_i), i = 1, 2, \ldots, m$. Write for each i,

$$y_i = \beta_{i1}w_1 + \beta_{i2}w_2 + \ldots + \beta_{in}w_n$$

for some scalars β_{ij}'s. Define for each $j = 1, 2, \ldots, n$,

$$y_j(x) = \sum_{i=1}^{m} \alpha_i\beta_{ij}, \quad x \in \mathbf{V},$$

where $x = \alpha_1x_1 + \alpha_2x_2 + \ldots + \alpha_mx_m$ is the unique representation of x in terms of the basis of **V**. It is clear that each $y_i(\cdot)$ is a linear functional on **V**. Finally, for any x in **V**,

$$\begin{aligned} T(x) &= T(\sum_{i=1}^{m} \alpha_ix_i) = \sum_{i=1}^{m} \alpha_iT(x_i) = \sum_{i=1}^{m} \alpha_iy_i = \sum_{i=1}^{m} \alpha_i \sum_{j=1}^{n} \beta_{ij}w_j \\ &= \sum_{j=1}^{n}(\sum_{i=1}^{m} \alpha_i\beta_{ij})w_j = \sum_{j=1}^{n} y_j(x)w_j. \end{aligned}$$

Thus we are able to write T as a combination of linear functionals! This completes the proof.

Linear transformations from a vector space $\mathbf{V}$ into itself are of special interest. For any two transformations T and S from $\mathbf{V}$ to $\mathbf{V}$, one can define the product TS as the composition of the maps T and S taken in this order. Moreover, one can define the identity transformation $\mathbf{I}$ from $\mathbf{V}$ to $\mathbf{V}$ by $I(x) = x$ for every x in $\mathbf{V}$. It is clear that for any transformation T from $\mathbf{V}$ to $\mathbf{V}, TI = IT = T$. Thus the space $L(\mathbf{V}, \mathbf{V})$ has an additional structure which arises from the operation of product of transformations defined above. The following proposition provides the necessary details what this additional structure entails.

P 3.2.6 The space $L(\mathbf{V}, \mathbf{V})$ of all transformations from a vector space $\mathbf{V}$ into itself is an algebra with an identity.

PROOF. We have already seen that the space $L(\mathbf{V}, \mathbf{V})$ is a vector space. The additional structure on $L(\mathbf{V}, \mathbf{V})$ comes from the binary operation of product or composition of transformations. We need to identify the identity of this new binary operation. The obvious candidate is the identity transformation I. In order to show that the space $L(\mathbf{V}, \mathbf{V})$ is an algebra, we need to verify the following.

(1) For any transformation $T, 0T = T0 = 0$ holds, where 0 is the transformation which maps every vector of $\mathbf{V}$ into the zero vector. The map 0 is the additive identity of the additive operation of the vector space $L(\mathbf{V}, \mathbf{V})$.
(2) For every transformation $T, IT = TI = T$ holds.
(3) (Associative law). For any three transformations T, S and U, $(TS)U = T(SU)$ holds.
(4) (Distributive laws). For any three transformations T, S and U, $(T + S)U = TU + SU$ and $T(S + U) = TS + TU$ hold.

These properties are easy to establish.

As for the distributive law and associative law, they are valid in a more general framework. For example, suppose T is a transformation from a vector space $\mathbf{V}_1$ into a vector space $\mathbf{V}_2$, and S and U are transformations from the vector space $\mathbf{V}_2$ into a vector space $\mathbf{V}_3$, the transformations $(S + U)T$, ST and UT are all well-defined from the vector space $\mathbf{V}_1$ to $\mathbf{V}_3$. Moreover, we have the following distributive law:

$$(S + U)T = ST + UT.$$

One could write down any number of identities of the above type. As an example of another variety, suppose T is a transformation from a vector space $\mathbf{V}_1$ to a vector space $\mathbf{V}_2$, S is a transformation from the vector space $\mathbf{V}_2$ to a vector space $\mathbf{V}_3$, and U a transformation from the vector space $\mathbf{V}_3$ to a vector space $\mathbf{V}_4$. Then the transformations $U(ST)$ and $(US)T$ make sense and they are indeed transformations from the vector space $\mathbf{V}_1$ to $\mathbf{V}_4$. Further, they are identical, i.e., $U(ST) = (US)T$. It is customary to denote this transformation by UST.

Complements

3.2.1 Let $\mathbf{Q}$ be the field of all rational numbers. Let $\mathbf{V} = \mathbf{Q}^2$ and T a transformation from $\mathbf{V}$ to $\mathbf{V}$. The only clues we have about T are that

$$T(1,0) = (2,-3),\ T(0,1) = (3,1).$$

Determine $T(x,y)$ for any $(x,y) \in \mathbf{V}$. Is T an isomorphism? Justify your answer.

3.2.2 Let $\mathbf{V} = \mathbf{Q}^2, \mathbf{W} = \mathbf{Q}^3$, and T a transformation from $\mathbf{V}$ to $\mathbf{W}$. The only clues we have about T are that

$$T(1,0) = (2,3,-2),\ T(1,1) = (4,-7,8).$$

Determine $T(x,y)$ for any $(x,y) \in \mathbf{V}$.

3.2.3 Let $\mathbf{V} = \mathbf{R}^3, \mathbf{W} = \mathbf{R}^4$, and T a transformation from $\mathbf{V}$ to $\mathbf{W}$. The only clues we have about T are that

$$T(1,0,0) = (1,-1,2,1),\ T(0,1,0) = (0,1,1,0),\ T(0,0,1) = (2,0,0,0).$$

Determine the range and kernel of the transformation T along with their dimensions.

3.2.4 Let $\mathbf{P}$ be the collection of all polynomials viewed as a vector space over the field of real numbers. (Note that $\mathbf{P}$ is an infinite-dimensional vector space.) Let T be the differential operator on $\mathbf{P}$ defined by

$$T(\sum_{i=0}^{n-1} \xi_i x^i) = (\sum_{i=1}^{n-1} i\xi_i x^{i-1}),$$

and S the transformation on $\mathbf{P}$ defined by

$$S(\sum_{i=0}^{n-1} \xi_i x^i) = \sum_{i=0}^{n-1} (\xi_i/(i+1))x^{i+1},$$

$\xi_0, \xi_1, \ldots, \xi_n$ real, $n \geq 1$. Compute TS and ST. Show that $ST \neq TS$.

3.3. Inverse Transformations

Let $T : \mathbf{V} \to \mathbf{W}$. We raise the question whether given any vector y in $\mathbf{W}$, is it possible to recover the vector x in $\mathbf{V}$ through a linear transformation, related to T in some way. The answer to the question depends on the nature of the transformation T. The following are two crucial properties that the answer depends on.

(1) The map is injective or one-to-one: recall that the map T is injective if x_1 and x_2 are any two distinct vectors in $\mathbf{V}$, then the vectors $T(x_1)$ and $T(x_2)$ are distinct.

(2) The map is surjective or onto: recall that the map T is surjective if for every vector y in $\mathbf{W}$, there exists at least one vector x in $\mathbf{V}$ such that $T(x) = y$, i.e., $\mathbf{R}(T) = \mathbf{W}$.

P 3.3.1 Let $T : \mathbf{V} \to \mathbf{W}$.

(1) If the map T is injective, then there exists a linear transformation $S : \mathbf{W} \to \mathbf{V}$ such that $ST = I$, the identity transformation from $\mathbf{V}$ to $\mathbf{V}$. (Such a transformation S is called the left inverse of T and is denoted by T_L^{-1}.)

(2) If the map T is surjective, then there exists a linear transformation $S : \mathbf{W} \to \mathbf{V}$ such that $TS = I$, the identity transformation from $\mathbf{W}$ to $\mathbf{W}$. (Such a transformation S is called the right inverse of T and is denoted by T_R^{-1}.)

(3) If the map T is bijective, i.e., T is injective and surjective, then there exists a transformation $S : \mathbf{W} \to \mathbf{V}$ such that $ST = I$ and $TS = I$ with the identity transformation I operating on the appropriate vector space. The transformation S is unique and is called the inverse of T. It is denoted by T^{-1}. A bijective map is also called invertible.

(4) There always exists a transformation $S : \mathbf{W} \to \mathbf{V}$ such that $TST = T$. Such a transformation S is called a generalized inverse (g-inverse) of T and is denoted by T^-.

PROOF. Before we embark on a proof, let us keep in line the entities we need. Let $\mathbf{R}(T)$ be the range of the transformation T. Choose and fix a complement $\mathbf{R}^c$ of $\mathbf{R}(T)$ in $\mathbf{W}$. See **P 1.5.5** for details. Let $\mathbf{K}(T)$ be the kernel of the transformation T. Choose and fix a complement $\mathbf{K}^c$ of $\mathbf{K}(T)$ in $\mathbf{V}$. We now proceed to prove every part of the above proposition. We basically make use of the above subspaces, their complements, and the associated projections.

Let y be any vector in $\mathbf{W}$. We can write $y = y_0 + y_1$ uniquely with $y_0 \in \mathbf{R}(T)$ and $y_1 \in \mathbf{R}^c$. Since T is injective, there exists a unique x_0 in $\mathbf{V}$ such that $T(x_0) = y_0$. Define $T_L^{-1}(y) = x_0$. Thus T_L^{-1} is a well-defined map from $\mathbf{W}$ to $\mathbf{V}$. It is clear that $T_L^{-1}T = I$, the identity transformation from $\mathbf{V}$ to $\mathbf{V}$. It remains to show that the map T_L^{-1} is a transformation. This essentially follows from the property that any projection is a linear operation. Let α and β be two scalars and u and w be any two vectors in $\mathbf{W}$. If we decompose $u = u_0 + u_1$ and $w = w_0 + w_1$ with $u_0, w_0 \in \mathbf{R}(T)$ and $u_1, w_1 \in \mathbf{R}^c$, then we can identify the decomposition of $\alpha u + \beta w$ as

$$\alpha u + \beta w = (\alpha u_0 + \beta w_0) + (\alpha u_1 + \beta w_1)$$

with $\alpha u_0 + \beta w_0 \in \mathbf{R}(T)$ and $\alpha u_1 + \beta w_1 \in \mathbf{R}^c$. Let x_0 and v_0 be the unique vectors in $\mathbf{V}$ such that $T(x_0) = u_0$ and $T(v_0) = w_0$. It is now clear that

$$T_L^{-1}(\alpha u + \beta w) = \alpha x_0 + \beta v_0 = \alpha T_L^{-1}(u) + \beta T_L^{-1}(w).$$

This shows that the map T_L^{-1} is a transformation. This proves (1).

We now work with $\mathbf{K}(T)$ and a complement $\mathbf{K}^c$ of $\mathbf{K}(T)$. Let y be any vector in $\mathbf{W}$. Since T is surjective, there exists a vector x in $\mathbf{V}$ such that $T(x) = y$. But there could be more than one vector x satisfying $T(x) = y$. But there is one and only one vector x_0 in $\mathbf{K}^c$ such that $T(x_0) = y$. This is not hard to see. Define

$$T_R^{-1}(y) = x_0.$$

By the very nature of the definition of the map T_R^{-1}, $TT_R^{-1} = I$, the identity map from $\mathbf{W}$ to $\mathbf{W}$. It remains to show that T_R^{-1} is a transformation. In order to show this, one can craft an argument similar to the one used in Part (1) above, which proves (2).

If the map is bijective, both the definitions of the maps T_L^{-1} and T_R^{-1} coincide. Let the common map be denoted by T^{-1}. Of course, we have $TT^{-1} = I$ and $T^{-1}T = I$. As for uniqueness, suppose S is a map such that $TS = I$, then

$$S = IS = T^{-1}TS = T^{-1}I = T^{-1}.$$

Another interesting feature is that if S and U are two maps from $\mathbf{W}$ to $\mathbf{V}$ satisfying $TS = I$ and $UT = I$, then we must have $S = U = T^{-1}$. This proves (3).

Let $y \in \mathbf{W}$. Write uniquely $y = y_0 + y_1$ with $y_0 \in \mathbf{R}(T)$ and $y_1 \in \mathbf{R}^c$. Determine $x \in \mathbf{V}$ such that $T(x) = y_0$. Decompose uniquely $x = x_0 + x_1$ with $x_0 \in \mathbf{K}^c$ and $x_1 \in \mathbf{K}(T)$. Define $T^-(y) = x_0$. T^- is well-defined and indeed is a transformation from $\mathbf{W}$ to $\mathbf{V}$. It is clear that $TT^-T = T$, which proves (4).

The transformations T_L^{-1}, T_R^{-1} and T^- are not, in general, unique. Different possible choices arise from different choices of the complements $\mathbf{R}^c$ and $\mathbf{K}^c$. We will exploit this fact when we explore the world of g-inverses in one of the subsequent chapters.

We record a few facts about inverse transformations for future reference. These results can easily be verified.

P 3.3.2 Let T be a bijective transformation (isomorphism) from a vector space $\mathbf{V}$ to a vector space $\mathbf{W}$, and S a bijective transformation from the vector space $\mathbf{W}$ to a vector space $\mathbf{U}$. Let α be a non-zero scalar. Then the following are valid.

(1) The transformation ST is a bijective transformation from $\mathbf{V}$ to $\mathbf{U}$ and $(ST)^{-1} = T^{-1}S^{-1}$.

(2) The transformation αT is a bijective transformation from $\mathbf{V}$ to $\mathbf{W}$ and $(\alpha T)^{-1} = \alpha^{-1}T^{-1}$.

(3) The transformation T^{-1} is a bijective transformation from $\mathbf{W}$ to $\mathbf{V}$ and $(T^{-1})^{-1} = T$.

Let us specialize about transformations that operate from a vector space $\mathbf{V}$ into itself. If T is such a transformation, T^{-1} exists if and only if $\rho(T) = \dim(\mathbf{V})$. This follows from the identity $\nu(T) + \rho(T) = \dim(\mathbf{V})$. See **P 3.1.7**.

Complements

3.3.1 Let $\mathbf{V}$ be the set of all complex numbers viewed as a vector space over the field of real numbers. Let T be the transformation from $\mathbf{V}$ to $\mathbf{V}$ defined as

$$T(x + iy) = x - iy,$$

the complex conjugate of the complex number $x + iy \in \mathbf{V}$. Determine T^{-1}.

3.3.2 Let T be the transformation on the vector space $\mathbf{C}^2$ defined by

$$T(x_1, x_2) = (\alpha x_1 + \beta x_2, \gamma x_1 + \delta x_2),$$

for $(x_1, x_2) \in \mathbf{C}^2$ and α, β, γ, and $\delta \in \mathbf{C}$. Show that T is an isomorphism if $\alpha\delta - \beta\gamma \neq 0$. In such an event, determine T^{-1}. If $\alpha\delta - \beta\gamma = 0$, determine a g-inverse T^- of T.

3.3.3 Let $\mathbf{P}$ be the vector space of all polynomials with real coefficients, which is viewed as an infinite-dimensional vector space over the field of real numbers. Let T be the differential operator on $\mathbf{P}$ defined by

$$T(\sum_{i=0}^{n-1} \alpha_i x^i) = \sum_{i=1}^{n-1} i\alpha_i x^{i-1},$$

$\alpha_0, \alpha_1, \alpha_2, \ldots, \alpha_{n-1}$ real and $n \geq 1$. Show that T is surjective. Let S be the transformation on $\mathbf{P}$ defined by

$$S(\sum_{i=0}^{n-1} \alpha_i x^i) = \sum_{i=0}^{n-1} \frac{\alpha_i}{i+1} x^{i+1}.$$

Show that S is a right inverse of T. Show that there is no left inverse for T.

3.3.4 If $T_1, T_2, \ldots, T_r$ are invertible transformations from a vector space $\mathbf{V}$ into $\mathbf{V}$ itself, show that $T_1 T_2 \ldots T_r$ is also invertible.

3.3.5 Let T be a transformation from $\mathbf{V}$ to $\mathbf{W}$. Show that for the existence of a left inverse of T, it is necessary that T is injective. If T is injective, show that any left inverse of T is surjective.

3.3.6 Let T be a transformation from $\mathbf{V}$ to $\mathbf{W}$. Show that for the existence of a right inverse of T, it is necessary that T is surjective. If T is surjective, show that any right inverse of T is injective.

3.3.7 If T and S are two transformations from a finite-dimensional vector space $\mathbf{V}$ into itself such that $TS = I$, show that T is invertible and $S = T^{-1}$. (Why do we need the finite-dimensionality condition on the vector space?)

3.3.8 Let T be a transformation from a finite-dimensional vector space $\mathbf{V}$ into itself enjoying the property that $T(x_1), T(x_2), \ldots, T(x_r)$ are linearly independent in $\mathbf{W}$ whenever $x_1, x_2, \ldots, x_r$ are linearly independent in $\mathbf{V}$ for any $r \geq 1$. Show that T is invertible. Is the finite-dimensionality condition needed?

3.4. Matrices

Let **V** and **W** be two finite-dimensional vector spaces over the same field **F**. Let T be a transformation from **V** to **W**. Let $x_1, x_2, \ldots, x_m$ be a basis of the vector space **V** and $y_1, y_2, \ldots, y_n$ that of **W**. For each $i = 1, 2, \ldots, m$, note that $T(x_i) \in \mathbf{W}$, and consequently, we can write

$$T(x_i) = \alpha_{1i}y_1 + \alpha_{2i}y_2 + \ldots + \alpha_{ni}y_n, \tag{3.4.1}$$

for some scalars α_{ij}'s in **F**. These scalars and the transformation T can be regarded as two sides of the same coin. Knowing the transformation T is equivalent to knowing the bunch of scalars α_{ij}'s. We explain why. The transformation T provides a rule or a formula which associates every vector in **V** with a vector in **W**. This rule can be captured by the set of scalars α_{ij}'s. First of all, let us organize the scalars in the form of an $n \times m$ grid A_T, which we call a matrix.

$$A_T = \begin{bmatrix} \alpha_{11} & \alpha_{12} & \ldots & \alpha_{1m} \\ \alpha_{21} & \alpha_{22} & \ldots & \alpha_{2m} \\ \cdot & \cdot & \ldots & \cdot \\ \alpha_{n1} & \alpha_{n2} & \ldots & \alpha_{nm} \end{bmatrix}.$$

Let x be any vector in **V**. We can write

$$x = \beta_1 x_1 + \beta_2 x_2 + \ldots + \beta_m x_m$$

for some scalars $\beta_1, \beta_2, \ldots, \beta_m$ in **F**. In view of the uniqueness of the representation, the vector x can be identified with the m-tuple $(\beta_1, \beta_2, \ldots, \beta_m)$ in $\mathbf{F}^m$. The transformed value $T(x)$ of x under T can be written as

$$T(x) = \gamma_1 y_1 + \gamma_2 y_2 + \ldots + \gamma_n y_n \tag{3.4.2}$$

for some scalars $\gamma_1, \gamma_2, \ldots, \gamma_n$ in **F**. The transformed vector $T(x)$ can be identified with the n-tuple $(\gamma_1, \gamma_2, \ldots, \gamma_n)$ in $\mathbf{F}^n$. Thus the transformation T can be identified with a transformation from $\mathbf{F}^m$ to $\mathbf{F}^n$. Once we know the vector x, or equivalently, its m-tuple $(\beta_1, \beta_2, \ldots, \beta_m)$, the coordinates $(\gamma_1, \gamma_2, \ldots, \gamma_n)$ of the transformed vector $T(x)$ can be obtained as

$$\gamma_i = \sum_{j=1}^{m} \alpha_{ij}\beta_j, \; i = 1, 2, \ldots, n. \tag{3.4.3}$$

For, we note that

$$
\begin{aligned}
T(x) &= T(\beta_1 x_1 + \beta_2 x_2 + \ldots + \beta_m x_m) \\
&= \beta_1 T(x_1) + \beta_2 T(x_2) + \ldots + \beta_m T(x_m) \\
&= \beta_1 (\sum_{i=1}^{n} \alpha_{i1} y_i) + \beta_2 (\sum_{i=1}^{n} \alpha_{i2} y_i) + \ldots + \beta_m (\sum_{i=1}^{n} \alpha_{im} y_i) \\
&= (\sum_{j=1}^{m} \alpha_{1j}\beta_j) y_1 + (\sum_{j=1}^{m} \alpha_{2j}\beta_j) y_2 + \ldots + (\sum_{j=1}^{m} \alpha_{nj}\beta_j) y_n.
\end{aligned}
\tag{3.4.4}
$$

using (3.4.1) in step 3 above. Then identifying (3.4.4) with (3.4.2), we obtain (3.4.3), which transforms β_i's to γ_j's.

Equations (3.4.3) can be written symbolically as:

$$
\begin{bmatrix} \alpha_{11} & \alpha_{12} & \ldots & \alpha_{1m} \\ \alpha_{21} & \alpha_{22} & \ldots & \alpha_{2m} \\ \cdot & \cdot & \ldots & \cdot \\ \alpha_{n1} & \alpha_{n2} & \ldots & \alpha_{nm} \end{bmatrix} \begin{bmatrix} \beta_1 \\ \beta_2 \\ \cdot \\ \beta_m \end{bmatrix} = \begin{bmatrix} \gamma_1 \\ \gamma_2 \\ \cdot \\ \gamma_n \end{bmatrix},
$$

or, in short

$$A_T b = c, \tag{3.4.5}$$

where b is the column vector consisting of entries β_i's and c is the column vector consisting of entries γ_i's. The symbolic representation etched above can be made algebraically meaningful. Operationally, the symbolic representation given above can be implemented as follows. Start with any vector x in $\mathbf{V}$. Determine its coordinates b with respect to the given basis of $\mathbf{V}$. Combine the entries of the matrix A_T and the vector b as per the arithmetic set out in the equation (3.4.3) in order to obtain the coordinates c of the transformed vector $T(x)$. Once we know the coordinates of the vector $T(x)$, we can write down the vector $T(x)$ using the given basis of the vector space $\mathbf{W}$.

There are two ways of giving an algebraic meaning to the symbolic representation (3.4.5) of the transformation T. In the representation, we seem to multiply a matrix of order $n \times m$ (i.e., a matrix consisting nm entries from the underlying field arranged in n rows and m columns) and a matrix of order $m \times 1$ resulting in a matrix of order $n \times 1$. One could spell out the rules of multiplication in such a scenario by invoking the equations (3.4.3). Another way is to define formally the multiplication of two matrices and then exclaim that in the symbolic representation (3.4.5) we are actually carrying out the multiplication. We will now spend some time on matrices and some basic operations on matrices.

DEFINITION 3.4.1. Let $\mathbf{F}$ be a field. A matrix A of order $m \times n$ is an array of mn scalars from $\mathbf{F}$ arranged in m rows and n columns. The array is presented in the following form:

$$A = \begin{bmatrix} \alpha_{11} & \alpha_{12} & \dots & \alpha_{1n} \\ \alpha_{21} & \alpha_{22} & \dots & \alpha_{2n} \\ \cdot & \cdot & \dots & \cdot \\ \alpha_{m1} & \alpha_{m2} & \dots & \alpha_{mn} \end{bmatrix}$$

with the entries α_{ij}'s coming from the field $\mathbf{F}$. Frequently, we abbreviate the matrix A in the form $A = (\alpha_{ij})$, where α_{ij} is the generic entry located in the matrix at the junction of the i-th row and j-th column. The scalar α_{ij} is also called the (i, j)-th entry of the matrix A.

The origin of the word "matrix" makes an interesting reading. In Latin, the word "matrix" means - womb, pregnant animal. The use of the word "matrix" in Linear Algebra, perhaps, refers to the way a matrix is depicted - it is a womb containing objects in an orderly fashion. The Indo-European root of the word "matrix" is *ma* which means mother.

It is time to introduce some operations on matrices. In what follows, we assume that all matrices have scalars from one fixed field.

DEFINITION 3.4.2.

(1) Addition. Let $A = (\alpha_{ij})$ and $B = (\beta_{ij})$ be two matrices of the same order $m \times n$. The matrix $C = (\gamma_{ij})$ of order $m \times n$ is defined by the rule that the (i, j)-th entry γ_{ij} of C is given by $\gamma_{ij} = \alpha_{ij} + \beta_{ij}$ for all $1 \le i \le m$ and $1 \le j \le n$. The matrix C is called the sum of A and B, and is denoted by $A + B$.

(2) Scalar Multiplication. Let $A = (\alpha_{ij})$ be a matrix of order $m \times n$ and α a scalar. The matrix $D = (\delta_{ij})$ of order $m \times n$ is defined by the rule that the (i, j)-th entry δ_{ij} of D is given by $\delta_{ij} = \alpha\alpha_{ij}$ for all i and j. The matrix D is called a scalar multiple of A and is denoted by αA. If $\alpha = -1$, the matrix αA is denoted by $-A$.

(3) Multiplication. Let $A = (\alpha_{ij})$ and $B = (\beta_{ij})$ be two matrices of order $m \times n$ and $p \times q$, respectively. Say that A and B are

conformable for multiplication in the order they are written if the number of columns of A is the same as the number of rows of B, i.e., $n = p$. Suppose A and B are conformable for multiplication. The matrix $E = (e_{ij})$ of order $m \times q$ is defined by the rule that the (i,j)-th entry e_{ij} of E is given by $e_{ij} = \sum_{k=1}^{n} \alpha_{ik}\beta_{kj}$ for all $1 \le i \le m$ and $1 \le j \le q$. The matrix E is called the product of A and B, and is denoted by AB.

Of all the operations defined on matrices defined above, the multiplication seems to be the most baffling. The operation of multiplication occurs in a natural way when we consider a pair of transformations and their composition. If two matrices A and B are conformable for multiplication, it is not true that B and A should be conformable for multiplication. Even if B and A are conformable for multiplication with the orders of AB and BA being identical, it is not true that we must have $AB = BA$!

We introduce two special matrices. The matrix of order $m \times n$ in which every entry is zero is called the zero matrix and is denoted by $0_{m\times n}$. If the order of the matrix is clear from the context, we will denote the zero matrix simply by 0. The matrix of order $n \times n$ in which every diagonal entry is equal to 1 and every off-diagonal entry is 0 is called the identity matrix of order $n \times n$ and is denoted by I_n. We now record some of the properties of the operations we have introduced above.

P 3.4.3 (1) Let A, B and C be matrices of the same order. Then

$$(A + B) + C = A + (B + C),$$
$$A + B = B + A,$$
$$A + 0 = A,$$
$$A + (-A) = 0.$$

(2) Let A, B and C be three matrices of orders $m \times n, n \times p$, and $p \times q$, respectively. Then

$$A(BC) = (AB)C \text{ of order } m \times q,$$
$$I_m A = A,$$
$$AI_n = A,$$
$$0_{r\times m} A = 0_{r\times n},$$
$$A0_{n\times s} = 0_{m\times s}.$$

(3) Let A and B be two matrices of the same order $m \times n$, C a matrix of order $n \times p$ and D a matrix of order $q \times m$. Then

$$(A+B)C = AC + BC$$
$$D(A+B) = DA + DB.$$

The above properties are not hard to establish. Some useful pointers emerge from the above deliberations. If $\mathbf{M}_{m,n}$ denotes the collection of all matrices of order $m \times n$, then it is a vector space over the field $\mathbf{F}$. If $m = n$, then $\mathbf{M}_{m,m}$ is an algebra. The operations are addition, scalar multiplication and multiplication of matrices. This algebra has a certain peculiarity. For two matrices A and B in $\mathbf{M}_{m,m}$, it is possible that $AB = 0$ without each of A and B being the zero matrix. Construct an example yourself.

We introduce two other operations on an $m \times n$ matrix A. One is called *transpose*, which is obtained by writing the columns of A as rows (i-th column as the i-th row, $i = 1, \ldots, n$) and denoted by A'. It is seen that the order of A' is $n \times m$. The following is an example of A and A'.

$$A = \begin{pmatrix} 1 & 2 & 4 \\ 0 & 3 & 1 \end{pmatrix}, \quad A' = \begin{pmatrix} 1 & 0 \\ 2 & 3 \\ 4 & 1 \end{pmatrix}.$$

The following results concerning the transpose operation are easily established.

P 3.4.4

(1) $(AB)' = B'A'$, $(ABC)' = C'B'A'$.
(2) $(A+B)' = A' + B'$.
(3) $(A^2)' = (A')^2$.
(4) If $A = (a_{ij})$ is a square symmetric matrix of order n, then

$$\sum_1^n \sum_1^n a_{ij} x_i x_j = x'Ax$$

where $x' = (x_1, \ldots, x_n)$. Note that x is a column vector.

(5) $(A^{-1})' = (A')^{-1}$ if A is invertible, i.e., $AA^{-1} = A^{-1}A = I$ holds.

Another is called the *conjugate transpose*, applicable when the elements of A are from the field of complex numbers, which is obtained by first writing the columns of A as rows as in the transpose and replacing each element by its complex conjugate, and denoted by A^*. Thus if $A = A_1 + iA_2$, where A_1 and A_2 are real, then $A^* = A_1' - iA_2'$. The following results concerning the conjugate transpose are easily established.

P 3.4.5

(1) $(AB)^* = B^*A^*$, $(ABC)^* = C^*B^*A^*$.
(2) $(A + B)^* = A^* + B^*$.
(3) $(\alpha A)^* = \bar{\alpha}A^*$, ($\bar{\alpha}$ is the complex conjugate of α).
(4) $(A^{-1})^* = (A^*)^{-1}$ if A is invertible.

We showed that associated with a transformation $T : \mathbf{V} \to \mathbf{W}$, there exists an $n \times m$ matrix A_T which provides a transformation from $\mathbf{F}^m$ to $\mathbf{F}^n$ through the operation of matrix multiplication $A_T x$, $x \in \mathbf{F}^m$. As observed earlier, the transformations T and A_T are isomorphic and A_T does the same job as T. We quote some results which are easy to establish.

P 3.4.6 Let $T : \mathbf{V} \to \mathbf{W}$ and $S : \mathbf{V} \to \mathbf{W}$ be two transformations with the associated matrices A_T and A_S. Then

$$A_{\alpha T + \beta S} = \alpha A_T + \beta A_S.$$

Let $T : \mathbf{V} \to \mathbf{W}$ and $S : \mathbf{W} \to \mathbf{U}$. Then $A_{ST} = A_S A_T$, which justifies the matrix multiplication as introduced in Definition 3.4.2 (3).

Complements

3.4.1 Let $A = (a_{ij})$ be a matrix of order $n \times n$ with entries from a field. Define

$$\text{Trace } A = \text{Tr } A = \sum_{i=1}^{n} a_{ii}$$

i.e., the sum of diagonal elements. The following results are easily established:

(1) $\text{Tr}(A + B) = \text{Tr } A + \text{Tr } B$
(2) $\text{Tr}\alpha A = \alpha \text{ Tr } A$, for any scalar α.
(3) $\text{Tr}(AB) = \text{Tr}(BA)$ for A of order $m \times n$ and B of order $n \times m$.
(4) Let x be an n-vector and A be $n \times n$ matrix. Then $x'Ax = \text{Tr}(Axx')$.

Let us recall that in Section 3.1, we have introduced the range $\mathbf{R}(T)$ and the nullity $\mathbf{K}(T)$ of a transformation T and defined $\dim \mathbf{R}(T) = \rho(T)$ as the rank of T and $\dim \mathbf{K}(T) = \nu(T)$ as the nullity of T, satisfying the condition $\dim \mathbf{V} = \nu(T) + \rho(T)$. The matrix A_T associated with T, for chosen complements of $\mathbf{R}(T)$ and $\mathbf{K}(T)$, is a transformation on its own right from $\mathbf{F}^m$ to $\mathbf{F}^n$ as represented in (3.4.5) with rule of multiplication given in (3.4.3). So we have $\nu(A_T)$ and $\rho(A_T)$ associated with A_T. It is easy to establish the following proposition.

P 3.4.7 Let $T : \mathbf{V} \to \mathbf{W}$ and $A_T : \mathbf{F}^m \to \mathbf{F}^n$, the matrix associated with T be as described above. Then

(1) $\rho(T) = \rho(A_T)$
(2) $\nu(T) = \nu(A_T)$
(3) $\rho(A_T) = \dim Sp(A_T) = \dim Sp(A'_T)$
(4) $\rho(A_T) + \nu(A_T) = m$
(5) $\rho(A'_T) + \nu(A'_T) = n$

where $Sp(A_T)$ is the vector space generated by the column vectors of A_T and $Sp(A'_T)$ in the vector space generated the row vectors of A_T, i.e., the column vectors of A'_T.

In the rest of the chapters, we develop the algebra of matrices as a set of elements with the operations of addition, scalar multiplication, matrix multiplication, transpose and conjugate transpose as defined in this section. The results of Chapter two on spectral theory in the context bilinear forms are proved with special reference to matrices for applications to problems in statistics and econometrics.

Complements

3.4.2 Let $A \in \mathbf{M}_n$, i.e., a square matrix of order n and $p(\alpha) = a_0 + a_1\alpha + \ldots + a_k\alpha^k$ be a polynomial in α of degree k. Define $P(A) = \alpha_0 I + a_1 A + \ldots + a_k A^k \in \mathbf{M}_n$. Show that if $p(\alpha) + q(\alpha) = h(\alpha)$ and $p(\alpha)q(\alpha) = t(\alpha)$ for scalar polynomials, then $p(A) + q(A) = h(A)$ and $p(A)q(A) = t(A)$.

3.4.3 Let $p(\lambda) = (\lambda - 1)(\lambda + 2)$ and consider the matrix equation $P(A) = 0$. Show that $A = I$ and $-2I$ are roots of $P(A) = 0$. Construct an example to show that there can be roots other than I and $-2I$.

CHAPTER 4

CHARACTERISTICS OF MATRICES

In Chapter 3, linear transformations and their associated matrices have been at the center of attraction. The associated matrix itself can be viewed as a linear transformation in its own right. To bring matters into proper perspective, let us recapitulate certain features of Chapter 3. Let **V** and **W** be two vector spaces of dimensions n and m, respectively, over the same field **F**. Let T be a linear transformation from **V** to **W**. Let A_T be the matrix of order $m \times n$ associated with the transformation T. See Section 3.4. The entries of the matrix are all scalars belonging to the underlying field **F**. Once we arrive at the matrix A_T, the flesh and blood of the transformation T, we can ignore the underlying transformation T for what goes on, and concentrate solely on the matrix A_T. The matrix A_T can now be viewed as a linear transformation from the vector space $\mathbf{F}^n$ to $\mathbf{F}^m$ by the following operational device:

$$b \rightarrow A_T b, \ b \in \mathbf{F}^n,$$

where, as usual, we identify members of the vector spaces $\mathbf{F}^m$ and $\mathbf{F}^n$ by column vectors. The subscript T attached to the associated matrix now becomes superficial and we drop it.

In this chapter, we are solely concerned with matrices of order $m \times n$ with entries belonging to a field **F** of scalars. Let us recall that a matrix A of order $m \times n$ is an arrangement of mn elements in m rows and n columns with the (i, j)-th element indicated by a_{ij}. Addition and multiplication operations are as in Definition 3.4.2. Such matrices can be regarded as linear transformations from the vector space $\mathbf{F}^n$ into the vector space $\mathbf{F}^m$. A variety of decomposition theorems for matrices will be presented along with their usefulness. In the initial part of this chapter, we will rehash certain notions introduced in the environment of linear transformations for matrices.

4.1. Rank and Nullity of a Matrix

Let A be a matrix of order $m \times n$. Unless otherwise specified, the entries of the matrices are always from a fixed field $\mathbf{F}$ of scalars. The range space $\mathbf{R}(A)$ of A is defined by,

$$\mathbf{R}(A) = \{Ax : x \in \mathbf{F}^n\}. \tag{4.1.1}$$

The set (4.1.1) or the subspace (4.1.1) has the equivalent expression

$$\mathbf{R}(A) = \{\alpha_1 a_1 + \ldots + \alpha_n a_n : \alpha_1, \ldots, \alpha_n \in \mathbf{F}\}$$

where $a_1, \ldots, a_n$ are the n column vectors of A. It is also called the span of the column vectors of A, and alternatively written as $Sp(A)$.

It is time to move on to other entities. The rank $\rho(A)$ of the matrix A is defined by,

$$\rho(A) = \dim[\mathbf{R}(A)] = \dim[\mathbf{R}(A')]. \tag{4.1.2}$$

The number $\rho(A)$ can be interpreted as the maximal number of linearly independent column vectors of A, or equivalently, as the maximal number of linearly independent row vectors of A.

Another object of interest is the kernel $\mathbf{K}(A)$ of the matrix, which is defined by,

$$\mathbf{K}(A) = \{x \in \mathbf{F}^n : Ax = 0\}, \tag{4.1.3}$$

which can be verified to be a subspace of the vector space $\mathbf{F}^n$, and the nullity $\nu(A)$ of the matrix A by,

$$\nu(A) = \dim[\mathbf{K}(A)]. \tag{4.1.4}$$

The kernel is also called by a different name: null space of A. From **P 3.1.1**,

$$\dim[\mathbf{R}(A)] + \dim[\mathbf{K}(A)] = n = \dim[\mathbf{F}^n], \tag{4.1.5}$$

or equivalently,

$$\rho(A) + \nu(A) = n. \tag{4.1.6}$$

If we look at the identity (4.1.5), there is something odd about it. The range space $\mathbf{R}(A)$ is a subspace of the vector space $\mathbf{F}^m$ whereas the null space $\mathbf{K}(A)$ is a subspace of the vector space $\mathbf{F}^n$, and the dimensions of these two subspaces add up to n! Let us rewrite the identity (4.1.5) in a different form:

$$\dim[\mathbf{R}(A')] + \dim[\mathbf{K}(A)] = n. \tag{4.1.7}$$

Now, both $\mathbf{R}(A')$ and $\mathbf{K}(A)$ are subspaces of the vector space $\mathbf{F}^n$. The identity (4.1.7) reads better. Then one is immediately led to the question whether

$$\mathbf{R}(A') \cap \mathbf{K}(A) = \{0\}? \tag{4.1.8}$$

The answer to this question depends on the make-up of the field $\mathbf{F}$. If $\mathbf{F}$ is the field of real numbers, (4.1.8) certainly holds. There are fields for which (4.1.8) is not valid. For an example, let $\mathbf{F} = \{0, 1\}$ and

$$A = \begin{bmatrix} 1 & 1 \\ 1 & 1 \end{bmatrix}.$$

Note that $\dim[\mathbf{K}(A)] = 1$ and $\dim[\mathbf{R}(A')] = 1$. Further, $\mathbf{K}(A) \cap \mathbf{R}(A')$ is a one-dimensional subspace of $\mathbf{F}^2$. Even the field $\mathbf{F} = \mathbf{C}$ of complex numbers is an oddity. We can have (4.1.8) not satisfied for a matrix A. The following is a simple example. Fix two real numbers a and b, both not zero. Let

$$A = \begin{bmatrix} a+ib & b-ia \\ a+ib & b-ia \end{bmatrix}.$$

It is clear that the subspace $\mathbf{R}(A')$ of $\mathbf{F}^2$ is one-dimensional. Let $x' = (a+ib,\ b-ia)$. One can check that $Ax = 0$. By (4.1.7), $\dim[\mathbf{K}(A)] = 1$. But $\mathbf{K}(A) \cap \mathbf{R}(A')$ is a one-dimensional subspace providing a negative response to (4.1.8). Does it puzzle you why the answer to the question in (4.1.8) is in the affirmative when $\mathbf{F} = \mathbf{R}$ but not when $\mathbf{F} = \mathbf{C}$ or $\{0, 1\}$? It is time for some introspection. Let us search for a necessary and sufficient condition on a matrix A for which $\mathbf{K}(A) \cap \mathbf{R}(A') = \{0\}$.

Let A be a matrix of order $n \times k$ and $x \in \mathbf{R}(A')$. Let $a_1, a_2, \ldots, a_n$ be the columns of A'. We can write

$$x = \alpha_1 a_1 + \alpha_2 a_2 + \ldots + \alpha_n a_n \tag{4.1.9}$$

for some scalars $\alpha_1, \alpha_2, \ldots, \alpha_n$ in $\mathbf{F}$. Suppose $x \in \mathbf{K}(A)$. This is equivalent to:

$$0 = Ax = (a_1, a_2, \ldots, a_n)'x,$$
$$a_i'x = 0 \quad \text{for} \quad i = 1, 2, \ldots, n. \tag{4.1.10}$$

Combining (4.1.9) and (4.1.10), we obtain

$$\alpha_1 a_1' a_1 + \alpha_2 a_1' a_2 + \ldots + \alpha_n a_1' a_n = a_1' x = 0,$$
$$\alpha_1 a_2' a_1 + \alpha_2 a_2' a_2 + \ldots + \alpha_n a_2' a_n = a_2' x = 0,$$
$$\ldots \qquad \ldots \qquad \ldots$$
$$\alpha_1 a_n' a_1 + \alpha_2 a_n' a_2 + \ldots + \alpha_n a_n' a_n = a_n' x = 0.$$

Let us rewrite the above equations in a succinct matrix form: $AA'\alpha = 0$, where $\alpha' = (\alpha_1, \alpha_2, \ldots, \alpha_n)$. Note that (4.1.9) can be rewritten as $A'\alpha = x$. The deliberations carried out so far make us conclude that $\mathbf{K}(A) \cap \mathbf{R}(A') = \{0\}$ if and only if $AA'\alpha = 0$ for any vector α implies that $A'\alpha = 0$. We can enshrine this result in the form of a proposition.

P 4.1.1 Let A be any matrix of order $n \times k$. Then $\mathbf{K}(A) \cap \mathbf{R}(A') = \{0\}$ if and only if

$$AA'\alpha = 0 \quad \text{for any} \quad \alpha \Rightarrow A'\alpha = 0. \tag{4.1.11}$$

The condition (4.1.11) is the one that sorts out the fields. If $\mathbf{F}$ is the field of real numbers, then (4.1.11) is always true for any matrix A. To see this, suppose $AA'\alpha = 0$ for some α. Then $\alpha' AA'\alpha = 0$. Let $y = A'\alpha$. We can rewrite $\alpha AA'\alpha = 0$ as $y'y = 0$. But $y'y$ is a sum of squares of real numbers, which implies that $y = A'\alpha = 0$. If $\mathbf{F} = \{0, 1\}$, there are matrices for which (4.1.11) is not true. Take

$$A = \begin{bmatrix} 1 & 1 \\ 1 & 1 \end{bmatrix} \quad \text{and} \quad \alpha' = (1, 0).$$

If $\mathbf{F}$ is the field of complex numbers, there are matrices A for which (4.1.11) is not true. It all boils down to the following query. Suppose $y_1, y_2, \ldots, y_n$ are complex numbers such that $y_1^2 + y_2^2 + \ldots + y_n^2 = 0$. Does this mean that each $y_i = 0$?

Complements

4.1.1 Let A and B be matrices of orders $m \times n$ and $s \times m$, respectively. Show that

$$\mathbf{R}(A) = \mathbf{K}(B)$$

if and only if

$$\mathbf{R}(B') = \mathbf{K}(A').$$

4.1.2 Show that a subset $\mathbf{S}$ of the vector space $\mathbf{F}^n$ is a subspace of $\mathbf{F}^n$ if and only if $\mathbf{S}$ is the null space of a matrix.

4.2. Rank and Product of Matrices

Let us extend the range of discussion carried above to products of matrices. Suppose A and B are two matrices such that the product AB is meaningful. Observe that every column of AB is a linear combination of the columns of A. In a similar vein, every row of AB is a linear combination of the rows of B. This simple observation leads to the fruitful inclusion relations:

$$\mathbf{R}(AB) \subset \mathbf{R}(A), \ \mathbf{R}(B'A') \subset \mathbf{R}(B'). \tag{4.2.1}$$

$$\Leftrightarrow \rho(AB) \leq \rho(A), \ \rho(AB) \leq \rho(B). \tag{4.2.2}$$

We can combine both the inequalities into a proposition.

P 4.2.1

$$\rho(AB) \leq \min\{\rho(A), \rho(B)\}. \tag{4.2.3}$$

We are led to another inquiry: when does equality hold in (4.2.3)? We have a precise answer for this question. For this, we need to take a detour on inverses of one kind or the other.

In Section 3.3, we spoke about inverse transformations of various types. We need to refurbish these notions in the environment of matrices. There are two ways to do this. One way is to view each matrix as a linear transformation from one vector space to another, work out a relevant inverse transformation and then obtain its associated matrix. This process is a little tortuous. Another way is to define inverses of various kinds for a given matrix directly. We pursue the second approach.

DEFINITION 4.2.2. Let A be a matrix of order $m \times n$.

(1) A left inverse of A is a matrix B of order $n \times m$ such that $BA = I$, where I is the identity matrix of order $n \times n$, i.e., I is a diagonal matrix in which every diagonal entry is equal to the unit element 1 of the field **F**. (If B exists, it is usually denoted by A_L^{-1}.)

(2) A right inverse of A is a matrix C of order $n \times m$ such that $AC = I$, where I is the identity matrix of order $m \times m$. (If C exists, it is usually denoted by A_R^{-1}.)

(3) A regular inverse of A of order $n \times n$ is a matrix C such that $AC = I$. If C exists, it is denoted by A^{-1}. In such a case $AA^{-1} = I = A^{-1}A$.

We know precisely the conditions under which a transformation admits a left inverse, a right inverse, or an inverse. These conditions, in the realm of matrices, have a nice interpretation. The following results are easily established.

P 4.2.3 Let A be a matrix of order $m \times n$.

(1) A admits a left inverse if and only if $\rho(A) = n$.
(2) A admits a right inverse if and only if $\rho(A) = m$.
(3) A admits a regular inverse if and only if $m = n$ and $\rho(A) = n$.

P 4.2.4 Let A and B be two matrices of orders $m \times n$ and $n \times s$, respectively. Then:

(1) $\rho(AB) = \rho(B)$ if A_L^{-1} exists, i.e., $\rho(A) = n$.
(2) $\rho(AB) = \rho(A)$ if B_R^{-1} exists, i.e., $\rho(B) = n$.

PROOF. Suppose that the left inverse of A exists. Note that $B = (A_L^{-1}A)B$. Also, by (4.2.3),

$$\rho(AB) \leq \rho(B) = \rho(A_L^{-1}AB) \leq \rho(AB),$$

from which the professed equality follows. One can establish (2) in a similar vein.

The above proposition has very useful applications in a variety of contexts. A standard scenario can be described as follows. Suppose B is any arbitrary matrix. If we pre-multiply or post-multiply B by any non-singular matrix A, the rank remains unaltered, i.e., $\rho(AB) = \rho(B)$ or $\rho(BA) = \rho(B)$, as the case may be.

Once equality in (4.2.3) holds, it sets a chain reaction, as the following proposition exemplifies.

P 4.2.5 Let A and B be two matrices of order $m \times n$ and $n \times s$, respectively.

(1) If $\rho(AB) = \rho(A)$, then $\rho(CAB) = \rho(CA)$ for any matrix C for which the multiplication makes sense.
(2) If $\rho(AB) = \rho(B)$, then $\rho(ABD) = \rho(BD)$ for any matrix D for which the multiplication makes sense.

PROOF. (1) Each column of the matrix AB is a linear combination of the columns of A. The condition $\rho(AB) = \rho(A)$ implies that the subspaces spanned by the columns of AB and of A individually are identical. What this means is that every column of A is a linear combination of the columns of AB. (Why?) Consequently, we can write $A = ABT$ for some suitable matrix T. Now, for any arbitrary matrix C for which the multiplication involved in the following makes sense,

$$\rho(CA) = \rho(CABT) \leq \rho(CAB) \leq \rho(CA),$$

from which the desired equality follows.

The result (2) follows in a similar vein.

The condition given in **P 4.2.5** for the equality of ranks in (4.2.3) is sufficient but not necessary. Let $\mathbf{F}$ be the field of real numbers. Further let

$$A = B = \begin{bmatrix} 1 & 0 & 0 \\ 0 & 1 & 0 \\ 0 & 0 & 0 \end{bmatrix}.$$

Then $\rho(AB) = \rho(B)$ but $\rho(A) \neq 3$. We are still in search of a necessary and sufficient condition for the equality in (4.2.3). Let us pinpoint the exact relationship between $\rho(AB)$ and $\rho(B)$.

P 4.2.6 Let A and B be two matrices of order $m \times n$ and $n \times s$, respectively. Then

$$(1) \qquad \rho(B) = \rho(AB) + \dim[\mathbf{K}(A) \cap \mathbf{R}(B)],$$

$$(2) \qquad \rho(A) = \rho(AB) + \dim[\mathbf{R}(A') \cap \mathbf{K}(B')].$$

PROOF. It suffices to prove the first part. The second part is a consequence of the first part. (Why?) To prove (1), we observe that

$$\{Bx \in \mathbf{F}^n : ABx = 0\} \ = \ \mathbf{K}(A) \cap \mathbf{R}(B).$$

We will now find the dimension of the subspace $\{Bx \in \mathbf{F}^n : ABx = 0\}$ of $\mathbf{F}^n$. Note that

$$\{Bx \in \mathbf{F}^n : ABx = 0\} = \{Bx \in \mathbf{F}^n : x \in \mathbf{K}(AB)\}.$$

Let $\mathbf{V} = \mathbf{K}(AB)$ and $\mathbf{W} = \mathbf{F}^n$. The matrix B can be viewed as a transformation from the vector space $\mathbf{V}$ to the vector space $\mathbf{W}$. The set $\{Bx \in \mathbf{F}^n : x \in \mathbf{K}(AB)\}$ then becomes the range of this transformation. By **P 3.1.1** (3),

$$\begin{aligned}
&\dim(\{Bx \in \mathbf{F}^n : x \in \mathbf{K}(AB)\}) + \dim(\{x \in \mathbf{K}(AB) : Bx = 0\}) \\
&\quad = \dim(\mathbf{K}(AB)), \text{ which gives} \\
&\dim(\{Bx \in \mathbf{F}^n : x \in \mathbf{K}(AB)\}) \\
&\quad = \dim(\mathbf{K}(AB)) - \dim(\{x \in \mathbf{K}(AB) : Bx = 0\}) \\
&\quad = s - \dim(\mathbf{R}(AB)) - [s - \dim(\mathbf{R}(B))] = \rho(B) - \rho(AB).
\end{aligned}$$

This completes the proof.

One can officially close the search for a necessary and sufficient condition if one is satisfied with the following result.

P 4.2.7 Let A and B be two matrices of orders $m \times n$ and $n \times s$, respectively. Then

$$\begin{aligned}
&(1) \qquad \rho(AB) = \rho(B) \quad \text{if and only if} \quad \mathbf{K}(A) \cap \mathbf{R}(B) = \{0\}; \\
&(2) \qquad \rho(AB) = \rho(A) \quad \text{if and only if} \quad \mathbf{R}(A') \cap \mathbf{K}(B') = \{0\}.
\end{aligned}$$

PROOF. (1) This is an immediate consequence of **P 4.2.5**. The second statement (2) is a consequence of (1).

We now spend some time on a certain celebrated inequality on ranks, namely, Frobenius inequality.

P 4.2.8 (**Frobenius Inequality**) Let A, B and C be three matrices such that the products AC and CB are defined. Then

$$\rho(ACB) + \rho(C) \geq \rho(AC) + \rho(CB). \tag{4.2.4}$$

PROOF. Consider the product of partitioned matrices,

$$\begin{bmatrix} I & -A \\ 0 & I \end{bmatrix} \begin{bmatrix} 0 & AC \\ CB & C \end{bmatrix} \begin{bmatrix} I & 0 \\ -B & I \end{bmatrix} = \begin{bmatrix} -ACB & 0 \\ 0 & C \end{bmatrix}.$$

Since the matrices

$$\begin{bmatrix} I & -A \\ 0 & I \end{bmatrix} \quad \text{and} \quad \begin{bmatrix} I & 0 \\ -B & I \end{bmatrix}$$

are non-singular, it follows that

$$\rho\begin{bmatrix} 0 & AC \\ CB & C \end{bmatrix} = \rho\begin{bmatrix} -ACB & 0 \\ 0 & C \end{bmatrix} \quad = \quad \rho(ACB) + \rho(C).$$

But

$$\begin{aligned} \rho\begin{bmatrix} 0 & AC \\ CB & C \end{bmatrix} &\leq \rho\begin{bmatrix} 0 \\ CB \end{bmatrix} + \rho\begin{bmatrix} AC \\ C \end{bmatrix} \\ &= \quad \rho(CB) + \rho(AC), \quad \text{(why?)} \end{aligned}$$

from which the desired inequality follows.

The following inequality is a special case of Frobenius inequality.

P 4.2.9 (Sylvester's Inequality) Let A and B be two matrices of orders $m \times n$ and $n \times s$, respectively. Then

$$\rho(AB) \geq \rho(A) + \rho(B) - n,$$

with equality if and only if $\mathbf{K}(A) \subset \mathbf{R}(B)$.

PROOF. In the Frobenius inequality, take $C = I_n$. The inequality follows. From **P 4.2.6** (1), $\rho(B) = \rho(AB) + \dim[\mathbf{K}(A) \cap \mathbf{R}(B)]$. If equality holds in the Sylvester's inequality, then we have $\dim[\mathbf{K}(A) \cap \mathbf{R}(B)] = n - \rho(A) = n - \dim[\mathbf{R}(A)] = \dim[\mathbf{K}(A)]$. See (4.1.5). Consequently, equality holds in Sylvester's inequality if and only if $\mathbf{K}(A) \subset \mathbf{R}(B)$.

Complements

4.2.1 Reprove **P 4.2.3** using results of Sections 3.3 and 3.4.

4.2.2 Let A and B be square matrices of the same order. Is $\rho(AB) = \rho(BA)$?

4.2.3 Let A and B be two matrices of orders $m \times n$ and $m \times s$, respectively. Let $(A|B)$ be the augmented matrix of order $m \times (n + s)$. Show that

$$\rho(A|B) = \rho(A)$$

if and only if $B = AC$ for some matrix C.

4.3. Rank Factorization and Further Results

In a variety of applications in engineering and statistics, we need to express a matrix as a product of simple and elegant matrices. In this section, we consider one of the simplest such factorizations of matrices.

P 4.3.1 (Rank Factorization Theorem) Let A be a matrix of order $m \times n$ with rank a. Then A can be factorized as

$$A = RF, \tag{4.3.1}$$

where R is of order $m \times a$, F is of order $a \times n$, $Sp(A) = Sp(R)$, $Sp(A') = Sp(F')$, and $\rho(R) = \rho(F) = a$. Alternatively,

$$A = SDG, \tag{4.3.2}$$

where S and G are non-singular matrices of order $m \times m$ and $n \times n$, respectively, and D is a block matrix of the following structure:

$$D = \begin{bmatrix} \underset{a \times a}{I_a} & \underset{a \times (n-a)}{0} \\ \underset{(m-a) \times a}{0} & \underset{(m-a) \times (n-a)}{0} \end{bmatrix}.$$

PROOF. The main goal in the above factorization is to write the matrix A of rank a as a product of two matrices one of them has a full column rank and the other full row rank. This is like removing the chaff from the grain so as to get to the kernel. A proof can be built by imitating the winnowing process. Take any basis of the vector space $Sp(A)$. The column vectors of the basis are taken as the column vectors of the matrix R of order $m \times a$. Every column vector of A is now a linear combination of the columns of R. Consequently, we can write $A = RF$ for some matrix F. Once R is chosen, there is only one matrix F which fulfills the above rank factorization. (Why?) It is obvious that $\rho(R) = a$. It is also true that $\rho(F) = a$. This can be seen as follows. Note that

$$a = \rho(A) = \rho(RF) \leq \rho(F) \leq \min\{a, n\} \leq a.$$

The fact that $Sp(A') = Sp(F')$ follows from the facts that every row vector of A is a linear combination of the rows of F and $\rho(F) = a$.

To obtain the form (4.3.2), one simply needs to augment matrices in (4.3.1) to get non-singular matrices. For example, let $S = (R : R_0)$, where the columns of the matrix R_0 is any basis of a complementary subspace of $Sp(R)$ in $\mathbf{F}^m$. The matrix G is obtained by adjoining F with $(n - a)$ rows so that G becomes non-singular. Take any basis of the complementary subspace of the vector space spanned by the rows of F. The basis vectors are adjoined to F to form the matrix G. A direct multiplication of (4.3.2) gives (4.3.1), which justifies the validity of (4.3.2).

That is all there is in rank factorization. Such a simple factorization is useful to answer some pertinent questions on ranks of matrices. One of the perennial questions is to know under what condition does the equality $\rho(A + B) = \rho(A) + \rho(B)$ holds for two given matrices A and B. The following result goes some way to answer this question.

P 4.3.2 Let A and B be two matrices of the same order with ranks a and b, respectively. Then the following statements are equivalent.

(1) $$\rho(A + B) = \rho(A) + \rho(B). \tag{4.3.3}$$

(2) $$\rho(A|B) = \rho \begin{bmatrix} A \\ B \end{bmatrix} = \rho(A) + \rho(B). \tag{4.3.4}$$

(3) $$Sp(A) \cap Sp(B) = \{0\} \quad \text{and} \quad Sp(A') \cap Sp(B') = \{0\}. \tag{4.3.5}$$

(4) The matrices A and B can be factorized in the following style:

$$A = S \begin{bmatrix} I_a & 0 & 0 \\ 0 & 0 & 0 \\ 0 & 0 & 0 \end{bmatrix} G; \quad B = S \begin{bmatrix} 0 & 0 & 0 \\ 0 & I_b & 0 \\ 0 & 0 & 0 \end{bmatrix} G, \tag{4.3.6}$$

where the zeros are matrices of appropriate order, and $|S| \neq 0$, $|G| \neq 0$.

PROOF. We start with rank factorization of A and B. Write $A = R_1 F_1$ and $B = R_2 F_2$ with $Sp(A) = Sp(R_1), Sp(A') = Sp(F_1'), Sp(B) = Sp(R_2)$ and $Sp(B') = Sp(F_2')$. It is obvious that

$$Sp(A|B) = Sp(R_1|R_2), \; Sp(A'|B') = Sp(F_1'|F_2'),$$

from which it follows that

$$\rho(A|B) = \rho(R_1|R_2), \; \rho\begin{pmatrix} A \\ B \end{pmatrix} = \rho\begin{pmatrix} F_1 \\ F_2 \end{pmatrix}.$$

We are now ready to prove the equivalence of the statements. Let us assume (1) is true. Then

$$\begin{aligned}\rho(A)+\rho(B) &= \rho(A+B) = \rho(R_1F_1+R_2F_2) = \rho((R_1|R_2)\begin{pmatrix}F_1\\F_2\end{pmatrix})\\ &\leq \rho(R_1|R_2) = \rho(A|B) \leq \rho(R_1)+\rho(R_2) = \rho(A)+\rho(B),\end{aligned}$$

from which $\rho(A|B)=\rho(A)+\rho(B)$ follows. The equality $\rho\binom{A}{B}=\rho(A)+\rho(B)$ follows in a similar vein. Thus (1) $\Rightarrow$ (2).

Since $\rho(A|B) = \dim[Sp(A|B)]$ = the maximal number of linearly independent columns of the augmented matrix $(A|B) = \rho(A)+\rho(B) = \dim[Sp(A)]+\dim[Sp(B)]$ = the maximal number of linearly independent columns of $A+$ maximal number of linearly independent columns of B, it follows that $Sp(A)\cap Sp(B)=\{0\}$. It also follows in a similar vein that $Sp(A')\cap Sp(B')=\{0\}$. Thus (3) follows from (2).

Let us augment the matrix $(R_1|R_2)$ of order $m\times(a+b)$ to a non-singular matrix of order $m\times m$. Since $Sp(A)=Sp(R_1)$, $Sp(B)=Sp(R_2)$, $Sp(R_1)\cap Sp(R_2)=\{0\}$ (by hypothesis), $\rho(R_1)=a$, and $\rho(R_2)=b$, all the columns of the matrix $(R_1|R_2)$ are linearly independent. Consequently, we can find a matrix R_0 of order $m\times(m-(a+b))$ such that the augmented matrix $S=(R_1|R_2|R_0)$ is nonsingular. Following the same line of reasoning, we can find a matrix F_0 of order $(n-(a+b))\times n$ such that

$$G = \begin{bmatrix}F_1\\F_2\\F_0\end{bmatrix}$$

is non-singular. The professed equalities of expressions in (4) follow now routinely. Finally, (4) $\Rightarrow$ (1) obviously.

The equality $\rho(A+B)=\rho(A)+\rho(B)$ is very hard to fulfill. What the above result indicates is that if no column vector of A is linearly dependent on the columns of B, no column vector of B is linearly dependent on the columns of A, no row vector of A is linearly dependent on the rows of B, and no row vector of B is linearly dependent on the rows of A, only then the purported equality in (1) will hold.

P 4.3.2 can be generalized. Looking at the proof a little critically, it seems that what is good for two matrices should be good for any finite number of matrices.

P 4.3.3 Let $A_1, A_2, \ldots, A_k$ be any finite number of matrices each of order $m\times n$ with ranks $a_1, a_2, \ldots, a_k$, respectively. Then the following statements are equivalent.

(1) $\rho(A_1 + A_2 + \ldots + A_k) = \rho(A_1) + \rho(A_2) + \ldots + \rho(A_k)$.

(2) $\rho(A_1|A_2|\ldots|A_k) = \rho(A_1'|A_2'|\ldots|A_k') = \rho(A_1) + \ldots + \rho(A_k)$.

(3) $Sp(A_i) \cap Sp(A_j) = \{0\}$ and $Sp(A_i') \cap Sp(A_j') = \{0\}$ for all $i \neq j$.

(4) There exist nonsingular matrices S and G of orders $m \times m$ and $n \times n$, respectively, such that

$$A_i = SD_iG, \ i = 1, 2, \ldots, k,$$

where D_i is a block matrix of order $m \times n$ with the following structure.

Row Block No. Column Block No.

$$\begin{array}{c} \\ 1 \\ 2 \\ \cdot \\ \cdot \\ i \\ \cdot \\ \cdot \\ k+1 \end{array} \quad \begin{array}{c} \begin{array}{ccccccc} 1 & 2 & \ldots & i & \ldots & (k+1) \end{array} \\ \begin{bmatrix} 0 & 0 & \ldots & 0 & \ldots & 0 \\ 0 & 0 & \ldots & 0 & \ldots & 0 \\ \cdot & \cdot & \ldots & \cdot & \ldots & \cdot \\ \cdot & \cdot & \ldots & \cdot & \ldots & \cdot \\ 0 & 0 & \ldots & I_{a_i} & \ldots & 0 \\ \cdot & \cdot & \ldots & \cdot & \ldots & \cdot \\ \cdot & \cdot & \ldots & \cdot & \ldots & \cdot \\ 0 & 0 & \ldots & 0 & \ldots & 0 \end{bmatrix} \end{array}$$

The zeros appearing in the above are zero matrices of appropriate order.

We now consider the problem of computing the rank of a matrix utilizing the notion of a Schur complement. Let A be a matrix of order $(n+s) \times (n+t)$ partitioned in the following style.

$$A = \begin{bmatrix} \underset{n\times n}{E} & \underset{n\times t}{F} \\ \underset{s\times n}{G} & \underset{s\times t}{H} \end{bmatrix}. \tag{4.3.7}$$

Assume that E is nonsingular. Define

$$A/E = H - GE^{-1}F.$$

The matrix A/E of order $s \times t$ is called the Schur complement of A with respect to the non-singular submatrix E of A. The Schur complement occurs in a wide variety of contexts in mathematics and statistics. Suppose that A is a square matrix, i.e., $s = t$. The determinant $|A|$ of A can be expressed in terms of the determinant of any Schur complement of A. More precisely,

$$|A| = (|E|)\,(|A/E|).$$

We are jumping ahead a little talking about determinants. A detailed discussion of determinants begins in Section 4.4 (see Complement 4.4.7 in this connection).

P 4.3.4 Let A be a matrix of order $(n+s) \times (n+t)$ partitioned in the style of (4.3.7) with E non-singular. Then

$$\rho(A) = \rho(E) + \rho(A/E).$$

PROOF. Consider the product.

$$\begin{bmatrix} I_n & 0 \\ -GE^{-1} & I_s \end{bmatrix} \begin{bmatrix} E & F \\ G & H \end{bmatrix} \begin{bmatrix} I_n & -E^{-1}F \\ 0 & I_t \end{bmatrix} = \begin{bmatrix} E & 0 \\ 0 & A/E \end{bmatrix}, \qquad (4.3.8)$$

where the zeros are zero matrices of appropriate order. The matrix that appears on the left of A is of order $(n+s) \times (n+s)$ and non-singular. The matrix that appears next to A is of order $(n+t) \times (n+t)$ and non-singular. Consequently,

$$\rho(A) = \rho \begin{bmatrix} E & 0 \\ 0 & A/E \end{bmatrix} = \rho(E) + \rho(A/E). \quad \text{Why?}$$

The above result has a practical significance. The computation of the rank of a matrix is painfully and computationally laborious. The above result breaks up the computation into manageable pieces of computation. As an example, consider the following matrix.

$$A = \begin{bmatrix} 1 & 1 & 0 & 1 & 0 & 0 \\ 1 & 1 & 0 & 0 & 1 & 0 \\ 1 & 1 & 0 & 0 & 0 & 1 \\ 1 & 0 & 1 & 1 & 0 & 0 \\ 1 & 0 & 1 & 0 & 1 & 0 \\ 1 & 0 & 1 & 0 & 0 & 1 \end{bmatrix}.$$

Partition the matrix A in the form

$$A = \begin{bmatrix} \underset{3\times 3}{E} & \underset{3\times 3}{F} \\ G & H \\ \underset{3\times 3}{G} & \underset{3\times 3}{H} \end{bmatrix}.$$

The matrix H is the identity matrix of order 3 whose inverse is itself. Therefore,

$$\rho(A) = \rho(H) + \rho(A/H),$$

where

$$A/H = E - FH^{-1}G = E - G = \begin{bmatrix} 0 & 1 & -1 \\ 0 & 1 & -1 \\ 0 & 1 & -1 \end{bmatrix}.$$

It is clear that $\rho(A/H) = 1$. Consequently, $\rho(A) = 3 + 1 = 4$.

In a variety of engineering applications, one comes across large dimensional matrices as big as the order of $5,000 \times 5,000$. An algorithm can be developed to determine the rank of such matrices based on the result of **P 4.3.4**.

Complements

4.3.1 For any two matrices A and B of the same order, show that

$$\rho(A) + \rho(B) - c - d \;\le\; \rho(A + B) \;\le\; \rho(A) + \rho(B) - \max\{c, d\},$$

where $c = \dim[Sp(A) \cap Sp(B)]$ and $d = \dim[Sp(A') \cap Sp(B')]$. Show that these inequalities are the best possible.

See Marsaglia (1967) and Marsaglia and Styan (1974).

4.3.2 Let A be a matrix of order $m \times n$. Show that $\rho(A) = 1$ if and only if $A = xy'$ for some non-zero vectors x and y of orders $m \times 1$ and $n \times 1$, respectively.

4.3.3 Obtain the rank factorization of the matrix

$$A = \begin{bmatrix} -1 & 2 & 4 \\ 2 & -1 & 2 \\ 0 & 3 & 10 \end{bmatrix}.$$

4.3.4 Let A be a matrix of order $m \times m$ and of rank $m - 1$. Show that A can be made non-singular by changing just one element of A.

4.3.5 Let A be a non-singular matrix of order $m \times m$. Show that A can be made singular by changing just one element of A. What can you say about the rank of the resultant matrix?

4.4. Determinants

We came across determinants in earlier chapters. We now introduce the notion of a determinant from an abstract angle. Let $\mathbf{M}_n$ denote the collection of all matrices of order $n \times n$ with entries coming from a fixed field $\mathbf{F}$ of scalars. The integer $n \geq 1$ is fixed. A matrix $A \in \mathbf{M}_n$ can be identified in three equivalent ways.

(1) $A = (\alpha_{ij})$
The (i, j)-th entry of A, i.e., the entry located at the junction of the i-th row and j-th column of A is α_{ij}.
(2) $A = (a_1, a_2, \ldots, a_n)$, where a_i is the i-th column vector of A.

Let I_n denote the identity matrix of order $n \times n$ in which every diagonal entry is 1, the unit element of the field $\mathbf{F}$, and every off-diagonal entry is zero. In this section, every column vector is taken to be of order $n \times 1$.

DEFINITION 4.4.1. The determinant of a matrix $A \in \mathbf{M}_n$, denoted by det A or $|A|$, is a map from $\mathbf{M}_n$ into the field $\mathbf{F}$ having the following properties.

(1) Multilinearity

For any $1 \leq i \leq n$ and $(n+1)$ column vectors $a_1, a_2, \ldots, a_{i-1}, a_{i+1}, \ldots, a_n, b_1$, and b_2, and scalars α and β,

$$\begin{aligned}&\det(a_1, a_2, \ldots, a_{i-1}, \alpha b_1 + \beta b_2, a_{i+1}, \ldots, a_n)\\&= \alpha \det(a_1, a_2, \ldots, a_{i-1}, b_1, a_{i+1}, \ldots, a_n)\\&+\beta \det(a_1, a_2, \ldots, a_{i-1}, b_2, a_{i+1}, \ldots, a_n).\end{aligned}$$

If $i = 1$, the above equality is taken to be

$$\begin{aligned}&\det(\alpha b_1 + \beta b_2, a_2 \ldots, a_n)\\&\quad = \alpha \det(b_1, a_2, \ldots, a_n) + \beta \det(b_2, a_2, \ldots, a_n).\end{aligned}$$

If $i = n$, the above equality is taken to be

$$\begin{aligned}&\det(a_1, a_2, \ldots, a_{n-1}, \alpha b_1 + \beta b_2)\\&\quad = \alpha \det(a_1, a_2, \ldots, a_{n-1}, b_1) + \beta \det(a_1, a_2, \ldots, a_{n-1}, b_2).\end{aligned}$$

(2) Alternating

For any n column vectors $a_1, a_2, \ldots, a_n$, $\det(a_1, a_2, \ldots, a_n) = 0$ if any two columns are identical.

(3) $\det(I_n) = 1$.

The common notation that is used for the determinant of a matrix A is $|A|$. We will use this notation as well as "det" depending upon which one makes a convenient reading. The following properties of the determinant map follow solely from the definition of the map.

P 4.4.2 Let $A = (a_1, a_2, \ldots, a_n)$ be any matrix.

(1) If a_1 is dependent on $a_2, a_3, \ldots, a_n$, i.e., a_1 is a linear combination of $a_2, a_3, \ldots, a_n$, then $|A| = 0$.
(2) Let B be the matrix obtained from A by interchanging two columns of A. Then $|B| = -|A|$.
(3) Let $A = (\alpha_{ij})$. Let Π be the collection of all permutations of $\{1, 2, \ldots, n\}$. Then

$$|A| = \sum_{\pi \in \Pi} \operatorname{sign}(\pi)\, \alpha_{1\pi(1)}\alpha_{2\pi(2)} \cdots \alpha_{n\pi(n)},$$

where

$$\operatorname{sign}(\pi) = \begin{cases} +1 & \text{if } \pi \text{ is an even permutation,} \\ -1 & \text{if } \pi \text{ is an odd premutation.} \end{cases}$$

(4) Let A be any matrix. Then $|A| = |A'|$. (Recall that A' is the transpose of the matrix A.)
(5) Let A and B be any two matrices. Then $|AB| = |A|\,|B|$.

The computation of the determinant of a matrix is important in many applications. The formula given in (3) of **P 4.4.2** is very labor-intensive. A simple recurrence formula should come handy. Some recurrence formulas will be presented in Section 4.5. We will now present a simple formula of expanding the determinant of a matrix in terms of lower order determinants. Let $A = (\alpha_{ij})$ be a matrix of order $n \times n$. For each $1 \leq i, j \leq n$, let A_{ij} be the submatrix of A obtained from A by deleting its i-th row and j-th column. The determinant of A_{ij}, $|A_{ij}|$ is called the

(i, j)-th minor of A. The (i, j)-th co-factor C_{ij} is defined by the formula: $C_{ij} = (-1)^{i+j}|A_{ij}|$. The following results are easily established.

P 4.4.3 (1) Let A be a diagonal matrix of order $n \times n$, i.e., every off-diagonal entry is zero. Then the determinant of A is the product of its diagonal entries.

(2) Let A be an upper triangular matrix of order $n \times n$, i.e., every entry below the main diagonal of A is zero. Then the determinant of A is the product of its diagonal entries.

(3) Let A be a lower triangular matrix of order $n \times n$, i.e., every entry above the main diagonal of A is zero. Then the determinant of A is the product of its diagonal entries.

(4) Let $A = (\alpha_{ij})$ be any matrix of order $n \times n$. Then

$$(a) \quad |A| = \alpha_{r1}C_{r1} + \alpha_{r2}C_{r2} + \ldots + \alpha_{rn}C_{rn}, \quad \text{for every } 1 \leq r \leq n$$
(expansion of the determinant by the r-th row of A),

$$(b) \quad |A| = \alpha_{1i}C_{1i} + \alpha_{2i}C_{2i} + \ldots + \alpha_{ni}C_{ni}, \quad \text{for every } 1 \leq i \leq n$$
(expansion of the determinant by the i-th column of A).

(5) Let $A = (\alpha_{ij})$ be any matrix of order $n \times n$. Then for any $r \neq s$,

$$\alpha_{r1}C_{1s} + \alpha_{r2}C_{2s} + \ldots + \alpha_{rn}C_{ns} = 0.$$

(6) Let A be a square matrix such that $|A| \neq 0$. Let $B = |A|^{-1}(C_{ij})'$, i.e., the (i, j)-th entry of B is $C_{ji}/|A|$. Then $AB = BA = I_n$.

(7) A square matrix A is invertible or non-singular if and only if $|A| \neq 0$.

(8) Let A be a square matrix and α a scalar. Then $|\alpha A| = \alpha^n|A|$.

(9) Let A be any square matrix. Then $|A^2| = (|A|)^2$.

Finally, we close this section with a couple of definitions. A square matrix A is said to be symmetric if $A' = A$, i.e., the (i, j)-th and (j, i)-th entries of A are identical for every i and j. The adjugate of a square matrix A is defined to be the matrix $(C_{ij})'$, where we recall that C_{ij} is the (i,j)-th co-factor of A. The adjugate of A is denoted by adj(A).

Complements

4.4.1 A matrix $A = (a_{ij})$ of order $n \times n$ is said to be skew-symmetric if $a_{ii} = 0$ for all i and $a_{ij} = -a_{ji}$ for all i and j. Show that:

$$(1)\quad |A| = \begin{cases} 0, & \text{if } n \text{ is odd,} \\ \text{a perfect square,} & \text{if } n \text{ is even} \end{cases}.$$

(2) $\rho(A)$ is always an even number.

(3) Every square matrix can be written as a sum of a symmetric matrix and a skew-symmetric matrix.

4.4.2 Let $\alpha_1, \alpha_2, \ldots, \alpha_n$ be n scalars. The Vandermond determinant based on the given scalars is defined to be the determinant of the matrix

$$A = \begin{bmatrix} 1 & 1 & \ldots & 1 \\ \alpha_1 & \alpha_2 & \ldots & \alpha_n \\ \alpha_1^2 & \alpha_2^2 & \ldots & \alpha_n^2 \\ \ldots & \ldots & \ldots & \ldots \\ \alpha_1^{n-1} & \alpha_2^{n-1} & \ldots & \alpha_n^{n-1} \end{bmatrix}.$$

Show that $|A| = \Pi_{i>j}(\alpha_i - \alpha_j)$.

4.4.3 Let A be a matrix of order $n \times n$ and x a column vector of order $n \times 1$. Let α be a scalar. Show that:

$$(1)\quad \begin{vmatrix} A & x \\ x' & \alpha \end{vmatrix} = \alpha|A| - x'(\text{adj}(A))x,$$

$$(2)\quad \begin{vmatrix} A & x \\ x' & -1 \end{vmatrix} = -|A + xx'| = -|A| - x'(\text{adj}(A))x,$$

$$= -|A|(1 + x'A^{-1}x) \quad \text{if } |A| \neq 0.$$

4.4.4 Let A be a matrix of order $n \times n$, U a matrix of order $n \times n$ every entry of which is equal to 1, V a column vector of order $n \times 1$ in which every entry is equal to 1, and α a scalar. Show that:

(1) $|A+\alpha U| = |A| + \alpha V'(\text{adj}(A))V$;
(2) $V'\text{adj}(A+\alpha U) = V'(\text{adj}(A))$;
(3) $(\text{adj}(A+\alpha U))V = (\text{adj}(A))V$.

4.4.5 Let A and B be two matrices of the same order $n \times n$. Determine a necessary and sufficient condition in order that $|A+B| = |A| + |B|$.
4.4.6 Let A be a matrix of order $n \times n$. Show that $\text{adj}(\text{adj}(A)) = |A|^{n-2}A$.
4.4.7 Let A be a matrix of order $(n+s)\times(n+s)$ partitioned as follows.

$$A = \begin{bmatrix} \underset{n\times n}{E} & \underset{n\times s}{F} \\ \underset{s\times n}{G} & \underset{s\times s}{H} \end{bmatrix}.$$

Assume that E is non-singular. Show that (using **P 4.3.4**)

$$|A| = (|E|)(|H - GE^{-1}F|).$$

4.5. Determinants and Minors

Another important problem is the computation of the determinant of a square matrix. If the matrix is big, the computation is tedious. One needs to find a simple way of computing determinants. We present one such method. For this we need to introduce a cogent notation for minors of a matrix.

Let A be a square matrix of order $m \times m$. Let $1 \le p \le m$ and $1 \le i_1 < i_2 < \ldots < i_p \le m$ and $1 \le j_1 < j_2 < \ldots < j_p \le m$. Let $\alpha = (i_1, i_2, \ldots, i_p)$ and $\beta = (j_1, j_2, \ldots, j_p)$. Consider the submatrix of A obtained by retaining only the i_1-th, i_2-th, $\ldots$, i_p-th rows and j_1-th, j_2-th, $\ldots$, j_p-th columns of A and deleting the rest from A. Suppose

$$A = \begin{bmatrix} 1 & 2 & 3 & 4 \\ -3 & 4 & 6 & -5 \\ 2 & 2 & 1 & -1 \\ 1 & -1 & 2 & 2 \end{bmatrix},$$

$\alpha = (1,3)$ and $\beta = (2,4)$. The submatrix that corresponds to the choice of α and β is given by

$$\begin{bmatrix} 2 & 4 \\ 2 & -1 \end{bmatrix}.$$

Let us get back to the generalities. For a given choice of α and β described above, let $A_{\alpha\beta}$ be the submatrix of A associated with α and β. The determinant $|A_{\alpha\beta}|$ of matrix $A_{\alpha\beta}$ is called a minor of order p. (Note that the minor A_{ij} introduced in Section 4.4 is the same as the minor $A_{\alpha\beta}$ of order $m-1$ with $\alpha = (1, 2, \ldots, i-1, i+1, \ldots, m)$ and $\beta = (1, 2, \ldots, j-1, j+1, \ldots, m)$.)

How many such minors can one work out? If $p = m$, there is only one minor of order m which is the determinant of A. If $p = 1$, there are m^2 minors of order 1. The basic computation of the number reduces to how many α's and β's one can find. There are $\binom{m}{p}$ each of such α's and β's. Let $M = \binom{m}{p}$, which is the number of combinations of selecting p objects out of m. Thus there are M^2 minors of order p. We would like to build a matrix based on these M^2 minors. Before we do this, we need to arrange the p-tuples α's and β's individually in some order. But it is customary to arrange α's as well as β's in lexicographic order. Let

$$\mathbf{A}_p = \{\alpha = (i_1, i_2, \ldots, i_p) : 1 \leq i_1 < i_2 < \ldots < i_p \leq m\}.$$

For distinct $\alpha = (i_1, i_2, \ldots, i_p)$ and $\alpha^* = (i_1^*, i_2^*, \ldots, i_p^*)$ in $\mathbf{A}_p$, say that $\alpha < \alpha^*$ lexicographically if $i_1 < i_1^*$, or $i_1 = i_1^*$ but $i_2 < i_2^*$, or $i_1 = i_1^*, i_2 = i_2^*$ but $i_3 < i_3^*, \ldots$, or $i_1 = i_1^*, i_2 = i_2^*, \ldots, i_{p-1} = i_{p-1}^*$ but $i_p < i_p^*$. The lexicographic order is also called dictionary order and it is indeed a linear order, i.e., in addition to being a partial order, any two members of $\mathbf{A}_p$ are comparable. For example, suppose $m = 4$ and $p = 2$. The cardinality of the set $\mathbf{A}_2$ is six and its members are laid out in the dictionary order as follows.

$$(1,2) < (1,3) < (1,4) < (2,3) < (2,4) < (3,4).$$

Now we are ready to define what is called a compound matrix of A.

DEFINITION 4.5.1. Let A be a matrix of order $m \times m$ and $1 \leq p \leq m$. The compound matrix of A of order p is a matrix $A_{[p]}$ of order $M \times M$ whose (α, β)-th entry is given by $|A_{\alpha\beta}|$, $\alpha, \beta \in \mathbf{A}_p$, where $M = \binom{m}{p}$.

The order in which α's and β's are written down is the lexicographical order enunciated above. We look at a numerical illustration. Suppose $m = 4$ and $p = 2$. Let

$$A = \begin{bmatrix} 1 & 2 & 3 & 4 \\ -3 & 4 & 6 & -5 \\ 2 & 2 & 1 & -1 \\ 1 & -1 & 2 & 2 \end{bmatrix}.$$

The compound matrix of order 2 is given by:

$$A_{[2]} = \begin{bmatrix} 10 & 15 & 7 & 0 & -26 & -39 \\ -2 & -5 & -9 & -4 & -10 & -7 \\ -3 & -1 & -2 & 7 & 8 & -2 \\ -14 & -15 & -7 & -8 & 6 & -1 \\ -1 & -12 & -1 & 14 & 3 & 22 \\ -4 & 3 & 5 & 5 & 3 & 4 \end{bmatrix}.$$

Let us go back to the general case. Note that in the extreme cases, $A_{[1]} = A$ and $A_{[m]} = (|A|)$. There are quite a number of properties enjoyed by the operation of compounding. We relegate these properties to the Complements Section.

We need to broaden the definition of a minor in two directions. First, we need not confine our attention to square matrices alone. Let A be any matrix of order $m \times n$. A p-th order minor of A makes sense with $1 \leq p \leq \min\{m, n\}$. We retain some p rows, p columns of A, discard the rest and then form the determinant. Even if p exceeds $\min\{m, n\}$, conventionally, we can define the p-th order minor of A to be equal to zero. Secondly, the definition of $\mathbf{A}_p$ seems to be too restrictive. It is not necessary to have the components of any vector α in $\mathbf{A}_p$ to be distinct and strictly increasing. Let us enlarge the set $\mathbf{A}_p$ to

$$\mathbf{A}_p^* = \{\alpha = (i_1, i_2, \ldots, i_p) : i_1, i_2, \ldots, i_p \in \{1, 2, \ldots, m\}\}.$$

For $\alpha = (i_1, i_2, \ldots, i_p)$ and $\beta = (j_1, j_2, \ldots, j_p)$ in A_p^*, define the submatrix $A_{\alpha\beta}$ of A to be

$$A_{\alpha\beta} = \begin{bmatrix} a_{i_1 j_1} & a_{i_1 j_2} & \cdots & a_{i_p j_p} \\ a_{i_2 j_1} & a_{i_2 j_2} & \cdots & a_{i_p j_p} \\ \cdots & \cdots & \cdots & \cdots \\ a_{i_p j_1} & a_{i_p j_2} & \cdots & a_{i_p j_p} \end{bmatrix}.$$

The determinant $|A_{\alpha\beta}|$ of the matrix $A_{\alpha\beta}$ is called a minor of order p. If two components of α are identical or two components of β are identical, it is clear that $|A_{\alpha\beta}| = 0$. Further, if α and α^* belong to $\mathbf{A}_p^*$, and α is a permutation of α^*, then the minors $|A_{\alpha\beta}|$ and $|A_{\alpha^*\beta}|$ differ only in sign. There are several advantages in extending the definition of a minor as we will see in some of the proofs we present below.

We now spend some time on the Cauchy-Binet formula. Let A, B and C be three matrices of order $m \times m$ such that $A = BC$. We have already seen that the determinant of A has a simple relationship with the determinants of B and C. More precisely, we have $|A| = (|B|)(|C|)$. In the Cauchy-Binet formula this simple relationship is extended to cover minors of A. Recall the definition of $\mathbf{A}_p = \{\alpha = (i_1, i_2, \ldots, i_p) : 1 \le i_1 < i_2 < \ldots < i_p \le m\}$, where $1 \le p \le m$.

P 4.5.2 (Cauchy-Binet Formula). Let $A = (a_{ij}), B = (b_{ij})$ and $C = (c_{ij})$ be three matrices each of order $m \times m$ such that $A = BC$. Let $1 \le p \le m$. Let $\alpha = (i_1, i_2, \ldots, i_p)$ and $\beta = (j_1, j_2, \ldots, j_p) \in \mathbf{A}_p$. Then the p-th order minor $|A_{\alpha\beta}|$ is given by

$$|A_{\alpha\beta}| = \sum_{\gamma \in \mathbf{A}_p} |B_{\alpha\gamma}| \quad |C_{\gamma\beta}|.$$

We omit the proof which follows from definitions.

Let A be a matrix of order $m \times m$. The compound matrix $A_{[m-1]}$ of order $m \times m$ is of special interest. This is related to the adjugate matrix adj(A) of A. See Section 4.4. The (i, j)-th entry a^{ij} of adj(A) is given by,

$$a^{ij} = (-1)^{i+j} |A_{(1,2,\ldots,j-1,j+1,\ldots,m),(1,2,\ldots,i-1,i+1,\ldots,m)}|.$$

The members of the set $\mathbf{A}_{(m-1)}$ can be arranged in the lexicographic order as follows:

$$\alpha_1 < \alpha_2 < \ldots < \alpha_m,$$

where $\alpha_1 = (1, 2, \ldots, m-1)$; $\alpha_2 = (1, 2, \ldots, m-2, m)$; $\alpha_3 = (1, 2, \ldots, m-3, m-1, m)$; $\ldots$; $\alpha_{m-2} = (1, 2, 4, \ldots, m)$; $\alpha_{m-1} = (1, 3, 4, \ldots, m)$; and $\alpha_m = (2, 3, \ldots, m)$.

The (i, j)-th entry of $A_{[m-1]}$ is given by $|A_{\alpha_i \alpha_j}|$, which is equal to $(-1)^{i+j} a^{m+1-j, m+1-i}$. More transparently, the $(1, 1)$-th entry of adj(A) is the (m, m)-th entry of $A_{[m-1]}$. Let us look at a simple example. Let

$$A = \begin{bmatrix} a_{11} & a_{12} & a_{13} \\ a_{21} & a_{22} & a_{23} \\ a_{31} & a_{32} & a_{33} \end{bmatrix}.$$

Then adj (A)

$$= \begin{bmatrix} \det\begin{bmatrix} a_{22} & a_{23} \\ a_{32} & a_{33} \end{bmatrix} & -\det\begin{bmatrix} a_{12} & a_{13} \\ a_{32} & a_{33} \end{bmatrix} & \det\begin{bmatrix} a_{12} & a_{13} \\ a_{22} & a_{23} \end{bmatrix} \\ -\det\begin{bmatrix} a_{21} & a_{23} \\ a_{31} & a_{33} \end{bmatrix} & \det\begin{bmatrix} a_{11} & a_{13} \\ a_{31} & a_{33} \end{bmatrix} & -\det\begin{bmatrix} a_{11} & a_{13} \\ a_{21} & a_{23} \end{bmatrix} \\ \det\begin{bmatrix} a_{21} & a_{22} \\ a_{31} & a_{32} \end{bmatrix} & -\det\begin{bmatrix} a_{11} & a_{12} \\ a_{31} & a_{32} \end{bmatrix} & \det\begin{bmatrix} a_{11} & a_{12} \\ a_{21} & a_{22} \end{bmatrix} \end{bmatrix},$$

and $A_{[2]}$

$$= \begin{bmatrix} \det\begin{bmatrix} a_{11} & a_{12} \\ a_{21} & a_{22} \end{bmatrix} & \det\begin{bmatrix} a_{11} & a_{13} \\ a_{21} & a_{23} \end{bmatrix} & \det\begin{bmatrix} a_{12} & a_{13} \\ a_{22} & a_{23} \end{bmatrix} \\ \det\begin{bmatrix} a_{11} & a_{12} \\ a_{31} & a_{32} \end{bmatrix} & \det\begin{bmatrix} a_{11} & a_{13} \\ a_{31} & a_{33} \end{bmatrix} & \det\begin{bmatrix} a_{12} & a_{13} \\ a_{32} & a_{33} \end{bmatrix} \\ \det\begin{bmatrix} a_{21} & a_{22} \\ a_{31} & a_{32} \end{bmatrix} & \det\begin{bmatrix} a_{21} & a_{23} \\ a_{31} & a_{33} \end{bmatrix} & \det\begin{bmatrix} a_{22} & a_{23} \\ a_{32} & a_{33} \end{bmatrix} \end{bmatrix}.$$

One can obtain $A_{[2]}$ from adj(A) in a simple manner. First, ignore all the negative signs present in front of the determinants. Keeping the (1,3)-th position in adj(A) as fixed, move the first row 90^0 anticlockwise. Keeping the (2,2)-th position as fixed, move the second row 90^0 anticlockwise. Finally, keeping the (3,1)-th position as fixed, move the third row 90^0 anticlockwise. We have $A_{[2]}$. These operations can be executed purely in a mathematical fashion. Let

$$E = \begin{bmatrix} 1 & 0 & 0 \\ 0 & -1 & 0 \\ 0 & 0 & 1 \end{bmatrix},$$

and

$$F = \begin{bmatrix} 0 & 0 & 1 \\ 0 & 1 & 0 \\ 1 & 0 & 0 \end{bmatrix}.$$

One can check that $A_{[2]} = FE[(\text{adj}(A))']EF = FE(\text{adj}(A'))EF$. The multiplication by the matrix E neutralizes the negative signs in adj(A). The multiplication by the matrix F moves the entries of $(\text{adj}(A))'$ the way outlined above. (Try it.) The apparent discrepancy in the location of the entries between the matrices $A_{[m-1]}$ and adj(A) stems from the fact we have agreed upon the way to write down the members of $\mathbf{A}_{m-1}$ in lexicographical order.

We now concentrate on minors of submatrices of compound matrices. Let A be a matrix of order $m \times m$. Let p be any positive integer with $p+1 \leq m$. For each i and $j \in \{p+1, p+2, \ldots, m\}$, let

$$b_{ij} = |A_{(1,2,\ldots,p,i),(1,2,\ldots,p,j)}|.$$

Let B be the matrix of order $(m-p) \times (m-p)$ whose (i,j)-th entry is given by b_{ij}, $i, j \in \{p+1, p+2, \ldots, m\}$. We can immediately recognize that the matrix B is a submatrix of the compound matrix $A_{[p+1]}$. We would like to compute minors of various orders for the matrix B. Let us make an assault. Let

$$p+1 \leq i_1 < i_2 < \ldots < i_q \leq m,$$

and

$$p+1 \leq j_1 < j_2 < \ldots < j_q \leq m$$

be two choices of integers for some $1 \leq q \leq m-p$. The following result known as Sylvester's determinantal identity provides a formula for the computation of q-th order minors of B in terms of minors of A.

P 4.5.3 (Sylvester's Determinantal Identity) With the notation set as above, we have

$$\begin{aligned}|B_{(i_1,i_2,\ldots,i_q),(j_1,j_2,\ldots,j_q)}| &= [|A_{(1,2,\ldots,p),(1,2,\ldots,p)}|]^{q-1} \\ &\times |A_{(1,2,\ldots,p,i_1,i_2,\ldots,i_q),(1,2,\ldots,p,j_1,j_2,\ldots,j_q)}|.\end{aligned}$$

PROOF. The case $q=1$ is clear. The result follows from the definition of the entries of B. It is not difficult to establish the identity for $q=2$. The general case $q>2$ follows in the foot-steps of the case for $q=2$.

The Sylvester's identity can be generalized in several directions. We will describe the generalization in general terms without proof.

Choose and fix two sets of p integers: $1 \leq \mu_1 < \mu_2 < \ldots < \mu_p \leq m$ and $1 \leq \nu_1 < \nu_2 < \ldots < \nu_p \leq m$, for some $1 \leq p < m$. Let $i \in$

$\{1,2,\ldots,m\}-\{\mu_1,\mu_2,\ldots,\mu_p\}$ and $j \in \{1,2,\ldots,m\}-\{\nu_1,\nu_2,\ldots,\nu_p\}$. Define

$$b_{ij} = |A_{(i_1,i_2,\ldots,i_p,i_{p+1}),(j_1,j_2,\ldots,j_p,j_{p+1})}|,$$

where $i_1 < i_2 < \ldots < i_{p+1}$ is a permutation of $i,\mu_1,\mu_2,\ldots,\mu_p$ and $j_1 < j_2 < \ldots < j_{p+1}$ is a permutation of $j,\nu_1,\nu_2,\ldots,\nu_p$. Let $B = (b_{ij})$ be the resultant matrix of order $(m-p)\times(m-p)$. The entries b_{ij}'s appear in the matrix B in the natural order of i's and j's consistent with the lexicographical order in which the minors of order $p+1$ are arranged in the compound matrix $A_{[p+1]}$. It is clear that the matrix B is a submatrix of $A_{[p+1]}$. As an illustration, let $m = 5$, $p = 2$, $\mu_1 = 2$, $\mu_2 = 4$, $\nu_1 = 1$, and $\nu_2 = 3$. The matrix B is of order 3×3 given by,

$$B = \begin{bmatrix} b_{12} & b_{14} & b_{15} \\ b_{32} & b_{34} & b_{35} \\ b_{52} & b_{54} & b_{55} \end{bmatrix}$$

$$= \begin{bmatrix} |A_{(1,2,4),(1,2,3)}| & |A_{(1,2,4),(1,3,4)}| & |A_{(1,2,4),(1,3,5)}| \\ |A_{(2,3,4),(1,2,3)}| & |A_{(2,3,4),(1,3,4)}| & |A_{(2,3,4),(1,3,5)}| \\ |A_{(2,4,5),(1,2,3)}| & |A_{(2,4,5),(1,3,4)}| & |A_{(2,4,5),(1,3,5)}| \end{bmatrix}.$$

(Check the natural order of the subscripts of the entries of B and the matching lexicographical order of the minors of A.)

Let us go back to the general case. Let $1 \le q \le m-p$. We want a formula for the q-th order minor of B. Let $k_1 < k_2 < \ldots < k_q$ and $\ell_1 < \ell_2 < \ldots < \ell_q$ be two choices of integers with $k_1,k_2,\ldots,k_q \in \{1,2,\ldots,m\} - \{\mu_1,\mu_2,\ldots,\mu_p\}$ and $\ell_1,\ell_2,\ldots,\ell_q \in \{1,2,\ldots,m\} - \{\nu_1,\nu_2,\ldots,\nu_p\}$. We are now ready to state the identity.

P 4.5.4 (Sylvester's Determinantal Identity for the q-th Order Minor of B)

$$|B_{(k_1,k_2,\ldots,k_q),(\ell_1,\ell_2,\ldots,\ell_q)}| = [|A_{(\mu_1,\mu_2,\ldots,\mu_p),(\nu_1,\nu_2,\ldots,\nu_p)}|]^{q-1} \times |A_{(\alpha_1,\alpha_2,\ldots,\alpha_{p+q}),(\beta_1,\beta_2,\ldots,\beta_{p+q})}|,$$

where $\alpha_1 < \alpha_2 < \ldots < \alpha_{p+q}$ is a permutation of $\mu_1,\mu_2,\ldots,\mu_p,k_1,k_2,\ldots,k_q$ and $\beta_1 < \beta_2 < \ldots < \beta_{p+q}$ is a permutation of $\nu_1,\nu_2,\ldots,\nu_p,\ell_1,\ell_2,\ldots,\ell_q$.

Consider the example presented above with $m = 5$. Let $q = 2$, $k_1 = 2$, $k_2 = 3$, $\ell_1 = 2$ and $\ell_2 = 4$. Then the second-order minor of B is given by

$$|B_{(1,3),(2,4)}| = \begin{vmatrix} b_{12} & b_{14} \\ b_{32} & b_{34} \end{vmatrix} = |A_{(2,4),(1,3)}| \cdot |A_{(1,2,3,4),(1,2,3,4)}|.$$

We now need to explore the relationship between minors of a matrix and its rank. We report some results without proof.

P 4.5.5 Let A be a matrix of order $m \times n$. Let $1 \leq r \leq \min\{m, n\}$. Then the following statements are equivalent.

(1) $\rho(A) = r$.

(2) There exists a non-zero minor of A of order r and every minor of order $(r + 1)$ is zero. (We are adopting the convention that if $r + 1$ exceeds $\min\{m, n\}$, minors of order $r + 1$ are set equal to zero.)

P 4.5.6 Let A be a matrix of order $m \times m$ with $m \geq 2$. Then

(1) $\rho(\text{adj}(A)) = m$ if and only if $\rho(A) = m$;
(2) $\rho(\text{adj}(A)) = 1$ if and only if $\rho(A) = m - 1$;
(3) $\rho(\text{adj}(A)) = 0$ if and only if $\rho(A) \leq m - 2$.

P 4.5.7 (Laplace Expansion) Let A be a matrix of order $m \times m$. Choose and fix $1 \leq p \leq m$.

$$|A| = \sum (-1)^{i_1+i_2+\ldots+i_p+j_1+j_2+\ldots+j_p} |A_{(i_1,i_2,\ldots,i_p),(j_1,j_2,\ldots,j_p)}| \cdot |A_{(\alpha_1,\alpha_2,\ldots,\alpha_{m-p}),(\beta_1,\beta_2,\ldots,\beta_{m-p})}|,$$

where the summation is taken over all p-tuples $1 \leq i_1 < i_2 < \ldots < i_p \leq m$, and $1 \leq j_1 < j_2 < \ldots < j_p \leq m, 1 \leq \alpha_1 < \alpha_2 < \ldots < \alpha_{m-p} \leq m$ is a permutation of the members of the set $\{1, 2, \ldots, m\} - \{i_1, i_2, \ldots, i_p\}$, and $1 \leq \beta_1 < \beta_2 < \ldots < \beta_{m-p} \leq m$ is a permutation of the members of the set $\{1, 2, \ldots, m\} - \{j_1, j_2, \ldots, j_p\}$. (If $p = m$, the second determinant in the above summation is taken equal to unity.)

Complements

4.5.1 Suppose A, B and C are three matrices each of order $m \times m$ such that $A = BC$. Show that for any $1 \leq p \leq m$,

$$A_{[p]} = B_{[p]} C_{[p]}.$$

(This is essentially the Cauchy-Binet formula.)

4.5.2 Suppose A is a matrix of order $m \times m$ and $\alpha \in \mathbf{F}$. Show that for any $1 \leq p \leq m$,

$$(\alpha A)_{[p]} = \alpha^p (A_{[p]}).$$

4.5.3 Suppose A is a matrix of order $m \times m$. Show that for any $1 \leq p \leq m$,

$$(A')_{[p]} = (A_{[p]})'.$$

4.5.4 Suppose A is a non-singular matrix. Show that for any $1 \leq p \leq m$, $A_{[p]}$ is non-singular, and

$$(A_{[p]})^{-1} = (A^{-1})_{[p]}.$$

4.5.5 There is no particular reason that the operation of compounding should be confined to square matrices. The notion makes sense for any matrix of any order. Reformulate Complements 4.5.1-4.5.3 above in the context of matrices of any order and solve them.

4.5.6 Let A, B, and C be three matrices of orders $m \times n$, $m \times r$, and $r \times n$, respectively, such that $A = BC$. Let $1 \leq p \leq \min\{m, r, n\}$. We need to take into account possible differences in m and n in the computation of minors of A. Let $\mathbf{A}_p^m = \{\alpha = (i_1, i_2, \ldots, i_p) : 1 \leq i_1 < i_2 < \ldots < i_p \leq m\}$. Let $\alpha \in \mathbf{A}_p^m$ and $\beta \in \mathbf{A}_p^n$. Show that the p-th order minor of A is given by

$$|A_{\alpha\beta}| = \sum_{\gamma \in \mathbf{A}_p^r} |B_{\alpha\gamma}| \quad |C_{\gamma\beta}|.$$

4.5.7 Let A be a matrix of order $m \times m$. Write the relationship between the compound matrix $A_{[m-1]}$ and $\operatorname{adj}(A)$ in mathematical terms. Establish this relationship.

4.5.8 Let

$$A = \begin{bmatrix} 1 & 2 & 3 \\ 4 & 5 & 6 \\ 7 & 8 & 9 \end{bmatrix}.$$

Evaluate $\operatorname{adj}(A)$ and $A_{[2]}$. Show that $\operatorname{adj}(A)$ is symmetric. (Does this surprise you?)

4.5.9 Let $T : \mathbf{V} \to \mathbf{W}$ and A_T be the associated matrix. Further let $\mathbf{K}(T) = \{x \in \mathbf{V} : T(x) = 0\}$ and $\mathbf{R}(T) = \{Tx : x \in \mathbf{V}\}$ be the kernel and range of T, respectively. Define $\rho(T) = \dim \mathbf{R}(T)$ and $\nu(T) = \dim \mathbf{K}(T)$. Then show that

$$\nu(T) = \nu(A_T) \text{ and } \rho(T) = \rho(A_T).$$

4.5.10 Let $T : \mathbf{V} \to \mathbf{V}, \dim \mathbf{V} = n$ and A_T be the associated matrix. Then show that there exists a matrix B of order $n \times n$ such that $A_T B = BA_T = I_n$, if and only if $\rho(A_T) = n$.

4.5.11 Let $T : \mathbf{V} \to \mathbf{W}$. Define a map T' from the dual space $\mathbf{V}'$ of $\mathbf{V}$ to the dual space $\mathbf{W}'$ of $\mathbf{W}$ as follows. Let f be a linear functional on $\mathbf{W}$ and $f'(x) = f(T(x)), x \in \mathbf{V}$. Then f' is a linear functional on $\mathbf{V}$, i.e., $f' \in \mathbf{V}'$. Note that $T'(f) = f',\ f \in \mathbf{W}'$. The map T' is called the transpose of T. Show that

$$T'(\alpha f_1 + \beta f_2) = \alpha T'(f_1) + \beta T'(f_2)$$

i.e., T' is a transformation.

If T and S are two transformations from $\mathbf{V} \to \mathbf{W}$, then show that $(T+S)' = T' + S', (\alpha T)' = \alpha T', (ST)' = T'S'$.

4.5.12 Let T' be as defined in Complement 4.5.11. Then show that

$$A_{T'} = (A_T)'$$

where A_T and $A_{T'}$ are the matrices associated with T and T' respectively.

4.5.13 Let $\mathbf{V}$ be a vector space over the field $\mathbf{C}$ of complex numbers equipped with an inner product $< \cdot, \cdot >$. Consider for each fixed $y \in \mathbf{V}$, the linear functional $f_y : \mathbf{V} \to \mathbf{C}$ defined by

$$f_y(x) = < T(x), y >,\ x \in \mathbf{V}.$$

Then by **P 2.4.1**, there exists a unique vector $z \in \mathbf{V}$ such that

$$f_y(x) = < x, z >, x \in \mathbf{V}.$$

Let us denote z by $T^*(y)$. Thus we have a map $T^* : \mathbf{V} \to \mathbf{V}$ which we call the adjoint of T.

Show that T^* is a transformation and the matrix A_{T^*} is the conjugate transpose of A_T.

4.5.14 Let $T : \mathbf{V} \to \mathbf{W}$ and $S : \mathbf{W} \to \mathbf{U}$. Show that

$$\rho(ST) \leq \min\{\rho(S), \rho(T)\}.$$

4.5.15 Let A be an $m \times n$ matrix of rank k. Show the following:

(1) There is a $k \times k$ submatrix of A with nonzero determinant, but all $(k+1) \times (k+1)$ submatrices of A have determinant zero.

(2) There is a set k, but not more than k, of linearly independent vectors b such that the linear equation $Ax = b$ is consistent.

4.5.16 Let A be an $m \times n$ matrix. Then $\rho(A^*A) = \rho(A)$.

4.5.17 Let A and B be $m \times n$ matrices. Then $\rho(A) = \rho(B)$ if and only if there exist nonsingular matrices X of order m and Y of order n such that $B = XAY$.

4.5.18 Let A be a square Hermitian matrix of order n, i.e., $A = A^*$. Show that $\rho(A) \geq (\operatorname{tr} A)^2/\operatorname{tr} A^2$, with equality if and only if there exists an $n \times r$ matrix $U = (u_1, \ldots, u_r)$ with orthonormal columns and some real number a such that $A = aUU^*$.

4.5.19 Let $A = (a_{ij}) = (a_1|\cdots|a_n)$ be a square matrix of order n obtained by writing n column vectors $a_1, \ldots, a_n$ one after the other to form a square grid. Show that

$$\rho(A) \geq \sum_{i=1}^{n} |a_{ii}|^2/\|a_i\|_2^2,$$

where $\|a_i\|_2^2 = a_i^* a_i$.

CHAPTER 5

FACTORIZATION OF MATRICES

In Chapter 4, we have demonstrated how a matrix A of order $m \times n$ with rank a has the factorization $A = RF$, where R is an $m \times a$ matrix of rank a and F a matrix of order $a \times n$ with rank a. In this chapter, we consider a number of other factorizations of A in terms of special matrices. These factorizations are useful in theoretical investigations and practical applications.

5.1. Elementary Matrices

Consider the following operations on a matrix A of order $m \times n$.

Operations on the rows of the matrix

(R_1) Multiply each of the entries in the r-th row by a scalar $\alpha \neq 0$.

(R_2) Replace the r-th row by "r-th row $+\beta$(s-th row)," where β is a scalar.

(R_3) Interchange two rows.

Operations on the columns of the matrix

(C_1) Multiply each of the entries in the r-th column by a scalar $\alpha \neq 0$.

(C_2) Replace the r-th column by "r-th column $+\beta$(s-th column)," where β is a scalar.

(C_3) Interchange two columns.

We show that all the above row and column operations on A are equivalent to pre and post-multiplying A respectively, with what we call appropriate elementary matrices. Let $E_r(\alpha) = (\delta_{ij})$ be the matrix of order $m \times m$, where

$$\delta_{ij} = \begin{cases} \alpha, & \text{if } i = j = r, \\ 1, & \text{if } i = j \neq r, \\ 0, & \text{if } i \neq j. \end{cases}$$

Let B be the matrix obtained from A after performing the operation (R_1) on A. One can show that $B = E_r(\alpha)A$. If C is the matrix obtained from A after the performance of the operation (C_1), then this operation is equivalent to postmultiplying A by $E_r(\alpha)$, i.e., $C = AE_r(\alpha)$, with the matrix $E_r(\alpha)$ being of order $n \times n$.

As per operation (R_2), assume that $r \neq s$. If $r = s$, the operation (R_2) is similar to the operation (R_1). Let $E_{rs}(\beta) = (\gamma_{ij})$ be the matrix of order $m \times m$, where

$$\gamma_{ij} = \begin{cases} 1, & \text{if } i = j, \\ \beta, & \text{if } i = r,\ j = s, \\ 0, & \text{otherwise.} \end{cases}$$

Note that $E_{rs}(\beta)$ is either an upper-triangular matrix or a lower-triangular matrix depending upon the relationship between r and s. Let D be the matrix obtained from A after performing the operation (R_2) on A. One can check that $D = E_{rs}(\beta)A$. Let us look at the column operation (C_2). Let F be the matrix obtained from A after performing the operation (C_2) on A. Then one can check that $F = AE_{sr}(\beta)$. But now the matrix $E_{sr}(\beta)$ is of order $n \times n$. (Do you see the difference in how the matrices $E_{rs}(\beta)$'s work out to be in the operations (R_2) and (C_2)?)

One can look at the row operation (R_1) as a special case of the operation (R_2). One can take $r = s$ in the operation (R_2), and define $E_{rr}(\beta)$ as $E_r(1 + \beta)$.

Suppose we interchange the r-th and s-th rows at A. Let F be the resultant matrix. Let $E_{rs} = (\nu_{ij})$ be the matrix of order $m \times m$, where

$$\nu_{ij} = \begin{cases} 1 & \text{if } i = j,\ i \neq r,\ i \neq s, \\ 1 & \text{if } i = r,\ j = s, \\ 1 & \text{if } i = s,\ j = r, \\ 0 & \text{otherwise.} \end{cases}$$

One can check that $F = E_{rs}A$. It is instructive to observe that the matrix E_{rs} can be obtained as a product of matrices of the type $E_{rs}(\beta)$. More precisely,

$$E_{rs} = E_s(-1)\, E_{sr}(-1)\, E_{rs}(1)\, E_{sr}(-1).$$

This follows from the observation:

$$\begin{bmatrix} 1 & 0 \\ 0 & -1 \end{bmatrix} \begin{bmatrix} 1 & 0 \\ -1 & 1 \end{bmatrix} \begin{bmatrix} 1 & 1 \\ 0 & 1 \end{bmatrix} \begin{bmatrix} 1 & 0 \\ -1 & 1 \end{bmatrix} = \begin{bmatrix} 0 & 1 \\ 1 & 0 \end{bmatrix}.$$

As mentioned earlier, one can take, conventionally, that $E_s(-1) = E_{ss}(-2)$. Let G be the matrix obtained from A after performing the operation (C_3) on A. Then $G = AE_{rs}$ with the understanding that E_{rs} is of order $n \times n$.

The matrices $E_r(\alpha)$, $E_{rs}(\beta)$, and E_{rs} of whatever order are called elementary matrices. It is obvious that they are all nonsingular. Consequently, the matrix obtained after performing any of the row operations will have the same rank as that of A.

The elementary matrices can be obtained by a different route. Start with the identity matrix I_m. Perform the operation (R_1) on I_m. The resultant matrix is $E_r(\alpha)$. If we perform the operation (R_2) on I_m, we will obtain the matrix $E_{rs}(\beta)$. If we perform the operation (R_3) on I_m, we will obtain E_{rs}.

Upper-triangular and lower-triangular matrices have nice properties. Some of these will be touched upon in the complements. There are other special forms of matrices called echelon matrices which play an important role in the factorization of matrices.

DEFINITION 5.1.1. A matrix A of order $m \times n$ is said to be in echelon form if each row of A has *one* of the following properties:

(1) All the entries in the row are zeros.
(2) If the row is non-zero, then all the entries in the column below the first non-zero entry of the row are zeros.

The following matrices are all in echelon form:

$$\begin{bmatrix} 2 & 1 & 1 \\ 0 & 0 & 3 \\ 0 & 0 & 0 \end{bmatrix}, \begin{bmatrix} 0 & 1 & 2 \\ 0 & 0 & 1 \\ 2 & 0 & 0 \end{bmatrix}, \begin{bmatrix} 0 & 1 & 2 & 0 & 1 \\ 1 & 0 & 1 & 2 & 3 \\ 0 & 0 & 2 & 0 & 3 \end{bmatrix}.$$

DEFINITION 5.1.2. Let A be a matrix of order $m \times n$. For each $1 \leq i \leq m$, let a_i be the number of zeros preceding the first non-zero element of the i-th row. The matrix A is said to be in an upper echelon form if $a_1 < a_2 < \ldots < a_m$. Let b_i be the number of zeros above the first non-zero element of the i-th column, $1 \leq i \leq n$. The matrix A is said to be in a lower echelon form if $b_1 < b_2 < \ldots < b_n$.

The following matrices are in upper echelon form:

$$\begin{bmatrix} 0 & 1 & 2 & 0 \\ 0 & 0 & 1 & 1 \\ 0 & 0 & 0 & 0 \end{bmatrix}, \begin{bmatrix} 1 & 0 & 2 & 3 \\ 0 & 0 & 1 & 2 \\ 0 & 0 & 0 & 1 \end{bmatrix}.$$

Complements

5.1.1 Let A and B be two upper-triangular matrices of the same order. Show that AB is upper-triangular.

5.1.2 Let A be a non-singular upper-triangular matrix. Show that its inverse is also upper-triangular.

5.1.3 Determine the inverse of each of the elementary matrices $E_r(\alpha)$, $E_{rs}(\beta)$, and E_{rs}.

5.1.4 Let

$$A = \begin{bmatrix} 1 & -3 & 0 & 7 \\ 0 & 1 & 3 & 3 \\ 0 & 0 & -1 & -2 \end{bmatrix}.$$

Which row operation on A will render (2,4)-th element of A zero?

5.2. Reduction of General Matrices

First, we consider matrices with elements belonging to any field of scalars and derive a number of factorizations of a matrix obtained by pre- and post-multiplying it by elementary matrices. Before this, we would like to introduce the notion of a unit lower-triangular matrix. A square matrix A is said to be a unit lower-triangular matrix if it is lower triangular and each of its diagonal entries is equal to unity. If $r > s$, the elementary matrix $E_{rs}(\beta)$ is a unit lower-triangular matrix. The notion of a unit upper-triangular matrix is analogous. If $r < s$, the elementary matrix $E_{rs}(\beta)$ is a unit upper-triangular matrix. If A and B are two unit upper-triangular matrices of the same order, it can be checked that the product AB is also a unit upper-triangular matrix. A similar assertion is valid for unit lower-triangular matrices.

P 5.2.1 Let $A = (\alpha_{ij})$ be a matrix of order $m \times n$. Then there exists a unit lower-triangular matrix B of order $m \times m$ such that BA is in echelon form.

PROOF. Start with the first non-zero row of the matrix A. Assume, without loss of generality, that the first row of A is non-zero. Let α_{1i} be the first non-zero entry in the first row. Take any $2 \leq s \leq m$. If $\alpha_{si} = 0$, we do nothing. If $\alpha_{si} \neq 0$, we multiply the first row by $-\alpha_{si}/\alpha_{1i}$ and add this to the s-th row. This operation makes the (s,i)-th entry zero. This operation is equivalent to pre-multiplying A by the elementary matrix $E_{s1}(-\alpha_{si}/\alpha_{1i})$ which is a unit lower-triangular matrix. Thus we have used the $(1,i)$-th entry, namely α_{1i}, as a pivot to liquidate or sweep out all the other entries in the i-th column. The resultant matrix or reduced matrix is mathematically obtainable by pre-multiplying A successively by a finite number of unit lower-triangular matrices whose product is again a unit lower-triangular matrix. Now start with the reduced matrix. Look at the second row. If all of its entries are equal to zero, move on to the third row. If we are unable to find any non-zero vectors among the second, third, ... , m-th rows, the process stops. The reduced matrix is clearly in echelon form. Otherwise, locate the first non-zero vector among the $m-1$ rows of the reduced matrix starting from the second. Repeat the process of sweeping out all the entries below the first non-zero entry (pivot) of the chosen non-zero vector. Repeat this process until we could find no more non-zero vectors in the reduced matrix. The reduced matrix is clearly in echelon form. The promised matrix B is simply the product of all pre-multiplying unit lower-triangular matrices employed in the sweep-out process. Clearly, B is a unit lower-triangular matrix. This completes the proof.

COROLLARY 5.2.2. Let A be a matrix of order $m \times n$. Let B be the unit lower-triangular matrix obtained in the proof of **P 5.2.1** such that BA is in echelon form. Then the rank of A is equal to the total number of non-zero row vectors in BA, or equivalently to the total number of pivots.

We have already established the rank factorization of A in **P 4.3.1**. This result can be obtained as a corollary of **P 5.2.1**. Let $BA = C$. Write $A = B^{-1}C$. Let $\rho(A) = a$. The matrix C will have exactly a non-zero rows. Eliminate all zero rows. Let F denote the resultant matrix. The matrix F will be of order $a \times n$. Eliminate the corresponding columns from B^{-1}. Let R be the resultant matrix which will be of order $m \times a$. It is clear that $\rho(R) = \rho(F) = a$. Further, $A = B^{-1}C = RF$. We will jot this result as a corollary.

COROLLARY 5.2.3. (**Rank Factorization**) Any given matrix A of order $m \times n$ and of rank $a \neq 0$ admits a rank factorization

$$A = RF \tag{5.2.1}$$

where R is of order $m \times a$ with rank a and F is of order $a \times n$ with rank a.

We can polish the echelon form a little. For the given matrix A, determine a lower-triangular matrix B such that $BA = C$ is in echelon form. By interchanging the columns of C, one can produce a matrix in which all the entries below the leading diagonal are zeros, For example, if

$$C = \begin{bmatrix} 0 & 1 & 2 & 0 & 1 \\ 1 & 0 & 1 & 2 & 3 \\ 0 & 0 & 2 & 0 & 1 \end{bmatrix},$$

interchange columns 1 and 2 to produce the matrix

$$\begin{bmatrix} 1 & 0 & 2 & 0 & 1 \\ 0 & 1 & 1 & 2 & 3 \\ 0 & 0 & 2 & 0 & 0 \end{bmatrix},$$

in which we notice that all the entries below the leading diagonal are zeros. In some cases, one may have to do more than one interchange of columns. Any interchange of columns in a matrix is equivalent to post-multiplying the matrix with an elementary matrix of the type E_{rs}. When the matrix C in echelon form is changed into another matrix D in echelon form with the additional property that all entries below the leading diagonal of D are zeros, one can write $D = CG$, where G is a product of matrices of the type E_{rs}. In the final analysis, we will have $BAG = CG = D$. Let $\tilde{A} = AG$. Observe that $\tilde{A}$ is obtainable from A by some interchanges of columns of A. Let us put down all these deliberations in the form of a corollary.

COROLLARY 5.2.4. Let A be a given matrix and B a unit lower-triangular matrix such that $BA = C$ is in echelon form. Then one can construct a matrix $\tilde{A}$ by interchanging columns of A such that $B\tilde{A}$ is in echelon form with the additional property that all entries below the leading diagonal are zeros.

COROLLARY 5.2.5. **(Upper Echelon Form)** For any given matrix A there exists a non-singular matrix B such that BA is in upper echelon form.

PROOF. If A is the zero matrix, it is already in upper echelon form. Assume that A is a non-zero matrix. Identify the i-th column such that it is a non-zero vector and all the first $(i-1)$ columns are zero vectors. Identify the first entry a_{ji} in the i-th column which is non-zero. Add j-th row to the first row. In the resulting matrix, the $(1,i)$-th element is non-zero. Using this element as a pivot, sweep out the rest of the entries in the i-th column. Disregard the first row and the first i columns of the reduced matrix. Repeat the above operation on the submatrix created out of the reduced matrix. Continue this process until no non-zero column is left. All these matrices may not be unit lower-triangular matrices. The addition of a row at a lower level of the matrix to a row at a higher level of the matrix cannot be done by pre-multiplying the matrix with a unit lower-triangular matrix. In any case, the final reduced matrix is in upper echelon form and the product of all elementary matrices involved is the desired non-singular matrix B.

For the next result, we need the notion of a principal minor of a square matrix. Let A be a square matrix of order $m \times m$. Let $1 \leq i_1 < i_2 < \ldots < i_p \leq m$. The determinant of the submatrix of A obtained from A by deleting the i_1-th, i_2-th, $\ldots$, i_p-th rows and i_1-th, i_2-th, $\ldots$, i_p-th columns from A is called a principal minor of A. The submatrix itself is called a principal submatrix. If $i_1, i_2, \ldots, i_p$ are consecutive integers and $i_p = m$, the associated principal minor is called a leading principal minor of A. The corresponding submatrix of A is called a leading principal submatrix.

P 5.2.6 (LU Triangular Factorization) Let A be an $m \times m$ matrix such that all its leading principal minors are non-zero. Then A can be factorized as

$$A = LU, \tag{5.2.2}$$

where L is a unit lower-triangular matrix and U is a non-singular upper-triangular matrix each of order $m \times m$.

PROOF. Let $A = (\alpha_{ij})$. Since $\alpha_{11} \neq 0$, it can be used as a pivot to sweep out all the other elements in the first column. Let us identify the

(2,2)-th element $\alpha_{22}^{(1)}$ in the reduced matrix. Observe that

$$\begin{vmatrix} \alpha_{11} & \alpha_{12} \\ \alpha_{21} & \alpha_{22} \end{vmatrix} = \begin{vmatrix} \alpha_{11} & \alpha_{12} \\ 0 & \alpha_{22}^{(1)} \end{vmatrix} = \alpha_{11}\alpha_{22}^{(1)} .$$

The determinant on the left-hand side of the above expression is a leading principal minor of A and by hypothesis, it is non-zero. The second determinant above is the result of the sweep-out process and this does not change the value of the minor. Now it follows that $\alpha_{22}^{(1)} \neq 0$. Further, all the leading principal minors of the reduced matrix are non-zero. (Why?) Using $\alpha_{22}^{(1)}$ as a pivot, sweep out all the entries below the second row in the second column of the reduced matrix. Continue this process until we end up with an upper triangular matrix U. All the operations involved are equivalent to pre-multiplying the matrix A with a series of unit lower-triangular matrices whose product B is clearly a unit lower-triangular matrix. Thus we have $BA = U$, from which we have $A = B^{-1}U = LU$, say. Observe that B^{-1} is a unit lower-triangular matrix. It is clear that U is non-singular. This completes the proof.

COROLLARY 5.2.7. Let A be a matrix of order $m \times m$ for which every leading principal minor is non-zero. Then it can be factorized as

$$A = LDU, \tag{5.2.3}$$

where L is unit lower-triangular, U is unit upper-triangular, and D is a non-singular diagonal matrix.

PROOF. First, we write $A = LU_1$ following (5.2.2). Divide each row of U_1 by its diagonal element. This operation results in a matrix U which is unit upper-triangular. This operation is also equivalent to pre-multiplying U_1 by a non-singular diagonal matrix D^{-1}. Thus we can write $U = D^{-1}U_1$. The result now follows.

The assumption that the matrix A should have leading principal minors non-zero looks a little odd. Of course, every such matrix is non-singular. Consider the following matrix:

$$A = \begin{bmatrix} 0 & 1 \\ 1 & 1 \end{bmatrix}.$$

The matrix A is non-singular but the assumption of **P 5.2.6** on leading principal minors is not met. There is no way we could write $A = LU$ with L unit lower-triangular and U non-singular upper-triangular. (Try.) In the following corollary we record that this assumption is inviolable.

COROLLARY 5.2.8. Let A be a non-singular matrix. Then A admits a factorization of the form (5.2.3) if and only if all the leading principal minors of A are non-zero.

PROOF. The critical observation is that if A can be factorized in the form (5.2.3), then any leading principal submatrix of A can be factorized in the form (5.2.3). Now the result becomes transparent.

The notion of leading minors and leading submatrices can be defined for any general matrix A not necessarily square. A result analogous to **P 5.2.6** should work out for rectangular matrices too.

COROLLARY 5.2.9. Let A be a matrix of order $m \times n$ for which all its leading principal minors are non-zero. Then we have the factorization

$$A = LU, \tag{5.2.4}$$

where L is unit lower-triangular of order $m \times m$ and U is in upper echelon form.

The hypothesis that all leading principal minors are non-zero can be relaxed, but the conclusion will have to be diluted.

COROLLARY 5.2.10. Let A be matrix of order $m \times m$ such that its i_1-th, i_2-th, ... , i_r-th columns are dependent on the previous columns for some $1 < i_1 < i_2 < \ldots < i_r \leq m$. Let B be the matrix obtained from A by deleting the i_1-th, i_2-th, ... , i_r-th rows and i_1-th, i_2-th, ... , i_r-th columns of A. Suppose all the leading principal minors of B are non-zero. Then we have the factorization

$$A = LU, \tag{5.2.5}$$

where L is unit lower-triangular and U is upper-triangular (with some diagonal elements possibly zero).

PROOF. The factorization (5.2.5) can be established by following the same argument that is used in the proof of **P 5.2.6**. If we encounter a zero leading element at any stage of the reduction process, skip the row containing the leading element and then move on to the next row. The hypothesis of the corollary ensures that when a leading element is zero at any stage, all the elements below it in the same column are zero.

When the matrix involved is special, the factorization becomes special too. In the following, we look at symmetric matrices.

COROLLARY 5.2.11. Let A be a symmetric matrix of order $m \times m$ with all its leading principal minors to be non-zero. Then we have the factorization

$$A = L\Delta L', \tag{5.2.6}$$

where L is unit lower-triangular and Δ is diagonal with non-zero diagonal entries.

PROOF. First, obtain the factorization $A = LU$ following **P 5.2.6**, where L is unit lower-triangular and U non-singular and upper-triangular. Since A is symmetric,

$$A = LU = A' = U'L',$$

from which we note that $U = L^{-1}U'L'$. Let $\Delta = L^{-1}U'$. If we can show that Δ is diagonal, then we can write $A = LU = L\Delta L'$ and the result follows. The diagonality of Δ follows from the fact that UL^{-1} is upper-triangular and $L^{-1}U'$ is lower-triangular. (Check.)

5.3. Factorization of Matrices with Complex Entries

In addition to the factorization results given in the last section, some special results are available when the elements of the matrices belong to the field of complex numbers. Let A be a square matrix with complex numbers as entries. The complex conjugate of A is defined to be that matrix A^* obtained from A by replacing the entries of A by their complex conjugates and then taking the transpose of the matrix. If the entries of A are real, the complex conjugate of A is merely the transpose A' of A. A matrix A is said to be Hermitian if $A^* = A$. If A has real entries, recall that A is said to be symmetric if $A' = A$. Two column vectors a_1 and a_2 of the same order with complex entries are said to be orthogonal if $a_1^*a_2 = a_2^*a_1 = 0$. They are said to be orthonormal if, in addition, $a_1^*a_1 = 1 = a_2^*a_2$. These notions are not different from what we already know about orthogonal and orthonormal vectors in an inner product space. The relevant inner product space in this connection is $\mathbf{C}^n$ equipped with the standard inner product. A square matrix A is said to be unitary if $A^*A = AA^* = I$. A square matrix A with real entries is said to be orthogonal if $A'A = AA' = I$. We now introduce formally Householder matrices.

DEFINITION 5.3.1. Let ω be a column vector of order $n \times 1$ with complex entries satisfying $\omega^*\omega = 1$. Let

$$E(\omega) = I_n - 2\omega\omega^*.$$

The matrix $E(\omega)$ is called a *Householder matrix.*

The following are some of the properties of Householder matrices.

P 5.3.2 (1) Every Householder matrix is Hermitian and unitary.

(2) Let the column vectors a and b be such that $a^*a = b^*b,\ a^*b = b^*a$ (which is automatically true if the vectors are real), and distinct. Then there exists a vector ω of unit length such that $E(\omega)a = b$.

(3) Let a and b be two distinct column vectors of the same order $n\times 1$ with the following structure: $a' = (a_1', a_2')$ and $b' = (a_1', b_2')$, where a_1 is a column vector of order $t\times 1$. (What this means is that the first t entries of a and b are identical.) Suppose that $b_2^*b_2 = a_2^*a_2$ and $b_2^*a_2 = a_2^*b_2$. Then there exists a vector ω of unit length such that $E(\omega)a = b$ and $E(\omega)(c', 0') = (c', 0')$ for any column vector c of order $t \times 1$.

PROOF. The verification of (1) is easy. For (2), take $\omega = r(a - b)$, where $r > 0$ is such that the length of ω is unity. In fact $1/r^2 = (a - b)^*(a - b)$. The given conditions on the vectors a and b imply that the vectors $(a - b)$ and $(a + b)$ are orthogonal, i.e., $(a - b)^*(a + b) = 0$. We are now ready to prove the assertion:

$$\begin{aligned} E(\omega)a = (I_n - 2\omega\omega^*)a &= [I_n - 2r^2(a-b)(a-b)^*][\frac{a+b}{2} + \frac{a-b}{2}] \\ &= \frac{a+b}{2} + \frac{a-b}{2} - (a-b) = b. \end{aligned}$$

As for (3), take $\omega = r(a - b)$, where $r > 0$ is such that the vector ω has unit length. This would do the trick.

The following is one of the most useful factorizations of a matrix. It is similar to the rank factorization of a matrix presented earlier.

P 5.3.3 **(QR Factorization)** Let A be a matrix of order $m \times n$ with complex entries. Let the rank of A be a. Then A can be factorized as

$$A = QR, \tag{5.3.1}$$

where Q is an $m \times a$ matrix of rank a such that $Q^*Q = I_a$ and R is an $a \times n$ matrix in upper echelon form.

PROOF. If $A = 0$, the avowed factorization of A can be set down easily. Assume that A is non-zero. Identify all non-zero columns of A. Let $1 \leq i_1 < i_2 < \ldots < i_r \leq n$ be such that $a_{i_1}, a_{i_2}, \ldots, a_{i_r}$ are all those columns of A which are non-zero. Let d be the first entry in the column a_{i_1}. If d is real, let b_1 be the column vector given by $b_1' = (\sqrt{a_{i_1}^* a_{i_1}}, 0, \ldots, 0)$. If d is complex, let $b_1' = (\frac{d}{|d|}\sqrt{a_{i_1}^* a_{i_1}}, 0, .., 0)$. Note that $b_1^* b_1 = a_{i_1}^* a_{i_1}$ and $b_1^* a_{i_1} = a_{i_1}^* b_1$. By **P 5.3.2** (2), there exists a vector ω_1 of unit length such that $E(\omega_1) a_{i_1} = b_1$. Let $E(\omega_1)A = A_1$. Let us examine closely the structure of A_1. The first $(i_1 - 1)$ columns of A_1 are zero vectors. The i_1-th column of A_1 is b_1. Let us work on A_1. Identify all non-zero columns of A_1. It turns out that i_1-th, i_2-th, $\ldots, i_r$-th columns of A_1 are the only ones which are non-zero. Let us work with i_2-th column of A_1. Let us denote this column by c. Partition $c' = (c_1 | c_2, c_3, \ldots, c_m) = (c_1 | \tilde{c}')$, say. Let $b_2' = (c_1 | \sqrt{\tilde{c}^* \tilde{c}}, 0, \ldots, 0)$ if c_2 is real. If c_2 is complex, b_2 is modified in the same way as in b_1. The vectors c and b_2 satisfy the conditions of **P 5.3.2**(3) with $t = 1$. There exists a vector ω_2 of length one such that $E(\omega_2)c = b_2$. Also, $E(\omega_2)b_1 = b_1$. Let $A_2 = E(\omega_2)A_1$. Let us look at the structure of A_2. The i_1-th, i_2-th, $\ldots, i_r$-th columns are the only vectors of A_2 which are non-zero. In the i_1-th column, all entries below the first element are zeros. In the i_2-th column, all entries below the first two elements are zeros. The trend and strategy should be clear by now. We now have to work with the matrix A_2 and its i_3-th column. Continuing this way, we will have Householder matrices $E(\omega_1), E(\omega_2), \ldots, E(\omega_r)$ such that

$$E(\omega_r)E(\omega_{r-1}) \ldots E(\omega_2)E(\omega_1)\, A = R_1, \quad \text{say,}$$

is in upper echelon form.

Let $Q_1 = E(\omega_r)E(\omega_{r-1}) \ldots E(\omega_2)E(\omega_1)$. It is clear that Q_1 is unitary and $A = Q_1^* R_1$. Since $\rho(A) = a$, exactly a rows of R_1 will be non-zero. Delete these rows from R_1 and let R be the resultant matrix. Delete the correspondingly numbered columns from Q_1^* and let the resultant matrix be Q. Note that $A = QR$ and this is the desired factorization.

The QR factorization of a matrix is computationally very important. It can be used to determine the rank of the matrix, a g-inverse of the matrix, and a host of other features of the matrix. The method outlined in the proof of **P 5.3.3** is called Householder method and used in many computational routines. There is another method of obtaining QR factorization of a matrix based on the Gram-Schmidt orthogonalization process which was outlined in Chapter 2.

Gram-Schmidt Method: Let us denote the column vectors of A by $a_1, a_2, \ldots, a_n$. The Gram-Schmidt process is designed to provide vectors $b_1, b_2, \ldots, b_n$ such that $b_r^* b_s = 0$ for every $r \neq s$, and

$$Sp(a_1, a_2, \ldots, a_i) = Sp(b_1, b_2, \ldots, b_i)$$

for each $i = 1, 2, \ldots, n$, where $Sp(a_1, \ldots, a_i)$ denotes the subspace spanned by $a_1, \ldots, a_i$. The relationship between $a_1, a_2, \ldots, a_n$ and $b_1, b_2, \ldots, b_n$ can be written in the form

$$\begin{aligned} a_1 &= b_1 \\ a_2 &= r_{12}b_1 + b_2 \\ a_3 &= r_{13}b_1 + r_{23}b_2 + b_3 \\ \ldots & \quad \ldots \quad \ldots \\ a_n &= r_{1n}b_1 + r_{2n}b_2 + \ldots + r_{n-1,n}b_{n-1} + b_n, \end{aligned} \tag{5.3.2}$$

where r_{ij}'s and b_i's are determined so as to satisfy $b_i^* b_j = 0$ for all $i \neq j$.

One computational procedure for obtaining $b_1, b_2, \ldots, b_n$ is as follows. Suppose $b_1, b_2, \ldots, b_{i-1}$ are determined. We will determine $r_{1i}, r_{2i}, \ldots, r_{i-1,i}$, and b_i. Consider the equation

$$a_i = r_{1i}b_1 + r_{2i}b_2 + \ldots, r_{i-1,i}b_{i-1} + b_i.$$

Since $b_r^* b_s = 0$ for every $r \neq s$, $r \in \{1, 2, \ldots, i-1\}$,

$$b_r^* a_i = r_{ri} b_r^* b_r, \; r = 1, 2, \ldots, i-1. \tag{5.3.3}$$

If $b_r^* b_r = 0$, set $r_{ri} = 0$. If $b_r^* b_r \neq 0$, set $r_{ri} = b_r^* a_i / b_r^* b_r$, $r = 1, \ldots, i-1$. Finally, b_i is obtained from

$$b_i = a_i - r_{1i}b_1 - r_{2i}b_2 - \ldots - -r_{i-1,i}b_{i-1}. \tag{5.3.4}$$

Now that $r_{1i}, r_{2i}, \dots, r_{i-1,i}$, and b_i are determined, we can proceed to the next step of determining $r_{1,i+1}, r_{2,i+1}, \dots, r_{i,i+1}$, and b_{i+1} in an analogous fashion. The process is continued until all $b_1, b_2, \dots, b_n$ are determined.

Equations (5.3.2) can, indeed, be written in a succinct form. Let $A = (a_1, a_2, \dots, a_n)$, $B = (b_1, b_2, \dots, b_n)$, and

$$C = \begin{bmatrix} 1 & r_{12} & r_{13} & \dots & r_{1n} \\ 0 & 1 & r_{23} & \dots & r_{2n} \\ \cdot & \cdot & \cdot & \dots & \cdot \\ 0 & 0 & 0 & \dots & 1 \end{bmatrix}.$$

Note that $A = BC$ and C is upper triangular. If some of the b_i's are zero, we can omit these columns from B and the correspondingly numbered rows from C resulting in matrices Q_1 and R_1, respectively. Thus we will have $A = Q_1 R_1$, where the columns of Q_1 are orthogonal and R_1 is in upper echelon form. We can normalize the column vectors of Q_1 so that the resultant vectors are orthonormal. Let Q be the matrix so obtained from Q_1 by the process of normalization. The matrix R_1 is also modified to absorb the constants involved in the normalization process with the resultant matrix denoted by R. Thus we have the desired QR decomposition of A.

Some computational caution is in order in implementing the Gram-Schmidt method. From the nature of the formulae (5.3.3) and (5.3.4), some problems may arise if $b_i^* b_i = 0$ or close to zero indicating that a_i depends on or closely related to $a_1, a_2, \dots, a_{i-1}$. In such a case it may be advisable to shift a_i to the last position and consider a_{i+1} after a_{i-1}. Such a rearrangement of a_i decided upon at different stages of the Gram-Schmidt process does not alter the nature of the problem. We can always restore the order by shifting the columns of A and the corresponding rows of R under these circumstances, the resulting matrix Q will still be orthonormal and R will be in echelon form.

Modified Gram-Schmidt Method: This is a slightly different way of transforming a given set $a_1, a_2, \dots, a_n$ of vectors into an orthogonal set of vectors. The procedure is carried out in n stages as detailed in the following.

Stage 1

Set $b_1 = a_1$. Compute

$$r_{12} = \frac{b_1^* a_2}{b_1^* b_1} \quad \text{and} \quad a_2^{(1)} = a_2 - r_{12} b_1;$$

$$r_{13} = \frac{b_1^* a_3}{b_1^* b_1} \quad \text{and} \quad a_3^{(1)} = a_3 - r_{13} b_1;$$

$$\cdots \quad \cdots \quad \cdots \quad \cdots$$

$$r_{1n} = \frac{b_1^* a_n}{b_1^* b_1} \quad \text{and} \quad a_n^{(1)} = a_n - r_{1n} b_1.$$

Stage 2

Set $b_2 = a_2^{(1)}$. Compute

$$r_{23} = \frac{b_2^* a_3^{(1)}}{b_2^* b_2} \quad \text{and} \quad a_3^{(2)} = a_3^{(1)} - r_{23} b_2;$$

$$r_{24} = \frac{b_2^* a_4^{(1)}}{b_2^* b_2} \quad \text{and} \quad a_4^{(2)} = a_4^{(1)} - r_{24} b_2;$$

$$\cdots \quad \cdots \quad \cdots \quad \cdots$$

$$r_{2n} = \frac{b_2^* a_n^{(1)}}{b_2^* b_2} \quad \text{and} \quad a_n^{(2)} = a_n^{(1)} - r_{2n} b_2.$$

$$\cdots \quad \cdots \quad \cdots \quad \cdots.$$

Stage $(n-1)$

Set $b_{n-1} = a_{n-1}^{(n-2)}$. Compute

$$r_{n-1,n} = \frac{b_{n-1}^* a_n^{(n-2)}}{b_{n-1}^* b_{n-1}} \quad \text{and} \quad a_n^{(n-1)} = a_n^{(n-2)} - r_{n-1,n} b_{n-1}.$$

Stage n

Set $b_n = a_n^{(n-1)}$.

If, at any stage, $b_i = 0$, set $r_{i,i+1} = r_{i,i+2} = \ldots = r_{i,n} = 0$. One can check that $b_i^* b_j = 0$ for all $i \neq j$. The set $b_1, b_2, \ldots, b_n$ of vectors and the coefficients r_{ik}'s can be used to set out the QR decomposition of the matrix $A = (a_1, a_2, \ldots, a_n)$.

The singular value decomposition (SVD) of a matrix of order $m \times n$ is a basic result in matrix algebra.

P 5.3.4 (SVD) A matrix A of order $m \times n$ with rank a can be factorized as

$$A = P\Delta Q^* \tag{5.3.5}$$

where P is of order $m \times a$ such that $P^*P = I_a$, Q is of order $n \times a$ such that $Q^*Q = I_a$ and Δ is a diagonal matrix of order $a \times a$ with all positive entries.

PROOF. A simple proof of (5.3.5) depends on what is called SD (spectral decomposition) of a Hermitian matrix derived later in (5.3.9). Note that AA^* is Hermitian of order $m \times m$ and using (5.3.9), we have

$$\begin{aligned} AA^* &= \lambda_1^2 p_1 p_1^* + \ldots + \lambda_a^2 p_a p_a^* \\ I &= p_1 p_1^* + \ldots + p_a p_a^* + p_{a+1} p_{a+1}^* + \ldots + p_m p_m^* \end{aligned}$$

where $p_1, \ldots, p_m$ are orthonormal and

$$\begin{aligned} p_i^* AA^* p_j &= \lambda_i^2 \text{ if } i = j, i = 1, \ldots, a \\ &= 0 \text{ if } i \neq j, i = 1, \ldots, m, j = 1, \ldots, m \end{aligned}$$

As a consequence of the above two equations, we have

$$\begin{aligned} p_i^* A &= 0, \quad i = a+1, \ldots, m \\ \text{and with } q_i^* &= \lambda_i^{-1} p_i^* A, i = 1, \ldots, a \\ q_i^* q_j &= 1 \text{ if } i = j \\ &= 0 \text{ if } i \neq j. \end{aligned}$$

Now consider

$$\begin{aligned} A &= (p_1 p_1' + \ldots + p_m p_m')A \\ &= p_1 p_1^* A + \ldots + p_a p_a^* A + p_{a+1} p_{a+1}^* A + \ldots + p_m p_m^* A \\ &= \lambda_1 p_1 q_1^* + \ldots + \lambda_a p_a q_a^* \\ &= P\Delta Q^* \end{aligned}$$

where $P = (p_1 : \ldots : p_a)$, $Q = (q_1 : \ldots : q_1)$. This completes the proof.

If A is $m \times m$ Hermitian, i.e., $A = A^*$, and non-negative definite (nnd), i.e., $x^*Ax \geq 0$ or positive definite (pd), i.e., $x^*Ax > 0$ for all complex vectors x, then the decomposition takes a finer hue as established in **P 5.3.5** and **P 5.3.9**.

P 5.3.5 (Cholesky Decomposition) Let A be a non-negative definite matrix of order $m \times m$. Then A can be factorized as

$$A = KK^*, \tag{5.3.6}$$

where K is a lower-triangular matrix with non-negative entries in the diagonal. If A is positive definite, then all the entries in the diagonal of K are positive and the decomposition is unique.

PROOF. The result is obviously true when $m = 1$. We will establish the result by induction. Assume that the result is true for any non-negative definite matrix of order $(m-1) \times (m-1)$. Let A be a non-negative definite matrix of order $m \times m$. Partition A as

$$A = \begin{bmatrix} \alpha^2 & \alpha a^* \\ \alpha a & A_1 \end{bmatrix},$$

where α is a real scalar, a is a column vector of order $(m-1) \times 1$ and the matrix A_1 is of order $(m-1) \times (m-1)$. (Check that it is possible to partition A the way we did.) Suppose $\alpha \neq 0$. Let x be any vector of order $m \times 1$ partitioned as $x' = (x_1 | \tilde{x}')$, where x_1 is a scalar and $\tilde{x}$ is a column vector of order $(m-1) \times 1$. Since A is non-negative definite,

$$\begin{aligned} 0 \leq x^* A x &= \alpha^2 |x_1|^2 + \alpha x_1 \tilde{x}^* a + \alpha \bar{x}_1 a^* \tilde{x} + \tilde{x}^* A_1 \tilde{x} \\ &= |\alpha x_1 + a^* \tilde{x}|^2 + \tilde{x}^* (A_1 - a a^*) \tilde{x}. \end{aligned}$$

Since this inequality is valid for all complex vectors x, it follows that $A_1 - aa^*$ is non-negative definite. By the induction hypothesis, there exists a lower-triangular matrix L of order $(m-1) \times (m-1)$ with non-negative diagonal entries such that $A_1 - aa^* = LL^*$. Observe that

$$\begin{bmatrix} \alpha & 0 \\ a & L \end{bmatrix} \begin{bmatrix} \alpha & a^* \\ 0 & L^* \end{bmatrix} = \begin{bmatrix} \alpha^2 & \alpha a^* \\ \alpha a & aa^* + LL^* \end{bmatrix} = \begin{bmatrix} \alpha^2 & \alpha a^* \\ \alpha a & A_1 \end{bmatrix} = A.$$

Take

$$K = \begin{bmatrix} \alpha & 0 \\ a & L \end{bmatrix}.$$

If $\alpha = 0$, choose

$$K = \begin{bmatrix} 0 & 0 \\ 0 & M \end{bmatrix},$$

where $A_1 = MM^*$ with M being a lower-triangular matrix with non-negative diagonal elements.

If A is positive definite, $|K| \neq 0$ which means that every diagonal entry of A is positive. In this case, suppose HH^* is an alternative factorization of A. By directly comparing the elements of H and K, one can show that $H = K$.

P 5.3.6 (A General Decomposition Theorem) Let A be a square matrix of order $m \times m$. Then A can be factorized as

$$A = P\Gamma P^*, \tag{5.3.7}$$

where P is unitary and Γ is upper-triangular.

PROOF. We prove this result by induction. The case $m = 1$ is clear. Assume that the result is true for all matrices of order $(m-1) \times (m-1)$. Let A be a matrix of order $m \times m$. Consider the equation $Ax = \lambda x$ in unknown λ, a scalar, and x, a vector. Choose a λ satisfying the determinantal equation $|A - \lambda I_m| = 0$, which is a polynomial in λ of m-th degree. Let λ_1 be one of the roots of the polynomial. Let x_1 be a vector of unit length satisfying $Ax_1 = \lambda_1 x_1$. Choose a matrix X of order $m \times (m-1)$ such that the partitioned matrix $(x_1|X)$ is unitary. One can verify that

$$\begin{bmatrix} x_1^* \\ X^* \end{bmatrix} A(x_1|X) = \begin{bmatrix} \lambda_1 & x_1^*AX \\ 0 & X^*AX \end{bmatrix}.$$

Note that X^*AX is of order $(m-1) \times (m-1)$. By the induction hypothesis, we can write

$$X^*AX = Q\Gamma_1 Q^*$$

with $Q^*Q = I_{m-1}$ and Γ_1 upper-triangular. Note that

$$\begin{bmatrix} \lambda_1 & x_1^*AX \\ 0 & X^*AX \end{bmatrix} = \begin{bmatrix} 1 & 0 \\ 0 & Q \end{bmatrix} \begin{bmatrix} \lambda_1 & x_1^*AXQ \\ 0 & \Gamma_1 \end{bmatrix} \begin{bmatrix} 1 & 0 \\ 0 & Q^* \end{bmatrix}.$$

Consequently,

$$A = (x_1|X) \begin{bmatrix} \lambda_1 & x_1^*AX \\ 0 & X^*AX \end{bmatrix} \begin{bmatrix} x_1^* \\ X^* \end{bmatrix}$$

$$= (x_1|X) \begin{bmatrix} 1 & 0 \\ 0 & Q \end{bmatrix} \begin{bmatrix} \lambda_1 & x_1^*AXQ \\ 0 & \Gamma_1 \end{bmatrix} \begin{bmatrix} 1 & 0 \\ 0 & Q^* \end{bmatrix} \begin{bmatrix} x_1^* \\ X^* \end{bmatrix}.$$

Let

$$P = (x_1|X)\begin{bmatrix} 1 & 0 \\ 0 & Q \end{bmatrix}$$

and

$$\Gamma = \begin{bmatrix} \lambda_1 & x_1^*AXQ \\ 0 & \Gamma_1 \end{bmatrix}.$$

Clearly, P is unitary and Γ is upper-triangular.

COROLLARY 5.3.7. Let A be of order $m \times m$. In the factorization $A = P\Gamma P^*$ in (5.3.7), the diagonal entries of Γ are the roots of the polynomial $|A - \lambda I_m| = 0$ in λ.

The diagonal entries of Γ in Corollary 5.3.7 have a special name. They are the eigenvalues of A. See Section 5.4 that follows. **P 5.3.6** is also called Schur's Decomposition Theorem.

We will discuss the roots of the polynomial $|A - \lambda I| = 0$ in λ in the next section. Some inequalities concerning these roots will be presented in a later chapter.

We want to present two more decompositions of matrices before we close this section. First, we need a definition. A square matrix A is said to be normal if $A^*A = AA^*$.

P 5.3.8 (Decomposition of a Normal Matrix) A normal matrix A can be factorized as

$$A = P\Gamma P^* \tag{5.3.8}$$

where Γ is diagonal and P unitary.

PROOF. Let us use the general decomposition result (5.3.7) on A. Write $A = P\Gamma P^*$, where Γ is upper-triangular and P unitary. Note that

$$\begin{aligned} AA^* &= P\Gamma P^* P\Gamma^* P^* = P\Gamma\Gamma^* P^* \\ &= A^*A = P\Gamma^* P^* P\Gamma P^* = P\Gamma^*\Gamma P^*. \end{aligned}$$

Consequently, $\Gamma\Gamma^* = \Gamma^*\Gamma$. If Γ is upper-triangular, then Γ has to be diagonal. (Why?)

P 5.3.9 (Decomposition of a Hermitian Matrix) A Hermitian matrix A can be factorized as

$$A = P\Gamma P^*, \tag{5.3.9}$$

where Γ is diagonal with real entries and P unitary.

PROOF. Let us use the general decomposition result (5.3.7) again. Write $A = P\Gamma P^*$, where Γ is upper-triangular and P unitary. Note that $A^* = P\Gamma^* P^* = A = P\Gamma P^*$. Consequently, $\Gamma = \Gamma^*$ and hence Γ is diagonal with real entries.

Complements

5.3.1 Let $E(\omega)$ be the Householder matrix based on the vector ω of unit length. Show that: $E(\omega)a = a$, if a and ω are orthogonal, i.e., $a^*\omega = 0$, and $E(\omega)\omega = -\omega$.

5.3.2 Let

$$A = \begin{bmatrix} 0 & 1 & 0 & 2 & 1 \\ 0 & 2 & 0 & 4 & 1 \\ 0 & 3 & 0 & 6 & 1 \end{bmatrix}.$$

Obtain QR factorization of A following the method outlined in the proof of **P 5.3.3**. Obtain also QR factorization of A following the Gram-Schmidt as well as the modified Gram-Schmidt process. Comment on the computational stability of the two methods.

5.3.3 Let

$$A = \begin{bmatrix} a_{11} & a_{12} \\ a_{21} & a_{22} \end{bmatrix}$$

be non-negative definite. Spell out explicitly a Cholesky decomposition of A. If A is non-negative definite but not positive definite, explore the source of non-uniqueness in the Cholesky decomposition of A.

5.3.4 The QR factorization is very useful in solving systems of linear equations. Suppose $Ax = b$ is a system of linear equations in unknown x of order $n \times 1$, where A of order $m \times n$ and b of order $m \times 1$ are known. Suppose a QR decomposition of A, i.e., $A = QR$ is available. Rewriting the equations as $Rx = Q^*b$, demonstrate a simple way of solving the linear equations.

5.3.5 (SQ Factorization) If A is an $m \times n$ matrix, $m \leq n$, show that A can be factorized as $A = SQ$ where S is $m \times m$ lower triangular and Q is $m \times n$ matrix with orthonormal rows.

5.3.6 Work out the SVD (singular value decomposition) of the matrix A in the example **5.3.2** above.

5.4. Eigenvalues and Eigenvectors

In many of the factorization results on matrices, the main objective is to reduce a given matrix to a diagonal matrix. The question is whether one can attach some meaning to the numbers that appear in the diagonal matrix. In this section, we embark on such an investigation.

DEFINITION 5.4.1. Let A be a square matrix of order $m \times m$ with complex entries. A complex number λ is said to be an eigenvector of A if there exists a non-zero vector x such that

$$Ax = \lambda x. \tag{5.4.1}$$

We have come across the word "eigenvalue" before in a different context in Chapter 2. There is a connection between what was presented in Chapter 2 in the name of "eigenvalue" and what we are discussing now. This connection is explored in the complements at the end of the section.

Rewriting (5.4.1) as

$$(A - \lambda I_m)x = 0, \tag{5.4.2}$$

we see that any λ for which $A - \lambda I_m$ is singular would produce a non-zero solution to the system $(A - \lambda I_m)x = 0$ of homogeneous linear equations in x. The matrix $A - \lambda I_m$ being singular would imply that the determinant $|A - \lambda I_m| = 0$. But $|A - \lambda I_m|$ is a polynomial in λ of degree m. Consequently, one can conclude that every matrix of order $m \times m$ has m eigenvalues which are the roots of the polynomial equation $|A - \lambda I_m| = 0$ in λ.

Suppose λ is an eigenvalue of a matrix A. Let x be an eigenvector of A corresponding to λ. If α is a non-zero scalar, αx is also an eigenvector corresponding to the eigenvalue λ. For some special matrices eigenvalues are real.

P 5.4.2 Let A be a Hermitian matrix. Then all its eigenvalues are real.

PROOF. Let λ be an eigenvalue and x a corresponding eigenvector of $A = (a_{ij})$, i.e., $Ax = \lambda x$. Take x to be of unit length. This implies that

$$x^* Ax = \sum_{i=1}^{m} \sum_{j=1}^{m} a_{ij}\, x_i \bar{x}_j = \lambda x^* x = \lambda,$$

where $x^* = (x_1, x_2, \ldots, x_m)$. We show that x^*Ax is real. It is clear that $a_{ii}x_i\bar{x}_i$ is real for all i. If $i \neq j$,

$$a_{ij}x_i\bar{x}_j + a_{ji}x_j\bar{x}_i = a_{ij}x_i\bar{x}_j + \bar{a}_{ij}\bar{x}_i x_j$$

is real. Hence λ is real.

In Section 5.3, we presented a decomposition of a Hermitian matrix. The entries in the diagonal matrix involved have a special meaning.

P 5.4.3 Let A be a Hermitian matrix and $A = P\Gamma P^*$ its decomposition with P unitary and $\Gamma = \text{Diag}\{\lambda_1, \lambda_2, \ldots, \lambda_m\}$. Then $\lambda_1, \lambda_2, \ldots, \lambda_m$ are the eigenvalues of A.

PROOF. Look at the determinantal equation $|A - \lambda I_m| = 0$ for eigenvalues of A. Note that

$$\begin{aligned} |A - \lambda I_m| &= |P\Gamma P^* - \lambda I_m| = |P\Gamma P^* - \lambda PP^*| \\ &= |P(\Gamma - \lambda I_m)P^*| = |P|\ |\Gamma - \lambda I_m|\ |P^*| \\ &= |\Gamma - \lambda I_m| = (\lambda_1 - \lambda)(\lambda_2 - \lambda)\cdots(\lambda_m - \lambda). \end{aligned}$$

The proof is complete.

It can also be verified that the i-th column vector of P is an eigenvector corresponding to the eigenvalue λ_i of A.

We now spend some time on eigenvectors. Let λ and μ be two distinct eigenvalues of a Hermitian matrix A and x and y the corresponding eigenvectors. Then x and y are orthogonal. Observe that

$$Ax = \lambda x \Rightarrow y^*Ax = \lambda y^*x,$$
$$Ay = \mu y \Rightarrow x^*Ay = \mu x^*y.$$

Since $x^*Ay = y^*Ax$ and $x^*y = y^*x$, we have

$$(\lambda - \mu)y^*x = 0,$$

from which we conclude that $y^*x = 0$, i.e., x and y are orthogonal.

If A is any matrix and λ and μ are two distinct eigenvalues of A with corresponding eigenvectors x and y, the good thing we can say about x and y is that they are linearly independent. Suppose they are linearly dependent. Then there exist two scalars α and β such that

$$\alpha x + \beta y = 0. \tag{5.4.3}$$

Since the vectors x and y are non-zero, both α and β have to be non-zero. Note that

$$0 = A(\alpha x + \beta y) = \alpha Ax + \beta Ay = \alpha\lambda x + \beta\mu y. \tag{5.4.4}$$

Multiplying (5.4.3) by μ and then subtracting it from (5.4.4), we note that

$$\alpha\lambda x - \alpha\mu x = 0 \Leftrightarrow \alpha(\lambda - \mu)x = 0.$$

Since $\alpha \neq 0$, $\lambda \neq \mu$, and $x \neq 0$, we have a contradiction to the assumption of linear dependence of x and y.

We will now discuss the notion of multiplicity in the context of eigenvalues and eigenvectors. The multiplicity a_0 of a root λ_0 of the equation $|A - \lambda I| = 0$ is called its algebraic multiplicity. The number g_0 of linearly independent solutions of the system $(A - \lambda_0 I)x = 0$ of equations is called the geometric multiplicity of the root λ_0. If A is Hermitian, we show that $g_0 = a_0$. This result follows from the decomposition theorem for Hermitian matrices reported in **P 5.3.8**. From this theorem, we have a unitary matrix $P = (x_1, x_2, \ldots, x_m)$ such that $A = P\Delta P^*$, where $\Delta = \text{Diag}\{\lambda_1, \lambda_2, \ldots, \lambda_m\}$, x_i is the i-th column of P, and λ_i's are the eigenvalues of A. Let $\lambda_{(1)}, \lambda_{(2)}, \ldots, \lambda_{(s)}$ be the distinct values among $\lambda_1, \lambda_2, \ldots, \lambda_m$ with multiplicities $a_1, a_2, \ldots, a_s$ respectively. Clearly, $a_1 + a_2 + \ldots + a_s = m$. Assume, without loss of generality, that the first a_1 numbers among $\lambda_1, \lambda_2, \ldots, \lambda_m$ are each equal to $\lambda_{(1)}$, the next a_2 numbers are each equal to $\lambda_{(2)}$ and so on. Since

$$Ax_i = \lambda_{(1)}x_i, \; i = 1, 2, \ldots, a_1,$$

and x_i, $i = 1, 2, \ldots, a_1$ are orthogonal, it follows that $a_1 \leq g_1$. A similar argument yields that $a_i \leq g_i$ for $i = 2, 3, \ldots, s$. Consequently,

$$m = \sum_{i=1}^{s} a_i \leq \sum_{i=1}^{s} g_i \leq m.$$

(We cannot have more than m linearly independent vectors.) Hence $a_i = g_i$ for all i.

This result is not true for any matrix. As an example, let

$$A = \begin{bmatrix} 0 & 1 \\ 0 & 0 \end{bmatrix}.$$

Zero is an eigenvalue of A of multiplicity 2. The corresponding eigenvector is of the form

$$\begin{bmatrix} x \\ 0 \end{bmatrix}$$

with $x \neq 0$. Consequently, the geometric multiplicity of the zero eigenvalue is one.

We will now present what is called the spectral decomposition of a Hermitian matrix A. This is essentially a rehash of **P 5.3.8**. From the decomposition $A = P\triangle P^*$ we can write

$$A = \lambda_1 x_1 x_1^* + \lambda_2 x_2 x_2^* + \ldots + \lambda_m x_m x_m^* \tag{5.4.5}$$

where $P = (x_1, x_2, \ldots, x_m)$, x_i is the i-th column of P, and

$$\triangle = \text{Diag}\{\lambda_1, \lambda_2, \ldots, \lambda_m\}.$$

Without loss of generality, assume that $\lambda_i \neq 0$, $i = 1, 2, \ldots, r$ and $\lambda_i = 0$, $i = r+1, r+2, \ldots, m$. Let $\lambda_{(1)}, \lambda_{(2)}, \ldots, \lambda_{(s)}$ be the distinct values among $\lambda_1, \ldots, \lambda_m$. We can rewrite (5.4.5) as

$$A = \lambda_{(1)} E_1 + \lambda_{(2)} E_2 + \ldots + \lambda_{(s)} E_s, \tag{5.4.6}$$

where E_i is the sum of all the matrices $x_j x_j^*$ associated with the same eigenvalue λ_i. Note that $E_i^2 = E_i$, $E_i^* = E_i$ for all i, and $E_i E_j = 0$ for all $i \neq j$. Further, $\rho(E_1)$ is the multiplicity of the root $\lambda_{(1)}$. The form (5.4.6) is the spectral decomposition of A.

The spectral decomposition (5.4.6) is unique. We will demonstrate its uniqueness as follows.

Suppose

$$A = \lambda_{(1)} F_1 + \lambda_{(2)} F_2 + \ldots + \lambda_{(s)} F_s \tag{5.4.7}$$

is another decomposition of A with the properties $F_i^2 = F_i$ for all i and $F_i F_j = 0$ for all $i \neq j$. Subtracting (5.4.7) from (5.4.6), we note that

$$\lambda_{(1)}(E_1 - F_1) + \lambda_{(2)}(E_2 - F_2) + \ldots + \lambda_{(s)}(E_s - F_s) = 0. \tag{5.4.8}$$

Multiplying (5.4.8) by E_i on the left and F_j on the right, for $i \neq j$, we have $E_i F_j = 0$. Multiplying (5.4.8) by E_i on the left, we note that $E_i = E_i F_i$. Using a similar argument, we can show that $E_i F_i = E_i$. Thus we have $E_i = F_i$ for all i.

Now we take up the case of an important class of matrices, namely non-negative definite (also called positive semi-definite) matrices.

DEFINITION 5.4.4. A Hermitian matrix A is said to be non-negative definite (nnd) if $x^*Ax \geq 0$ for all column vectors $x \in \mathbf{C}^m$.

Non-negative definite matrices are like non-negative numbers. For example, we can take the square root of an nnd matrix which is also nnd. We shall now identify a special subclass of nnd matrices.

DEFINITION 5.4.5. A Hermitian matrix A is said to be positive definite (abbreviated as pd) if A is nnd and $x^*Ax = 0$ if and only if $x = 0$.

What we need is a tangible criterion to check whether or not a given Hermitian matrix is nnd. In the following result, we address this problem.

P 5.4.6 Let A be a Hermitian matrix.

(1) The matrix A is nnd if and only if all its eigenvalues are non-negative.
(2) For the matrix A to be nnd, it is necessary (not sufficient) that all its leading principal minors are non-negative.
(3) The matrix A is positive definite if and only if all its eigenvalues are positive.
(4) The matrix A is positive definite if and only if all its leading principal minors are positive.

PROOF. (1) Let λ be an eigenvalue of A and x a corresponding eigenvector, i.e., $Ax = \lambda x$. Note that $0 \leq x^*Ax = \lambda x^*x$ from which we have $\lambda \geq 0$. Conversely, suppose every eigenvalue of A is non-negative. By **P 5.3.8**, there exists a unitary P such that $P^*AP = \Delta = \text{Diag}\{\lambda_1, \lambda_2, \dots, \lambda_m\}$, where $\lambda_1, \lambda_2, \dots, \lambda_m$ are the eigenvalues of A. Let $x \in \mathbf{C}^m$ be any given vector. Let $y = P^*x = (y_1, \dots, y_m)'$. Then

$$\begin{aligned} x^*Ax &= y^*P^*APy = y^*\Delta y \\ &= \sum_{i=1}^{m} \lambda_i |y_i|^2 \geq 0. \end{aligned}$$

(2) Suppose A is nnd. Then the determinant $|A|$ of A is non-negative. This follows from **P 5.3.8**. As a matter of fact,

$$|A| = \Pi_{i=1}^m \lambda_i \geq 0.$$

Observe that any principal submatrix of A is also non-negative definite. Consequently, the leading principal minors of A are non-negative.

(3) A proof of this assertion can be built based on the proof of (1).

(4) If A is positive definite, it is clear that every leading principal minor of A is positive. Conversely, suppose every leading principal minor of A is positive. We want to show that A is positive definite. Let $A = (a_{ij})$. By hypothesis, $a_{11} > 0$. Using a_{11} as a pivot, sweep out the first column and first row of A. Let B be the resultant matrix. Write

$$B = \begin{bmatrix} a_{11} & 0 \\ 0 & B_1 \end{bmatrix}.$$

We note the following. 1. Every principal minor of A and the corresponding principal minor of B which includes the first row are equal. 2. Any leading principal minor of B_1 of order $k \times k$ is equal to $(\frac{1}{a_{11}})$ times the leading principal minor of A_1 of order $(k+1) \times (k+1)$. These facts are useful in proving the result. We use induction. The result is obviously true for 1×1 matrices. Assume that the assertion holds for any matrix of order $(m-1) \times (m-1)$. The given matrix A is assumed to be of order $m \times m$ and for which all the leading principal minors are positive. Let B be the matrix obtained from A as above. It now follows that every leading principal minor of B_1 is positive. By the induction hypothesis, B_1 is positive definite. Consequently, B is positive definite. Hence A is positive definite. (Why?) This completes the proof.

Note the difference between the assertions (2) and (4) of **P 5.4.6**. If all the principal leading minors of A are non-negative, it does not follow that A is non-negative definite. For an example, look at

$$A = \begin{bmatrix} 0 & 0 \\ 0 & -1 \end{bmatrix}.$$

However, if *all* the principal minors of the Hermitian matrix are non-negative, then A is non-negative definite.

Complements

5.4.1 Let $\lambda_1, \lambda_2, \ldots, \lambda_m$ be the eigenvalues of a matrix A of order $m \times m$. Show that

$$\operatorname{Tr} A = \sum_{i=1}^{m} \lambda_i \text{ and } |A| = \Pi_{i=1}^{m} \lambda_i.$$

5.4.2 If λ is an eigenvalue of A, show that λ^2 is an eigenvalue of A^2.

5.4.3 Let A be a Hermitian matrix of order $m \times m$ with complex entries. For row vectors $x, y \in \mathbf{C}^m$, let $K(x, y) = xAy^*$. Show that $K(\cdot, \cdot)$ is a Hermitian conjugate bilinear functional on the vector space $\mathbf{C}^m$.

5.4.4 Let $K(\cdot, \cdot)$ be as defined in Complement 5.4.3. Let $< \cdot, \cdot >$ be the usual inner product on $\mathbf{C}^m$. Let $\lambda_1 \geq \lambda_2 \geq \ldots \geq \lambda_m$ be the eigenvalues of $K(\cdot, \cdot)$. Show that $\lambda_1, \lambda_2, \ldots, \lambda_m$ are the eigenvalues of A.

5.4.5 Let A be a Hermitian matrix of order $m \times m$ and A_i the i-th leading principal minor of A, $i = 1, 2, \ldots, m$. If $A_1 > 0, \ldots, A_{m-1} > 0$ and $A_m \geq 0$, show that A is nnd.

5.4.6 Let $A = (a_{ij})$ be an nnd matrix. If $a_{ii} = 0$, show that $a_{ij} = 0$ for all j.

5.4.7 Let A be an nnd matrix. Show that there exists an nnd matrix B such that $B^2 = A$. Show also that B is unique. (It is customary to denote B by $A^{1/2}$.)

5.4.8 If A is nnd, show that A^2 is nnd.

5.4.9 If A is pd, show that A^{-1} is pd.

5.4.10 If A is positive definite, show that these exists a non-singular matrix C such that $C^*AC = I$.

5.4.11 If $A = (a_{ij})$ is nnd, show that $|A| \leq \Pi_{i=1}^m a_{ii}$.

5.4.12 For any matrix B of order $m \times n$, show that BB^* is nnd. What is the relationship between eigenvalues of BB^* and singular values of B?

5.4.13 Let $B = (b_{ij})$ be a matrix of order $m \times m$ with real entries. Show that

$$|B|^2 \leq \Pi_{j=1}^m (b_{1j}^2 + b_{2j}^2 + \ldots + b_{mj}^2).$$

Hint. Look at $A = B'B$.

5.4.14 Let A be an $n \times n$ nonsingular matrix with complex entries. Show that $(A^{-1})' = A^{-1}$ if $A' = A$, where A' is the transpose of A.

5.4.15 Show that a complex symmetric matrix need not be diagonalizable.

5.4.16 Show that every square matrix is similar to its transpose.

5.4.17 Let A be a matrix of order $m \times m$ given by

$$A = \begin{bmatrix} 1 & \rho & \rho & \dots & \rho \\ \rho & 1 & \rho & \dots & \rho \\ \rho & \rho & 1 & \dots & \rho \\ \rho & \rho & \rho & \dots & 1 \end{bmatrix}.$$

Show that A is positive definite if and only if $-\frac{1}{m-1} < \rho < 1$.

5.4.18 Let A and B be two Hermitian matrices of the same order. Say $A \geq B$ if $A - B$ is nnd. Prove the following.

(1) If $A \geq B$ and $B \geq C$, then $A \geq C$.
(2) If $A \geq B$ and $B \geq A$, then $A = B$.
(3) If A and B are nnd and $A \geq B$, then $|A| \geq |B|$.
(4) If A and B are positive definite, $A \geq B$, and $|A| = |B|$, then $A = B$.

5.4.19 A Hermitian matrix A is said to be negative semi-definite if $x^* Ax \leq 0$ for all column vectors $x \in \mathbf{C}^m$. A Hermitian matrix A is said to be negative definite if A is negative semi-definite and $x^* Ax = 0$ only if $x = 0$. Formulate **P 5.4.6** for these matrices.

5.5. Simultaneous Reduction of Two Matrices

The principal goal of this section is to investigate under what circumstances two given matrices can be factorized in such a way that some factors are common. First, we tackle Hermitian matrices. Before this, we need a result which is useful in our quest.

P 5.5.1 Let A be a square matrix of order $n \times n$ with complex entries. Let x be a non-zero column vector of order $n \times 1$. Then there exists an eigenvector y of A belonging to the span of $x, Ax, A^2x, \dots$.

PROOF. The vectors $x, Ax, A^2x, \dots$ cannot all be linearly independent. Let k be the smallest integer such that

$$A^k x + b_{k-1}A^{k-1}x + b_{k-2}A^{k-2}x + \dots + b_1 Ax + b_0 x = 0 \qquad (5.5.1)$$

for some scalars $b_0, b_1, \dots, b_{k-1}$. (Why?) Let $\mu_1, \mu_2, \dots, \mu_k$ be the roots of the polynomial,

$$z^k + b_{k-1}z^{k-1} + b_{k-2}z^{k-2} + \dots + b_1 z + b_0 = 0,$$

of degree k in z, i.e., we can write

$$z^k + b_{k-1}z^{k-1} + b_{k-2}z^{k-2} + \ldots + b_1 z + b_0$$
$$= (z - \mu_1)(z - \mu_2)\ldots(z - \mu_k).$$

Consequently,

$$0 = A^k x + b_{k-1}A^{k-1}x + b_{k-2}A^{k-2}x + \ldots + b_1 Ax + b_0 x$$
$$= (A - \mu_1 I)(A - \mu_2 I)\ldots(A - \mu_k I)x.$$

Let $y = (A - \mu_2 I)(A - \mu_3 I)\ldots(A - \mu_k I)x$. It is clear that $y \neq 0$ (Why?) and y is an eigenvector of A. Further, y belongs to the span of the vectors x, Ax, $A^2x, \ldots , A^{k-1}x$. This completes the proof.

P 5.5.2 Let A and B be two Hermitian matrices of the same order $n \times n$. Then a necessary and sufficient condition that A and B have factorizations, $A = P\Delta_1 P^*$ and $B = P\Delta_2 P^*$, with P unitary and Δ_1 and Δ_2 diagonal matrices is that A and B commute, i.e., $AB = BA$.

PROOF. If A and B have the stipulated factorizations, it is clear that $AB = BA$. On the other hand, let A and B commute. Let y_1 be an eigenvector of B. For any positive integer r, we show that $A^r y_1$ is also an eigenvector of B provided that it is non-zero. Since $By_1 = \lambda y_1$ for some scalar λ, we have $B(A^r y_1) = A^r B y_1 = \lambda A^r y_1$, from which the avowed assertion follows. We now look at the sequence $y_1, Ay_1, A^2y_1, \ldots$ of vectors. There is a vector p_1 in the span of the sequence which is an eigenvector of A. The vector p_1 is obviously an eigenvector of B. Thus we are able to find a common eigenvector of both A and B. We can take p_1 to be of unit length. Let y_2 be an eigenvector of B orthogonal to p_1. We claim that p_1 is orthogonal to every vector in the span of $y_2, Ay_2, A^2y_2, \ldots$. For any $r \geq 1$,

$$(A^r y_2)^* p_1 = y_2^* A^r p_1 = y_2^*(\alpha^r p_1) = \alpha^r y_2^* p_1 = 0,$$

where α is the eigenvalue of A associated with the eigenvector p_1 of A. Following the argument used earlier in the proof, we can find a vector p_2 in the span of $y_2, Ay_2, A^2y_2, \ldots$ which is a common eigenvector of both A and B. It is clear that p_2 and p_1 are orthogonal. Take p_2 to be of unit length. Continuing this way, we obtain orthonormal eigenvectors

$p_1, p_2, \ldots, p_n$ common to both A and B. Let $P = (p_1, p_2, \ldots, p_n)$. Note that P is unitary and both P^*AP and P^*BP are diagonal. This completes the proof.

Thus the above result clearly identifies the situation in which two Hermitian matrices are diagonalizable by the same unitary matrix. This result can be extended for any number of matrices. The diagonal matrices involved in the decomposition consist of eigenvalues of their respective matrices in the diagonals.

COROLLARY 5.5.3. Let $A_1, A_2, \ldots, A_k$ be a finite number of Hermitian matrices of the same order. Then there exists a unitary matrix P such that P^*A_iP is diagonal for every i if and only if the matrices commute pairwise, i.e., $A_iA_j = A_jA_i$ for all $i \neq j$.

Finally, we present a result similar to the one presented in **P 5.5.2** for special matrices.

P 5.5.4 Let A and B be two Hermitian matrices at least one of which is positive definite. Then there exists a non-singular matrix C such that C^*AC and C^*BC are both diagonal matrices.

PROOF. Assume that A is positive definite. Then there exists a non-singular matrix D such that $D^*AD = I$. See Complements 5.4.10. Since D^*BD is Hermitian, there exists a unitary matrix U such that $U^*(D^*BD)U$ is diagonal. Take $C = DU$. Then $C^*AC = U^*D^*ADU = U^*IU = I$, which is obviously diagonal. This completes the proof.

It will be instructive to explore the nature of entries in the diagonal matrix C^*BC in **P 5.5.4**. Let $C^*BC = \Delta = \text{diag}\{\alpha_1, \alpha_2, \ldots, \alpha_m\}$. The equation $C^*AC = I$ implies that $C^* = C^{-1}A^{-1}$. Consequently,

$$\Delta = C^*BC = C^{-1}A^{-1}BC$$

from which we have

$$A^{-1}BC = C\Delta. \tag{5.5.2}$$

Let x_i be the i-th column vector of C which is obviously non-zero. From (5.5.2), we have

$$A^{-1}Bx_i = \alpha_i x_i,$$

which means that α_i is an eigenvalue of $A^{-1}B$ and x_i the corresponding eigenvector for every i. It remains to be seen that all the eigenvalues of A^{-1} are accounted by $\alpha_1, \alpha_2, \ldots, \alpha_m$, i.e., $\alpha_1, \alpha_2, \ldots, \alpha_m$ are the roots of the determinantal equation $|A^{-1}B - \alpha I| = 0$. The linear independence of x_i's settles this question. The x_i's have an additional nice property: $x_i^* A x_j = 0$ for all $i \neq j$.

The determinantal equation $|A^{-1}B - \alpha I| = 0$ makes an interesting reading. This equation is equivalent to the equation $|B - \alpha A| = 0$. The roots of this equation can be called the eigenvalues of B with respect to the positive definite matrix A! The usual eigenvalues of B as we know them traditionally can now be called the eigenvalues of B with respect to the positive definite matrix I.

Complements

5.5.1 If A is positive definite and B is Hermitian, show that the eigenvalues of B with respect to A are real.

5.5.2 Let $A_1, A_2, \ldots, A_k$ be k Hermitian matrices with A_1 positive definite. Show that there exists a non-singular matrix C such that $C^* A_i C$ is diagonal for every i if and only if $A_i A_1^{-1} A_j = A_j A_1^{-1} A_i$ for all i and j.

5.5.3 Let

$$A = \begin{bmatrix} 5 & 1 & 3 \\ 1 & 3 & -1 \\ 3 & -1 & 13 \end{bmatrix} \quad \text{and } B = \begin{bmatrix} 10 & 2 & 6 \\ 2 & 3 & -8 \\ 6 & -8 & 14 \end{bmatrix}.$$

Determine a non-singular matrix which diagonalizes A and B simultaneously.

5.5.4 Suppose A and B are Hermitian and commute. Let $\lambda_1, \lambda_2, \ldots, \lambda_m$ be the eigenvalues of A and $\mu_1, \mu_2, \ldots, \mu_m$ those of B. Show that the eigenvalues of $A + B$ are

$$\lambda_1 + \mu_{i_1}, \lambda_2 + \mu_{i_2}, \ldots, \lambda_m + \mu_{i_m}$$

for some permutation $i_1, i_2, \ldots, i_m$ of $1, 2, \ldots, m$.

5.5.5 (**Polar Decomposition**) Let A be an $n \times n$ nonsingular matrix. Then there exist a positive definite matrix G and an orthogonal matrix H such that $A = GH$.

(*Hint:* Consider the positive definite matrix AA' and take $G = (AA')^{\frac{1}{2}}$ the positive definite square root of AA'. Then take $H = G^{-1}A$. Verify that H is orthogonal.)

5.6. A Review of Matrix Factorizations

Because of the importance of matrix factorization theorems in applications, all major results of Chapter 5 and some additional propositions not proved in the chapter are recorded in this section for ready reference. The reader may consult the references given to books and papers for further details. We use the following notations. The class of $m \times n$ matrices is represented by $M_{m,n}$ and the matrices of order $n \times n$ by M_n.

Triangular Matrices. A matrix $A = (a_{ij}) \in M_n$ is called upper triangular if $a_{ij} = 0$ whenever $j < i$ and lower triangular if $a_{ij} = 0$ whenever $j > i$. A unit triangular matrix is a triangular matrix which has unities in the diagonal.

Permutation Matrices. A matrix $P \in M_n$ is called a permutation matrix if exactly one entry in each row and column is equal to 1, and all other entries are zero. Premultiplication by such a matrix interchanges the rows and postmultiplication, the columns.

Hessenberg Matrices. A matrix $A \in M_n$ is said to be an upper Hessenberg matrix if $a_{ij} = 0$ for $i > j + 1$, and its transpose is called a lower Hessenberg matrix.

Tridiagonal Matrices. $A \in M_n$ is said to be a tridiagonal matrix if $a_{ij} = 0$, whenever $|i - j| > 1$. For example

$$\begin{bmatrix} a_{11} & a_{12} & a_{13} & a_{14} \\ a_{21} & a_{22} & a_{23} & a_{24} \\ 0 & a_{32} & a_{33} & a_{34} \\ 0 & 0 & a_{43} & a_{44} \end{bmatrix} \text{ and } \begin{bmatrix} a_{11} & a_{12} & 0 & 0 \\ a_{21} & a_{22} & a_{23} & 0 \\ 0 & a_{32} & a_{33} & a_{34} \\ 0 & 0 & a_{43} & a_{44} \end{bmatrix}$$

are upper Hessenberg and tridiagonal matrices respectively.

Givens Matrices. A matrix $A(l, m; c, s) \in M_n$ is said to be a Givens matrix if

$$a_{ii} = 1, \; i \neq l, \; i \neq m; \; a_{ll} = a_{mm} = c$$
$$a_{lm} = -s, \; a_{ml} = s, \text{ and } a_{ij} = 0 \text{ elsewhere.}$$

We may choose $c = \cos\theta$ and $s = \sin\theta$. Geometrically, the Givens matrix $A = (l, m; c, s)$ rotates the l-th and m-th coordinate axes in the (l, m)-th plane through an angle θ.

Other matrices. $A \in M_n$ is said to be Hermitian if $A = A^*$, positive (negative) definite if $x^*Ax > 0 (< 0)$ for all nonzero $x \in \mathbf{C}^n$ and non-negative (nonpositive) definite if $x^*Ax \geq 0 (x^*Ax \leq 0)$ for all $x \in \mathbf{C}^n$. The abbreviations pd is used for positive definite and nnd for non-negative definite. An alternative term for non-negative definite used in books on algebra is positive semi-definite abbreviated as psd.

P 5.6.1 **(Rank Factorization)** Let $A \in M_{m,n}$ and rank $\rho(A) = k$. Then there exist matrices $R \in M_{m,k}$, $F \in M_{k,n}$ and $\rho(R) = \rho(F) = k$ such that $A = RF$.

In the following propositions $A \in M_n$ represents a general matrix, $L \in M_n$, a lower triangular matrix and $U \in M_n$, an upper triangular matrix, all with complex entries unless otherwise stated.

P 5.6.2 **(LU Factorization Theorems)**

(1) If all the leading principal minors of A are nonzero, then A can be factorized as $A = LU$, where the diagonal entries of L can all be chosen as unities.
(2) If A is nonsingular, there exists a permutation matrix $P \in M_n$ such that $PA = LU$, where the diagonal entries of L can all be chosen as unities.
(3) In any case there exist permutation matrices $P_1, P_2 \in M_n$ such that $A = P_1LUP_2$. If A is nonsingular, it may be written as $A = P_1LU$.

P 5.6.3 **(Schur's Triangulation Theorems)**

(1) Let $\lambda_1, \ldots, \lambda_n$ be eigenvalues of A in any prescribed order. Then there exists a unitary matrix $Q \in M_n$ such that $Q^*AQ = U$ or $A = QUQ^*$, where U is upper triangular with the eigenvalues $\lambda_1, \ldots, \lambda_n$ as diagonal entries. That is, every square matrix is unitarily equivalent to a triangular matrix whose diagonal entries are in a prescribed order. If A is real and if all its eigenvalues are real, then U may be chosen to be real orthogonal.

(2) Given a real $A \in M_n$ with k real eigenvalues, $\lambda_1, \ldots, \lambda_k$, and $x_j + iy_j$ as complex eigenvalues for $j > k$, there exists a real orthogonal matrix $Q \in M_n$ such that $A = QRQ'$ where R is a quasi diagonal $n \times n$ matrix

$$R = \begin{bmatrix} \lambda_1 & \cdots & R_{1,k} & R_{1k+1} & \cdots & R_{1m} \\ & \ddots & \vdots & \vdots & & \vdots \\ & & \lambda_k & R_{k,k+1} & \cdots & R_{k,m} \\ & & & Z_{k+1} & \cdots & R_{k+1,m} \\ & & & & \ddots & \vdots \\ & & & & & Z_m \end{bmatrix}$$

and $m = (n+k)/2$, with blocks

$$R_{ij} \text{ of size } \begin{cases} 1 \times 1 & \text{if } i \leq k, j \leq k \\ 1 \times 2 & \text{if } i \leq k, j > k \\ 2 \times 1 & \text{if } i > k, j \leq k \\ 2 \times 2 & \text{if } i > k, j > k \end{cases}$$

$$Z_j = \begin{bmatrix} x_j & b_j \\ -c_j & x_j \end{bmatrix}, \ \sqrt{b_j c_j} = y_j, \text{ for } j > k; \ b_j \geq c_j, \ b_j c_j > 0.$$

Note that $R_{ii} = \lambda_i$, $i = 1, \ldots, k$. For an application of this result to probability theory see Edelman (1997).

P 5.6.4 (QR Factorization). Let $A \in M_{m,n}$. Then A can be factorized as $A = QR$, where $Q \in M_m$ is unitary and

$$R = \begin{cases} \begin{bmatrix} R_0 \\ 0 \end{bmatrix} & \text{if } m > n, \\ [R_1 : S_1] & \text{if } m \leq n, \end{cases}$$

where $R_0 \in M_n$ and $R_1 \in M_m$ are upper triangular and $S_1 \in M_{m,n-m}$.

P 5.6.5 (Upper Hessenberg Reduction) For any A, there exists a unitary matrix $Q \in M_n$ such that $QAQ^* = H_u$ (upper Hessenberg).

P 5.6.6 (Tridiagonal Reduction). If A is Hermitian, there exists a unitary matrix $Q \in M_n$ such that $QAQ^* = H_T$ (Tridiagonal).

P 5.6.7 (Normal Matrix Decomposition) If A is normal, i.e., $AA^* = A^*A$, then there exists a unitary matrix $Q \in M_n$ such that $A = Q\Lambda Q^*$, where $\Lambda \in M_n$ is a diagonal matrix with the eigenvalues of A as diagonal elements.

P 5.6.8 (Spectral Decomposition) Let A be Hermitian. Then $A = Q\Lambda Q^*$, where $Q \in M_n$ is unitary and $\Lambda \in M_n$ is diagonal with real entries, which are the eigenvalues of A. If A is real symmetric, then

$A = Q\Lambda Q'$ where Q is orthogonal.

P 5.6.9 (**Singular Value Decomposition**) For $A \in M_{m,n}$, we have $A = V\Delta W^*$, where $V \in M_m$ and $W \in M_n$ are unitary matrices and $\Delta \in M_{m,n}$ has non-negative elements in the main diagonal and zeros elsewhere. If $\rho(A) = k$, then $A = V_0\Delta_0 W_0^*$, where $V_0 \in M_{m,k}$ and $W_0 \in M_{n,k}$ are such that $V_0^* V_0 = I_k, W_0^* W_0 = I_k$ and $\Delta_0 \in M_k$ is a diagonal matrix with positive elements in the main diagonal.

P 5.6.10 (**Hermitian Matrix Decomposition**) If A is Hermitian, we have the factorization $A = S\Delta S^*$, where $S \in M_n$ is nonsingular and $\Delta \in M_n$ is diagonal with +1,−1 or 0 as diagonal entries. The number of +1's and −1's are same as the number of positive and negative eigenvalues of A and the number of zeros is $n - \rho(A)$.

P 5.6.11 (**Symmetric Matrix Decomposition**) If A is real symmetric, then A has the factorization $A = S\Delta S'$, where $S \in M_n$ is nonsingular and Δ is diagonal with +1 or 0 as diagonal entries and $\rho(\Delta) = \rho(A)$.

P 5.6.12 (**Cholesky Decomposition**) If A is Hermitian and nnd, then it can be factorized as $A = LL^*$, where $L \in M_n$ is lower triangular with non-negative diagonal entries. The factorization is unique if A is nonsingular.

P 5.6.13 (**General Matrix Decomposition**) Any A can be factorized as $A = SQ\Sigma Q^* S^{-1}$, where $S \in M_n$ is nonsingular, $Q \in M_n$ is unitary and Σ is diagonal with non-negative entries.

P 5.6.14 (**Polar Decomposition**) Any A with rank k can be factorized as $A = SQ$, where $S \in M_n$ is nnd with rank k and $Q \in M_n$ is unitary. If A is nonsingular, then A can be factorized as $A = GQ$, where $QQ' = I$ and $G = G'$.

P 5.6.15 (**Jordan Canonical Form**) Let A be a given complex matrix. Then there is a nonsingular matrix $S \in M_n$ such that $A = SJS^{-1}$ where J is a block diagonal matrix with the r-th diagonal block as $J_{n_r}(\lambda_r) \in M_{n_r}, r = 1, \ldots, k$ and $n_1 + \ldots + n_k = n$. The λ_i's are eigenvalues of A which are not necessarily distinct. The matrix $J_{n_r}(\lambda_r) = (a_{ij})$ is defined as follows (see Horn and Johnson (1985) for a detailed proof):

$a_{ii} = \lambda_r,\ i = 1, \ldots, n_r; a_{i,i+1} = 1,\ i = 1, \ldots, n_r - 1; a_{ij} = 0$ elsewhere.

P 5.6.16 **(Takagi Factorization)** If $A \in M_n$ is symmetric (i.e., $A = A'$), then there exists a unitary $Q \in M_n$ and a real non-negative diagonal matrix Σ such that $A = Q\Sigma Q'$. The columns of Q are an orthogonal set of eigenvectors of $A\bar{A}$ and the corresponding diagonal entries of Σ are the non-negative square roots of the corresponding eigenvalues of $A\bar{A}$. [If $A = (a_{ij})$, $\bar{A} = (\bar{a}_{ij})$, where $\bar{a}_{ij}$ is the complex conjugate of a_{ij}.]

P 5.6.17 Let $A \in M_n$ be given. There exists a unitary $U \in M_n$ and an upper triangular $\Delta \in M_n$ such that $A = U\Delta U'$ if and only if all the eigenvalues of $A\bar{A}$ are real and non-negative. Under this condition all the main diagonal entries of Δ may be chosen to be non-negative.

P 5.6.18 **(Complete Orthogonal Theorem)** Given $A \in M_{m,n}$ with $\rho(A) = k$, there exist unitary matrices $Q \in M_m$ and $W \in M_n$ such that $Q^*AW = \begin{pmatrix} U & 0 \\ 0 & 0 \end{pmatrix}$ where $U \in M_k$ is upper triangular.

P 5.6.19 **(Similarity of Matrices)** Every $A \in M_n$ is similar to a symmetric matrix. [A is similar to B if there exists a nonsingular $S \in M_n$ such that $B = S^{-1}AS$.]

P 5.6.20 **(Simultaneous Singular Value Decomposition)** Let $A, B \in M_{m,n}$. Then there exist unitary matrices $P \in M_m$ and $Q \in M_n$ such that $A = P\Sigma_1 Q^*$ and $B = P\Sigma_2 Q^*$ with both Σ_1 and $\Sigma_2 \in M_{m,n}$ and diagonal if and only if AB^* and B^*A are both normal [C is said to be normal if $CC^* = C^*C$].

P 5.6.21 For a set $\mathcal{F} = \{A_i, i \in \mathcal{I}\} \subset M_{m,n}$, there exist unitary matrices P and Q such that $A_i = P\Lambda_i Q^*$ for all $i \in \mathcal{I}$ and Λ_i are all diagonal if and only if each $A_i^* A_j \in M_n$ is normal and $\mathcal{G} = \{A_i A_j^* : i, j \in \mathcal{I}\} \subset M_m$ is a commuting family.

P 5.6.22 Let $A = A^*, B = B^*$ and $AB = BA$. Then there exists a unitary matrix U such that UAU^* and UBU^* are both diagonal.

P 5.6.23 The Hermitian matrices, $A_1, A_2, \ldots$, are simultaneously diagonalizable by the same unitary matrix U if they commute pairwise.

Note: The main references for this Chapter are: Bhatia (1991), Datta (1995), Golub and van Loan (1989), Horn and Johnson (1985, 1990), and Rao (1973c).

CHAPTER 6

OPERATIONS ON MATRICES

Matrix multiplication is at the core of a substantial number of developments in Matrix Algebra. In this chapter, we look at other multiplicative operations on matrices and their applications.

6.1. Kronecker Product

Let $A = (a_{ij})$ and $B = (b_{ij})$ be two matrices of order $m \times n$ and $p \times q$, respectively. The Kronecker product of A and B is denoted by $A \otimes B$ and defined by

$$A \otimes B = \begin{bmatrix} a_{11}B & a_{12}B & \dots & a_{1n}B \\ a_{21}B & a_{22}B & \dots & a_{2n}B \\ \cdot & \cdot & \dots & \cdot \\ \cdot & \cdot & \dots & \cdot \\ a_{m1}B & a_{m2}B & \dots & a_{mn}B \end{bmatrix}.$$

For a specific example, take $m = 3$, $n = 2$, $p = 2$, and $q = 3$. The Kronecker product of A and B spreads out to be a 6×6 matrix,

$$\underset{3\times 2}{A} \otimes \underset{2\times 3}{B} = \left[\begin{array}{ccc|ccc} a_{11}b_{11} & a_{11}b_{12} & a_{11}b_{13} & a_{12}b_{11} & a_{12}b_{12} & a_{12}b_{13} \\ a_{11}b_{21} & a_{11}b_{22} & a_{11}b_{23} & a_{12}b_{21} & a_{12}b_{22} & a_{12}b_{23} \\ \hline a_{21}b_{11} & a_{21}b_{12} & a_{21}b_{23} & a_{22}b_{11} & a_{22}b_{12} & a_{22}b_{13} \\ a_{21}b_{21} & a_{21}b_{22} & a_{21}b_{23} & a_{22}b_{21} & a_{22}b_{22} & a_{22}b_{23} \\ \hline a_{31}b_{11} & a_{31}b_{12} & a_{31}b_{13} & a_{32}b_{11} & a_{32}b_{12} & a_{32}b_{13} \\ a_{31}b_{21} & a_{31}b_{22} & a_{31}b_{23} & a_{32}b_{21} & a_{32}b_{22} & a_{32}b_{23} \end{array}\right].$$

In the general case, the Kronecker product, $A \otimes B$, of A and B is of order $mp \times nq$. From the very definition of the Kronecker product, it is clear that there are no restrictions on the numbers m, n, p, and q for the product to be meaningful. Notice also that the equality $A \otimes B = B \otimes A$ rarely holds just as it is the case for the usual multiplication of matrices. However, the Kronecker product has one distinctive feature. The matrix $B \otimes A$ can be obtained from $A \otimes B$ by interchanging rows and columns of $A \otimes B$. This feature is absent in the usual multiplication of matrices. We list some of the salient properties of this operation in the following proposition. Most of these properties stem directly from the definition of the product.

P 6.1.1 (1) The operation of performing Kronecker product on matrices is associative. More precisely, if A, B, and C are any three matrices, then

$$(A \otimes B) \otimes C = A \otimes (B \otimes C).$$

It is customary to denote $(A \otimes B) \otimes C$ by $A \otimes B \otimes C$.

(2) If A, B, and C are three matrices with B and C being of the same order, then

$$A \otimes (B + C) = A \otimes B + A \otimes C.$$

(3) If α is a scalar and A is any matrix, then

$$\alpha \otimes A = \alpha A = A \otimes \alpha.$$

(In the Kronecker product operation, view α as a matrix of order 1×1.)

(4) If A, B, C, and D are four matrices such that each pair A and C and B and D is conformable for the usual multiplication, then

$$(A \otimes B)(C \otimes D) = AC \otimes BD.$$

(5) If A and B are any two matrices, then

$$(A \otimes B)' = A' \otimes B'.$$

(6) If A and B are any two matrices, then

$$(A \otimes B)^* = A^* \otimes B^*.$$

(7) If A and B are square matrices not necessarily of the same order, then

$$\operatorname{tr}(A \otimes B) = [\operatorname{tr}(A)][\operatorname{tr}(B)].$$

(8) If A and B are non-singular matrices not necessarily of the same order, then

$$(A \otimes B)^{-1} = A^{-1} \otimes B^{-1}.$$

One of the most important questions concerning Kronecker products is about the relationship of the eigenvalues of the Kronecker product of two matrices and the eigenvalues of the constituent matrices of the product. We will address this question now.

P 6.1.2 Let A and B be two square matrices with eigenvalues $\lambda_1, \lambda_2, \dots, \lambda_m$ and $\mu_1, \mu_2, \dots, \mu_n$, respectively. Then the $\lambda_i\mu_j$, $i = 1, 2, \dots, m$ and $j = 1, 2, \dots, n$ are the eigenvalues of $A \otimes B$.

PROOF. If λ is an eigenvalue of A with a corresponding eigenvector x, and μ is an eigenvalue of B with a corresponding eigenvector y, it is easy to show that $\lambda\mu$ is an eigenvalue of $A \otimes B$ with a corresponding eigenvector $x \otimes y$. As a matter of fact, note that

$$(A \otimes B)(x \otimes y) = (Ax) \otimes (By) = (\lambda x) \otimes (\mu y) = \lambda\mu(x \otimes y),$$

which settles the avowed assertion. This does not prove that the eigenvalues of $A \otimes B$ are precisely $\lambda_i\mu_j$, $i \in \{1, 2, \dots, m\}$ and $j \in \{1, 2, \dots, n\}$. (Why?) Let us prove the assertion invoking the General Decomposition Theorem (**P 5.3.6**). There exist unitary matrices U of order $m \times m$ and V of order $n \times n$ such that $UAU^* = \Delta_1$ and $VBV^* = \Delta_2$, where Δ_1 and Δ_2 are upper-triangular, the diagonal entries of Δ_1 are the eigenvalues of A, and the diagonal entries of Δ_2 are the eigenvalues of B. Note that

$$\begin{aligned}(U \otimes V)(A \otimes B)(U \otimes V)^* &= (U \otimes V)(A \otimes B)(U^* \otimes V^*) \\ &= (UAU^*) \otimes (VBV^*) = \Delta_1 \otimes \Delta_2.\end{aligned}$$

We also note that $U \otimes V$ is unitary and $\Delta_1 \otimes \Delta_2$ upper-triangular. The diagonal entries of $\Delta_1 \otimes \Delta_2$ should exhaust all the eigenvalues of $A \otimes B$. (Why?) This completes the proof.

We now look at the status of eigenvectors. A proof of the following result has already been included in the proof of **P 6.1.2**.

P 6.1.3 Let x be an eigenvector corresponding to some eigenvalue of a matrix A and y an eigenvector corresponding to some eigenvalue of a matrix B. Then $x \otimes y$ is an eigenvector of $A \otimes B$.

A word of warning. Not all eigenvectors of $A \otimes B$ do arise the way it was described in **P 6.1.3**. (See the contrast in the statements of **P 6.1.2** and **P 6.1.3**.) As a counterexample, look at the following 2×2 matrices:

$$A = B = \begin{bmatrix} 0 & 1 \\ 0 & 0 \end{bmatrix}.$$

Let $x' = (1, 0)$. The eigenvalues of A are $\lambda_1 = \lambda_2 = 0$ and the eigenvalues of B are $\mu_1 = \mu_2 = 0$. Only non-zero multiples of x are the eigenvectors of A. But $A \otimes B$ has only three linearly independent eigenvectors, written as rows,

$$u' = (1, 0, 0, 0), \quad v' = (0, 1, 0, 0) \text{ and } w' = (0, 0, 1, 0).$$

There is no way we can write u, v and w in the way expostulated in **P 6.1.3**!

P 6.1.2 has a number of interesting implications. We chronicle some of these in the following Proposition.

P 6.1.4 (1) If A and B are non-negative definite matrices, so is $A \otimes B$.

(2) If A and B are positive definite, so is $A \otimes B$.

(3) If A and B are matrices of order $m \times m$ and $n \times n$, respectively, then (using the notation $|\cdot|$ for determinant),

$$|A \otimes B| = (|A|)^n (|B|)^m.$$

(4) If A and B are two matrices not necessarily square, then

$$\text{rank}(A \otimes B) = [\text{rank}(A)][\text{rank}(B)].$$

Hint: Look at $(AA^*) \otimes (BB^*)$.

It is time to look at the usefulness of Kronecker products. Consider linear equations

$$AX = B, \tag{6.1.1}$$

where A and B are known matrices of orders $m \times n$ and $m \times p$, respectively, and X is of order $n \times p$ and unknown. As an example, look at

$$\begin{bmatrix} a_{11} & a_{12} \\ a_{21} & a_{22} \end{bmatrix} \begin{bmatrix} x_1 & x_2 \\ x_3 & x_4 \end{bmatrix} = \begin{bmatrix} b_{11} & b_{12} \\ b_{21} & b_{22} \end{bmatrix}. \tag{6.1.2}$$

There are two ways we can write these equations in the format we are familiar with. One way is:

$$\begin{bmatrix} a_{11} & a_{12} & 0 & 0 \\ a_{21} & a_{22} & 0 & 0 \\ 0 & 0 & a_{11} & a_{12} \\ 0 & 0 & a_{21} & a_{22} \end{bmatrix} \begin{bmatrix} x_1 \\ x_3 \\ x_2 \\ x_4 \end{bmatrix} = \begin{bmatrix} b_{11} \\ b_{21} \\ b_{12} \\ b_{22} \end{bmatrix}. \tag{6.1.3}$$

Let $x' = (x_1, x_2, x_3, x_4)$ and $b' = (b_{11}, b_{21}, b_{12}, b_{22})$. The system (6.1.3) can be rewritten as

$$(I_2 \otimes A)x = b. \tag{6.1.4}$$

Another way is:

$$\begin{bmatrix} a_{11} & 0 & a_{12} & 0 \\ 0 & a_{11} & 0 & a_{12} \\ a_{21} & 0 & a_{22} & 0 \\ 0 & a_{21} & 0 & a_{22} \end{bmatrix} \begin{bmatrix} x_1 \\ x_2 \\ x_3 \\ x_4 \end{bmatrix} = \begin{bmatrix} b_{11} \\ b_{12} \\ b_{21} \\ b_{22} \end{bmatrix}. \tag{6.1.5}$$

Let $y' = (x_1, x_2, x_3, x_4)$ and $c' = (b_{11}, b_{12}, b_{21}, b_{22})$. The system (6.1.5) can be rewritten as

$$(A \otimes I_2)y = c. \tag{6.1.6}$$

In the general case of (6.1.1), let x be the column vector of order $np \times 1$ obtained from $X = (x_{ij})$ by stacking the rows of X one by one, i.e.,

$$x' = (x_{11}, x_{12}, \ldots, x_{1n}, x_{21}, x_{22}, \ldots, x_{2n}, \ldots, x_{m1}, x_{m2}, \ldots, x_{mn}).$$

This vector has a special name. We will introduce this concept in the next section. Let b be the column vector of order $mp \times 1$ obtained from $B = (b_{ij})$ in the same way x was obtained from X. The system of equations (6.1.1) can be rewritten as

$$(A \otimes I_p)x = b. \tag{6.1.7}$$

Suppose $m = n$, i.e., A is a square matrix. Then the system (6.1.7) has a unique solution if A is non-singular. In the general case, a discussion of the consistency of the system $AX = B$ of matrix equation now becomes easy in (6.1.7), courtesy of Kronecker products!

Another matrix equation of importance is given by

$$AX + XB = C, \tag{6.1.8}$$

where A is of order $m \times m$, B of order $n \times n$, C of order $m \times n$, and X of order $m \times n$. The matrices A, B, and C are known and X is unknown. Let x be the column vector of order $mn \times 1$ obtained from X by stacking the rows of X, and c is obtained analogously from C. It can be shown that the system (6.1.8) is equivalent to

$$(A \otimes I_n + I_m \otimes B')x = c. \tag{6.1.9}$$

Now we can say that the system (6.1.8) admits a unique solution if and only if the matrix

$$D = A \otimes I_n + I_m \otimes B' \tag{6.1.10}$$

of order $mn \times mn$ is non-singular. The matrix D is very special. It will be certainly of interest to know when it is non-singular. First, we would like to say something about the eigenvalues of D.

P 6.1.5 Let D be as specified in (6.1.10) and $\lambda_1, \lambda_2, \ldots, \lambda_m$ be the eigenvalues of A and $\mu_1, \mu_2, \ldots, \mu_n$ be those of B. Then the eigenvalues of D are $\lambda_i + \mu_j$, $i = 1, 2, \ldots, m$ and $j = 1, 2, \ldots, n$.

PROOF. Let $\epsilon > 0$ be any number. Look at the product

$$\begin{aligned}(I_m + \epsilon A) \otimes (I_n + \epsilon B') &= I_m \otimes I_n + \epsilon(A \otimes I_n + I_m \otimes B') + \epsilon^2 A \otimes B' \\ &= I_m \otimes I_n + \epsilon D + \epsilon^2 A \otimes B'.\end{aligned}$$

The eigenvalues of $I_m + \epsilon A$ are $1+\epsilon\lambda_1,\ 1+\epsilon\lambda_2, \ldots, 1+\epsilon\lambda_m$ and those of $I_n + \epsilon B$ are $1+\epsilon\mu_1,\ 1+\epsilon\mu_2, \ldots, 1+\epsilon\mu_n$. Consequently, the eigenvalues $(I_m + \epsilon A) \otimes (I_n + \epsilon B')$ are all given by

$$(1 + \epsilon\lambda_i)(1 + \epsilon\mu_j) = 1 + \epsilon(\lambda_i + \mu_j) + \epsilon^2\lambda_i\mu_j$$

for $i = 1, 2, \ldots, m$ and $j = 1, 2, \ldots, n$. Since ϵ is arbitrary, it now follows that the eigenvalues of D are all given by $(\lambda_i + \mu_j)$ for $i = 1, 2, \ldots, m$ and $j = 1, 2, \ldots, n$. (Why?)

We can reap some benefits out of **P 6.1.5**. The non-singularity of D can be settled.

COROLLARY 6.1.6. Let D be as defined in (6.1.9) and λ_i's be the eigenvalues of A and μ_j's those of B. Then D is non-singular if and only if $\lambda_i + \mu_j \neq 0$ for all i and j.

Complements

6.1.1 The matrices A and B are of the same order and so are the matrices C and D. Show that

$$(A + B) \otimes (C + D) = A \otimes C + A \otimes D + B \otimes C + B \otimes D.$$

6.1.2 Let x and y be two column vectors not necessarily of the same order. Show that

$$x' \otimes y = yx' = y \otimes x'.$$

6.1.3 Let A and B be matrices of orders $m \times n$ and $n \times p$, respectively. Let x be a column vector of order $q \times 1$. Show that

$$(A \otimes x)B = (AB) \otimes x.$$

6.1.4 Let A be a matrix of order $m \times n$. Define the Kronecker powers of A by

$$A^{[2]} = A \otimes A, A^{[3]} = A \otimes A^{[2]}, \quad \text{etc.}$$

Let A and C be two matrices of orders $m \times n$ and $n \times p$, respectively. Show that

$$(AC)^{[2]} = A^{[2]}C^{[2]}.$$

Hence show that $(AC)^{[k]} = A^{[k]}C^{[k]}$ for all positive integers k.

6.1.5 Let $AX = B$ be a matrix equation. Provide a criterion for the existence of a solution to the equation. If the matrix equation is consistent, describe the set of all solutions of the equation.

6.1.6 Obtain a solution to the matrix equiation $AX = B$, where

$$A = \begin{bmatrix} 4 & 2 & -1 \\ 5 & 3 & -1 \\ 3 & -1 & 4 \end{bmatrix} \text{ and } B = \begin{bmatrix} 1 & 3 \\ -1 & 1 \\ 6 & 4 \end{bmatrix}.$$

6.1.7 By looking at the eigenvalues of A and B, show that the following matrix equation has a unique solution:

$$\begin{bmatrix} 1 & -1 \\ 0 & 2 \end{bmatrix} X + X \begin{bmatrix} -3 & 4 \\ 1 & 0 \end{bmatrix} = \begin{bmatrix} 1 & 3 \\ -2 & 2 \end{bmatrix}.$$

Determine the unique solution.

6.1.8 By looking at the eigenvalues of A and B, show that the following matrix equation has more than one solution:

$$\begin{bmatrix} 1 & -1 \\ 0 & 2 \end{bmatrix} X + X \begin{bmatrix} -3 & 4 \\ 0 & -1 \end{bmatrix} = \begin{bmatrix} 0 & 5 \\ 2 & -9 \end{bmatrix}.$$

Determine all solutions of the equation.

6.1.9 Show that the matrix equation $AX - XB = C$ has a unique solution if and only if A and B have no common eigenvalues.

6.1.10 For what values of μ, the matrix equation

$$AX - XA = \mu X$$

has a non-trivial solution in X. If $\mu = -2$, obtain a non-trivial solution to the equation.

6.2. The Vec Operation

One of the most common problems that occurs in many fields of scientific endeavor is solving a system of linear equations. Typically, a system of linear equations can be written in the form $Ax = b$, where the matrices A and b of orders $m \times n$ and $m \times 1$, respectively, are known and x of order $n \times 1$ unknown. Another problem of similar nature is solving matrix equations of the form $AX = B$ or $AX + XB = C$, where A, B, and C are known matrices and X unknown. These matrix equations can be recast in the traditional linear equations format as explained in Section 6.1. The vec operation and Kronecker products play a key role.

Let $A = (a_{ij})$ be a matrix of order $m \times n$. One can create a single column vector comprising all the entries of A. This can be done in two ways. One way is to stack the entries of all the rows of A one after another starting from the first row. Another way, which is more popular, is to stack the columns of A one underneath the other. Let us follow the popular way. Let a_i be the i-th column of A, $i = 1, 2, \ldots, n$.

Formally, we define the vec of A as the column vector of order $mn \times 1$ given by

$$\text{vec}(A) = \begin{bmatrix} a_1 \\ \cdot \\ \cdot \\ a_n \end{bmatrix}.$$

The notation vec A is an abbreviation of the operation of creating a single column vector comprising all the entries of the matrix A in a systematic way as outlined above. In the new notation, the matrix equation $AX = B$, where A is of order $m \times n$, X of order $n \times p$, and B of order $m \times p$, can be rewritten as

$$(I_p \otimes A)\text{vec}X = \text{vec}B.$$

This is just a system of linear equations. If the system is consistent, i.e., admits a solution, one can write down all solutions to the system.

We will now examine some properties of the vec operation. One thing we would like to emphasize is that the vec operation can be defined for any matrix not necessarily square. Another point to note is that if $\text{vec}(A) = \text{vec}(B)$, it does not mean that $A = B$. The matrices A and B could be of different orders and yet $\text{vec}(A) = \text{vec}(B)$ is possible.

P 6.2.1 (1) If x and y are two column vectors not necessarily of the same order, then

$$\text{vec}(xy') = y \otimes x.$$

(2) If A and B are matrices of the same order, then

$$\text{tr}(A'B) = [\text{vec}(A)]'\text{vec}(B).$$

(3) If A and B are matrices of the same order, then

$$\text{vec}(A + B) = \text{vec}(A) + \text{vec}(B).$$

(4) If A, B, and C are three matrices such that the product ABC makes sense, then

$$\text{vec}(ABC) = (C' \otimes A)\text{vec}B.$$

(5) If A and B are two matrices of orders $m \times n$ and $n \times p$, respectively, then

$$\text{vec}(AB) = (B' \otimes I_m)\text{vec}A = (I_p \otimes A)\text{vec}B.$$

PROOF. The assertions (1), (2), and (3) are easy to verify. We tackle (4). Let $A = (a_{ij})$, $B = (b_{ij})$, and $C = (c_{ij})$ be of orders $m \times n$, $n \times p$, and $p \times q$, respectively. Let $b_1, b_2, \ldots, b_p$ be the columns of B. Let $e_1, e_2, \ldots, e_p$ be the columns of the identity matrix I_p. We can write

$$B = BI_p = (b_1, b_2, \ldots, b_p)(e_1, e_2, \ldots, e_p)' = \sum_{j=1}^{p} b_j e_j'.$$

Consequently, by (3) and (4), vec (ABC)

$$\begin{aligned}
&= \text{vec}(A(\sum_{j=1}^{p} b_j e_j')C) = \sum_{j=1}^{p} \text{vec}(Ab_j e_j' C) = \sum_{j=1}^{p} \text{vec}((Ab_j)(C'e_j)') \\
&= \sum_{j=1}^{p} (C'e_j) \otimes (Ab_j) = \sum_{j=1}^{p} (C' \otimes A)(e_j \otimes b_j) \\
&= (C' \otimes A) \sum_{j=1}^{p} (e_j \otimes b_j) \\
&= (C' \otimes A) \sum_{j=1}^{p} \text{vec}(b_j e_j') \\
&= (C' \otimes A)\text{vec}(\sum_{j=1}^{p} b_j e_j') = (C' \otimes A)\text{vec}B.
\end{aligned}$$

The assertion (5) follows from (4) directly if we note that the matrix AB can be written in two ways, namely, $AB = I_m AB = ABI_p$.

Complements

6.2.1 Let A, B, C, and D be four matrices such that $ABCD$ is square. Show that

$$\text{tr}(ABCD) = (\text{vec}D')'(C' \otimes A)\text{vec}B = (\text{vec}(D'))'(A \otimes C')\text{vec}B'.$$

6.2.2 Give a necessary and sufficient condition for the existence of a solution to the matrix equation $AXB = C$, where A, B, and C are all matrices of the same order $m \times m$.

Hint: Use **P 6.2.1**(4).

6.3. The Hadamard-Schur Product

Let $A = (a_{ij})$ and $B = (b_{ij})$ be two matrices of the same order. The Hadamard-Schur (hereafter abbreviated as HS) product of A and B is again a matrix of the same order whose (i, j)-th entry is given by $a_{ij}b_{ij}$. Symbolically, the product is denoted by $A \cdot B = (a_{ij}b_{ij})$. The HS product is precisely the entry-wise product of A and B.

Let $\mathbf{M}_{m,n}$ be the collection of all matrices of order $m \times n$ with complex entries. We have already known that $\mathbf{M}_{m,n}$ is a vector space with respect to the operations of addition of matrices and scalar multiplication of matrices. (If $m = n$, we denote $\mathbf{M}_{m,n}$ by $\mathbf{M}_n$.) The HS product is associative and distributive over matrix addition. The identity element with respect to HS multiplication is the matrix J in which every entry is equal to 1. In short, $\mathbf{M}_{m,n}$ is a commutative algebra, i.e., a commutative ring with a multiplicative identity. If m and n are different, the usual matrix multiplication of matrices does not make sense. However, if $m = n$, the usual matrix multiplication is operational in $\mathbf{M}_n$, and $\mathbf{M}_n$ is indeed a non-commutative algebra. In this section, we study some of the properties of HS multiplication and present some statistical applications. Some of the properties mentioned earlier are chronicled below.

P 6.3.1 (1) If A and B are matrices of the same order, then

$$A \cdot B = B \cdot A.$$

(2) If A, B, and C are three matrices of the same order, then

$$A \cdot (B \cdot C) = (A \cdot B) \cdot C.$$

(Now the brackets in multiplication involving three or more matrices can be deleted.)

(3) If A, B, C, and D are four matrices of the same order, then

$$(A + B) \cdot (C + D) = A \cdot C + A \cdot D + B \cdot C + B \cdot D.$$

(4) If A is any matrix and 0 is the zero matrix of the same order, then $A \cdot 0 = 0$.

(5) If A is any matrix and J is the matrix of the same order each entry of which is 1, then $A \cdot J = A$.

(6) If $m = n$, $A = (a_{ij})$ is any matrix, and I_m is the identity matrix, then $A \cdot I_m = \text{diag}(a_{11}, a_{22}, \ldots, a_{mm})$.

(7) If A and B are any two matrices of the same order, then

$$(A \cdot B)' = A' \cdot B'.$$

(8) If $A = (a_{ij})$ is any matrix with the property that each $a_{ij} \neq 0$, and $B = (1/a_{ij})$, then $A \cdot B = J$. (The matrix B is the (HS) multiplicative inverse of A.)

There seems to be no universal agreement on an appropriate name for entry-wise product of matrices of the same order. In some research papers and books, the product is called Schur product. In 1911, Schur conducted a systematic study of what we call HS multiplication. In 1899, Hadamard studied properties of three power series $f(z) = \Sigma a_n z^n$, $g(z) = \Sigma b_n z^n$, and $h(z) = \Sigma a_n b_n z^n$, and obtained some remarkable results. Even though he never mentioned entry-wise multiplication of matrices in his study but the idea was mute when he undertook the study of coefficient-wise multiplication of two power series.

The following is one of the celebrated results of Schur. It can be proved in a number of different ways. We will concentrate just on the one which is statistical!

P 6.3.2 **(Schur's Theorem)** If A and B are two non-negative definite matrices of the same order, then $A \cdot B$ is also non-negative definite. If A and B are both positive definite, then so is $A \cdot B$.

PROOF. Let X and Y be two independent random vectors with mean vector 0 and dispersion matrices A and B, respectively. The random vector $X \cdot Y$ has mean vector 0 and dispersion matrix $A \cdot B$. It is clear that every dispersion matrix is non-negative definite. The nonstatistical proof is as follows using Kronecker product $A \otimes B$.

The HS product $A \cdot B$ of two matrices can be regarded as a submatrix of $A \otimes B$ of A and B. Let A and B be two square matrices of the same order m. Consider the submatrix of $A \otimes B$ by retaining its rows by numbers $1, m+2, 2m+3, \ldots, (m-1)m+m = m^2$ and the columns by numbers $1, m+2, 2m+3, \ldots, (m-1)m+m = m^2$ and chucking out the rest. This submatrix is precisely $A \cdot B$ and moreover, it is indeed a principal submatrix of $A \otimes B$. If A and B are non-negative definite, then so is $A \otimes B$. Consequently, any principal submatrix of $A \otimes B$ is also non-negative definite. This is another proof of **P 6.3.2**. In the general case when m and n are different, the HS product $A \cdot B$ is still a submatrix of the Kronecker product $A \otimes B$, and the same proof holds.

Schur's theorem has a converse! If A is a non-negative definite matrix, is it possible to write A as a HS product of two non-negative definite matrices? The answer is, trivially, yes. Write $A = A \cdot J$. Recall that J is the matrix in which every entry is equal to 1. If A is positive definite, is it possible to write A as a HS product of two positive definite matrices? The answer is yes. See the complements at the end of the section.

Rank and HS product are the next items to be considered jointly. The following result provides an inequality.

P 6.3.3 Let A and B be two matrices of the same order $m \times n$. Then $\text{rank}(A \cdot B) \leq [\text{rank}(A)][\text{rank}(B)]$.

PROOF. Let us use rank factorization of matrices. Let A and B have ranks a and b, respectively. Then there exist matrices $X = (x_1, x_2, \ldots, x_a)$ of order $m \times a$, $Y = (y_1, y_2, \ldots, y_a)$ of order $n \times a$, $Z = (z_1, z_2, \ldots, z_b)$ of order $m \times b$, and $U = (u_1, u_2, \ldots, u_b)$ of order $n \times b$ such that $A = XY' = \sum_{i=1}^{a} x_i y_i'$ and $B = ZU' = \sum_{i=1}^{b} z_i u_i'$. The matrices X and Y each has rank a and Z and U each has rank b. Note that

$$
\begin{aligned}
A \cdot B &= (\sum_{i=1}^{a} x_i y_i') \cdot (\sum_{j=1}^{b} z_j u_j') = \sum_{i=1}^{a} \sum_{j=1}^{b} (x_i y_i') \cdot (z_j u_j') \\
&= \sum_{i=1}^{a} \sum_{j=1}^{b} (x_i \cdot z_j)(y_i \cdot u_j)'.
\end{aligned}
$$

Consequently, $\rho(A \cdot B) \leq ab = [\rho(A)][\rho(B)]$. See Complement 6.3.1 at the end of this section. The stipulated inequality follows if we observe that each matrix within the summation symbols is of rank 1 at most.

The inequality stated in **P 6.3.3** seems to be very crude. On one hand, the rank of $A \cdot B$ cannot exceed $\min\{m, n\}$ and on the other hand, $[\text{rank}(A)][\text{rank}(B)]$ could be $[\min[m, n]]^2$. However, equality in **P 6.3.3** is possible. Let

$$A = \begin{bmatrix} 1 & 1 & 0 & 0 \\ 1 & 1 & 0 & 0 \\ 0 & 0 & 1 & 1 \\ 0 & 0 & 1 & 1 \end{bmatrix} \text{ and } B = \begin{bmatrix} 1 & 0 & 1 & 0 \\ 0 & 1 & 0 & 1 \\ 1 & 0 & 1 & 0 \\ 0 & 1 & 0 & 1 \end{bmatrix}.$$

Note that $\text{rank}(A) = \text{rank}(B) = 2$ and $\text{rank}(A \cdot B) = 4$.

Next we concentrate in obtaining some bounds for the eigenvalues of HS products of matrices. For any Hermitian matrix A of order $m \times m$, let $\lambda_1(A) \geq \lambda_2(A) \geq \ldots \geq \lambda_m(A)$ be the eigenvalues of A arranged in decreasing order.

P 6.3.4 Let A and B be two non-negative definite matrices of order $m \times m$. Let b_1 and b_m be the largest and smallest entries respectively among the diagonal entries of B. Then

$$b_m \lambda_m(A) \leq \lambda_j(A \cdot B) \leq b_1 \lambda_1(A), \ j = 1, 2, \ldots, m.$$

PROOF. This is virtually a consequence of variational characterization of the largest and smallest eigenvalues of a Hermitian matrix. We will see more of this in a later chapter. We need to note right away that for any Hermitian matrix A and vector x,

$$[\lambda_m(A)]x^*x \leq x^*Ax \leq [\lambda_1(A)]x^*x. \tag{6.3.1}$$

One can establish this inequality by appealing to **P 5.3.8**. Note that

$$B \cdot A = B \cdot (A - \lambda_m(A)I_m) + \lambda_m(A)B \cdot I_m.$$

Next we note that $A - \lambda_m(A)I_m$ is non-negative definite. Use (6.3.1). By **P 6.3.2**, both $B \cdot (A - \lambda_m(A)I_m)$ and $[\lambda_m(A)]B \cdot I_m$ are non-negative definite. If x is a vector of unit length, then

$$x^*(B \cdot A)x \geq [\lambda_m(A)]x^*(B \cdot I_m)x \geq [\lambda_m(A)][\lambda_m(B \cdot I_m)] = [\lambda_m(A)]b_m.$$

(Why?) Since x is arbitrary, it follows that $\lambda_m(A \cdot B) \geq b_m[\lambda_m(A)]$. In a similar vein, the inequality $\lambda_1(A \cdot B) \leq b_1[\lambda_2(A)]$ follows. This completes the proof.

Let us see what we can get out of this inequality. If $B = (b_{ij})$ is Hermitian, then $\lambda_m(B) \leq b_{ii} \leq \lambda_1(B)$ for all i. This follows from (6.3.1). Take x to be the unit vector with the i-th component being 1 and all other components being zeros. This simple observation yields the following result.

COROLLARY 6.3.5. If A and B are two non-negative definite matrices of the same order m, then for all j

$$\lambda_m(A)\lambda_m(B) \leq \lambda_j(A \cdot B) \leq \lambda_1(A)\lambda_1(B).$$

This corollary resembles the result on the rank of an HS product (see **P 6.3.3**). For the next result, we need the notion of a correlation matrix. A non-negative definite matrix $R = (\rho_{ij})$ is said to be a correlation matrix if $\rho_{ii} = 1$ for all i, i.e., every diagonal entry of R is equal to unity. If R is a correlation matrix, then $|\rho_{ij}| \leq 1$ for all i and j. (Why?) If the correlation matrix R is non-singular, then $|\rho_{ij}| < 1$ for all $i \neq j$. Correlation matrices arise naturally in Multivariate Analysis. Let X be a random vector with dispersion matrix $\Sigma = (\sigma_{ij})$. Define $\rho_{ij} =$ correlation between X_i and $X_j = \sigma_{ij}/[\sigma_{ii}\sigma_{jj}]^{1/2} =$ [covariance between X_i and X_j]/[(standard deviation of X_i) (standard deviation of X_j)], where X_i is the i-th component of X. Let $R = (\rho_{ij})$. Then R is a correlation matrix. The manner in which we arrived at the correlation matrix begs an apology. What happens when one of the variances σ_{ii}'s is zero. If σ_{ii} is zero, the entire i-th row of Σ is zero. In such an event, we could define $\rho_{ij} = 1$ for all j. With this convention, we can proclaim that every correlation matrix arises this way. The following is a trivial corollary of **P 6.3.4**.

COROLLARY 6.3.6. Let A be a non-negative definite matrix of order $m \times m$ and R any correlation matrix. Then for all j

$$\lambda_m(A) \leq \lambda_j(A \cdot R) \leq \lambda_1(A).$$

Let us examine some implications of this corollary. If we take $R = I_m$, we get the result that every diagonal entry of A is sandwiched between the smallest and largest eigenvalues of A. Let $\Sigma = (\sigma_{ij})$ be any non-negative definite matrix. It can be regarded as a dispersion matrix of some random vector X with components $X_1, X_2, \ldots, X_m$.

Let $R = (\rho_{ij})$ be the correlation matrix associated with Σ. Write $\sigma_{ii} =$ variance of $X_i = \sigma_i^2$ for all i. Observe that $\sigma_{ij} = \rho_{ij}\sigma_i\sigma_j$ for all i and j. Define $A = \sigma\sigma'$, where the vector $\sigma' = (\sigma_1, \ldots, \sigma_m)$. Note that A is of rank 1. Further, $\Sigma = A \cdot R$. Corollary 6.3.6 provides bounds on the eigenvalues of Σ in terms of the eigenvalues of A. Since A is of rank 1, $(m-1)$ eigenvalues of A are all equal to zero. The other eigenvalue is $\sum_{i=1}^{m} \sigma_i^2$. (Why?) Corollary 6.3.6 offers the inequality $0 \leq \lambda_j(\Sigma) \leq \sum_{i=1}^{m} \sigma_i^2$.

Corollary 6.3.6 offers a good insight as to the magnitude of the eigenvalues when the correlations associated with a dispersion matrix are modified. Suppose we have a dispersion matrix $\Sigma = (\sigma_{ij})$ with the associated correlation matrix $R = (\rho_{ij})$. Suppose we reduce the correlations in R in absolute value by a systematic factor keeping the variances the same. The question is how the eigenvalues of the modified dispersion matrix are affected. Let us make this a little more concrete. Let

$$R_0 = \begin{bmatrix} 1 & \rho & \rho & \ldots & \rho \\ \rho & 1 & \rho & \ldots & \rho \\ \rho & \rho & 1 & \ldots & \rho \\ \cdot & \cdot & \cdot & \ldots & \cdot \\ \rho & \rho & \rho & \ldots & 1 \end{bmatrix},$$

where $-1/(m-1) \leq \rho \leq 1$ is fixed. Clearly, R_0 is a correlation matrix. Let $\Sigma_0 = \Sigma \cdot R_0$. It is clear that Σ_0 is non-negative definite and can be regarded as a dispersion matrix of some random vector. The variances in Σ and Σ_0 are identical. The correlations ρ_{ij}'s associated with the dispersion matrix Σ_0 are a constant multiple of the correlations ρ_{ij}'s associated with Σ. The gist of Corollary 6.3.6 is that the eigenvalues of the modified dispersion matrix Σ_0 are sandwiched between the smallest and largest eigenvalues of the dispersion matrix Σ.

Let us now deal with some determinantal inequalities. The following result is very useful in this connection.

P 6.3.7 Let $A = (a_{ij})$ be a non-negative definite matrix of order $m \times m$ and A_1 the submatrix of A obtained from A by deleting the first row and first column of A. Let $e' = (1, 0, 0, \ldots, 0)$; $\alpha = 0$, if $|A| = 0$; $\alpha = |A|/|A_1|$ if $|A| \neq 0$ and

$$A_2 = A - \alpha ee'.$$

Then the following are valid.

(1) The matrix A_2 is non-negative definite.
(2) If A is positive definite, A^{-1} satisfies $A_2 A^{-1} A_2 = A_2$. (In the nomenclature of a later chapter, A^{-1} is a g-inverse of A_2.)
(3) If A is positive definite and $A^{-1} = (a^{ij})$, then $a_{11}a^{11} \geq 1$.
(4) $|A| \leq a_{11}a_{22}\cdots a_{mm}$.

PROOF. (1) and (2). If $|A| = 0$, $A_2 = A$ and hence A_2 is non-negative definite. Suppose $|A| \neq 0$. Let us determine what α precisely is. Write

$$A = \begin{bmatrix} a_{11} & a_1' \\ a_1 & A_1 \end{bmatrix},$$

where (a_{11}, a_1') is the first row of A. Observe that

$$|A| = |A_1|(a_{11} - a_1' A_1^{-1} a_1).$$

Check the material on Schur complements. Consequently, $\alpha = a_{11} - a_1' A_1^{-1} a_1$. There is another way to identify α. Recall how the inverse of a matrix is computed using its minors. The determinant of A_1 is the cofactor of a_{11}. Therefore, the (1, 1)-th entry a^{11} in A^{-1} is given by $a^{11} = |A_1|/|A| = 1/\alpha = e'A^{-1}e$. Note that

$$\begin{aligned} A_2 A^{-1} A_2 &= (A - \alpha ee')A^{-1}(A - \alpha ee') \\ &= A + \alpha^2 ee'A^{-1}ee' - \alpha ee' - \alpha ee' \\ &= A + \alpha ee' - \alpha ee' - \alpha ee' = A_2. \end{aligned}$$

It is clear that A_2 is symmetric and hence $A_2 A^{-1} A_2$ is non-negative definite. This means that A_2 is non-negative definite. With one stroke, we are able to establish both (1) and (2).

(3) Since A_2 is non-negative definite, its (1,1)-th element must be non-negative, i.e., $a_{11} - \alpha \geq 0$. But $\alpha = 1/a^{11}$. Consequently, $a_{11}a^{11} \geq 1$. As a matter of fact, $a_{ii}a^{ii} \geq 1$ for all i.

(4) The inequality $a_{11}a^{11} \geq 1$ can be rewritten as $|A| \leq a_{11}|A_1|$. Let B be the submatrix of A obtained by deleting the first two rows and first two columns of A. In an analogous way, we find that we have $|A_1| \leq a_{22}|B|$. If we keep pushing this inequality to its utmost capacity, we have the inequality that $|A| \leq a_{11}a_{22}\cdots a_{mm}$. This inequality goes under the name of HS determinantal inequality.

If A is positive definite, it is not necessary that A_2 is positive definite. For a counterexample, take $A = I_2$.

The correlation matrix is an important landmark in Multivariate Analysis. The above result provides a good understanding on the makeup of a correlation matrix.

COROLLARY 6.3.8. If $R = (\rho_{ij})$ is a non-singular correlation matrix then the diagonal entries of $R^{-1} = (\rho^{ij})$ satisfy the inequality $\rho^{ii} \geq 1$. Further, $|R| \leq 1$.

The case when $|R| = 1$ is of interest. If $R = I_m$, then $|R| = 1$. In fact, this is the only situation we have determinant equal to 1. This can be shown as follows. Let $\lambda_1, \lambda_2, \dots, \lambda_m$ be the eigenvalues of R. By the Arithmetic-Geometric mean inequality,

$$(\Pi_{i=1}^m \lambda_i)^{1/m} \leq \frac{1}{m} \sum_{i=1}^m \lambda_i.$$

The equality holds if and only of all λ_i's are equal. In our case, $1 = |R| = \Pi_{i=1}^m \lambda_i$ and $\sum_{i=1}^m \lambda_i = \operatorname{tr}(R) = m$. Thus equality holds in the Arithmetic-Geometric mean inequality. Hence all λ_i's are equal and in fact, they are all equal to unity. Hence $R = I_m$. (Why?)

Now we come to an interesting phase of HS multiplication. The following result involves determinants.

P 6.3.9 Let $A = (a_{ij})$ and $B = (b_{ij})$ be two non-negative definite matrices of the same order $m \times m$. Then

$$|A \cdot B| \geq |A|\ |B|. \tag{6.3.2}$$

PROOF. First, we establish the following inequality:

$$|A \cdot B| \geq |A| b_{11} b_{22} \cdots b_{mm}. \tag{6.3.3}$$

If A is singular or one of the diagonal entries of B is zero, the inequality (6.3.3) is crystal clear. Assume that A is non-singular and none of the diagonal entries of B is zero. Let R be the correlation matrix associated with the dispersion matrix B. Now observe that

$$|A \cdot B| = b_{11} b_{22} \cdots b_{mm} |A \cdot R|.$$

(Why?) In order to establish (6.3.3), it suffices to prove that

$$|A \cdot R| \geq |A| \tag{6.3.4}$$

for any correlation matrix $R = (\rho_{ij})$. Let $A_2 = A - \alpha ee'$, where α and e are as defined in **P 6.3.7**. Let $A^{-1} = (a^{ij})$. Observe that

$$0 \leq |A_2 \cdot R| = |(A - \alpha ee') \cdot R| = |A \cdot R - (1/a^{11})ee'|.$$

The computation of the last determinant requires some tact. We need to borrow a trick or two from the theory of determinants. We note that the determinant $|A \cdot R - a_{11}^{-1} ee'| = |A \cdot R| - a_{11}^{-1}|A_1 \cdot R_1|$ where A_1 is the submatrix of A obtained by deleting the first row and first column of A and R_1 is created analogously. From these deliberations, we obtain the inequality,

$$|A \cdot R| \geq (1/a^{11})|A_1 \cdot R_1| = (|A|/|A_1|)|A_1 \cdot R_1|.$$

This is an interesting inequality begging for a continuation of the process. Let B be the submatrix obtained from A by deleting the first two rows and first two columns. Let R_2 be the correlation matrix obtained from R in a similar fashion. Continuing the work, we have

$$|A \cdot R| \geq (|A|/|A_1|)|A \cdot R_1| \geq (|A|/|A_1|)(|A_1|/|B|)|B \cdot R_2|.$$

Pushing the chain of inequalities to the end, we obtain $|A \cdot R| \geq |A|$. This establishes (6.3.3). Now from (6.3.3), it is clear that $|A \cdot B| \geq |A|\ |B|$. Use **P 6.3.7** (4).

Let us examine what HS multiplication means in certain quarters of Statistics. Suppose $X^{(1)}, X^{(2)}, \ldots$ is a sequence of independent identically distributed random vectors with common mean vector 0 and dispersion matrix Σ. Assume that Σ is non-singular. Let R be the correlation matrix associated with Σ. Let, for each $n \geq 1$,

$$Y^{(n)} = X^{(1)} \cdot X^{(2)} \cdot \ldots \cdot X^{(n)},$$

i.e., $Y^{(n)}$ is the HS product of the random vectors $X^{(1)}, X^{(2)}, \ldots, X^{(n)}$. The correlation matrix $R^{(n)}$ of the random vector $Y^{(n)}$ is precisely the HS product of R with itself n times. Let us denote this product by

$R^{(n)} = (\rho_{ij}^{(n)})$. If $|R| = 1$, the components of $Y^{(n)}$ are clearly uncorrelated. If $0 < |R| < 1$, the components of $Y^{(n)}$ are nearly uncorrelated if n is large. (Why?) The determinantal inequality referred to in **P 6.3.9**, namely $|R^{(n)}| \geq (|R|)^n$, is not informative. The determinant $|R^{(n)}|$ is nearly equal to one if n is large, whereas the quantity $(|R|)^n$ is nearly equal to zero.

One can easily improve the lower bound provided by **P 6.3.9** on the determinant of the HS product of two non-negative definite matrices. This is what we do next.

P 6.3.10 If $A = (a_{ij})$ and $B = (b_{ij})$ are two non-negative definite matrices of order $m \times m$, then

$$|A \cdot B| + |A|\ |B| \geq b_{11}b_{22}\cdots b_{mm}|A| + a_{11}a_{22}\cdots a_{mm}|B|. \qquad (6.3.5)$$

PROOF. Note that the determinantal inequality (6.3.2) is a special case of (6.3.5). The inequality (6.3.5) leads to

$$|A{\cdot}B|+|A||B| \geq b_{11}b_{22}\cdots b_{mm}|A|+a_{11}a_{22}\cdots a_{mm}|B| \geq |B||A|+|A||B|,$$

from which (6.3.2) follows. As for the validity of (6.3.5), if A or B is singular, (6.3.5) is essentially the inequality (6.3.3). Assume that both A and B are non-singular. Let Q and R be the correlation matrices associated with A and B, respectively. It suffices to prove that

$$|Q \cdot R| + |Q|\ |R| \geq |Q| + |R|.$$

To prove this inequality, we can employ the trick we have used in the proof of **P 6.3.9** by looking at the relationship between the minors of a matrix. Let Q_i be the submatrix of Q obtained by deleting the first i rows and i columns of Q, $i = 0, 1, 2, \ldots, m-1$. Let R_i stand for the submatrix of R likewise. Let

$$\ell_{i+1} = |Q_i \cdot R_i| + |Q_i|\ |R_i| - |Q_i|,\ i = 0, 1, 2, \ldots, m-1.$$

Our objective is to show that $\ell_1 \geq 0$. We shall scrutinize ℓ_1 and ℓ_2 a little closely. We need to put in some additional work before the scrutinization. Let $R^{-1} = (\rho^{ij})$. Recall the vector $e' = (1, 0, 0, \ldots, 0)$

we have used in the proof of **P 6.3.9**. By **P 6.3.7**, $\tilde{R} = R-(1/\rho^{11})ee'$ is non-negative definite and hence $Q \cdot \tilde{R}$ is non-negative definite. Further, by (6.3.3),

$$|Q|(1 - 1/\rho^{11}) \leq |Q \cdot \tilde{R}| = |Q \cdot R|\left(1 - \frac{|Q_1 \cdot R_1|}{|Q \cdot Q|\rho^{11}}\right), \tag{6.3.6}$$

from which we have

$$|Q \cdot R| - |Q_1 \cdot R_1|/\rho^{11} \geq |Q| - |Q|/\rho^{11}. \tag{6.3.7}$$

(The equality in (6.3.6) is fashioned after (6.3.6).) By (6.3.7),

$$\begin{aligned}
\ell_1 - \ell_2/\rho^{11} &= |Q \cdot R| - |Q_1 \cdot R_1|/\rho^{11} + |Q|\ |R| - |Q_1|\ |R_1|/\rho^{11} \\
&\quad + |Q_1|/\rho^{11} - |Q| + |R_1|/\rho^{11} - |R| \\
&\geq |Q| - |Q|/\rho^{11} + |Q|\ |R| - |Q_1|\ |R_1|/\rho^{11} \\
&\quad + |Q_1|/\rho^{11} - |Q| + |R_1|/\rho^{11} - |R| \\
&= (1/\rho^{11})(|Q_1| - |Q|) + |Q|\ |R| - |Q_1|\ |R| + |R| - |R| \\
&= (1/\rho^{11} - |R|)(|Q_1| - |Q|).
\end{aligned}$$

In these deliberations, we have used the fact that $\rho^{11} = |R_1|/|R|$. Observe that $(1/\rho^{11}) - |\rho| = (1 - |\rho_1|)/\rho^{11} \geq 0$ as the determinant of a correlation matrix is ≤ 1. Note also that $|Q_1| - |Q| = q^{11}|Q| - |Q| = (q^{11} - 1)|Q| \geq 0$, where $Q^{-1} = (q^{ij})$. Consequently, we observe that $\ell_1 - \ell_2/\rho^{11} \geq 0$, which implies that $\ell_1 \geq \ell_2(|R|/|R_1|)$. This inequality sets a chain reaction. It now follows that $\ell_2 \geq \ell_3(|R_1|/|R_2|)$. Proceeding inductively, we achieve that $\ell_1 \geq 0$. This completes the proof.

Let us spend some time on HS multiplication and ranks. For any two matrices A and B, we have seen that $\text{rank}(A \cdot B) \leq [\text{rank}(A)][\text{rank}(B)]$. If the matrices are non-negative definite, we can do a better job.

P 6.3.11 If A is positive definite and B is non-negative definite with r non-zero diagonal entries, then $\text{rank}(A \cdot B) = r$.

PROOF. Observe that $A \cdot B$ is non-negative definite and has r non-zero diagonal entries. Consequently, $\text{rank}(A \cdot B) \leq r$. Consider the principal submatrix of $A \cdot B$ of order $r \times r$ whose diagonal entries are precisely these non-zero numbers. By (6.3.3), the determinant of this submatrix is non-zero. Thus we now have a minor of order r which is non-zero. Hence $\text{rank}(A \cdot B) \geq r$. This completes the proof.

From **P 6.3.11**, the following interesting results emerge.

(1) If A is pd and B is nnd with diagonal entries nonzero, then $A \cdot B$ is nnd even if B is singular.

(2) If $\rho(A) = \rho(B) = 1$, then $\rho(A \cdot B) = 1$.

(3) If A is pd and B is nd, then $A \cdot B$ is nd.

(4) It is feasible for $A \cdot B$ to have full rank even if A and B are not of full rank. For instance, each of the matrices

$$A = \begin{bmatrix} 2 & 1 & 1 \\ 1 & 1 & 1 \\ 1 & 1 & 1 \end{bmatrix} \text{ and } B = \begin{bmatrix} 2 & 1 & 1 \\ 1 & 1 & 0 \\ 1 & 0 & 1 \end{bmatrix}$$

has rank 2, but $A \cdot B$ has rank 3.

P 6.3.12 **(Fejer's Theorem).** Let $A = (a_{ij})$ be an $n \times n$ matrix. Then A is nnd if and only if $\text{tr}(A \cdot B) \geq 0$ for all nnd matrices B of order $n \times n$.

PROOF. To establish the only if part, let A and B be both nnd and consider a vector $x \in C^n$, with all its components unity. Then $A \cdot B$ is nnd and $x^*(A \cdot B)x \geq 0$, i.e., $\text{tr}(A \cdot B) \geq 0$. Conversely let $\text{tr}(A \cdot B) \geq 0$ for all nnd B. Choose $B = (b_{ij}) = (\bar{x}_i x_j)$ for any $x \in C^n$. Then B is nnd and $\text{tr}(A \cdot B) = x^* A x \geq 0$ which implies that A is nnd.

As a corollary to Schur's product theorem we have the following.

COROLLARY 6.3.13. Let A be an $n \times n$ nnd matrix. Then: (1) The matrix $A \cdot A \cdot \ldots \cdot A$, with any number of terms is nnd. (2) If $f(z) = a_0 + a_1 z + a_2 z^2 + \ldots$ is an analytic function with non-negative coefficients and radius of convergence $R > 0$, then the matrix $(f(a_{ij}))$ is nnd if all $|a_{ij}| < R$.

Complements

6.3.1 Let A and B be matrices of the same order $m \times n$ and the same unit rank, i.e., $A = xy'$ and $B = uv'$ for some non-zero column vectors

x and u each of order $m \times 1$ and non-zero column vectors y and v each of order $n \times 1$. Show that

$$A \cdot B = (x \cdot u)(y \cdot v)'.$$

Show also that $A \cdot B$ is at most of rank 1.

6.3.2 Let $A = (a_{ij})$ be a square matrix of order $m \times m$. Show that

$$\text{Diag}(a_{11}, a_{22}, \ldots, a_{mm}) = A \cdot I_m.$$

6.3.3 (An alternative proof of Schur's Theorem **P 6.3.2**). Since B is non-negative definite, write $B = TT^*$. Let t_k be the k-th column of $T = (t_{ij})$, $k = 1, 2, \ldots, m$. Let $x' = (x_1, x_2, \ldots, x_m)$ be any vector with complex entries. Then

$$\begin{aligned} x^*(A \cdot B)x &= \sum_{i=1}^{m}\sum_{j=1}^{m} a_{ij} b_{ij} \bar{x}_i x_j = \sum_{i=1}^{m}\sum_{j=1}^{m} a_{ij} \left(\sum_{k=1}^{m} t_{ik}\bar{t}_{jk}\right)\bar{x}_i x_j \\ &= \sum_{k=1}^{m} (x \cdot t_k)^* A (x \cdot t_k) \geq 0, \quad \text{as } A \text{ is non-negative definite.} \end{aligned}$$

6.3.4 Let A and B be two matrices of the same order $m \times m$. Let 1_m be a column vector of order $m \times 1$ in which each entry is equal to 1. Show that

$$\text{tr}(AB) = 1'_m(A \cdot B')1_m.$$

6.3.5 Show that every positive definite matrix is the HS product of two positive definite matrices. Explore the uniqueness of the factorization.

6.3.6 If A and B are positive definite, show that $A \cdot B = AB$ if and only if A and B are both diagonal matrices.

Hint: Use **P 6.3.7** (4) and (6.3.3). Is it necessary that A and B have to be positive definite?

6.3.7 Let A be a symmetric non-singular matrix. Show that 1 is an eigenvalue of $A \cdot A^{-1}$ with corresponding eigenvector e, where $e' = (1, 1, \ldots, 1)$.

Hint: Observe that each row sum of $A \cdot A^{-1}$ is unity.

6.4. Khatri-Rao Product

In this section, we introduce another product of matrices known as Khatri-Rao product and examine some of its ramifications. Let A and B be two matrices of orders $p \times n$ and $m \times n$, respectively. Let α_i be the i-th column of A and β_i, be the i-th column of B, $i = 1, 2, \ldots, n$. The Khatri-Rao product $A \odot B$ of A and B is the partitioned matrix of order $pm \times n$ given by

$$A \odot B = (\alpha_1 \otimes \beta_1 | \alpha_2 \otimes \beta_2 | \ldots | \alpha_n \otimes \beta_n). \tag{6.4.1}$$

We will establish some useful results stemming out of this type of matrix.

P 6.4.1 Let A, B, C, and D be four matrices of orders $p \times n$, $m \times n$, $m \times p$, and $n \times m$, respectively. Then

$$(C \otimes D)(A \odot B) = (CA) \odot (DB),$$

PROOF. Let α_i be the i-th column of A and β_i, the i-th column of B, $i = 1, 2, \ldots, n$. Then the i-th column of CA is $C\alpha_i$ and that of DB is $D\beta_i$. Consequently, the i-th column of $(CA) \odot (DB)$ is $C\alpha_i \otimes D\beta_i = (C \otimes D)(\alpha_i \otimes \beta_i)$, which is precisely the i-th column of $(C \otimes D)(A \odot B)$.

In the next result, we rope HS multiplication into the process of Khatri-Rao product.

P 6.4.2 Let A and B be two non-negative definite matrices each of order $n \times n$. Let $A = \Gamma'\Gamma$ and $B = \Omega'\Omega$ be the Gram-matrix representations of A and B, respectively, for some matrices Γ of order $r \times n$ and Ω of order $s \times n$. Then the HS product of A and B is related to the Khatri-Rao product by

$$A \cdot B = (\Gamma \odot \Omega)'(\Gamma \odot \Omega).$$

PROOF. Let $\alpha_1, \alpha_2, \ldots, \alpha_n$ be the columns of Γ and $\beta_1, \beta_2, \ldots, \beta_n$ be those of Ω. If $A = (a_{ij})$ and $B = (b_{ij})$, note that $a_{ij} = \alpha_i'\alpha_j$ and $b_{ij} = \beta_i'\beta_j$ for all i and j. The (i, j)-th entry of $(\Gamma \odot \Omega)'(\Gamma \odot \Omega)$ is given by $(\alpha_i \otimes \beta_i)'(\alpha_j \otimes \beta_j) = \alpha_i'\alpha_j \otimes \beta_i'\beta_j = (\alpha_i'\alpha_j)(\beta_i'\beta_j) = a_{ij}b_{ij}$, which is the (i, j)-th entry of the HS product $A \cdot B$.

From **P 6.4.2**, we can draw the important conclusion that the HS product $A \cdot B$ is also non-negative definite. We have arrived at this conclusion in Section 6.3 by a different route.

P 6.4.3 Let A and B be two non-negative definite matrices of the same order $n \times n$. Let $A = \Gamma'\Gamma$ and $B = \Omega'\Omega$ be the Gram-matrix representations of A and B, respectively, for some matrices Γ of order $r \times n$ and Ω of order $s \times n$. If the HS product $A \cdot B$ is not of full rank, then there exists a non-null diagonal matrix Δ such that $\Gamma\Delta\Omega' = 0$.

PROOF. By **P 6.4.2**, we can write $A \cdot B = (\Gamma \odot \Omega)'(\Gamma \odot \Omega)$. Since $A \cdot B$ is not of full rank, there exists a non-null vector x such that $(A \cdot B)x = 0$. This implies that $(\Gamma \odot \Omega)x = 0$. (Why?) Let $\Gamma = (\gamma_{ij}), \Omega = (\omega_{ij})$, and $x' = (x_1, x_2, \dots, x_n)$. Let $\Delta = \operatorname{diag}(x_1, x_2, \dots, x_n)$. The statement that $(\Gamma \odot \Omega)x = 0$ implies that

$$x_1\gamma_{i1}\omega_{j1} + x_2\gamma_{i2}\omega_{j2} + \dots + x_n\gamma_{in}\omega_{jn} = 0$$

for $1 \leq i \leq r$ and $1 \leq j \leq s$. This is equivalent to $\Gamma\Delta\Omega' = 0$. This completes the proof.

Let us explore the world of estimation of heteroscedastic variances in linear models as an application of the results presented in this section. Suppose $Y_1, Y_2, \dots, Y_n (n > 2)$ are pairwise uncorrelated random variables all with the same mean μ but with variances $\sigma_1^2, \sigma_2^2, \dots, \sigma_n^2$ respectively. The objective is to estimate each of the variances using the data $Y_1, Y_2, \dots, Y_n$. Let $\bar{Y} = \frac{1}{n}\sum_{i=1}^{n} Y_i$ and $S^2 = \frac{1}{n-1}\sum_{i=1}^{n}(Y_i - \bar{Y})^2$. The estimation problem arises when one wants to ascertain precisions of some n instruments. All instruments measure the same quantitative phenomenon μ but with error characterized by the variances $\sigma_1^2, \sigma_2^2, \dots, \sigma_n^2$. Based on the measurements $Y_1, Y_2, \dots, Y_n$ made one on each instrument, the objective is to estimate the precisions $\sigma_1^2, \sigma_2^2, \dots, \sigma_n^2$ of the instruments. Suppose we want an unbiased estimator of σ_1^2. It is natural to seek a quadratic function of the data as an unbiased estimator of σ_1^2. Any quadratic function of the data can be written as $\sum_{i=1}^{n}\sum_{j=1}^{n} a_{ij}Y_iY_j$ for some constants a_{ij}'s. Setting

$$\sigma_1^2 = E\sum_{i=1}^{n}\sum_{j=1}^{n} a_{ij}Y_iY_j = \mu^2\sum_{i=1}^{n}\sum_{j=1}^{n} a_{ij} + \sum_{i=1}^{n} a_{ii}\sigma_i^2,$$

we note that the coefficients a_{ij}'s have to satisfy the conditions:

$$a_{11} = 1, a_{ii} = 0,\ i = 2, 3, \ldots, n; \text{ and } \sum_{i=1}^{n}\sum_{j=1}^{n} a_{ij} = 0.$$

As an example, $Y_1^2 - Y_1Y_2$ is an unbiased estimator of σ_1^2. One can jot down any number of estimators of σ_1^2. We would like to focus on one particular estimator of σ_1^2 and demonstrate that it is optimal in a sense to be made precise later. First, we note the following:

$$\begin{aligned}
E(Y_1 - \bar{Y})^2 &= E((Y_1 - \mu) - (\bar{Y} - \mu))^2 \\
&= E[(\frac{n-1}{n})(Y_1 - \mu) - \frac{1}{n}(Y_2 - \mu) - \ldots - \frac{1}{n}(Y_n - \mu)]^2 \\
&= (\frac{n-1}{n})^2\sigma_1^2 + (\frac{1}{n})^2\sum_{i=2}^{n}\sigma_i^2; \\
ES^2 &= (\frac{1}{n-1})(\frac{n-1}{n})^2\sum_{i=1}^{n}\sigma_i^2 + (\frac{1}{n-1})(n-1)(\frac{1}{n})^2\sum_{i=1}^{n}\sigma_i^2 \\
&= \frac{1}{n}\sum_{i=1}^{n}\sigma_i^2.
\end{aligned}$$

Now it is simple to verify that

$$T_1 = \frac{n}{n-2}(Y_1 - \bar{Y})^2 - \frac{1}{n-2}S^2$$

is an unbiased estimator of σ_1^2. What is special about this estimator? One thing is that it uses all the data. One could provide other estimators of σ_1^2 which make use of all the data. The estimator T_1 enjoys a certain invariance property. More precisely, if each Y_i is replaced by $Y_i + c$ for some fixed constant c, the value of the statistic T_1 computed using the new data $Y_1 + c, Y_2 + c, \ldots, Y_n + c$ is the same as the value of the statistic T_1 computed using the data $Y_1, Y_2, \ldots, Y_n$. This invariance property is a desirable property for an estimator to possess because the parameters σ_i^2's enjoy such a property. One could provide different invariant estimators of σ_1^2. Let us try a different track. The estimator T_1 is a quadratic form in $Y_1, Y_2, \ldots, Y_n$. More precisely,

$$T_1 = Y'AY,$$

where $Y' = (Y_1, Y_2, \ldots, Y_n)$, and $A = (a_{ij})$ where

$$a_{11} = 1, a_{i1} = a_{1i} = -(n-1)^{-1}, \ i = 2, \ldots, n$$
$$a_{i2} = \ldots = a_{nn} = 0,$$
$$a_{ij} = (n-1)^{-1}(n-2)^{-1} \text{ for } i \neq j, \ i, j = 2, \ldots, n. \qquad (6.4.2)$$

Let Q be the collection of all quadratic unbiased estimators of σ_1^2 which are invariant under translations of the data. Any quadratic estimator is of the form $Y'BY$ for some symmetric matrix $B = (b_{ij})$ of order $n \times n$. The estimator $Y'BY$ is invariant means that

$$(Y + c1)'B(Y + c1) = Y'BY$$

for all real numbers c and for all realizations of the data vector Y, where 1 is a column vector of order $n \times 1$ in which every entry is equal to one. The data vector $Y + c1$ is simply a translation of the data vector Y. The demand that the estimator be invariant is equivalent to the condition that

$$c^2 1'B1 + 2c1'BY = 0$$

for all real numbers and for all realizations of the data vector Y. This is equivalent to the condition that

$$B1 = 0. \qquad (6.4.3)$$

This condition can be rephrased as the condition that every row sum of B is equal zero. In the presence of invariance condition, the estimator $Y'BY$ is unbiased for σ_1^2 if

$$b_{11} = 1 \quad \text{and} \quad b_{ii} = 0, \ i = 2, 3, \ldots, n. \qquad (6.4.4)$$

(Why?) The matrix A spelled out in (6.4.2) meets all these requirements (6.4.3) and (6.4.4). The matrix A has the optimum property that among all matrices $B = (b_{ij})$ satisfying (6.4.3) and (6.4.4), $\|B\|^2 = \sum_{i=1}^{n} \sum_{j=1}^{n} b_{ij}^2$ is a minimum when $B = A$. One can easily verify this using the Lagrangian multipliers method. There is yet another way of deriving the estimator T_1. The random vector Y falls into the mould of a linear model, i.e.,

$$Y = X\mu + \epsilon \tag{6.4.5}$$

with $E(\epsilon) = 0$ and $\text{Disp}(\epsilon) = \text{Diag}(\sigma_1^2, \sigma_2^2, \dots, \sigma_n^2)$, where $\epsilon' = (\epsilon_1, \epsilon_2, \dots, \epsilon_n)$, the unobservable error random vector associated with the data vector Y, and $X' = (1, 1, \dots, 1)$. The least squares estimator of μ is $\bar{Y}$. The projection matrix F associated with the linear model is given by

$$F = (m_{ij}) = (I_n - X(X'X)^{-1}X') = I_n - \frac{1}{n}J_n,$$

where J_n is the matrix of order $n \times n$ in which every entry is equal to one. The vector $\hat{\epsilon}$ of residuals is given by

$$\hat{\epsilon}' = (FY)' = (Y_1 - \bar{Y}, Y_2 - \bar{Y}, \dots, Y_n - \bar{Y}).$$

Let $\hat{\sigma}^{2\prime} = (\hat{\sigma}_1^2, \hat{\sigma}_2^2, \dots, \hat{\sigma}_n^2)$. Look at the system of linear equations

$$(F \cdot F)\hat{\sigma}^2 = \hat{\epsilon} \cdot \hat{\epsilon} \tag{6.4.6}$$

in the unknown $\hat{\sigma}^2$, where the symbol $\cdot$ stands for HS multiplication. The matrix $F \cdot F$ is non-singular and its inverse is given by

$$(F \cdot F)^{-1} = \begin{bmatrix} a & b & \dots & b \\ b & a & \dots & b \\ \cdot & \cdot & \dots & \cdot \\ b & b & \dots & a \end{bmatrix},$$

where $b = [-1/(n-1)(n-2)]$ and $a - b = n/(n-2)$. The solution to the linear equations (6.4.6) is given by $\hat{\sigma}^2 = (F \cdot F)^{-1}(\hat{\epsilon} \cdot \hat{\epsilon})$. After a round of routine simplification, we note that

$$\hat{\sigma}_i^2 = \frac{n}{n-2}(Y_i - \bar{Y})^2 - \frac{S^2}{n-2}, \quad i = 1, 2, \dots, n.$$

For $i = 1$, $\hat{\sigma}_1^2$ is precisely equal to the statistic T_1 we have been harboring and nurturing all along! There are a couple of things to be sorted out before we can conclude this section. The first is about the significance of the linear equations (6.4.6) we have jotted down. The second is the significance of the optimization problem in arriving at the estimator $T_1 = Y'AY$. Let us explain why T_1 is optimal. When we think about optimality of a certain estimator, we would like to phrase the optimality of the estimator in terms of variance. Declare an unbiased estimator of a parameter to be optimal if its variance is the least among all unbiased estimators of the same parameter. If the variance were to be the criterion of optimality, we need to assume some structure on the fourth moments of the random variables $Y_1, Y_2, \ldots, Y_n$. If we do not want to assume beyond what we imposed on the data, namely pairwise uncorrelatedness and finite second moments, variance criterion is beyond our reach. We need to seek other optimality criteria intuitively justifiable and acceptable. Suppose μ is known. A reasonable estimator of σ_1^2 is $(Y_1 - \mu)^2$. If μ is known, the residuals $\epsilon_1, \epsilon_2, \ldots, \epsilon_n$ in the linear model (6.4.5) are observable. In terms of the residuals, the reasonable estimator can be rewritten as $\epsilon' C \epsilon$, where $C = \text{diag}(1, 0, 0, \ldots, 0)$. Can we do better than this? Is it possible to find an invariant unbiased estimator $Y'BY$ of σ_1^2 which is close to the reasonable estimator? The conditions on the estimator, especially the invariance property, imply that

$$Y'BY = (Y - \mu 1)'B(Y - \mu 1) = \epsilon' B \epsilon.$$

The problem is to determine the matrix B such that $\epsilon' C \epsilon - \epsilon' B \epsilon$ is as small as possible. This is tantamount to choosing the matrix B with all the constraints so that $\|C - B\|$ is minimum. This is equivalent to minimizing $\|B\|$ subject to the constraints of invariance and unbiasedness. This is the story behind the estimator $T_1 = Y'AY$. One can also justify the estimator T_1 on the ground that the variation exhibited by the invariant unbiased estimator $Y'BY$ be reduced as much as possible. We can achieve this by choosing B as small as possible. This is the so-called Minimum Norm Quadratic Unbiased Estimation, with the acronym MINQUE principle of, C.R. Rao (1972a).

Now we come to the significance of the linear equations (6.4.6). Let

$$Y = X\beta + \epsilon$$

be a general linear model with

$$\text{Disp}(\epsilon) = \text{Diag}(\sigma_1^2, \sigma_2^2, \ldots, \sigma_n^2),$$

where the matrix X of order $n \times m$ is known and the parameter vector β and variances σ_i^2 are unknown. We need good estimators of the variances based on the data vector Y. Assume that the variances are all distinct and the rank of X is m. (These assumptions can be relaxed.) When we say that the variances are all distinct we mean that the vector $(\sigma^2)' = (\sigma_1^2, \sigma_2^2, \dots, \sigma_n^2)$ of variances has the parameter space $(0, \infty) \times (0, \infty) \times \dots \times (0, \infty)$. Let $p_1\sigma_1^2 + p_2\sigma_2^2 + \dots + p_n\sigma_n^2 = p'\sigma^2$ be a linear function of the variances, where the vector $p' = (p_1, p_2, \dots, p_n)$ is known. As per the MINQUE principle, we seek a quadratic estimator $Y'BY$ of $p'\sigma^2$ such that $B = (b_{ij})$ satisfies the conditions

$$BX = 0, \tag{6.4.7}$$

$$\sum_{i=1}^{n} b_{ii}\sigma_i^2 = \sum_{i=1}^{n} p_i\sigma_i^2, \tag{6.4.8}$$

and

$$\|B\|^2 = \sum_{i=1}^{n}\sum_{j=1}^{n} b_{ij}^2$$

is a minimum. The condition (6.4.7) implies that the estimator $Y'BY$ is invariant, i.e.,

$$Y'BY = (Y - X\beta_0)'B(Y - X\beta_0)$$

for all vectors β_0, and condition (6.4.8) implies that the estimator $Y'BY$ is unbiased for $p'\sigma^2$. A solution to this problem has been discussed in C.R. Rao (1972). Let $F = (I_n - X(X'X)^{-1}X')$ be the projection matrix and $\hat{\epsilon} = (I_n - X(X'X)^{-1}X')Y$ be the vector of residuals. Let $(\hat{\sigma}^2)' = (\hat{\sigma}_1^2, \hat{\sigma}_2^2, \dots, \hat{\sigma}_n^2)$. Consider the system $(F \cdot F)\hat{\sigma}^2 = \hat{\epsilon} \cdot \hat{\epsilon}$ of linear equations in the unknown vector $\hat{\sigma}^2$. If $F \cdot F$ is non-singular, the MINQUE of σ^2 is given by $\hat{\sigma}^2 = (F \cdot F)^{-1}(\hat{\epsilon} \cdot \hat{\epsilon})$. This is the story behind (6.4.6).

The next line of inquiry is to understand when the HS product $F \cdot F$ is non-singular. Hartley, Rao, and Kiefer (1969) and Rao (1972a) throw some light on this problem.

Complements

6.4.1 Let $Y_1, Y_2, \dots, Y_n$ be n pairwise uncorrelated random variables with the same mean μ. Let $k_1, k_2, \dots, k_r$ be positive integers such that

$k_1 + k_2 + \ldots + k_r = n$. Suppose

$$\text{Var}(Y_1) = \text{Var}(Y_2) = \ldots = \text{Var}(Y_{k_1}) = \sigma_1^2,$$
$$\text{Var}(Y_{k_1+1}) = \text{Var}(Y_{k_1+2}) = \ldots = \text{Var}(Y_{k_1+k_2}) = \sigma_2^2,$$
$$\ldots \qquad \ldots \qquad \ldots$$
$$\text{Var}(Y_{k_1+\ldots+k_{r-1}+1}) = \text{Var}(Y_{k_1+\ldots+k_{r-1}+2}) = \ldots = \text{Var}(Y_n) = \sigma_r^2.$$

The mean and variances are all unknown. Develop MINQUE's of the variances.

6.4.2 Let Y_{ij}, $i = 1, 2, \ldots, p$ and $j = 1, 2, \ldots, q$ be pairwise uncorrelated random variables with the following structure.

$$E(Y_{ij}) = \alpha_i + \beta_j \quad \text{for all } i \text{ and } j,$$
$$\text{Var}(Y_{ij}) = \sigma_i^2, \; j = 1, 2, \ldots, q \quad \text{and for all } i.$$

The variances, α_i's and β_j's are all unknown. Show that the residuals are given by

$$\hat{\epsilon}_{ij} = Y_{ij} - \bar{Y}_{i.} - \bar{Y}_{.j} + \bar{Y}_{..}, \; i = 1, 2, \ldots, p \quad \text{and} \quad j = 1, 2, \ldots, q,$$

where $\bar{Y}_{i.} = \frac{1}{q}\sum_{j=1}^{q} Y_{ij}, \bar{Y}_{.j} = \frac{1}{p}\sum_{i=1}^{p} Y_{ij}$, and $\bar{Y}_{..} = \frac{1}{pq}\sum_{i=1}^{p}\sum_{j=1}^{q} Y_{ij}$. Show that the MINQUE of σ_1^2 is given by

$$a\left(\sum_{j=1}^{q} \hat{\epsilon}_{ij}^2\right) + b\left(\sum_{i=1}^{p}\sum_{j=1}^{q} \hat{\epsilon}_{ij}^2\right),$$

where $a = [(p-1)(q-2)]^{-1}$ and $b = -[(p-1)(q-1)(q-2)]^{-1}$.

6.5. Matrix Derivatives

Suppose f is a real valued function of mn variables x_{ij}, $i = 1, 2, \ldots, m$ and $j = 1, 2, \ldots, n$. Suppose these variables are arranged in the form of a matrix $X = (x_{ij})$ of order $m \times n$. Assume that the partial derivatives of f exist with respect to each of its variables. The *matrix* derivative $\partial f/\partial X$ of f with respect to X is a matrix of order $m \times n$ given by

$$\frac{\partial f}{\partial X} = \left(\frac{\partial f}{\partial x_{ij}}\right),$$

i.e., the (i,j)-th element of $\partial f/\partial X$ is $\partial f/\partial X_{ij}$. If $n=1$, X is a column vector and it is denoted by x with components $x_1, x_2, \ldots, x_m$. The corresponding derivative $\partial f/\partial x$ is called the *vector* derivative of f with respect to x. More generally, suppose $F=(f_{ij})$ is a matrix function of a matrix variable X. What we mean by this is that each entry f_{ij} of the matrix F is a real valued function of the variables in X. Let F be of order $p\times q$ and X of order $m\times n$. Assume that each of the entries of F has partial derivatives with respect to all the variables in X. The *matrix* derivative $\partial F/\partial X$ of F with respect to X is defined to be the matrix

$$\frac{\partial F}{\partial X}=\left(\frac{\partial f_{ij}}{\partial X}\right) \tag{6.5.1}$$

of order $pm\times qn$ broken up into pq partitions or compartments strung along p rows and q columns. Each partition of the matrix derivative is of order $m\times n$. As an illustration, suppose F is of order 2×4 and X is of order 3×2. The matrix (6.5.1) comports itself as

$$\frac{\partial F}{\partial X}=\left[\begin{array}{cc|cc|cc|cc}
\frac{\partial f_{11}}{\partial x_{11}} & \frac{\partial f_{11}}{\partial x_{12}} & \frac{\partial f_{12}}{\partial x_{11}} & \frac{\partial f_{12}}{\partial x_{12}} & \frac{\partial f_{13}}{\partial x_{11}} & \frac{\partial f_{13}}{\partial x_{12}} & \frac{\partial f_{14}}{\partial x_{11}} & \frac{\partial f_{14}}{\partial x_{12}} \\
\frac{\partial f_{11}}{\partial x_{21}} & \frac{\partial f_{11}}{\partial x_{22}} & \frac{\partial f_{12}}{\partial x_{21}} & \frac{\partial f_{12}}{\partial x_{22}} & \frac{\partial f_{13}}{\partial x_{21}} & \frac{\partial f_{13}}{\partial x_{22}} & \frac{\partial f_{14}}{\partial x_{21}} & \frac{\partial f_{14}}{\partial x_{22}} \\
\frac{\partial f_{11}}{\partial x_{31}} & \frac{\partial f_{11}}{\partial x_{32}} & \frac{\partial f_{12}}{\partial x_{31}} & \frac{\partial f_{12}}{\partial x_{32}} & \frac{\partial f_{13}}{\partial x_{31}} & \frac{\partial f_{13}}{\partial x_{32}} & \frac{\partial f_{14}}{\partial x_{31}} & \frac{\partial f_{14}}{\partial x_{32}} \\
\hdashline
\frac{\partial f_{21}}{\partial x_{11}} & \frac{\partial f_{21}}{\partial x_{12}} & \frac{\partial f_{22}}{\partial x_{11}} & \frac{\partial f_{22}}{\partial x_{12}} & \frac{\partial f_{23}}{\partial x_{11}} & \frac{\partial f_{23}}{\partial x_{12}} & \frac{\partial f_{24}}{\partial x_{11}} & \frac{\partial f_{24}}{\partial x_{12}} \\
\frac{\partial f_{21}}{\partial x_{21}} & \frac{\partial f_{21}}{\partial x_{22}} & \frac{\partial f_{22}}{\partial x_{21}} & \frac{\partial f_{22}}{\partial x_{22}} & \frac{\partial f_{23}}{\partial x_{21}} & \frac{\partial f_{23}}{\partial x_{22}} & \frac{\partial f_{24}}{\partial x_{21}} & \frac{\partial f_{24}}{\partial x_{22}} \\
\frac{\partial f_{21}}{\partial x_{31}} & \frac{\partial f_{21}}{\partial x_{32}} & \frac{\partial f_{22}}{\partial x_{31}} & \frac{\partial f_{22}}{\partial x_{32}} & \frac{\partial f_{23}}{\partial x_{31}} & \frac{\partial f_{23}}{\partial x_{32}} & \frac{\partial f_{24}}{\partial x_{31}} & \frac{\partial f_{24}}{\partial x_{32}}
\end{array}\right]$$

There is some criticism mooted against the way the partial derivatives are strung out in $\partial F/\partial X$. Suppose the matrix function is the identity function, i.e., $F(X)=X$, or equivalently, $f_{ij}(X)=x_{ij}$ for all i and j. If we want to use the matrix of partial derivatives to build the Jacobian of the transformation, the entity $\partial F/\partial X$ is in for a disappointment. Suppose X is of order 2×3 and $F(X)=X$. Then $\partial f/\partial X=((\text{vec}I_2)\otimes I_3)'$, which is of order 4×9. It is clear that the rank of the matrix $\partial F/\partial X$ is one. The Jacobian of the transformation $F(X)=X$ is I_0. The derivative $\partial F/\partial X$ is nowhere near the Jacobian. Even the order of the matrix $\partial F/\partial X$ is wrong for the Jacobian. To ameliorate the standard definition of the matrix derivative (6.5.1) to meet the needs of the Jacobian, one could define the matrix derivative $^*\partial F/\partial X$ of F of order $p \times q$ with respect to X of order $m \times n$ as

$$\frac{^*\partial F}{\partial X} = \frac{\partial \text{vec} F}{\partial (\text{vec} X)'}, \tag{6.5.2}$$

which is of order $pq \times mn$. In order to work out the new matrix of partial derivatives, to begin with, one has to stack the variables in X column by column in one long vector, takes it transpose, stack the entries of F column by column into one long column vector, and then take the partial derivatives of each and every entry with respect to $(\text{vec}X)'$. Note that the order of $\text{vec}F$ is $pq \times 1$ and that of $(\text{vec}X)' = 1 \times mn$. Consequently, the order of the matrix (6.5.2) is $pq \times mn$. For example, the case of F with order 2×4 and X of order 3×2 gives rise to the following matrix derivative in its new incarnation:

$$\frac{^*\partial F}{\partial X} = \begin{bmatrix} \frac{\partial f_{11}}{\partial x_{11}} & \frac{\partial f_{11}}{\partial x_{21}} & \frac{\partial f_{11}}{\partial x_{31}} & \frac{\partial f_{11}}{\partial x_{12}} & \frac{\partial f_{11}}{\partial x_{22}} & \frac{\partial f_{11}}{\partial x_{32}} \\ \frac{\partial f_{21}}{\partial x_{11}} & \frac{\partial f_{21}}{\partial x_{21}} & \frac{\partial f_{21}}{\partial x_{31}} & \frac{\partial f_{21}}{\partial x_{12}} & \frac{\partial f_{21}}{\partial x_{22}} & \frac{\partial f_{21}}{\partial x_{32}} \\ \frac{\partial f_{12}}{\partial x_{11}} & \frac{\partial f_{12}}{\partial x_{21}} & \frac{\partial f_{12}}{\partial x_{31}} & \frac{\partial f_{12}}{\partial x_{12}} & \frac{\partial f_{12}}{\partial x_{22}} & \frac{\partial f_{12}}{\partial x_{32}} \\ \frac{\partial f_{22}}{\partial x_{11}} & \frac{\partial f_{22}}{\partial x_{21}} & \frac{\partial f_{22}}{\partial x_{31}} & \frac{\partial f_{22}}{\partial x_{12}} & \frac{\partial f_{22}}{\partial x_{22}} & \frac{\partial f_{22}}{\partial x_{32}} \\ \frac{\partial f_{13}}{\partial x_{11}} & \frac{\partial f_{13}}{\partial x_{21}} & \frac{\partial f_{13}}{\partial x_{31}} & \frac{\partial f_{13}}{\partial x_{12}} & \frac{\partial f_{13}}{\partial x_{22}} & \frac{\partial f_{13}}{\partial x_{32}} \\ \frac{\partial f_{23}}{\partial x_{11}} & \frac{\partial f_{23}}{\partial x_{21}} & \frac{\partial f_{23}}{\partial x_{31}} & \frac{\partial f_{23}}{\partial x_{12}} & \frac{\partial f_{23}}{\partial x_{22}} & \frac{\partial f_{23}}{\partial x_{32}} \\ \frac{\partial f_{14}}{\partial x_{11}} & \frac{\partial f_{14}}{\partial x_{21}} & \frac{\partial f_{14}}{\partial x_{31}} & \frac{\partial f_{14}}{\partial x_{12}} & \frac{\partial f_{14}}{\partial x_{22}} & \frac{\partial f_{14}}{\partial x_{32}} \\ \frac{\partial f_{24}}{\partial x_{11}} & \frac{\partial f_{24}}{\partial x_{21}} & \frac{\partial f_{24}}{\partial x_{31}} & \frac{\partial f_{24}}{\partial x_{12}} & \frac{\partial f_{24}}{\partial x_{22}} & \frac{\partial f_{24}}{\partial x_{32}} \end{bmatrix}$$

As one can see, the partial derivatives in $^*\partial F/\partial X$ are set out in the style of evaluating the Jacobian of a transformation. The entries of $^*\partial F/\partial X$ are simply a rearrangement of the entries of $\partial F/\partial X$. More precisely, in the special case, first we form transposes of vecs of each partition of

$\partial F/\partial X$ as

$$\begin{bmatrix} \left(\text{vec}\frac{\partial f_{11}}{\partial X}\right)' & \left(\text{vec}\frac{\partial f_{12}}{\partial X}\right)' & \left(\text{vec}\frac{\partial f_{13}}{\partial X}\right)' & \left(\text{vec}\frac{\partial f_{14}}{\partial X}\right)' \\ \left(\text{vec}\frac{\partial f_{21}}{\partial X}\right)' & \left(\text{vec}\frac{\partial f_{22}}{\partial X}\right)' & \left(\text{vec}\frac{\partial f_{23}}{\partial X}\right)' & \left(\text{vec}\frac{\partial f_{24}}{\partial X}\right)' \end{bmatrix},$$

and then treating each vec as a single entity arrange them in vec form in order to obtain $^*\partial F/\partial X$. For a more precise relationship, see Complement 6.5.3.

There is a minor conflict between the standard practice of writing the vector derivative and the version (6.5.2) in the case of a scalar valued function f of a vector variable x. On one hand, $\partial f/\partial x$ is a column vector and on the other hand, $^*\partial f/\partial x = \partial f/\partial x' = (\partial f/\partial x)'$, which is a row vector. When we provide a list of some derivatives of some standard functions, we follow the standard form of arranging the partial derivatives. The formulas for the modified form can be jotted down in a simple manner.

A critical result which is useful in deriving some formulas for matrix derivatives is the following. Let f be scalar valued function of a matrix variable X of order $m \times n$. Let Y be a constant matrix of order $m \times n$. Assume that f is differentiable, i.e., all its partial derivatives exist and are continuous. Then the directional derivative of f in the direction of Y as defined by (6.5.3) exists and

$$\lim_{t \to 0} \frac{f(X + tY) - f(X)}{t} = \text{tr}(Y' \frac{\partial f}{\partial X}). \tag{6.5.3}$$

In some problems, it may be relatively easy to evaluate the limit on the left-hand side of (6.5.3). Once we know what it is, $\partial f/\partial X$ can be figured out from (6.5.3). As an example, consider the following problem. Let A be a matrix of order $m \times m$ and for x in R^m, let $f(x) = x'Ax$. Observe that for any constant vector y,

$$\lim_{t \to 0} \frac{f(x + ty) - f(x)}{t} = \lim_{t \to 0} \frac{x'Ax + t^2 y'Ay + ty'Ax + tx'Ay - x'Ax}{t}$$
$$= y'Ax + x'Ay = y'(A + A')x = \text{tr}(y' \frac{\partial f}{\partial x}).$$

But $\text{tr}(y'(\partial f/\partial x)) = y'(\partial f/\partial x)$. Hence $\partial f/\partial x = (A + A')x$. This can also be obtained by a straightforward evaluation of the vector derivative.

The formula (6.5.3) is also useful in deriving some identities involving matrix derivatives. Some of them are jotted down below.

P 6.5.1 Let f and g be two differentiable real valued functions of a matrix variable X. Then the following are valid:

(1) $\frac{\partial fg}{\partial X} = f\frac{\partial g}{\partial X} + g\frac{\partial f}{\partial X}$.

(2) $\frac{\partial (f/g)}{\partial X} = \frac{1}{g}\frac{\partial f}{\partial X} - \frac{f}{g^2}\frac{\partial g}{\partial X}$ provided g is not zero.

(3) For a scalar valued function f of a matrix valued function $H = (h_{ij})$ of a matrix variable X,

$$\frac{\partial f(H)}{\partial H} = \sum_i \sum_j \frac{\partial f}{\partial h_{ij}} \frac{\partial h_{ij}}{\partial X}.$$

We now focus on vector derivatives. All our functions f are real valued functions defined on the vector space R^m.

P 6.5.2 (1) If $f(x) = a'x$ for some constant vector $a \in R^m$, then $\frac{\partial f}{\partial x} = a$.

(2) If $f(x) = x'x$, then $\frac{\partial f}{\partial x} = 2x$.

(3) If $f(x) = x'Ax$ for some constant matrix A of order $m \times m$ with real entries, then $\frac{\partial f}{\partial x} = (A + A')x$.

(4) If $f(x) = x'Ax$ for some constant symmetric matrix A of order $m \times m$ with real entries, then $\frac{\partial f}{\partial x} = 2Ax$.

We seek two important applications of these results. Let A be a non-negative definite matrix of order $m \times m$, B a matrix of order $r \times m$, and p a column vector of order $r \times 1$, all with constant real entries. The objective is to minimize the function f given by $f(x) = x'Ax$, $x \in R^m$, subject to the restriction that $Bx = p$. Introduce the vector λ of Lagrange multipliers of order $r \times 1$ and consider the function

$$L(x, \lambda) = x'Ax + 2\lambda'(Bx - p), \quad x \in R^m, \lambda \in R^r.$$

The stationary values of the function L are obtained by setting separately the vector derivatives of L with respect to x and λ equal to zero. Using **P 6.5.2**, we have

$$\frac{\partial L}{\partial x} = 2Ax + 2B'\lambda = 0, \frac{\partial L}{\partial \lambda} = 2(Bx - p) = 0.$$

These equations which are linear in x and λ can be rewritten as

$$\begin{bmatrix} A & B' \\ B & 0 \end{bmatrix} \begin{bmatrix} x \\ \lambda \end{bmatrix} = \begin{bmatrix} 0 \\ p \end{bmatrix}.$$

Solving these equations is another story. If rank$(B) = r$ and A is positive definite, then the system of equations admits a unique solution. From the equations, the following series of computations follow:

$$x = -A^{-1}B'\lambda, \ Bx = -BA^{-1}B'\lambda = p;$$
$$\lambda = -(BA^{-1}B')^{-1}p, \ x = A^{-1}B'(BA^{-1}B')^{-1}p.$$

This type of optimization problem arises in Linear Models. Suppose Y is a random vector of m components whose distribution could be any one of the distributions indexed by a finite dimensional parameter $\theta \in R^r$. Suppose under each $\theta \in R^r$, Y has the same dispersion matrix A but the expected value is given by $E_\theta Y = B'\theta$ for some known matrix B of order $r \times m$. (The expected value of each component of Y is a known linear combination of the components of θ.) One of the important problems in Linear Models is to estimate a linear function $p'\theta$ of the parameter vector unbiasedly with minimum variance, where the vector p of order $r \times 1$ is known. In order to make the estimation problem simple, we seek only linear functions of the data Y which are unbiased estimators of $p'\theta$ and in this collection of estimators we search for one with minimum variance. One can show that a linear function $x'Y$ of the data Y is unbiased for $p'\theta$ if $Bx = p$. For such x, the variance of $x'Y$ is $x'Ax$. Now the objective is to minimize $x'Ax$ over all x but subject to the condition $Bx = p$. If B is of rank r and A is of full rank, then the linear unbiased estimator of $p'\theta$ with minimum variance (Best Linear Unbiased Estimator with the acronym BLUE) is given by

$$x'Y = p'(BA^{-1}B')^{-1}BA^{-1}Y.$$

Let us look at another optimization problem. Let A be a symmetric and B a positive definite matrix with real entries. Let $f(x) = x'Ax$ and $g(x) = x'Bx$, $x \in R^m$. We would like to determine the stationary values of the function $f(x)/g(x)$, $x \in R^m - \{0\}$. We equate the vector derivative of this ratio with respect to x to zero. Using **P 6.5.2**, we have

$$\frac{\partial(f/g)}{\partial x} = \frac{2}{x'Bx}Ax - \frac{2x'Ax}{(x'Bx)^2}Bx = 0.$$

This equation leads to the equation

$$(A - \lambda B)x = 0,$$

where $\lambda = x'Ax/x'Bx$. Thus the stationary value x in $R^m - \{0\}$ of the ratio of quadratic forms has to satisfy the equation $(A - \lambda B)x = 0$ for some λ. (But λ will be automatically equal to $x'Ax/x'Bx$. Why?) A non-zero solution to the equation exists if the determinant $|A-\lambda B| = 0$. This determinantal equations has exactly m roots. Thus the stationary values of the ratio of the quadratic forms of interest are at most m in number.

We now focus on matrix derivatives. The function f is a real valued function of a matrix variable X of order $m \times m$. The domain of definition of f need not be the space of all matrices. For the determinant function, we will consider the collection $\mathbf{M}_m(ns)$ of all non-singular matrices of order $m \times m$ with real entries. This set is an open subset of the collection of all matrices of order $m \times m$ in its usual topology. The set $\{X \in \mathbf{M}_m(ns) : |X| > 0\}$ is also an open set and we will consider functions having this set as their domain. Differentiability of the determinant function $|X|$ of X in its domain should pose no problems.

P 6.5.3 (1) If $f(X) = |X|, X \in \mathbf{M}_m(ns)$, then $\partial f/\partial X = |X|(X^{-1})'$.
(2) If $f(X) = \log|X|, |X| > 0$, then $\frac{\partial f}{\partial X} = (X^{-1})'$.
(3) If $f(X) = |X|^r, |X| > 0$ fixed, then $\partial f/\partial X = r|X|^r|(X^{-1})'$.

PROOF. (1) We use (6.5.3). Let $X = (x_{ij}) \in \mathbf{M}_m(ns)$. Let $Y = (y_{ij})$ be any arbitrary matrix of order $m \times m$. For small values of t, $X+tY$ will be non-singular. Let us embark on finding the determinant of $X + tY$. Let $X^c = (x^{ij})$ be the matrix of cofactors of X. After expanding $|X + tY|$ and omitting terms of the order t^2, we have

$$|X + tY| = |X| + t\sum_{i=1}^{m}\sum_{j=1}^{m} y_{ij}x^{ij} = |X| + t\ \mathrm{tr}(Y'X^c).$$

Consequently,

$$\lim_{t\to 0}\frac{|X+tY|-|X|}{t}=\operatorname{tr}(Y'X^c)=\operatorname{tr}(Y'\frac{\partial f}{\partial X})$$

$$\text{i.e., } \frac{\partial f}{\partial X}=X^c=|X|(X^{-1})'.$$

This completes the proof. The proofs of (2) and (3) are now trivial.

The case of symmetric matrices requires some caution. The space $\mathbf{M}_m(s)$ of all symmetric matrices of order $m\times m$ is no longer an m^2-dimensional vector space. In fact, it is an $m(m+1)/2$-dimensional vector space. Now we consider the subset $\mathbf{M}_m(s,ns)$ of all non-singular matrices in $\mathbf{M}_m(s)$. [The letters ns stand for nonsingular.] This subset is an open set in $\mathbf{M}_m(s)$ in its usual topology. The determinant function on this subset is under focus. As a simple example, consider the case of $m=2$. Any matrix X in $\mathbf{M}_m(s,ns)$ is of the form,

$$X=\begin{bmatrix} x_{11} & x_{12}\\ x_{12} & x_{22}\end{bmatrix},$$

with the determinant $x_{11}x_{22}-x_{12}^2\neq 0$. Observe that

$$\frac{\partial|X|}{\partial X}=\begin{bmatrix}\frac{\partial|X|}{\partial x_{11}} & \frac{\partial|X|}{\partial x_{12}}\\ \frac{\partial|X|}{\partial x_{12}} & \frac{\partial|X|}{\partial x_{22}}\end{bmatrix}=\begin{bmatrix} x_{22} & -2x_{12}\\ -2x_{12} & x_{11}\end{bmatrix}$$

$$=2\begin{bmatrix} x_{22} & -x_{12}\\ -x_{12} & x_{11}\end{bmatrix}-\begin{bmatrix} x_{22} & 0\\ 0 & x_{11}\end{bmatrix}=|X|[2X^{-1}-\operatorname{diag}(X^{-1})].$$

This formula holds in general too. Before we jot it down let us discuss the problem of taking derivatives of functions whose domain of definition is the set of all symmetric matrices.

Let f be a scalar valued function of a matrix variable X. It is clear that $\partial f/\partial X'=(\partial f/\partial X)'$. Let f be a scalar valued function of a matrix variable X, where X is symmetric. What we need is a formula analogous to (6.5.3) operational in the case of a symmetric argument. We do have a direct formula which in conjunction with (6.5.3) can be used to solve the symmetric problem. The formula for X symmetric is

$$\frac{\partial f}{\partial X}=\left\{\frac{\partial f(Y)}{\partial Y}+\frac{\partial f(Y)}{\partial Y'}-\operatorname{diag}\left(\frac{\partial f(Y)}{\partial Y}\right)\right\}\Big|_{Y=X}. \qquad (6.5.4)$$

In working out the derivative $\partial f(Y)/\partial Y$, the function $f(\cdot)$ is pretended to have been defined on the class of all matrices Y, i.e., all the components of Y are regarded as independent variables, and then the derivative formed. Let us illustrate the mechanics of this formula with a simple example. Let $f(X) = |X|$, where X is of order 2×2, $|X| \neq 0$, and X symmetric. Regard $f(\cdot)$ as a function of $Y = (y_{ij})$, where Y is of order 2×2 and $|Y| \neq 0$. More precisely, $f(Y) = |Y| = y_{11}y_{22} - y_{12}y_{21}$. Note that

$$\frac{\partial f(Y)}{\partial Y} = \begin{bmatrix} \frac{\partial |Y|}{\partial y_{11}} & \frac{\partial |Y|}{\partial y_{12}} \\ \frac{\partial |Y|}{\partial y_{21}} & \frac{\partial |Y|}{\partial y_{22}} \end{bmatrix} = \begin{bmatrix} y_{22} & -y_{21} \\ -y_{12} & y_{11} \end{bmatrix},$$

$$\frac{\partial f(Y)}{\partial Y'} = \begin{bmatrix} \frac{\partial |Y|}{\partial y_{11}} & \frac{\partial |Y|}{\partial y_{21}} \\ \frac{\partial |Y|}{\partial y_{12}} & \frac{\partial |Y|}{\partial y_{22}} \end{bmatrix} = \begin{bmatrix} y_{22} & -y_{12} \\ -y_{21} & y_{11} \end{bmatrix},$$

$$\operatorname{diag}\left(\frac{\partial f(Y)}{\partial Y}\right) = \begin{bmatrix} \frac{\partial |Y|}{\partial y_{11}} & 0 \\ 0 & \frac{\partial |Y|}{\partial y_{22}} \end{bmatrix} = \begin{bmatrix} y_{22} & 0 \\ 0 & y_{11} \end{bmatrix},$$

and for X symmetric,

$$\frac{\partial f}{\partial X} = \left\{ \frac{\partial f(Y)}{\partial Y} + \frac{\partial f(Y)}{\partial Y'} - \operatorname{diag}\left(\frac{\partial f(Y)}{\partial Y}\right) \right\}\Big|_{Y=X}$$

$$= \begin{bmatrix} x_{22} & -x_{12} \\ -x_{12} & x_{11} \end{bmatrix} + \begin{bmatrix} x_{22} & -x_{12} \\ -x_{12} & x_{11} \end{bmatrix} - \begin{bmatrix} x_{22} & 0 \\ 0 & x_{11} \end{bmatrix}$$

$$= |X|[2X^{-1} - \operatorname{diag}(X^{-1})].$$

P 6.5.4 (1) If $f(X) = |X|$, $X \in \mathbf{M}_m(s, ns)$, then

$$\frac{\partial f}{\partial X} = |X|[2X^{-1} - \operatorname{diag}(X^{-1})].$$

(2) If $f(X) = \log|X|,\ X \in \mathbf{M}_m(s, ns), |X| > 0$, then

$$\frac{\partial f}{\partial X} = [2X^{-1} - \text{diag}(X^{-1})].$$

(3) If $f(X) = |X|^r,\ X \in \mathbf{M}_m(s, ns), |X| > 0$, then

$$\frac{\partial f}{\partial X} = r|X|^r[2X^{-1} - \text{diag}(X^{-1})].$$

We will now outline some useful formulas on matrix derivatives. Let U and V be two matrix functions of a matrix variable X, where $U = (u_{ij})$ and $V = (v_{ij})$ are of orders $p \times q$ and X is of order $m \times n$. Applying **P 6.5.1** (1) to each term $u_{ij}(X)v_{ji}(X)$, we deduce

$$\frac{\partial}{\partial X}\text{tr}(U(X)V(X))$$
$$= \frac{\partial}{\partial X}\text{tr}(U(X)V(Y))|_{Y=X} + \frac{\partial}{\partial X}\text{tr}(U(Y)V(X))|_{Y=X}. \quad (6.5.5)$$

Instead of the trace function dealt in (6.5.5), we could deal with any scalar valued function f of $U(X)$ and $V(X)$. Accordingly, we have

$$\frac{\partial}{\partial X}f(U(X), V(X)) = \left[\frac{\partial}{\partial U}f(U, V)\right]\left[\frac{\partial}{\partial X}U(X)\right]$$
$$+ \left[\frac{\partial}{\partial V}f(U, V)\right]\left[\frac{\partial}{\partial X}V(X)\right]. \quad (6.5.6)$$

Using (6.5.5) or (6.5.6), one can establish the validity of the following proposition.

P 6.5.5 (1) Let $U(X)$ be a matrix valued function of a matrix variable X, where $U(X)$ is of order $p \times p$, non-singular, and X is of order $m \times n$. Then

$$\frac{\partial}{\partial X}\text{tr}(U^{-1}(X)) = -\frac{\partial}{\partial X}\text{tr}(U^{-2}(Y)U(X))|_{Y=X}.$$

(2) Let A be a constant matrix of order $p \times p$ and $U(X)$ a matrix valued function of a matrix argument X, where $U(X)$ is of order $p \times p$, non-singular, and X is of order $m \times n$. Then

$$\frac{\partial}{\partial X}\text{tr}(U^{-1}(X)A) = -\frac{\partial}{\partial X}\text{tr}(U^{-1}(Y)AU^{-1}(Y)U(X))|_{Y=X}.$$

(3) Let A and B be constant matrices each of order $m \times m$ and $f(X) = \text{tr}(AX^{-1}B)$, $X \in \mathbf{M}_m(ns)$. Then

$$\frac{\partial f}{\partial X} = -(X^{-1}BAX^{-1})'.$$

(4) Let $U(X)$ be a matrix valued function of a matrix variable X, where $U(X)$ is of order $p \times p$, non-singular, and X is of order $m \times n$. Then

$$\frac{\partial}{\partial X}|U(X)| = |U(X)|\frac{\partial}{\partial X}\text{tr}(U^{-1}(Y)U(X))|_{Y=X}.$$

(5) Let A be a constant matrix of order $m \times m$ and $f(X) = |AX|$, X is of order $m \times m$ and AX non-singular. Then

$$\begin{aligned}\frac{\partial f}{\partial X} &= |AX|\frac{\partial}{\partial X}\text{tr}((AY)^{-1}AX)|_{Y=X} \\ &= |AX|((AX)^{-1}A)'.\end{aligned}$$

At the beginning of this section, we toyed with another idea of writing the matrix of partial derivatives. More precisely, let $F(X)$ be a matrix valued function of a matrix variable X. We defined

$$\frac{{}^*\partial F}{\partial X} = \frac{\partial \text{vec}(F)}{\partial(\text{vec}X)'}.$$

Even though the entries of ${}^*\partial F/\partial X$ are simply a rearrangement of the entries of $\partial F/\partial X$, it is useful to compile ${}^*\partial F/\partial X$ for some standard functions F of X. This is what we do in the following proposition. All these results can be derived from first principles.

P 6.5.6 (1) Let $F(X) = AX$, where A is a constant matrix of order $p \times m$ and X of order $m \times n$. Then

$$\frac{{}^*\partial F}{\partial X} = I_n \otimes A.$$

(2) Let $F(X) = XB$, where B is a constant matrix of order $n \times q$ and X of order $m \times n$. Then

$$\frac{{}^*\partial F}{\partial X} = B' \otimes I_m.$$

(3) Let $F(X) = AXB$, where A and B are constant matrices of orders $p \times m$ and $n \times q$, respectively, and X of order $m \times n$. Then

$$\frac{{}^*\partial F}{\partial X} = B' \otimes A.$$

(4) Let $F(X) = AX'B$, where A and B are constant matrices of orders $p \times n$ and $m \times q$, respectively, and X of order $m \times n$. Then

$$\frac{{}^*\partial F}{\partial X} = (A \otimes B')P,$$

where P is the permutation matrix which transforms the vector vec(X) into vec(X'), i.e., vec(X') = Pvec(X).

(5) Let $U(X)$ and $V(X)$ be matrix valued functions of a matrix variable X, where $U(X)$ is of order $p \times q$, $V(X)$ of order $q \times r$, and X of order $m \times n$. Then

$$\frac{{}^*\partial}{\partial X} U(X)V(X) = (V(X) \otimes I_r)' \frac{{}^*\partial}{\partial X} U(X) + (I \otimes U(X)) \frac{{}^*\partial}{\partial X} V(X).$$

(6) Let $F(X) = X'AX$, where A is a constant matrix of order $m \times m$ and X of order $m \times n$. Then

$$\frac{{}^*\partial F}{\partial X} = (X'A' \otimes I_n)P + (I_n \otimes X'A).$$

(7) Let $F(X) = AX^{-1}B$, where A and B are constant matrices of orders $p \times m$ and $m \times q$, respectively, X of order $m \times m$ and non-singular. Then

$$\frac{{}^*\partial F}{\partial X} = -(X^{-1}B)' \otimes (AX^{-1}).$$

(8) Let $U(X)$ and $Z(X)$ be two matrix valued functions of a matrix variable X, where $U(\cdot)$ is of order $p \times q$, $Z(\cdot)$ of order 1×1 and X of order $m \times n$. Let $f(X) = Z(X)U(X)$. Then

$$\frac{{}^*\partial f}{\partial X} = \text{vec}(U(X)) \frac{{}^*\partial Z(X)}{\partial X} + Z(X) \frac{{}^*\partial U(X)}{\partial X}.$$

(9) Let $U(X)$ be a matrix valued function of a matrix variable X, where $U(\cdot)$ is of order $p \times p$ and non-singular, and X is of order $m \times n$. Let $f(X) = [U(X)]^{-1}$. Then

$$\frac{{}^*\partial f}{\partial X} = ((U^{-1}(X))' \otimes U^{-1}(X)) \frac{{}^*\partial U(X)}{\partial X}.$$

(10) Let $Y(X)$ be a matrix valued function of a matrix variable X, where $Y(\cdot)$ is of order $p \times q$ and X of order $m \times n$. Let $Z(V)$ be a matrix valued function of a matrix variable V, where $Z(\cdot)$ is of order $r \times s$ and V of order $p \times q$. Let $f(X) = Z(Y(X))$, $X \in \mathbf{M}_{m,n}$. Then

$$\frac{{}^*\partial f}{\partial X} = \left(\left. \frac{{}^*\partial Z(V)}{\partial V} \right|_{V=Y(X)} \right) \left(\frac{{}^*\partial Y(X)}{\partial X} \right).$$

(11) Let $Z(X)$ and $Y(X)$ be two matrix valued functions of a matrix variable X, where $Z(X)$ and $Y(X)$ are of the same order $p \times q$ and X of order $m \times n$. Let $f(X) = Z(X) \cdot Y(X)$, $X \in \mathbf{M}_{m,n}$, where the symbol $\cdot$ denotes HS multiplication. Then

$$\frac{{}^*\partial f}{\partial X} = D(Z(X)) \frac{{}^*\partial Y(X)}{\partial X} + D(Y(X)) \frac{{}^*\partial Z(X)}{\partial X},$$

where for any matrix $Z = (z_{ij})$ of order $p \times q$,

$D(Z) = \mathrm{diag}(z_{11}, z_{12}, \ldots, z_{1q}, z_{21}, z_{22}, \ldots, z_{2q}, \ldots, z_{p1}, z_{p2}, \ldots, z_{pq})$.

(12) Let $Z(X)$ be a matrix valued function of a matrix variable X, where $Z(X)$ is of order $p \times q$ and X of order $m \times n$. Let B be a constant matrix of order $p \times q$ and $f(X) = Z(X) \cdot B$, $X \in \mathbf{S}_{mn}$. Then

$$\frac{{}^*\partial f}{\partial X} = D(B) \frac{{}^*\partial Z(X)}{\partial X}.$$

As has been indicated earlier, the matrix derivative defined as ${}^*\partial f / \partial X$ is very useful in evaluating the Jacobian of a transformation. Suppose $f(X)$ is a matrix valued function of a matrix variable X, where both X and $f(X)$ are of the same order $m \times n$. The Jacobian J of the transformation $f(\cdot)$ is given by

$$J = \left| \frac{{}^*\partial f}{\partial X} \right|_+,$$

where the suffix + indicates the positive value of the determinant of the matrix $^*\partial f/\partial X$ of order $mn \times mn$. Suppose $f(X) = AXB$, where A and B are constant non-singular matrices of orders $m \times m$ and $n \times n$, respectively, and $X \in \mathbf{M}_{m,n}$. The Jacobian of the transformation $f(\cdot)$ is given by

$$J = \left|\frac{^*\partial f}{\partial X}\right|_+ = |B' \otimes A|_+ = |A|_+^n |B|_+^m.$$

Complements

6.5.1 Let $F(X) = X$ be the identity transformation of the matrix variable X of order $m \times n$. Show that $\partial F/\partial X = (\text{vec}(I_m)) \otimes (\text{vec}(I_n))'$.

6.5.2 Let $F(X) = X$ be the identity transformation of the matrix variable X of order $m \times n$. Show that $^*\partial F/\partial X = I$.

6.5.3 Let F be a matrix valued function of order $p \times q$ of a matrix variable $X = (x_{ij})$ of order $m \times n$. Show that

$$\frac{^*\partial F}{\partial X} = \sum_{i=1}^{m}\sum_{j=1}^{n}\left(\text{vec}\frac{\partial F}{\partial x_{ij}}\right)\left(\text{vec}E_{ij}\right)',$$

where E_{ij} is a matrix of order $m \times n$ whose (i,j)-th entry is unity and the rest of its entries zeros.

6.5.4 Let A be a constant matrix of order $m \times n$ and $f(X) = \text{tr}(AX)$, $X \in \mathbf{M}_{m,n}$, the vector space of all matrices X of order $n \times m$. Show that $(\partial f/\partial X) = A'$. If $m = n$ and the domain of definition of f is the collection $\mathbf{M}(s)$ of all $m \times m$ symmetric matrices, show that $\partial f/\partial X = 2A' - \text{diag}(A)$.

6.5.5 Let $f(X) = \text{tr}(X^2)$, $X \in \mathbf{M}_m$, the space of all $m \times m$ matrices. Show that $(\partial f/\partial X) = 2X'$. If the domain of definition of f is the collection of all symmetric matrices, how does the matrix of partial derivatives change?

6.5.6 Let A and B be two constant matrices of orders $m \times m$ and $n \times n$, respectively. Let $f(X) = \text{tr}(X'AXB)$, $X \in \mathbf{M}_{m,n}$, the space of all $m \times n$ matrices, Show that $(\partial f/\partial X) = AXB + A'XB'$. If $m = n$ and the domain of definition of f is the space of all $m \times m$ matrices, show that

$$\frac{\partial f}{\partial X} = AXB + A'XB' + BXA + B'XA' - \text{diag}(AXB + A'XB').$$

6.5.7 Let A and B be two constant matrices of the same order $m \times m$ and $f(X) = \text{tr}(XAXB)$, $X \in \mathbf{S}_m$. Show that

$$\frac{\partial f}{\partial X} = A'X'B' + B'X'A'.$$

If the domain of definition of f is the space of all symmetric matrices, show that

$$\frac{\partial f}{\partial X} = A'XB' + B'XA' + AXB + BXA - \text{diag}(A'XB' + B'XA').$$

6.5.8 Let A be a constant matrix of order $m \times m$ and $f(X) = \text{tr}(X'AX)$, $X \in \mathbf{M}_{m,n}$. Show that $(\partial f/\partial X) = (A + A')X$. If $m = n$ and the domain of definition of f is the collection of all symmetric matrices, show that

$$\frac{\partial f}{\partial X} = (A + A')X + X(A + A') - \text{diag}((A + A')X).$$

6.5.9 Let $f(X) = \text{tr}(X^n)$, $X \in \mathbf{M}_m$, $n \geq 1$. Show that $(\partial f/\partial X) = nX^{n-1}$. If the domain of definition of f is the space of all symmetric matrices, show that

$$\frac{\partial f}{\partial X} = 2nX^{n-1} - n\text{diag}(X^{n-1}).$$

6.5.10 Let x and y be two fixed column vectors of orders $m \times 1$ and $n \times 1$, and $f(X) = x'Xy$, $X \in \mathbf{M}_{m,n}$. Show that $(\partial f/\partial X) = xy'$. If $m = n$ and the domain of definition of f is the set of all symmetric matrices, show that $(\partial f/\partial X) = xy' + yx'$.

6.5.11 Let A be a constant matrix of order $m \times m$ and $f(X) = \text{tr}(AX^{-1}), X \in \mathbf{M}_m(ns)$, the set of all non-singular matrices of order $m \times m$. Show that

$$\frac{\partial f}{\partial X} = -(X^{-1}AX^{-1})'.$$

If the domain of definition of f is confined to the collection of all non-singular symmetric matrices, show that

$$\frac{\partial f}{\partial X} = -X^{-1}AX^{-1} - X^{-1}A'X^{-1} + \text{diag}(X^{-1}AX^{-1}).$$

6.5.12 Let $f(X) = |XX'|$, $X \in \mathbf{M}_{m,n}$ and $\text{rank}(X) = m$. Show that

$$\frac{\partial f}{\partial X} = 2|XX'|(XX')^{-1}X.$$

6.5.13 Let a and b be two constant column vectors of orders $m\times 1$ and $n\times 1$, respectively. Determine the matrix derivative of each of the scalar valued functions $f_1(X) = a'Xb$, and $f_2(X) = a'XX'a$, $X \in \mathbf{M}_{m,n}$, the collection of all matrices of order $m \times n$, with respect to X.

6.5.14 Let a be a constant column vector of order $m \times 1$ and $f(X) = a'X^{-1}a$, $X \in \mathbf{M}_m(ns)$, the collection of all $m\times m$ non-singular matrices of order $m \times m$. Determine the matrix derivative of the scalar valued function f with respect to X.

6.5.15 Let p be any positive integer and $f(X) = X^p$, $X \in \mathbf{M}_m$. Show that

$$\frac{{}^*\partial f}{\partial X} = \sum_{j=1}^{p}(X')^{p-j} \otimes X^{j-1}.$$

6.5.16 Find the Jacobian of each of the following transformations, where A and B are constant matrices of order $m \times m$, and $X \in \mathbf{M}_m$.

(1) $f(X) = AX^{-1}B$, X non-singular.
(2) $f(X) = XAX'$.
(3) $f(X) = X'AX$.
(4) $f(X) = XAX$, $X \in \mathbf{M}_m$.
(5) $f(X) = X'AX'$.

Notes: The following papers and books have been consulted for developing the material in this chapter. Hartley, Rao, and Kiefer (1969), Rao and Mitra (1971b), Rao (1973c), Srivastava and Khatri (1979), Rao and Kleffe (1980), Graham (1981), Barnett (1990), Rao (1985a), Rao and Kleffe (1988), Magnus and Neudecker (1991), Liu (1995), among others.

CHAPTER 7

PROJECTORS AND IDEMPOTENT OPERATORS

The notion of an orthogonal projection has been introduced in Section 2.2 in the context of inner product spaces. Look up Definition 2.2.11 and the ensuing discussion. In this chapter, we will introduce projectors in the general context of vector spaces. Under a particular mixture of circumstances, an orthogonal projection is seen to be a special kind of projector. We round up the chapter with some examples and complements.

7.1. Projectors

DEFINITION 7.1.1. Let a vector space $\mathbf{V}$ be the direct sum of two subspaces $\mathbf{V}_1$ and $\mathbf{V}_2, \mathbf{V}_1 \cap \mathbf{V}_2 = \{0\}$, i.e., $\mathbf{V} = \mathbf{V}_1 \oplus \mathbf{V}_2$. (See **P 1.5.5** and the discussion preceding **P 1.5.7**.) Then any vector x in $\mathbf{V}$ has a unique decomposition $x = x_1 + x_2$ with $x_1 \in \mathbf{V}_1$ and $x_2 \in \mathbf{V}_2$. The transformation $x \to x_1$ is called the projection of x on $\mathbf{V}_1$ along $\mathbf{V}_2$. The operator or map P defined on the vector space $\mathbf{V}$ by $Px = x_1$ is called a projector from the vector space $\mathbf{V}$ to the subspace $\mathbf{V}_1$ along the subspace $\mathbf{V}_2$.

The first thing we would like to point out is that the map P is well-defined. Further, the map P is an onto map from $\mathbf{V}$ to $\mathbf{V}_1$. It is also transparent that the projector P restricted to the subspace $\mathbf{V}_1$ is the identity transformation from $\mathbf{V}_1$ to $\mathbf{V}_1$, i.e., $Px = x$ if $x \in \mathbf{V}_1$. If $\mathbf{V}$ is an inner product space and $x \in \mathbf{V}_1, y \in \mathbf{V}_2$ implies that x and y are orthogonal, i.e., $\mathbf{V}_2$ is the orthogonal complement of $\mathbf{V}_1$, the map P is precisely the orthogonal projection from the space $\mathbf{V}$ to the space $\mathbf{V}_1$ as enunciated in Definition 2.2.11. Suppose $\mathbf{V}_1$ is a subspace of $\mathbf{V}$. There could be any number of subspaces $\mathbf{V}_2$ of $\mathbf{V}$ such that $\mathbf{V}_1 \oplus \mathbf{V}_2 = \mathbf{V}$. Each such subspace $\mathbf{V}_2$ gives a projector P from $\mathbf{V}$ onto $\mathbf{V}_1$ along $\mathbf{V}_2$.

No two such projectors are the same! In the following proposition, we record a simple property of projectors.

P 7.1.2 A projector P is a linear transformation.

PROOF. Let $x = x_1 + x_2$ and $y = y_1 + y_2$ be the unique decompositions of two vectors x and y in $\mathbf{V}$, respectively, with respect to the subspaces $\mathbf{V}_1$ and $\mathbf{V}_2$ of $\mathbf{V}$. The decomposition of the vector $x + y$ works out to be

$$x + y = (x_1 + y_1) + (x_2 + y_2)$$

with $(x_1 + y_1) \in \mathbf{V}_1$ and $(x_2 + y_2) \in \mathbf{V}_2$. By definition,

$$P(x + y) = x_1 + y_1 = Px + Py.$$

This shows that the map P is additive. For a scalar $\alpha, \alpha x = \alpha x_1 + \alpha x_2$ with $\alpha x_1 \in \mathbf{V}_1$ and $\alpha x_2 \in \mathbf{V}_2$. Consequently, $P(\alpha x) = \alpha x_1 = \alpha(Px)$. Hence P is linear.

The definition of a projector involves two subspaces with only zero vector common between them. In the following proposition, we characterize projectors abstractly without alluding to the underlying subspaces.

P 7.1.3 A linear transformation P from a vector space $\mathbf{V}$ into itself is a projector from $\mathbf{V}$ onto some subspace of $\mathbf{V}$ along some complementary subspace of $\mathbf{V}$ if and only if it is idempotent, i.e., $P^2 = P$.

PROOF. Let P be a projector from $\mathbf{V}$ onto a subspace $\mathbf{V}_1$ along a subspace $\mathbf{V}_2$ of $\mathbf{V}$. Let $x = x_1 + x_2$ be the unique decomposition of any vector x in $\mathbf{V}$ with x_1 in $\mathbf{V}_1$ and x_2 in $\mathbf{V}_2$. The unique decomposition of x_1 in $\mathbf{V}$ is given by $x_1 = x_1 + 0$ with $x_1 \in \mathbf{V}_1$ and $0 \in \mathbf{V}_2$. Consequently,

$$P^2 x = P(Px) = Px_1 = x_1 = Px.$$

Hence it follows that P is idempotent. Conversely, let $P^2 = P$ and

$$\mathbf{V}_1 = \{x \in \mathbf{V} : Px = x\}$$
$$\mathbf{V}_2 = \{x \in \mathbf{V} : Px = 0\}.$$

Since P is a linear transformation, $\mathbf{V}_1$ and $\mathbf{V}_2$ are subspaces of $\mathbf{V}$. Further, $\mathbf{V}_1 \bigcap \mathbf{V}_2 = \{0\}$. For any given $x \in \mathbf{V}$, write $x = Px + (I -$

$P)x$, where I is the identity transformation from $\mathbf{V}$ to $\mathbf{V}$. Since P is idempotent, $Px = P^2x = 0$, from which we have $(I - P)x \in \mathbf{V}_2$. Thus $x = x_1 + x_2$, where $x_1 = Px$ and $x_2 = (I - P)x \in \mathbf{V}_2$, is the unique decomposition of x with respect to the subspaces $\mathbf{V}_1$ and $\mathbf{V}_2$. Hence P is a projector from $\mathbf{V}$ onto $\mathbf{V}_1$ along $\mathbf{V}_2$.

In view of **P 7.1.3**, we can safely omit mentioning the subspaces that define a projector. Once we recognize the projector as an idempotent operator, the associated subspaces $\mathbf{V}_1$ and $\mathbf{V}_2$ of the projector can be recovered via the formulas presented in the proof of **P 7.1.3**. These subspaces are explicitly identified in **P 7.1.4** below. In order to show that a particular linear map is a projector, in many cases, it is easier to show that it is an idempotent operator. We now jot down several equivalent characterizations of projectors. Let P be a linear transformation from a vector space $\mathbf{V}$ to $\mathbf{V}$. Let $\mathbf{R}(P)$ and $\mathbf{K}(P)$ be the range and kernel of the transformation P, respectively. See Section 3.1.

P 7.1.4 The following statements are equivalent.

(1) The map P is a projector.
(2) The map $(I - P)$ is a projector.
(3) The range $\mathbf{R}(P)$ of P is given by $\mathbf{R}(P) = \{x \in \mathbf{V} : Px = x\}$.
(4) $\mathbf{R}(P) = \mathbf{K}(I - P)$.
(5) $\mathbf{R}(I - P) = \mathbf{K}(P)$.
(6) $\mathbf{R}(P) \bigcap \mathbf{R}(I - P) = \{0\}$.
(7) $\mathbf{K}(P) \bigcap \mathbf{K}(I - P) = \{0\}$.

Proving the equivalence of these statements is left to the reader. In view of **P 7.1.4**, if P is a projector we can say that P is a projector from $\mathbf{V}$ onto $\mathbf{R}(P)$ along $\mathbf{R}(I - P)$. In the following we look at sums of projectors.

P 7.1.5 Let $P_1, P_2, \ldots, P_k$ be projectors such that $P_iP_j = 0$ for all $i \neq j$. Then:

(1) $P = P_1 + P_2 + \ldots + P_k$ is a projector.
(2) $\mathbf{R}(P_i) \bigcap \mathbf{R}(P_j) = \{0\}$ for all $i \neq j$ and $\mathbf{R}(P) = \mathbf{R}(P_1) \oplus \mathbf{R}(P_2) \oplus \ldots \oplus \mathbf{R}(P_k)$.

PROOF. (1) It is easy to establish that P is idempotent.

(2) Let $i \neq j$ and $z \in \mathbf{R}(P_i) \bigcap \mathbf{R}(P_j)$. Then $z = P_ix = P_jy$ for some vectors x and y in $\mathbf{V}$. Observe that $P_ix = P_i^2x = P_i(P_ix) = P_i(P_jy) =$

$P_iP_jy = 0$. Hence $z = 0$. This proves that $\mathbf{R}(P_i) \bigcap \mathbf{R}(P_j) = \{0\}$. By the very definition of the projector P, $\mathbf{R}(P) \subset \mathbf{R}(P_1) \oplus \ldots \oplus \mathbf{R}(P_k)$. On the other hand, note that $\mathbf{R}(P_i) \subset \mathbf{R}(P)$ for each i. For, if $x \in \mathbf{R}(P_i)$, then $x = P_iy$ for some y in $\mathbf{V}$ and $PP_iy = P_i^2y = P_iy = x$, from which it follows that $x \in \mathbf{R}(P)$. Consequently, $\mathbf{R}(P_1) \oplus \mathbf{R}(P_2) \oplus \ldots \oplus \mathbf{R}(P_k) \subset \mathbf{R}(P)$. This completes the proof.

The following result is complimentary to **P 7.1.5**. A given projector can be written as a sum of projectors under the right mixture of circumstances.

P 7.1.6 Let P be a projector defined on a vector space $\mathbf{V}$ onto a subspace $\mathbf{V}_1$ along a subspace $\mathbf{V}_2$ of $\mathbf{V}$. Suppose the subspace $\mathbf{V}_1$ is a direct sum of subspaces, i.e., $\mathbf{V}_1 = \mathbf{V}_{11} \oplus \mathbf{V}_{12} \oplus \ldots \oplus \mathbf{V}_{1r}$ for some subspaces $\mathbf{V}_{1j}$'s of $\mathbf{V}$. Then there exist unique projectors P_i from $\mathbf{V}$ onto $\mathbf{V}_{1i}$ along an appropriate subspace of $\mathbf{V}$ such that $P = P_1 + P_2 + \ldots + P_r$ and $P_iP_j = 0$ for all $i \neq j$.

PROOF. One can always bring into existence a projector as long as we have two subspaces whose direct sum is the underlying vector space. In order to identify P_i we need two subspaces. We already have one, namely, $\mathbf{V}_{1i}$. In order to avoid an appropriate subspace complementary to the subspace $\mathbf{V}_{1i}$, let us define the map P_i directly. Let $x \in \mathbf{V}$. We can write $x = x_{11} + x_{12} + \ldots + x_{1r} + y$ with $x_{1i} \in \mathbf{V}_{1i}$ and $y \in \mathbf{V}_2$. Define $P_ix = x_{1i}$. The map P_i is obviously a linear transformation and idempotent. Consequently, P_i is a projector. (Can you identify the subspace $\mathbf{V}_{2i}$ such that P_i is a projector from $\mathbf{V}$ onto $\mathbf{V}_{1i}$ along $\mathbf{V}_{2i}$?) It is clear that $P = P_1 + P_2 + \ldots + P_r$ and $P_iP_j = 0$ for all $i \neq j$. To prove uniqueness, let $P = Q_1 + Q_2 + \ldots + Q_r$ be an alternative representation of P as a sum of projectors. Then for any x in $\mathbf{V}$, $0 = Px - Px = (P_1 - Q_1)x + (P_2 - Q_2)x + \ldots + (P_r - Q_r)x$. This implies that $(P_i - Q_i)x = 0$ for each i in view of the fact that $(P_i - Q_i)x \in \mathbf{V}_{1i}$. If $(P_i - Q_i)x = 0$ for every x, then $P_i = Q_i$. This proves **P 7.1.6**.

We now look at a familiar problem that crops up in Statistics. Suppose $Y_1, Y_2, \ldots, Y_m$ are m real random variables whose joint distribution depends on a vector parameter $\theta' = (\theta_1, \theta_2, \ldots, \theta_k) \in \mathbf{R}^k$, with $m > k$. Suppose

$$E_\theta Y_i = x_{i1}\theta_1 + x_{i2}\theta_2 + \ldots + x_{ik}\theta_k, \; i = 1, 2, \ldots, m,$$

where x_{ij}'s are known. In the language of linear models, the random variables $Y_1, Y_2, \ldots, Y_m$ constitute a linear model. These models lie at the heart of multiple regression analysis and design of experiments problems. Let $\mathbf{V}$ be the collection of all linear functions of $Y_1, Y_2, \ldots, Y_m$. It is clear that $\mathbf{V}$ is a real space of dimension m. As a matter of fact, we can identify the vector space $\mathbf{V}$ with $\mathbf{R}^m$ in the obvious manner. Let $\mathbf{V}$ be the collection of all linear unbiased estimators of zero. A linear unbiased estimator of zero is any linear function $\ell_1 Y_1 + \ell_2 Y_2 + \ldots + \ell_m Y_m$ of $Y_1, Y_2, \ldots, Y_m$ such that $E_\theta(\ell_1 Y_1 + \ell_2 Y_2 + \ldots + \ell_m Y_m) = 0$ for all θ in $\mathbf{R}^k$. Such linear functions constitute a subspace $\mathbf{V}_1$ of $\mathbf{V}$. The space $\mathbf{V}_1$ can be identified explicitly. Let $X = (x_{ij})$. The matrix X of order $m \times k$ is called the design matrix of the linear model. One can check that

$$\mathbf{V}_1 = \{\ell' = (\ell_1, \ell_2, \ldots, \ell_m) \in \mathbf{R}^m : \ell' X = 0\}.$$

Every vector ℓ' in $\mathbf{V}_1$ is orthogonal to every column vector of the matrix X. Let $\mathbf{V}_2$ be the collection of all vectors in $\mathbf{R}^m$ each of which is a linear combination of the columns of X. The space $\mathbf{V}_2$ is clearly a subspace of $\mathbf{V}$. Further, $\mathbf{V} = \mathbf{V}_1 \oplus \mathbf{V}_2$. As a matter of fact, each vector in $\mathbf{V}_1$ and each vector in $\mathbf{V}_2$ are orthogonal. The next target is to identify explicitly the projector from the vector space $\mathbf{V}$ onto $\mathbf{V}_1$ along $\mathbf{V}_2$. To simplify the argument, assume that the rank of the matrix X is k. This ensures the matrix $X'X$ to be non-singular. Let

$$A = X(X'X)^{-1}X'.$$

It is clear that the matrix A is of order $m \times m$. One can also check that it is symmetric and idempotent, i.e., $A^2 = A$. Let ℓ' be any vector in $\mathbf{V} = \mathbf{R}^m$. Observe that $\ell = (I_m - A)\ell + A\ell$, where I_m is the identity matrix of order $m \times m$. We claim that the vector $((I_m - A)\ell)' = \ell'(I_m - A)$ belongs to $\mathbf{V}_1$. For $\ell'(I_m - A)X = \ell'(X - X) = 0$. Further, it is clear that $A\ell = X(X'X)^{-1}X'\ell$ is a linear combination of the columns of X. Thus we have demonstrated practically how the vector space $\mathbf{V}$ is the direct sum of the subspaces $\mathbf{V}_1$ and $\mathbf{V}_2$. Let P be the projector from $\mathbf{V}$ onto $\mathbf{V}_1$ along $\mathbf{V}_2$. The explicit formula for the computation of P is given by

$$P\ell = (I_m - X(X'X)^{-1}X')\ell. \tag{7.1.1}$$

If $\mathbf{V} = \mathbf{R}^m$ is equipped with its usual inner product, the projector P is indeed an orthogonal projection.

There is one benefit that accrues from the explicit formula (7.1.1) of the projector P. Suppose $Y_1, Y_2, \ldots, Y_m$ are pairwise uncorrelated with common variance $\sigma^2 > 0$. Let $\ell_1 Y_1 + \ell_2 Y_2 + \ldots + \ell_m Y_m = \ell' Y$ be a linear function of $Y_1, Y_2, \ldots, Y_m$, where $\ell' = (\ell_1, \ell_2, \ldots, \ell_m)$ and $Y' = (Y_1, Y_2, \ldots, Y_m)$. Let $\ell = \ell_{(1)} + \ell_{(2)}$ be the decomposition of ℓ with respect to the subspaces $\mathbf{V}_1$ and $\mathbf{V}_2$. One can verify that under each $\theta \in \mathbf{R}^k$,

$$\text{Variance}_\theta(\ell' Y) = \text{Variance}_\theta(\ell'_{(1)} Y) + \text{Variance}_\theta(\ell'_{(2)} Y),$$

i.e., $\ell'_{(1)} Y$ and $\ell'_{(2)} Y$ are uncorrelated.

The celebrated Gauss-Markov theorem unfolds in a simple way in this environment. Let $f(\cdot)$ be a linear parametric function, i.e., $f(\theta) = p_1\theta_1 + p_2\theta_2 + \ldots + p_k\theta_k$ for some known numbers $p_1, p_2, \ldots, p_k$, $\theta \in \mathbf{R}^k$. We now seek the best linear unbiased estimator (BLUE) of $f(\cdot)$. The estimator should be of the form $\ell_1 Y_1 + \ell_2 Y_2 + \ldots + \ell_m Y_m$, unbiased, and has minimum variance among all linear unbiased estimators of $f(\cdot)$. To begin with, cook up any linear unbiased estimator $\ell_1 Y_1 + \ell_2 Y_2 + \ldots + \ell_m Y_m = \ell' Y$ of $f(\cdot)$. Obtain the decomposition $\ell = \ell_{(1)} + \ell_{(2)}$, with respect to the subspaces $\mathbf{V}_1$ and $\mathbf{V}_2$. Then $\ell'_{(2)} Y$ is the desired BLUE of $f(\cdot)$. To see this, let $s'Y$ be any linear unbiased estimator of $f(\cdot)$. Write

$$s'Y = (s - \ell_{(2)})' Y + \ell'_{(2)} Y.$$

Note that $s - \ell_{(2)} \in \mathbf{V}_1$. (How?) Consequently, for each $\theta \in \mathbf{R}^k$,

$$\text{Variance}_\theta(s'Y) = \text{Variance}_\theta((s - \ell_{(2)})' Y) + \text{Variance}_\theta(\ell'_{(2)} Y),$$
$$\Rightarrow \text{Variance}_\theta(s'Y) \geq \text{Variance}(\ell'_{(2)} Y).$$

Complements

7.1.1 If P is a projector defined on a vector space $\mathbf{V}$ onto a subspace $\mathbf{V}_1$ of $\mathbf{V}$ along a subspace $\mathbf{V}_2$ of $\mathbf{V}$, identify the subspaces $\mathbf{V}_{1*}$ and $\mathbf{V}_{2*}$ such that the operator $I - P$ is a projector from $\mathbf{V}$ onto $\mathbf{V}_{1*}$ along $\mathbf{V}_{2*}$.

7.1.2 Let $\mathbf{V} = \mathbf{R}^2, \mathbf{V}_1 = \{(x_1, 0) \in \mathbf{R}^2 : x_1 \text{ real}\}$, and $\mathbf{V}_2 = \{(x_1, x_2) \in \mathbf{R}^2 : x_1 + x_2 = 0\}$. Identify the projector P_1 from $\mathbf{V}$ onto $\mathbf{V}_1$ along $\mathbf{V}_2$. Under the usual inner product on the real vector space $\mathbf{V}$, is P_1 an orthogonal projector? Explain fully.

7.1.3 Let $\mathbf{V} = \mathbf{R}^2, \mathbf{V}_1 = \{(x_1, 0) \in \mathbf{R}^2 : x_1 \text{ real}\}$, and $\mathbf{V}_2 = \{(x_1, x_2) \in \mathbf{R}^2 : 2x_1 + x_2 = 0\}$. Identify the projector P_2 from $\mathbf{V}$ onto $\mathbf{V}_1$ along $\mathbf{V}_2$.

7.1.4 Let $P = P_1 + P_2$, where P_1 is the projector identified in Complement 7.1.2 and P_2 the projector in Complement 7.1.3. Is P a projector? Explain fully.

7.1.5 Let $P_1, P_2, \ldots, P_k$ be projectors defined on a vector space $\mathbf{V}$ such that $P_iP_j = 0$ for all $i \neq j$. Identify the subspaces $\mathbf{V}_1$ and $\mathbf{V}_2$ such that $P = P_1 + P_2 + \ldots + P_k$ is projector from $\mathbf{V}$ onto $\mathbf{V}_1$ along $\mathbf{V}_2$.

7.1.6 Let $\mathbf{F} = \{0, 1\}$ be the two-element field and $\mathbf{V} = \mathbf{F}^2$, a vector space over the field $\mathbf{F}$. Let $\mathbf{V}_1 = \{(0,0), (1,0)\}$ and $\mathbf{V}_2 = \{(0,0), (0,1)\}$. Show that $\mathbf{V} = \mathbf{V}_1 \oplus \mathbf{V}_2$. Let P_1 be the projector from $\mathbf{V}$ onto $\mathbf{V}_1$ along $\mathbf{V}_2$. Show that $P_1 + P_1 = 0$, the map which maps every element of $\mathbf{V}$ to the vector $(0, 0)$. Let $\mathbf{P}_2$ be the projector from $\mathbf{V}$ onto $\mathbf{V}$ along the subspace $\{(0,0)\}$. Show that $P_1 + P_2$ is a projector but $P_1P_2 \neq 0$. Comment on **P 7.1.5** in this connection.

7.1.7 Let $\mathbf{V}$ be a vector space over a field $\mathbf{F}$. Suppose that the field $\mathbf{F}$ has the property that $1 + 1 \neq 0$ in $\mathbf{F}$. Let P_1 and P_2 be two projectors defined on $\mathbf{V}$. Show that $P = P_1 + P_2$ is a projector if and only if $P_1P_2 = P_2P_1 = 0$. If P is a projector, show that P is a projector from $\mathbf{V}$ onto $\mathbf{R}(P_1) \oplus \mathbf{R}(P_2)$ along $\mathbf{K}(P_1) \bigcap \mathbf{K}(P_2)$.

Hint: First, show that $P_1P_2 + P_2P_1 = 0$ and then $P_1P_2P_1 + P_1P_2P_1 = 0$.

7.1.8 Let P_1 and P_2 be two projectors. Show that $P_1 - P_2$ is a projector if and only if $P_1P_2 = P_2P_1 = P_2$, in which case $P_1 - P_2$ is a projector from $\mathbf{V}$ onto $\mathbf{R}(P_1) \bigcap \mathbf{K}(P_2)$ along $\mathbf{K}(P_1) \oplus \mathbf{R}(P_2)$. The condition on the underlying field $\mathbf{F}$ stipulated in Complement 7.1.7 is still operational.

7.1.9 Let P_1 and P_2 be two projectors such that $P_1P_2 = P_2P_1$. Show that $P = P_1P_2$ is a projector. Identify the subspaces $\mathbf{V}_1$ and $\mathbf{V}_2$ so that P is a projector from $\mathbf{V}$ onto $\mathbf{V}_1$ along $\mathbf{V}_2$.

7.2. Invariance and Reducibility

In this section we explore the world of invariant subspaces. We begin with some basic definitions and notions. Let $\mathbf{V}$ represent a generic symbol for a vector space in the following deliberations.

DEFINITION 7.2.1. A subspace $\mathbf{W}$ of $\mathbf{V}$ is said to be invariant under a linear transformation T from $\mathbf{V}$ if $Tx \in \mathbf{W}$ whenever $x \in \mathbf{W}$.

In other words, what this definition indicates is that if the map T is restricted to the space $\mathbf{W}$, then it is a linear map from $\mathbf{W}$ to $\mathbf{W}$. The notion of invariance can be extended to cover two subspaces as in the following definition.

DEFINITION 7.2.2. A linear transformation T from $\mathbf{V}$ to $\mathbf{V}$ is said to be reduced by a pair of subspaces $\mathbf{V}_1$ and $\mathbf{V}_2$ if $\mathbf{V}_1$ and $\mathbf{V}_2$ are both invariant under T and $\mathbf{V} = \mathbf{V}_1 \oplus \mathbf{V}_2$.

It will be instructive to examine the notion of invariance in the realm of projectors. Suppose P is a projector from $\mathbf{V}$ onto a subspace $\mathbf{V}_1$ along a subspace $\mathbf{V}_2$. It is clear that $\mathbf{V}_1$ is invariant under the linear transformation P. It not only maps elements of $\mathbf{V}_1$ into $\mathbf{V}_1$ but also all the elements of $\mathbf{V}$. It is also clear that $\mathbf{V}_2$ is also invariant under P. As a matter of fact, every element of $\mathbf{V}_2$ is mapped into 0. We will now determine the structure of the matrix associated with a linear transformation with respect to a basis in the context of invariance.

P 7.2.3 Let $\mathbf{W}$ be a subspace of $\mathbf{V}$ which is invariant under a given transformation T from $\mathbf{V}$ to $\mathbf{V}$. Then there exists a basis of $\mathbf{V}$ with respect to which the matrix A of the transformation T can be written in the triangular form

$$\underset{m\times m}{A} = \begin{pmatrix} \underset{k\times k}{A_1} & \underset{k\times(m-k)}{A_2} \\ \underset{(m-k)\times k}{0} & \underset{(m-k)\times(m-k)}{A_3} \end{pmatrix}, \tag{7.2.1}$$

where $m = \dim(\mathbf{V})$ and $k = \dim(\mathbf{W})$.

PROOF. For an exposition on matrices that are associated with linear transformations, see Section 3.4. Let $x_1, x_2, \ldots, x_m$ be a basis of the vector space $\mathbf{V}$ so that $x_1, x_2, \ldots, x_k$ form a basis for the subspace $\mathbf{W}$. Let $A = (\alpha_{ij})$ be the matrix associated with the linear transformation T with respect to this basis. As a matter of fact,

$$Ax_i = \sum_{j=1}^{m} \alpha_{ji} x_j, \ i = 1, 2, \ldots, m.$$

Since $\mathbf{W}$ is an invariant subspace under T, we must have

$$Ax_i = \sum_{j=1}^{k} \alpha_{ji} x_j, \ i = 1, 2, \ldots, k.$$

This implies that $\alpha_{ji} = 0$ for $j = k+1, k+2, \ldots, m$ and $i = 1, 2, \ldots, k$. Hence the matrix A must be of the form (7.2.1).

If the linear transformation T is reduced by a pair of subspaces, then the matrix associated with T is more elegant as we demonstrate in the following proposition.

P 7.2.4 Suppose a linear transformation T is reduced by a pair of subspaces $\mathbf{V}_1$ and $\mathbf{V}_2$ of $\mathbf{V}$. Then there exists a basis of $\mathbf{V}$ such that the matrix A of the transformation T with respect to the basis is of the form

$$\underset{m\times m}{A} = \begin{pmatrix} \underset{k\times k}{A_1} & \underset{k\times(m-k)}{0} \\ \underset{(m-k)\times k}{0} & \underset{(m-k)\times(m-k)}{A_3} \end{pmatrix}, \tag{7.2.2}$$

where $m = \dim(\mathbf{V})$, $k = \dim(\mathbf{V}_1)$, and $m - k = \dim(\mathbf{V}_2)$.

PROOF. Let $x_1, x_2, \ldots, x_m$ be a basis of $\mathbf{V}$ such that $x_1, x_2, \ldots, x_k$ form a basis for $\mathbf{V}_1$ and $x_{k+1}, x_{k+2}, \ldots, x_m$ form a basis for $\mathbf{V}_2$. Following the argument presented in the proof of **P 7.2.3**, we can discern that A must be of the form (7.2.2).

Projectors onto an invariant subspace of some linear transformation have an intimate relationship with the transformation. In the following propositions we bring out the connection.

P 7.2.5 If a subspace $\mathbf{W}$ of $\mathbf{V}$ is invariant under a linear transformation T from $\mathbf{V}$ to $\mathbf{V}$, then $PTP = TP$ for every projector P from $\mathbf{V}$ onto $\mathbf{W}$. Conversely, if $PTP = TP$ for some projector P from $\mathbf{V}$ onto $\mathbf{W}$, then $\mathbf{W}$ is invariant under T.

PROOF. Let P be a projector from $\mathbf{V}$ onto $\mathbf{W}$. Then for every x in $\mathbf{V}$, $x = Px + (I - P)x$ with $Px \in \mathbf{W}$. If $\mathbf{W}$ is invariant under T, then $TPx = Py$ for some y in $\mathbf{V}$. Here we use the fact that $\mathbf{W} = \mathbf{R}(P)$. Further, $PTPx = P^2y = Py = TPx$. Consequently, $PTP = TP$. Conversely, let $PTP = TP$ for some projector P from $\mathbf{V}$ onto $\mathbf{W}$. For every x in $\mathbf{V}$, the statement that $PTPx = TPx$ implies that $TPx \in \mathbf{R}(P) = \mathbf{W}$. If $y \in \mathbf{W} = \mathbf{R}(P)$, then $y = Px$ for some x in $\mathbf{V}$. Consequently, $Ty = TPx \in \mathbf{W}$. This shows that $\mathbf{W}$ is invariant under T.

P 7.2.6 A linear transformation T from $\mathbf{V}$ to $\mathbf{V}$ is reduced by a pair subspaces $\mathbf{V}_1$ and $\mathbf{V}_2$ if and only if $PT = TP$, where P is the projector from $\mathbf{V}$ onto $\mathbf{V}_1$ along $\mathbf{V}_2$.

PROOF. Suppose $TP = PT$. If $x \in \mathbf{V}_1$, then $Px = x$. Note that $PTx = TPx = Tx$ which implies that $Tx \in \mathbf{R}(P) = \mathbf{V}_1$. This shows that $\mathbf{V}_1$ is invariant under T. If $y \in \mathbf{V}_2$, $PTy = TPy = T0 = 0$. This shows that $Ty \in \mathbf{V}_2$. Hence $\mathbf{V}_2$ is invariant under T. Conversely, suppose that T is reduced by $\mathbf{V}_1$ and $\mathbf{V}_2$. Since $\mathbf{V}_1$ is invariant under T, we have $PTP = TP$ by **P 7.2.5**. Since $\mathbf{V}_2$ is invariant under T and $(I - P)$ is a projector from $\mathbf{V}$ onto $\mathbf{V}_2$ along $\mathbf{V}_1$, we have $(I - P)T(I - P) = T(I - P)$ by the same proposition. This simplifies to $T + PTP - PT - TP = T - TP$, from which we have $PTP = PT$. The result now follows.

Complements

7.2.1 Develop a result analogous to **P 7.2.3** for projectors.
7.2.2 Develop a result analogous to **P 7.2.4** for projectors.
7.2.3 Let $\mathbf{V} = \mathbf{R}^2$, $\mathbf{V}_1 = \{(x_1, 0) \in \mathbf{V} : x_1 \text{ real}\}$, and $\mathbf{V}_2 = \{(x_1, x_2) : x_1 + x_2 = 0\}$. Let P be the projector from $\mathbf{V}$ onto $\mathbf{V}_1$ along $\mathbf{V}_2$. Let (1,0) and (0,1) constitute a basis for the vector space $\mathbf{V}$. Construct the matrix of the linear transformation P with respect to the given basis. Let (1,0) and (1,−1) constitute another basis for the vector space. Construct the matrix of the linear transformation P with respect to the new basis.
7.2.4 Let $\mathbf{W}$ be a subspace of a vector space $\mathbf{V}$. Let $\dim(\mathbf{V}) = m$ and $\dim(\mathbf{W}) = k$. Let T be a linear transformation from $\mathbf{V}$ to $\mathbf{V}$ and $\mathbf{W}$ be invariant under T. Choose a basis $x_1, x_2, \ldots, x_m$ such that $x_1, x_2, \ldots, x_k$ form a basis for $\mathbf{W}$. Let P be any projector from $\mathbf{V}$ onto $\mathbf{W}$. Let A_T and A_P be the matrices of the transformations T and P, respectively, with respect to the given basis. Show that $A_P A_T A_P = A_T A_P$ directly. Place this assertion vis-a-vis with **P 7.2.5**.

7.3. Orthogonal Projection

In Section 2.2 we touched upon the orthogonal projection briefly. See Definition 2.2.12 and the ensuing discussion. In this section, we will spend some time with the orthogonal projection and learn some more. We will be working in the environment of inner product spaces. Let

V be an inner product space equipped with an inner product $< \cdot, \cdot >$. Let **W** be a subspace of **V** and $\mathbf{W}^{\perp}$ its orthogonal complement. Recall that the projector P from **V** onto **W** along $\mathbf{W}^{\perp}$ is called the orthogonal projection on **W**. Before we proceed with some characteristic properties of orthogonal projections, we need to brush up our knowledge on the adjoint of a transformation. If T is a linear transformation from an inner product space **V** into itself, then there exists a linear transformation T^* from **V** to **V** such that

$$< x, Ty > = < T^*x, y > \quad \text{for all } x \text{ and } y \text{ in } \mathbf{V}.$$

The transformation T^* is called the adjoint of T. If $T^* = T$, T is called self-adjoint.

P 7.3.1 A linear map P from **V** to **V** is an orthogonal projection if and only if P is idempotent and self-adjoint, i.e.,

$$P^2 = P \quad \text{and } P^* = P.$$

PROOF. Suppose P is an orthogonal projection. Since it is a projector, it is idempotent. See **P 6.1.3**. We can identify the relevant subspaces involved. The map P is a projector from **V** onto $\mathbf{R}(P)$, the range of P, along $\mathbf{R}(I - P)$. Since P is an orthogonal projection, the subspaces $\mathbf{R}(P)$ and $\mathbf{R}(I - P)$ must be orthogonal, i.e.,

$$\begin{aligned} & < u, v > = 0 \text{ for all } u \in \mathbf{R}(P) \text{ and } v \in \mathbf{R}(I - P) \\ \Leftrightarrow\ & < (I - P)x, Py >= 0 \quad \text{for all } x \text{ and } y \text{ in } \mathbf{V} \\ \Rightarrow\ & < P^*(I - P)x, y >= 0 \quad \text{for all } x \text{ and } y \text{ in } \mathbf{V}. \end{aligned}$$

Consequently,

$$P^*(I - P) = 0 \quad \text{or } P^*P = P^*.$$

(Why?) Observe that $P^* = P^*P = (P^*P)^* = (P^*)^* = P$. See the complements under Section 3.6. Conversely, suppose that $P^2 = P$ and $P^* = P$. It is clear that P is a projector from $\mathbf{R}(P)$ along $\mathbf{R}(I - P)$. See **P 6.1.3**. We need to show that the subspaces $\mathbf{R}(P)$ and $\mathbf{R}(I - P)$ are orthogonal. For x and y in **V**, observe that

$$< (I-P)x, Py >=< P^*(I-P)x, y >=< P(I-P)x, y >=< 0x, y >= 0.$$

Hence $\mathbf{R}(P)$ and $\mathbf{R}(I - P)$ are orthogonal.

EXAMPLE 7.3.2. Let $\mathbf{V} = \mathbf{C}^m$, a complex vector space of dimension m. The standard inner product on $\mathbf{V}$ is given by

$$< x, y > = \sum_{i=1}^{m} x_i \bar{y}_i,$$

where $x = (x_1, x_2, \dots, x_m)$ and $y = (y_1, y_2, \dots, y_m) \in \mathbf{V}$. Let $P = (p_{ij})$ be a matrix of order $m \times m$ with complex entries. The matrix P can be regarded as a linear transformation from $\mathbf{V}$ to $\mathbf{V}$. The complex conjugate of P is the matrix $P^* = (q_{ij})$ (abuse of notation?), where $q_{ij} = \bar{p}_{ji}$ for all i and j. Recall that P is Hermitian if $P^* = P$. One can verify that for any two vectors x and y in $\mathbf{V}$,

$$< x, Py > = < P^* x, y >,$$

with the understanding that when we write Py we view y as a column vector and then carry out matrix multiplication of P and y. The matrix P^* is after all the adjoint of P when they are viewed as transformations. The matrix P viewed as a linear transformation is a projector if and only if $P^2 = P$ and P is Hermitian.

Sums of orthogonal projections are easy to handle. The following proposition handles this situation, which is easy to prove.

P 7.3.3 Let $\mathbf{V}_1, \mathbf{V}_2, \dots, \mathbf{V}_r$ be pairwise orthogonal subspaces of $\mathbf{V}$. Let $\mathbf{V}_0 = \mathbf{V}_1 \oplus \mathbf{V}_2 \oplus \dots \oplus \mathbf{V}_r$. Let P_i be the orthogonal projection on $\mathbf{V}_i$, $i = 1, 2, \dots, r$. Then $P = P_1 + P_2 + \dots + P_r$ is an orthogonal projection on $\mathbf{V}_0$.

7.4. Idempotent Matrices

Every linear transformation T from a finite-dimensional vector space $\mathbf{V}$ to $\mathbf{V}$ has a matrix associated with it under a given basis of $\mathbf{V}$. See Section 3.4. In particular, the matrices associated with projectors are of special interest. In this section, we focus on matrices with entries as complex numbers. A square matrix A is said to be idempotent if $A^2 = A$. This definition is analogous to the one we introduced for linear transformations. In brief, transformations and matrices associated with them are hardly distinguishable and using the same word "idempotent" in both the contexts should not cause confusion.

The definition of idempotent matrix is also operational when the entries of the matrix come from any field. Some of the results stated below make sense in the general framework of matrices with entries coming from any field.

P 7.4.1 Let A be an idempotent matrix of order $m \times m$. The following are valid.

(1) The eigenvalues of A can only be zeros and ones.

(2) The matrix A admits a factorization $A = QR^*$ with Q and R being of order $m \times k$ and $R^*Q = I_k$, where $k = \rho(A)$, the rank of A.

(3) The matrix A is diagonalizable, i.e., there exists a non-singular matrix L and a diagonal matrix Δ such that $A = L\Lambda L^{-1}$, the diagonal entries of Λ being zeros and ones.

(4) $\rho(A) = Tr(A)$. (The trace operation is discussed in Complement 3.4.7.)

(5) There exists a positive definite matrix C such that $A = C^{-1}A^*C$.

(6) A is a projection matrix, i.e., there exist two subspaces $\mathbf{V}_1$ and $\mathbf{V}_2$ on $\mathbf{C}^m$ such that $\mathbf{V}_1 \bigcap \mathbf{V}_2 = \{0\}$, $\mathbf{C}^m = \mathbf{V}_1 \oplus \mathbf{V}_2$, and if $x = x_1 + x_2$ with $x_1 \in \mathbf{V}_1$ and $x_2 \in \mathbf{V}_2$, then $Ax = x_1$. (If we view A as a transformation from $\mathbf{C}^m$ to $\mathbf{C}^m$, then A is a projector from $\mathbf{C}^m$ onto $\mathbf{V}_1$ along $\mathbf{V}_2$, in the usual jargon. As usual, members of $\mathbf{C}^m$ are viewed as column vectors.)

PROOF. (1) Let λ be an eigenvalue of A with an associated eigenvector x. Then $Ax = \lambda x$ implies that $\lambda x = Ax = A^2x = A(Ax) = \lambda(Ax) = \lambda(\lambda x) = \lambda^2 x$. Since $x \neq 0, \lambda = 0$ or 1.

(2) By the Singular Value Decomposition Theorem (**P 5.3.4**), we can write $A = Q\Delta P^*$, where Q is of order $m \times k$ with the property that $Q^*Q = I_k$, P is of order $m \times k$ with the property that $P^*P = I_k$, and Δ is a diagonal matrix of order $k \times k$ with positive entries in the diagonal. Since $A^2 = A$, we have $Q\Delta P^*Q\Delta P^* = Q\Delta P^*$, from which we have $\Delta P^*Q\Delta = \Delta$ or $\Delta P^*Q = I_k$. Take $R^* = \Delta P^*$. Thus we have $A = QR^*$ with $R^*Q = I_k$.

(3) Choose a matrix S of order $m \times (m-k)$ so that the augmented matrix $L = (Q|S)$ is non-singular and $R^*S = 0$, where Q and R are the matrices that appear in the representation (2) of A above. (How?) Now choose a matrix U of order $m \times (m-k)$ such that $U^*S = I_{m-k}$ and $U^*Q = 0$. (How?) One can verify that $L^{-1} = (R|U)^*$. (Verify that

$L^{-1}L = I_m$.) Observe that

$$A = (Q|S)\begin{bmatrix} I_k & 0 \\ 0 & 0 \end{bmatrix}(R|U)^* = L\Lambda L^{-1}, \tag{7.4.1}$$

where Λ is the diagonal matrix given by

$$\Lambda = \begin{bmatrix} I_k & 0 \\ 0 & 0 \end{bmatrix}.$$

(4) From (7.4.1), $Tr(A) = Tr(L\Lambda L^{-1}) = Tr(\Lambda L^{-1}L) = Tr(\Lambda) = k = \rho(A)$. See Complement 3.4.7.

(5) Note that $I_m - A$ is also idempotent. Consequently, $\rho(I_m - A) = m - k$. Consider the rank factorizations of A and $I_m - A$. See Corollary 5.2.3. Write $A = D_1E_1$ and $I_m - A = D_2E_2$, where D_1 is order $m \times k$ with rank k, E_1 of order $k \times m$ with rank k, D_2 of order $m \times (m-k)$ with rank $(m-k)$, and E_2 of order $(m-k) \times m$ with rank $m-k$. Let $F_1 = (D_1|D_2)$ and $F_2 = \left[\begin{smallmatrix} E_1 \\ E_2 \end{smallmatrix}\right]$. Then $F_1F_2 = D_1E_1 + D_2E_2 = A + (I_m - A) = I_m$. It now follows that $F_1 = F_2^{-1}$. Let $C = F_2^*F_2$. It is clear that C is non-singular and Hermitian. Further, C is positive definite. Note that $F_1 = (D_1|D_2) = F_2^{-1} = C^{-1}F_2^* = C^{-1}(E_1^*|E_2^*)$, from which we have $D_1 = C^{-1}E_1^*$ or equivalently, $CD_1 = E_1^*$. Finally,

$$A = D_1E_1 = C^{-1}E_1^*(CD_1)^* = C^{-1}E_1^*D_1^*C^* = C^{-1}A^*C.$$

This completes the proof.

(6) Take $\mathbf{V}_1 = \mathbf{R}(A)$, the range of the matrix A, and $\mathbf{V}_2 = \mathbf{R}(I_m - A)$. It is clear that $\mathbf{V}_1 \bigcap \mathbf{V}_2 = \{0\}$. For every x in $\mathbf{C}^m$, note that $x = Ax + (I_m - A)x$, $Ax \in \mathbf{V}_1 = \mathbf{R}(A)$, and $(I_m - A)x \in \mathbf{V}_2 = \mathbf{R}(I_m - A)$. This implies that $\mathbf{V}_1 \oplus \mathbf{V}_2 = \mathbf{C}^m$ and the projector P from $\mathbf{C}^m$ onto $\mathbf{V}_1$ along $\mathbf{V}_2$ is precisely equal to A.

COROLLARY 7.4.2. If A is idempotent and Hermitian, one can write

$$A = TT^* \quad \text{with} \quad T^*T = I_k,$$

where $k = \rho(A)$.

PROOF. We use **P 5.4.3**. Since A is Hermitian, there exists a unitary matrix P such that $A = P\Gamma P^*$, where Γ is a diagonal matrix with diagonal entries constituting the eigenvalues of A. Since A is idempotent, each of the eigenvalues of A is either zero or one. Assume, without loss of generality, that Γ is of the form

$$\Gamma = \begin{bmatrix} I_k & 0 \\ 0 & 0 \end{bmatrix}.$$

Write $P = (T|S)$, where T is of order $m \times k$. We can now check that $A = TT^*$ and $T^*T = I_k$.

The idempotency of a matrix can be characterized solely based on the additive property of the ranks of two matrices A and $I_m - A$. The following proposition is concerned with this feature.

P 7.4.3 Let A be a matrix of order $m \times m$. Then A is idempotent if and only if

$$\rho(A) + \rho(I_m - A) = m.$$

PROOF. We have already seen that if A is idempotent then $\rho(A) + \rho(I_m - A) = m$. The argument is buried somewhere in the proof of **P 7.4.1**. Suppose $\rho(A) + \rho(I_m - A) = m$. Observe that

$$m = \rho(I_m) = \rho(A+(I_m-A)) = \rho(A)+\rho(I_m-A)-\dim(\mathbf{R}(A)\cap\mathbf{R}(I_m-A)).$$

This identity requires some explanation. It is clear that $\mathbf{C}^m = \mathbf{R}(A) + \mathbf{R}(I_m - A)$. (Why?) A look at **P 1.5.7** might help the reader to understand the meaning of the symbol $+$ in the orientation defined above. We do not know that $\mathbf{C}^m = \mathbf{R}(A) \oplus \mathbf{R}(I_m - A)$. Consequently,

$$\begin{aligned} m &= \dim(\mathbf{C}^m) = \dim(\mathbf{R}(A) + \mathbf{R}(I_m - A)) \\ &= \dim(\mathbf{R}(A)) + \dim(\mathbf{R}(I_m - A)) - \dim(\mathbf{R}(A) \cap \mathbf{R}(I_m - A)) \\ &= \rho(A) + \rho(I_m - A) - \dim(\mathbf{R}(A) \cap \mathbf{R}(I_m - A)) \\ &= m - \dim(\mathbf{R}(A) \cap \mathbf{R}(I_m - A)). \end{aligned}$$

This implies that $\dim(\mathbf{R}(A) \cap \mathbf{R}(I_m - A)) = 0$ from which we have

$$\mathbf{R}(A) \cap \mathbf{R}(I_m - A) = \{0\}.$$

Thus $\mathbf{C}^m = \mathbf{R}(A) \oplus \mathbf{R}(I_m - A)$ indeed. We claim that $A(I_m - A) = 0$. Suppose not. Then there exist non-zero vectors x and y in $\mathbf{C}^m$ such that $A(I_m - A)x = y$. This implies that $y \in \mathbf{R}(A)$. Note that $A(I_m - A) =$

$(I_m - A)A$. It is true that $(I_m - A)Ax = y$. This implies that $y \in \mathbf{R}(I_m - A)$. This is a contradiction. Hence $A(I_m - A) = 0$ or $A^2 = A$. This completes the proof.

The following result is analogous to the one stated in Complement 7.1.8. The statement is couched in terms of matrices with complex entries.

P 7.4.4 Let A_1 and A_2 be two square matrices of the same order and $A = A_1 + A_2$. Then the following statements are equivalent.

(1) $A^2 = A$ and $\rho(A) = \rho(A_1) + \rho(A_2)$.
(2) $A_1^2 = A_1$, $A_2^2 = A_2$, and $A_1A_2 = A_2A_1 = 0$.

PROOF. Suppose (2) is true. It is obvious that $A^2 = A$. Since A, A_1, and A_2 are idempotent, $\rho(A) = Tr(A_1) + Tr(A_2) = \rho(A_1) + \rho(A_2)$. Suppose (1) is true. By **P 7.4.3**, $m = \rho(A)+\rho(I_m-A) = \rho(A_1)+\rho(A_2)+\rho(I_m - A) \geq \rho(A_1) + \rho(A_2 + I_m - A) = \rho(A_1) + \rho(I_m - A_1) \geq \rho(A_1 + I_m - A_1) = \rho(I_m) = m$. Consequently, $\rho(A_1) + \rho(I_m - A) = m$. Again, by **P 7.4.3**, A_1 is idempotent. In a similar vein, one can show that A_2 is idempotent. The fact that A, A_1, and A_2 are idempotent implies that $A_1A_2 + A_2A_1 = 0$. The information that $\rho(A) = \rho(A_1) + \rho(A_2)$ implies that $\mathbf{R}(A_1) \cap \mathbf{R}(A_2) = \{0\}$. This coupled with $A_1A_2 = -A_2A_1$ gives $A_1A_2 = 0$. Follow the argument crafted in the proof of **P 7.4.3**.

A generalization of **P 7.4.4** is in order involving more than two matrices.

P 7.4.5 Let $A_1, A_2, \ldots, A_k$ be any k square matrices of the same order and $A = A_1 + A_2 + \ldots + A_k$. Consider the following statements.

(1) Each A_i is idempotent.
(2) $A_iA_j = 0$ for all $i \neq j$ and $\rho(A_i^2) = \rho(A_i)$ for all i.
(3) A is idempotent.
(4) $\rho(A) = \rho(A_1) + \rho(A_2) + \ldots + \rho(A_k)$.

Then the validity of any two of the statements (1), (2), and (3) imply the validity of all the four statements. Further, the validity of statements (3) and (4) imply the validity of the rest of the statements.

PROOF. Suppose (1) and (2) are true. It is clear that (3) is true. Since A and $A_1, A_2, \ldots, A_k$ are all idempotent, $\rho(A) = Tr(A) = Tr(A_1) + Tr(A_2) + \ldots + Tr(A_k) = \rho(A_1) + \rho(A_2) + \ldots + \rho(A_k)$. Thus (4) is true.

Suppose (2) and (3) are true. A computation of A^2 yields $A^2 = \sum_{i=1}^{k} A_i^2$. Fix $1 \leq i \leq k$. We show that A_i is idempotent. Note that $AA_i = A_iA = A_i^2$ and $A^2A_i = A_iA^2 = A_i^3$. Since A is idempotent, we have $A_i^2 = A_i^3$, which implies that $A_i^2(I_m - A_i) = 0$. The condition $\rho(A_i) = \rho(A_i^2)$ is equivalent to the statement that $dim(\mathbf{R}(A_i)) = \dim(\mathbf{R}(A_i^2))$. Since $\mathbf{R}(A_i^2) \subset \mathbf{R}(A_i)$, we must have $\mathbf{R}(A_i) = \mathbf{R}(A_i^2)$. Consequently, there exists a nonsingular matrix D such that $A_i = DA_i^2$. Hence $A_i^2(I_m - A) = 0$ implies that $A_i(I_m - A) = 0$ from which we conclude that A_i is idempotent. Thus (1) is true. Now (4) follows.

Suppose (3) and (4) are valid. Fix $i \neq j$. Let $B = A_i + A_j$ and $C = A - B$. By (4),

$$\sum_{r=1}^{k} \rho(A_r) = \rho(A) = \rho(B + C) \leq \rho(B) + \rho(C) \leq \sum_{r=1}^{k} \rho(A_r).$$

From this, we have $\rho(A) = \rho(B) + \rho(C)$ and $\rho(B) = \rho(A_i) + \rho(A_j)$. (Why?) Observe that

$$\begin{aligned} m = \rho(I_m) &= \rho(B + I_m - B) \leq \rho(B) + \rho(I_m - B) \\ &= \rho(B) + \rho(I_m - A + C) = \rho(B) + \rho(I_m - A) + \rho(C) \\ &= \rho(A) + \rho(I_m - A) = m. \end{aligned}$$

Hence $\rho(B) + \rho(I_m - B) = m$. By **P 7.4.3**, B is idempotent. Thus we have $A_i + A_j$ idempotent and $\rho(B) = \rho(A_i) + \rho(A_j)$. By **P 7.4.4**, $A_iA_j = 0$ and A_i and A_j idempotent. Thus (1) and (2) follow in one stroke.

Suppose (1) and (2) are valid. It is obvious that (4) follows exploiting the connection between rank and trace for idempotent matrices. Since we have (4) valid, (2) follows now from what we have established above. This completes the proof.

The condition in (2) of **P 7.4.5** that $\rho(A_i) = \rho(A_i^2)$ is somewhat intriguing. It could happen that $\rho(B) \neq \rho(B^2)$ for a matrix B. As an example, try

$$B = \begin{bmatrix} 0 & 1 \\ 0 & 0 \end{bmatrix}.$$

This will not happen if B is Hermitian or nonsingular.

Complements

7.4.1 Let $Y_1, Y_2, \ldots, Y_m$ be m real random variables whose joint distribution depends on a vector parameter $\theta' = (\theta_1, \theta_2, \ldots, \theta_k) \in \mathbf{R}^k$. Suppose $E_\theta Y_i = x_{i1}\theta_1 + x_{i2}\theta_2 + \ldots + x_{ik}\theta_k$, $i = 1, 2, \ldots, m$, where x_{ij}'s are known. Let $X = (x_{ij})$. Assume that $\rho(X) = k$. Let $g(\theta) = \theta, \theta \in \mathbf{R}^k$. The random variables constitute a linear model and one can rewrite the expectations as $E_\theta Y = X\theta, \theta \in \mathbf{R}^k$, where $Y' = (Y_1, Y_2, \ldots, Y_m)$. The least squares estimator $\hat{g}$ of $g(\cdot)$ is given by $\hat{g} = (X'X)^{-1}X'Y$. The residual sum of squares (RSS) is given by

$$RSS = (Y - X\hat{g})'(Y - X\hat{g}).$$

Show that

(1) $X(X'X)^{-1}X'$ and $(I_m - X(X'X)^{-1}X')$ are idempotent;
(2) $\rho(X(X'X)^{-1}X') = k$;
(3) $E_\theta \hat{g} = \theta$ for every $\theta \in \mathbf{R}^k$;
(4) $E_\theta RSS = (m - k)\sigma^2$ by assuming that $Y_1, Y_2, \ldots, Y_m$ are pairwise uncorrelated with common variance $\sigma^2 > 0$;
(5) $\hat{g}$ and RSS are independently distributed by assuming that $Y_1, \ldots, Y_m$ have a multivariate normal distribution with variance covariance matrix $\sigma^2 I_m$. (*Hint:* A linear form LY and a quadratic form $Y'AY$ are independently distributed if $LA = 0$.)

7.4.2 If A is idempotent and non-singular, show that $A = I$.

7.4.3 Let

$$A = \begin{bmatrix} B & C \\ 0 & D \end{bmatrix},$$

where B and D are square matrices. Show that A is idempotent if and only if B and D are idempotent, $BCD = 0$, and $(I - B)C(I - D) = 0$.

7.5. Matrix Representation of Projectors

In this section, we consider a finite dimensional vector space $\mathbf{V}$ over the field of complex numbers equipped with an inner product $< \cdot, \cdot >$. Let

$$S = (a_1, a_2, \ldots, a_n)$$

be an ordered collection of vectors from $\mathbf{V}$. We define some algebraic operations on the space of all ordered sets of the form S.

Addition of ordered sets

Let $S_1 = (a_1, a_2, \ldots, a_n)$ and $S_2 = (b_1, b_2, \ldots, b_n)$ be two ordered collections of vectors from $\mathbf{V}$. Define the sum of S_1 and S_2 by

$$S_1 + S_2 = (a_1 + b_1, a_2 + b_2, \ldots, a_n + b_n). \tag{7.5.1}$$

Multiplication of ordered sets

Let $S_1 = (a_1, a_2, \ldots, a_m)$ and $S_2 = (b_1, b_2, \ldots, b_n)$ be two ordered collections of vectors from $\mathbf{V}$. Define the product of S_1 and S_2 to be the matrix of order $m \times n$

$$S_1 \circ S_2 = \begin{bmatrix} < b_1, a_1 > & \ldots & < b_n, a_1 > \\ \cdot & \ldots & \cdot \\ < b_1, a_m > & \ldots & < b_n, a_m > \end{bmatrix}. \tag{7.5.2}$$

Multiplication of an ordered set and a matrix

Let $S = (a_1, a_2, \ldots, a_n)$ be an ordered set of vectors from $\mathbf{V}$ and $M = (m_{ij})$ a matrix of order $n \times k$ with complex entries. Define the product of S and M to be the ordered set

$$S \times M = \left(\sum_{j=1}^{n} m_{j1} a_j, \sum_{j=1}^{n} m_{j2} a_j, \ldots, \sum_{j=1}^{n} m_{jk} a_j\right), \tag{7.5.3}$$

which is an ordered set of k vectors.

If M is a column vector with entries $m_1, m_2, \ldots, m_n$, then $S \times M$ is simply the linear combination $m_1 a_1 + m_2 a_2 + \ldots + m_n a_n$ of the vectors $a_1, a_2, \ldots, a_n$.

The operation $\circ$ in (7.5.2) is a little fascinating. Let us examine what this means when the vectors a_i's and b_i's come from the coordinate vector space $\mathbf{C}^k$. The set S_1 turns out to be a matrix of order $k \times m$ with complex entries and S_2 is of order $k \times n$ with complex entries. It can be verified that

$$S_1 \circ S_2 = S_1^* S_2.$$

A special case of this is when $S_1 = S_2 = S$, say, in which case

$$S \circ S = S^* S.$$

The operation $\times$ simplifies the usual matrix multiplication in coordinate vector spaces. It turns out that $S \times M = SM$, the usual matrix multiplication of S and M.

We record some of the properties of these operations in the following proposition which can be verified easily. In the following the symbol S with subscripts stand for ordered sets of vectors and M with subscripts stand for matrices.

P 7.5.1 The following are valid.

(1) If $S = (a_1, a_2, \dots, a_m), M_1 = (m_{ij}^{(1)})$ and $M_2 = (m_{ij}^{(2)})$ each of order $m \times k$, then $S \times (M_1 + M_2) = (S \times M_1) + (S \times M_2)$.

(2) If $S_1 = (a_1, a_2, \dots, a_m), S_2 = (b_1, b_2, \dots, b_m)$, and $M = (m_{ij})$ of order $m \times k$, then $(S_1 + S_2) \times M = (S_1 \times M) + (S_2 \times M)$.

(3) If $S_1 = (a_1, a_2, \dots, a_m), S_2 = (b_1, b_2, \dots, b_n)$, and $M = (m_{ij})$ of order $n \times k$, then $S_1 \circ (S_2 \times M) = (S_1 \circ S_2)M$.

(4) If $S_1 = (a_1, a_2, \dots, a_m), M = (m_{ij})$ of order $m \times k$, and $S_2 = (b_1, b_2, \dots, b_n)$, then $(S_1 \times M) \circ S_2 = M^*(S_1 \circ S_2)$.

(5) If $S_1 = (a_1, a_2, \dots, a_m), S_2 = (b_1, b_2, \dots, b_n)$, and $S_3 = (c_1, c_2, \dots, c_r)$, then $S_1 \circ (S_2 + S_3) = S_1 \circ S_2 + S_1 \circ S_3$.

Having an explicitly workable expression for the projector from a vector space $\mathbf{V}$ onto a subspace $\mathbf{V}_1$ along a subspace $\mathbf{V}_2$ seems to be difficult. Using the new operations developed at the beginning of this section, we will attempt to provide an explicit formula.

P 7.5.2 Let $S = (a_1, a_2, \dots, a_m)$ be an ordered collection of vectors from a vector space. Let $\mathbf{R}(S)$ be the vector space spanned by the vectors of S. Then the orthogonal projection P from $\mathbf{V}$ onto $\mathbf{R}(S)$ has the representation,

$$P = S \times M(S \circ \cdot) \tag{7.5.4}$$

so that the orthogonal projection of a vector x in $\mathbf{V}$ is given by

$$Px = S \times M(S \circ x), \tag{7.5.5}$$

where M is any matrix of order $m \times m$ satisfying $(S \circ S)M(S \circ S) = S \circ S$.

PROOF. Let $b_1, b_2, \dots, b_k$ be a basis to the $\mathbf{R}(S)^\perp$, the orthogonal complement of $\mathbf{R}(S)$. Take $S_1 = (b_1, b_2, \dots, b_k)$. It is clear that

$$S \circ S_1 = 0.$$

Let $x \in \mathbf{V}$. Since $\mathbf{R}(S) \oplus \mathbf{R}(S)^{\perp} = \mathbf{V}$, we can write $x = x_1 + x_2$ with $x_1 \in \mathbf{R}(S)$ and $x_2 \in \mathbf{R}(S)^{\perp}$. The vector x_1 is the one we are going after, i.e., $Px = x_1$. The vector x_1 is a linear combination of the vectors in S and x_2 is a linear combination of the vectors in S_1. Consequently, there are vectors y and z of orders $m \times 1$ and $k \times 1$, respectively, such that, in terms of our new algebraic operations,

$$x_1 = S \times y \quad \text{and } x_2 = S_1 \times z.$$

If we know y, we will know the projection x_1 of x. Thus we have

$$x = S \times y + S_1 \times z \tag{7.5.6}$$

for some column vectors of complex numbers. Premultiply (7.5.6) by S with respect to the operation $\circ$

$$S \circ x = S \circ (S \times y) + S \circ (S_1 \times z) = (S \circ S)y + (S \circ S_1)z = (S \circ S)y.$$

We can view $(S \circ S)y = S \circ x$ as a linear equation in unknown y. Let M be a generalized inverse of the matrix $(S \circ S)$, i.e., M satisfies the equation, $(S \circ S)M(S \circ S) = (S \circ S)$. A solution of the linear equation is given by $y = M(S \circ x)$. We are jumping the gun again! We will see in Chapter 8 why this is so. Thus

$$Px = x_1 = S \times y = S \times M(S \circ x).$$

This completes the proof.

We will specialize this result for coordinate spaces. Suppose $\mathbf{V} = \mathbf{C}^n$. Then S is an $n \times m$ matrix. Take the inner product $< \cdot, \cdot >$ to be the standard inner product in $\mathbf{V}$, i.e., for x and y in $\mathbf{V}, < x, y >= y^*x$. Note that $S \circ S = S^*S$ and $S \circ x = S^*x$. Finally,

$$S \times M(S \circ x) = S \times MS^*x = SMS^*x,$$

where M is a generalized inverse of the Hermitian matrix S^*S. Let us enshrine this result in the form of a corollary.

COROLLARY 7.5.3. Let $\mathbf{V} = \mathbf{C}^n$ and S a matrix of order $n \times m$ with complex entries. Let $\mathbf{R}(S)$ be the vector space spanned by the columns of S. Then the projection operator P from $\mathbf{V}$ into $\mathbf{R}(S)$ has the representation,

$$P = SMS^* \tag{7.5.7}$$

where M is any matrix satisfying $(S^*S)M(S^*S) = S^*S$.

The expression (7.5.7) for the projection operator first given by Rao (1967) was useful in the discussion of linear models under a very general setup.

Now we take up the general case of projectors. Let S_1 and S_2 be two ordered sets of vectors from an inner product space $\mathbf{V}$. Suppose $\mathbf{R}(S_1)$ and $\mathbf{R}(S_2)$, the vector spaces spanned by S_1 and S_2, respectively, satisfy the conditions that $\mathbf{R}(S_1) \cap \mathbf{R}(S_2) = \{0\}$ and $\mathbf{R}(S_1) \oplus \mathbf{R}(S_2)$. The spaces $\mathbf{R}(S_1)$ and $\mathbf{R}(S_2)$ need not be orthogonal. Let P be the projector from $\mathbf{V}$ onto $\mathbf{R}(S_1)$ along $\mathbf{R}(S_2)$. We need an explicit representation of P. Let S_3 be an ordered set of vectors from $\mathbf{V}$ such that $\mathbf{R}(S_2)$ and $\mathbf{R}(S_3)$ are orthogonal and $\mathbf{V} = \mathbf{R}(S_2) \oplus \mathbf{R}(S_3)$. (How will you find S_3?) In particular, we will have $S_3 \circ S_2 = 0$.

P 7.5.4 In the framework described above, the projector P has the representation

$$Px = S_1 \times M(S_3 \circ x)$$

for any x in $\mathbf{V}$, where M is a generalized inverse of the matrix $S_3 \circ S_1$, i.e., M satisfies $(S_3 \circ S_1)M(S_3 \circ S_1) = S_3 \circ S_1$.

PROOF. Let $x \in \mathbf{V}$. Write $x = x_1 + x_2$ with $x_1 \in \mathbf{R}(S_1)$ and $x_2 \in \mathbf{R}(S_2)$. Since x_1 is a linear combination of the vectors in S_1, we can write $x_1 = S_1 \times y$ for some column vector of complex entries. In a similar vein, we can write $x_2 = S_1 \times z$ for some column vector z. Premultiplying $x = S_1 \times y + S_2 \times z$ by S_3 under the operation $\circ$, we have

$$\begin{aligned} S_3 \circ x &= S_3 \circ (S_1 \times y) + S_3 \circ (S_2 \times z) \\ &= (S_3 \circ S_1)y + (S_3 \circ S_2)z = (S_3 \circ S_1)y. \end{aligned}$$

A solution to the system of linear equations $(S_3 \circ S_1)y = S_3 \circ x$ in unknown y is given by $y = M(S_3 \circ x)$. Thus

$$Px = S_1 \times y = S_1 \times M(S_3 \circ x).$$

This completes the proof.

In the above proof, we roped the ordered set S_3 in the representation of the projector. We could avoid this.

P 7.5.5 Let the framework of **P 7.5.4** be operational here. Let G be any generalized inverse of the matrix

$$\begin{bmatrix} S_1 \circ S_1 & S_1 \circ S_2 \\ S_2 \circ S_1 & S_2 \circ S_2 \end{bmatrix}.$$

Partitioned as

$$G = \begin{bmatrix} C_1 & C_2 \\ C_3 & C_4 \end{bmatrix}$$

where the order of the matrix C_1 is the same as the order of $S_1 \circ S_1$ and the order of C_4 is the same as the order of $S_2 \circ S_2$. Then for any x in $\mathbf{V}$,

$$Px = C_1(S_1 \circ x) + C_2(S_2 \circ x).$$

PROOF. As in the proof of **P 7.5.4**, write

$$x = S_1 \times y + S_2 \times z \tag{7.5.8}$$

for some column vectors y and z with complex entries. Premultiply (7.5.8) by S_1 under the operation $\circ$. We will have

$$\begin{aligned} S_1 \circ x &= S_1 \circ (S_1 \times y) + S_1 \circ (S_2 \times z) \\ &= (S_1 \circ S_1)y + (S_1 \circ S_2)z. \end{aligned} \tag{7.5.9}$$

Premultiplying (7.5.8) by S_2 under the operation $\circ$, we have

$$S_2 \circ x = (S_1 \circ S_1)y + (S_2 \circ S_2)z. \tag{7.5.10}$$

Equations (7.5.9) and (7.5.10) can be written as

$$\begin{bmatrix} S_1 \circ S_1 & S_1 \circ S_2 \\ S_2 \circ S_1 & S_2 \circ S_2 \end{bmatrix} \begin{bmatrix} y \\ z \end{bmatrix} = \begin{bmatrix} S_1 \circ x \\ S_2 \circ x \end{bmatrix}.$$

This is a system of linear equations in the unknowns y and z. A solution is given by

$$\begin{bmatrix} y \\ z \end{bmatrix} = G \begin{bmatrix} S_1 \circ x \\ S_2 \circ x \end{bmatrix} = \begin{bmatrix} C_1 & C_2 \\ C_3 & C_4 \end{bmatrix} \begin{bmatrix} S_1 \circ x \\ S_2 \circ x \end{bmatrix}.$$

Consequently,

$$y = C_1(S_1 \circ x) + C_2(S_2 \circ x), \ z = C_3(S_1 \circ x) + C_4(S_2 \circ x).$$

Finally, $Px = S_1 \times y = C_1(S_1 \circ x) + C_2(S_2 \circ x)$. This completes the proof.

Complements

7.5.1 For vectors x and y in a vector space $\mathbf{C}^n$, define the inner product by $< x, y >= y^*\Sigma x$, where Σ is a positive definite matrix. Then P is an orthogonal projector if and only if

$$(1) \quad P^2 = P, \quad \text{and} \quad (2) \quad \Sigma P \text{ is Hermitian.}$$

7.5.2 (Rao (1967)) Let C_1 be a subspace of C^n spanned by columns of an $n \times k$ matrix A. Show that the orthogonal projector onto C_1 is

$$P = A(A^*\Sigma A)^- A^*\Sigma$$

where $(A^*\Sigma A)^-$ is a generalized inverse of $(A^*\Sigma A)$, i.e., any matrix B satisfying the property $(A^*\Sigma A)B(A^*\Sigma A) = A^*\Sigma A$. The expression for P is unique for any choice of the generalized inverse. (For a discussion of generalized inverses, see Chapter 8.)

7.5.3 Let A be $n \times p$ and B be $n \times q$ real matrices and denote their Kronecker product by $A \otimes B$. Denote by P_A, P_B and $P_{A\otimes B}$ the orthogonal projectors on $R(A), R(B)$ and $R(A \otimes B)$ respectively. Then

(1) $P_{A\otimes B} = P_A \otimes P_B$
(2) $P_{A\otimes I} = P_A \otimes I$
(3) $Q_{A\otimes B} = Q_A \otimes Q_B + Q_A \otimes P_B + P_A \otimes Q_B$, where $Q_A = I - P_A$, $Q_B = I - Q_B$.

Note: The references for this Chapter are: Rao (1967), Rao and Mitra (1971b), Rao (1974), Rao and Mitra (1974), Rao (1978c), Rao and Yanai (1979) and standard books on Linear Algebra.

CHAPTER 8

GENERALIZED INVERSES

In Section 3.3, we explored the concept of an inverse of a linear transformation T from a vector space $\mathbf{V}$ to a vector space $\mathbf{W}$. We found that the nature of the inverse depends upon what kind of properties T has. A summary of the discussion that had been carried out earlier is given below.

(1) Suppose T is bijective, i.e., T is injective (one-to-one) and surjective (onto). Then there exists a unique transformation S from $\mathbf{W}$ to $\mathbf{V}$ such that

$$ST = I \quad \text{and} \quad TS = I$$

with the identity transformation acting on the appropriate vector space.

(2) Suppose T is surjective. Then there exist a transformation S (called a right inverse of T) from $\mathbf{W}$ to $\mathbf{V}$ such that $TS = I$.

(3) Suppose T is injective. Then there exists a transformation S (called a left inverse of T) from $\mathbf{W}$ to $\mathbf{V}$ such that $ST = I$.

(4) There always exists a transformation S from $\mathbf{W}$ to $\mathbf{V}$ such that $TST = T$. Such a transformation S is called a g-inverse of T.

In this chapter, we focus on matrices. We reenact the entire scenario detailing inverses of transformation in the realm of matrices. Special attention will be paid to finding simple criteria for the existence of every type of inverse. The source material for this Chapter is Rao and Mitra (1971).

Before we proceed with the details, we need to set up the notation. Let $\mathbf{M}_{m,n}$ denote the collection of all matrices A of order $m \times n$ with entries coming from the field of real or complex numbers. The symbol $\mathbf{M}_{m,n}(r)$ denotes the collection of all matrices A in $\mathbf{M}_{m,n}$ with rank r. The rank of a matrix A is denoted by $\rho(A)$. The vector space spanned by the columns of a matrix A is denoted by $Sp(A)$. An equivalent notation is $\mathbf{R}(A)$, the range of A when A is viewed as a transformation. See Section 4.1. [$Sp(A)$ is more suggestive when A is a matrix as the vector space generated by the column vectors of A.]

8.1. Right and Left Inverses

In this section we characterize right and left inverses of matrices. In addition, the structure of a right inverse as well as left inverse of a matrix is described.

P 8.1.1 Let $A \in \mathbf{M}_{m,n}$. There exists a matrix $G \in \mathbf{M}_{n,m}$ such that $AG = I_m$ if and only if $\rho(A) = m$. In such a case a choice of G is given by

$$G = A^*(AA^*)^{-1}. \tag{8.1.1}$$

A general solution of G is given by

$$G = VA^*(AVA^*)^{-1}, \tag{8.1.2}$$

where V is any arbitrary matrix satisfying $\rho(A) = \rho(AVA^*)$.

PROOF. Suppose $\rho(A) = m$. Then $m = \rho(A) = \rho(AA^*)$. The matrix AA^* is of order $m \times m$ and has rank m. Consequently, AA^* is nonsingular. The matrix $G = A^*(AA^*)^{-1}$ indeed satisfies $AG = AA^*(AA^*)^{-1} = I_m$. Conversely, suppose there exists a matrix $G \in \mathbf{M}_{n,m}$ such that $AG = I_m$. Note that $m = \rho(I_m) = \rho(AG) \leq \rho(A) \leq m$. Hence $\rho(A) = m$. As for the general structure of G, if V is any matrix satisfying $\rho(AVA^*) = \rho(A)$, then $G = VA^*(AVA^*)^{-1}$ certainly satisfies the condition $AG = I_m$. On the other hand, if G is any matrix satisfying $AG = I_m$, it can be put in the form $G = VA^*(AVA^*)^{-1}$ for some suitable choice of V. Take $V = GG^*$. (How?)

The matrix G that figures in **P 8.1.1** can rightly be called a right inverse of A. One can also say that a right inverse of A exists if the rows of A are linearly independent. Incidentally, $\rho(G) = m$. A similar result can be crafted for left inverses of A.

P 8.1.2 Let $A \in \mathbf{M}_{m,n}$. Then there exists a matrix $G \in \mathbf{M}_{n,m}$ such that $GA = I_n$ if and only if $\rho(A) = n$. In such a case one choice of G is given by

$$G = (A^*A)^{-1}A^*. \tag{8.1.3}$$

A general solution of G satisfying $GA = I_n$ is given by

$$G = (A^*VA)^{-1}A^*V \tag{8.1.4}$$

for any matrix V satisfying $\rho(A) = \rho(A^*VA)$.

The matrix G that figures in **P 8.1.2** can be rightly called a left inverse of A. The existence of a left inverse of A is guaranteed if the columns of A are linearly independent. Incidentally, $\rho(G) = n$.

The right and left inverses have some bearing on solving linear equations. Suppose $Ax = y$ is a consistent system of linear equations in an unknown vector x, where $A \in \mathbf{M}_{m,n}$ and $y \in \mathbf{M}_{m,1}$ are known. Consistency means that the system admits a solution in x. Suppose $\rho(A) = n$. Let G be any left inverse of A. Then $x = Gy$ is a solution to the linear equations. This can be seen as follows. Since $Ax = y$ is consistent, y must be a linear combination of the column of A. In other words, we can write $y = A\alpha$ for some column vector α. We now proceed to verify that Gy is indeed a solution to the system $Ax = y$ of equations. Note that $A(Gy) = AGA\alpha = AI_n\alpha = A\alpha = y$.

Let us look at the other possibility where we have a consistent system $Ax = y$ of linear equations with $\rho(A) = m$. Let G be a right inverse of A. Then Gy is a solution to the system $Ax = y$. This can be verified directly. Incidentally, note that if $\rho(A) = m$, $Ax = y$ is always consistent whatever may be the nature of the vector y!

Complements

8.1.1 Let A be a matrix of order $m \times 1$ with at least one non-zero entry. Exhibit a left inverse of A. Obtain a general form of left inverses of A.

8.1.2 Let A be a matrix of order $1 \times n$ with at least one non-zero entry. Exhibit a right inverse of A. Obtain a general form of right inverse of A. Show that $Ax = \beta$ is consistent for any number β.

8.1.3 Let $A \in \mathbf{M}_{m,n}$. Show that $Ax = y$ is consistent for all $y \in \mathbf{M}_{m,1}$ if and only if $\rho(A) = m$.

8.1.4 Let

$$A = \begin{bmatrix} 1 & 2 & 4 \\ 0 & 3 & 6 \end{bmatrix}.$$

Obtain a right inverse of A.

8.2. Generalized Inverse (g-inverse)

One of the basic problems in Linear Algebra is to determine solutions to a system $Ax = y$ of consistent linear equations, where $A \in \mathbf{M}_{m,n}$.

If the matrix is of full rank, i.e., $\rho(A) = m$ or n, we have seen in the last section how the right and left inverses of A, as the case may be, help to obtain a solution. It is time to make some progress in the case when A is not of full rank. Generalized Inverses or g-inverses of A are the matrices needed to solve consistent linear equations. They can be introduced in a variety of ways. We follow the linear equations angle.

DEFINITION 8.2.1. Let $A \in \mathbf{M}_{m,n}$ be of arbitrary rank. A matrix $G \in \mathbf{M}_{n,m}$ is said to be a generalized inverse (g-inverse) of A if $x = Gy$ is a solution of $Ax = y$ for any y for which the equation is consistent.

This is not a neat definition. It is a goal-oriented definition. Later, we will provide some characterizations of g-inverses, one of which could give us a crisp mathematical definition. The customary notation for a g-inverse of A is A^-, if it exists. First, we settle the question of existence.

P 8.2.2 For any matrix $A \in \mathbf{M}_{m,n}$, $A^- \in \mathbf{M}_{n,m}$ exists.

PROOF. If $A = 0$, take $G = 0$. Assume that $A \neq 0$. Let us make use of the rank factorization of A. Write $A = RF$, where R is of order $m \times a$ with rank a and F of order $a \times n$ with rank a, where $a = \rho(A)$. See Corollary 5.2.3. Let B be a left inverse of R, i.e., $BR = I_a$, and C a right inverse of F, i.e., $FC = I$. Let $A^- = CB$. We show that A^- is a g-inverse of A. Let $y \in Sp(A)$, the vector space spanned by the columns of A. Then the system $Ax = y$ is consistent. Also, $y = A\alpha$ for some column vector α. We show that A^-y is a solution of $Ax = y$.

$$\begin{aligned} AA^-y &= (RF)(CB)y = R(FC)By = RBy \\ &= RBA\alpha = (RB)(RF)\alpha = R(BR)F\alpha \\ &= RF\alpha = A\alpha = y, \end{aligned}$$

which shows that $x = A^-y$ satisfies the equation $Ax = y$.

P 8.2.3 Let $A \in \mathbf{M}_{m,n}$ and $G \in \mathbf{M}_{n,m}$. The following statements are equivalent.

(1) G is a g-inverse of A.
(2) AG is an identity on $Sp(A)$, i.e., $AGA = A$.
(3) AG is idempotent and $\rho(A) = \rho(AG)$.

PROOF. We show that (1) $\Rightarrow$ (2) $\Rightarrow$ (3) $\Rightarrow$ (1).

Suppose (1) is true. Let $a_1, a_2, \ldots, a_n$ be the columns of A. It is clear that $Ax = a_i$ is a consistent system of equations for each i. Since G is a g-inverse of A, Ga_i is a solution to the system $Ax = a_i$, i.e., $AGa_i = a_i$. This is true for each i. Combining all these equations, we obtain $AGA = A$. The statement that AG is an identity on $Sp(A)$ is a restatement of the fact that $AGy = y$ for $y \in Sp(A)$. Thus (2) follows.

Suppose (2) is true. Post-multiply the equation $AGA = A$ by G. Thus we have $AGAG = (AG)^2 = AG$, which means that AG is idempotent. Note that $\rho(A) = \rho(AGA) \leq \rho(AG) \leq \rho(A)$. Hence $\rho(A) = \rho(AG)$. Thus (3) follows.

Suppose (3) is true. Let $Ax = y$ be a consistent system of equations. Consistency means that $y \in Sp(A)$, i.e., $y = A\alpha$ for some column vector α. Since $\rho(A) = \rho(AG), Sp(A) = Sp(AG)$. Consequently, $y = AG\beta$ for some column vector β. We will show that Gy is a solution of $Ax = y$. Note that $AGy = AGAG\beta = AG\beta = y$, since AG is idempotent. Thus (1) follows.

The characteristic property that a g-inverse G of A satisfies $AGA = A$ can be taken as a definition of g-inverse. The fact that for any g-inverse G of A, AG is idempotent puts us right into the ambit of projectors. In fact, AG is a projector from $\mathbf{C}^m$ onto $Sp(A)(= Sp(AG))$ along $Sp(I_m - AG)$. (Why?) It is interesting to note that GA is also idempotent. It is also interesting to note that $G \in \mathbf{M}_{n,m}$ is a g-inverse of A if and only if $\rho(A) + \rho(I_n - GA) = n$. This is reminiscent of the result **P 7.4.3**.

Let $G \in \mathbf{M}_{n,m}$ be a g-inverse of $A \in \mathbf{M}_{m,n}$. The matrix AG behaves almost like an identity matrix. What we mean by this is $(AG)A = A$, i.e., AG behaves like an identity matrix when multiplying A from the left. In the following proposition we examine under what circumstances $AGB = B$.

P 8.2.4 (1) For a matrix B of order $m \times k$, $(AG)B = B$ if and only if $Sp(B) \subset Sp(A)$, i.e., $B = AD$ for some matrix D.

(2) For a matrix B of order $k \times n$, $B(GA) = B$ if and only if $Sp(B') \subset Sp(A')$, i.e. $B = DA$ for some matrix D.

PROOF. (1) We have already seen that AG is an identity on $Sp(A)$. Consequently, AG is an identity on any subspace of $Sp(A)$. Thus if $B = AD$, then $AGB = B$. Note that AG is a projector from $\mathbf{C}^m$ onto

$Sp(A)$ along $Sp(I_m - AG)$. Consequently, if y is a non-zero vector with $(AG)y = y$, then y better be a linear combination of the columns of A, i.e., $y \in Sp(A)$. Hence, if $(AG)B = B$, then $Sp(B) \subset Sp(A)$.

(2) This is similar to (1).

A number of corollaries can be deduced from **P 8.2.4**.

COROLLARY 8.2.5. Let $A \in \mathbf{M}_{m,n}$ and $G \in \mathbf{M}_{n,m}$ a g-inverse of A. If α is a column vector consisting of n entries such that $\alpha \in Sp(A')$ and β is a column vector consisting of m entries such that $\beta \in Sp(A)$, then $\alpha' G\beta$ is invariant (i.e., a constant) for any choice of G.

PROOF. The conditions on α and β mean that $\alpha = A'\gamma$ for some column vector γ and $\beta = A\delta$ for some column vector δ. Consequently, $\alpha' G\beta = \gamma' AGA\beta = \gamma' A\delta$, a constant independent of the choice of G.

Corollary 8.2.5 can be generalized.

COROLLARY 8.2.6. Let $G \in \mathbf{M}_{n,m}$ stand as a generic symbol for a g-inverse of A. Suppose B and C are matrices of orders $p \times n$ and $m \times q$, respectively, such that $Sp(B') \subset Sp(A')$ and $Sp(C) \subset Sp(A)$. Then BGC is invariant for any choice of G.

PROOF. A proof can be crafted along the lines of the proof of Corollary 8.2.5.

COROLLARY 8.2.7. Let $A \in \mathbf{M}_{m,n}$. Let $(A^*A)^- \in \mathbf{M}_n$ stand for a g-inverse of A^*A. Then $A(A^*A)^-(A^*A) = A$ and $(A^*A)(A^*A)^-A^* = A^*$.

PROOF. First, we observe that $Sp(A^*) = Sp(A^*A)$. Consequently, $A^* = (A^*A)D$ for some matrix D. Therefore, $A = D^*A^*A$ and $A(A^*A)^-(A^*A) = D^*A^*A(A^*A)^-(A^*A) = D^*A^*A = A$. As for the second identity, note that $(A^*A)(A^*A)^-A^* = (A^*A)(A^*A)^-A^*AD = A^*AD = A^*$.

Corollary 8.2.7 can be strengthened. Let V be any matrix such that $\rho(A^*VA) = \rho(A)$. If V is positive definite, this rank condition is definitely satisfied. Then $A(A^*VA)^-(A^*VA) = A$ and $(A^*VA)^-(A^*VA)A^* = A^*$.

The matrix $A(A^*A)^-A^*$ for any A plays a crucial role in Linear Models. Linear Models provide a very general framework embodying Multiple Regression Models and Analysis of Variance Models. In the following proposition, we demonstrate that this matrix has some special properties.

P 8.2.8 Let $A \in \mathbf{M}_{m,n}$. Then $A(A^*A)^-A^*$ is Hermitian, idempotent, and invariant whatever may be the choice of $(A^*A)^-$.

PROOF. Idempotency is easy to settle: $(A(A^*A)^-A^*)A(A^*A)^-A^* = (A(A^*A)^-A^*A)((A^*A)^-A^*) = A(A^*A)^-A^*$, by Corollary 8.2.7. Let us look at the invariance property. Since $Sp(A^*) = Sp(A^*A)$, we can write $A^* = (A^*A)D$ for some matrix D. Consequently,

$$A(A^*A)^-A^* = D^*A^*A(A^*A)^-A^*AD = D^*A^*AD,$$

which is a constant whatever g-inverse of A^*A we use. Incidentally, we note that D^*A^*AD is Hermitian. This completes the proof.

P 8.2.8 can be strengthened. Let V be any positive definite matrix such that $\rho(A^*VA) = \rho(A)$. Then $A(A^*VA)^-A^*$ is invariant for any choice of $(A^*VA)^-$. Further, if A^*VA is Hermitian, so is $A(A^*VA)^-A^*$.

We now focus on a consistant system $Ax = y$ of linear equations. Using a single g-inverse of A, we will demonstrate how all the solutions of the system of equations can be generated. (See Rao (1962).)

P 8.2.9 Let $A \in \mathbf{M}_{m,n}$ and $G \in \mathbf{M}_{n,m}$ be a fixed g-inverse of A.

(1) A general solution of the homogeneous system $Ax = 0$ of equations is given by $x = (I_n - GA)z$, where z is an arbitrary vector.
(2) A general solution to the system $Ax = y$ of consistent equations is given by $x = Gy + (I_n - GA)z$ where z is an arbitrary vector.
(3) Let q be any column vector consisting of n components. Then $q'x$ is a constant for all solutions x of the consistent system $Ax = y$ of equations if and only if $q'(I_n - GA) = 0$ or equivalently, $q \in Sp(A')$.
(4) A necessary and sufficient condition that the system $Ax = y$ is consistent is that $AGy = y$.

PROOF. (1) First we note that for any vector $z, (I_n - GA)z$ is a solution of the system $Ax = 0$ of equations. On the other hand, let x be any solution of the system $Ax = 0$. Then

$$x = GAx + (I_n - GA)x = 0 + (I_n - GA)x = (I_n - GA)z,$$

with $z = x$. Thus we are able to write x in the form $(I_n - GA)z$ for some vector z.

(2) This result follows since a general solution of $Ax = y$ is the sum of a particular solution of $Ax = y$ and a general solution of $Ax = 0$. Note that Gy can be taken as a particular solution.

(3) Suppose $q'(I_n - GA) = 0$. Any solution of $Ax = y$ is of the form $Gy + (I_n - GA)z$ for some vector z. We compute $q'(Gy + (I_n - GA)z) = q'Gy$. Since $Ax = y$ is consistent, $y \in Sp(A)$. Since $q'(I_n - GA) = 0, q \in Sp(GA) = Sp(A')$. (Why?) By Corollary 8.2.5, $q'Gy$ remains the same regardless of the choice of G. Conversely, suppose $q'x$ remains the same for all solutions x of $Ax = y$. This means that $q'(Gy + (I_n - GA)z)$ = constant for all z in $\mathbf{C}^n$. This implies that $q'(I_n - GA)z = 0$ for all $z \in \mathbf{C}^n$. (Why?) Hence $q'(I_n - GA) = 0$.

(4) If $Ax = y$ is consistent, we have seen that $x = Gy$ is a solution. Then $AGy = y$. The converse is obvious.

If A is a nonsingular matrix, g-inverse of A is unique and is given by A^{-1}. This means that $AA^{-1} = A^{-1}A = I$. If A is a singular square matrix or rectangular matrix, AA^- and A^-A are, in general, not identity matrices. We shall investigate how exactly AA^- and A^-A differ from identity matrices and what is common to different g-inverses of A. The answer is contained in the following result.

P 8.2.10 Let $A \in \mathbf{M}_{m,n}$ and $G \in \mathbf{M}_{n,m}$ be any two matrices. Partition A and G as

$$A = \begin{bmatrix} \underset{p \times n}{A_1} \\ \\ \underset{q \times n}{A_2} \end{bmatrix}, \quad G = (\underset{n \times p}{G_1} : \underset{n \times q}{G_2})$$

with $p + q = m$. Then

$$Sp(A_1') \cap Sp(A_2') = \{0\} \quad \text{and } G \text{ is a } g\text{-inverse of } A$$

if and only if

$$A_1G_1A_1 = A_1, \ A_2G_1A_1 = 0, \ A_2G_2A_2 = A_2, \ A_1G_2A_2 = 0. \quad (8.2.1)$$

PROOF. Suppose G is a g-inverse of A. The equation $AGA = A$ yields

$$A_1G_1A_1 + A_1G_2A_2 = A_1, \ A_2G_1A_1 + A_2G_2A_2 = A_2.$$

These equations can be rewritten as

$$(A_1G_1 - I)A_1 = -A_1G_2A_2, \ (A_2G_2 - I)A_2 = -A_2G_1A_1.$$

Taking transposes, we obtain

$$A_1'(I - A_1G_1)' = -A_2'G_2'A_1', \ A_2'(I - A_2G_2)' = -A_1'G_1'A_2'.$$

If $Sp(A_1') \cap Sp(A_2') = \{0\}$, the above equations cannot hold unless each expression is zero. Hence

$$A_1G_1A_1 = A_1, \ A_2G_1A_1 = 0, \ A_2G_2A_2 = A_2, \ A_1G_2A_2 = 0.$$

Thus (8.2.1) follows. Conversely, suppose (8.2.1) holds and

$$A_1'x = A_2'y \tag{8.2.2}$$

for some vectors x and y. We will show that $A_1'x = A_2'y = 0$ which would then imply that $Sp(A_1') \cap Sp(A_2') = \{0\}$. Pre-multiply (8.2.2) by $A_1'G_1'$. Note that

$$A_1'G_1'A_1'x = A_1'G_1'A_2'y = 0,$$

which implies that $A_1'x = 0$, since $A_1'G_1'A_2' = 0$. It is simple to check that G is a g-inverse of A when (8.2.1) holds. This completes the proof.

This result has a cousin. The matrices could be partitioned in a different way.

P 8.2.11 Let $A \in \mathbf{M}_{m,n}$ and $G \in \mathbf{M}_{n,m}$ be partitioned as

$$A = (\underset{m\times p}{A_1} : \underset{m\times q}{A_2}), \ G = \begin{bmatrix} \underset{p\times m}{G_1} \\ \\ \underset{q\times m}{G_2} \end{bmatrix} \tag{8.2.3}$$

with $p + q = n$. Then $Sp(A_1) \cap Sp(A_2) = \{0\}$ and G is a g-inverse of A if and only if

$$A_1G_1A_1 = A_1, \ A_1G_1A_2 = 0, \ A_2G_2A_2 = A_2, \ A_2G_2A_1 = 0.$$

These are useful results. Under some conditions, a g-inverse of a submatrix of A can be picked up from the corresponding submatrix of a

g-inverse G of A. Note that the condition that $Sp(A_1) \cap Sp(A_2) = \{0\}$ is equivalent to the condition that $\rho(A) = \rho(A_1) + \rho(A_2)$. There is another way to look at the result. Pick up some g-inverses G_1 and G_2 of A_1 and A_2, respectively. Build the matrix (8.2.3). We could wish that G be a g-inverse of A. In order to have our wish realized, we must have as a pre-condition that the ranks be additive, i.e., $\rho(A) = \rho(A_1) + \rho(A_2)$. We will now derive a number of corollaries from **P 8.2.10**.

COROLLARY 8.2.12. Let $A \in \mathbf{M}_{m,n}$ and $G \in \mathbf{M}_{n,m}$ be any two matrices. Partition

$$A = \begin{bmatrix} \underset{p\times n}{A_1} \\ \\ \underset{q\times n}{A_2} \end{bmatrix}, \quad G = (\underset{n\times p}{G_1} \,|\, \underset{n\times q}{G_2})$$

with $p + q = m$. Then $Sp(A_1') \cap Sp(A_2') = \{0\}$, G is a g-inverse of A, and $\rho(A_1) = p$ if and only if

$$A_1G_1 = I_p, \ A_2G_1 = 0, \ A_2G_2A_2 = A_2, \text{ and } A_1G_2A_2 = 0.$$

PROOF. Suppose $Sp(A_1') \cap Sp(A_2') = \{0\}$, G is a g-inverse of A, and $\rho(A_1) = p$. Then $A_1G_1A_1 = A_1$, $A_2G_1A_1 = 0$, $A_2G_2A_2 = A_2$, and $A_1G_2A_2 = 0$. See **P 8.2.10**. If $\rho(A_1) = p$, we can cancel A_1 on the right from both sides of the equations $A_1G_1A_1 = A_1$ and $A_2G_1A_1 = 0$. (How?) The converse is straightforward.

Look at the product under the stipulated conditions of Corollary 8.2.12.

$$\underset{m\times m}{AG} = \begin{bmatrix} A_1 \\ A_2 \end{bmatrix} (G_1|G_2) = \begin{bmatrix} A_1G_1 & A_1G_2 \\ A_2G_1 & A_2G_2 \end{bmatrix} = \begin{bmatrix} I_p & A_1G_2 \\ 0 & A_2G_2 \end{bmatrix}.$$

The matrices AG and I_m have the same first p columns.

COROLLARY 8.2.13. Let $A \in \mathbf{M}_{m,n}$ and $G \in \mathbf{M}_{n,m}$ be any g-inverse of A. A necessary and sufficient condition that i-th column vector of AG is the same as the i-th column vector of I_m is that the i-th row vector of A is non-zero, and cannot be written as a linear combination of the remaining row vectors of A.

PROOF. One simply checks the condition of Corollary 8.2.12.

One of the interesting implications of Corollary 8.2.13 is that if all the row vectors of A are linearly independent and G is a g-inverse of A, then $AG = I_n$. In other words, if all row vectors of A are linearly independent, then $\rho(A) = m$ and any g-inverse G of A is indeed a right inverse of A confirming the results of Section 8.1. Corollary 8.2.11 can be rehashed in a different way.

COROLLARY 8.2.14. Let $A \in \mathbf{M}_{m,n}$ be partitioned as

$$A = (\underset{m\times p}{A_1} : \underset{m\times q}{A_2})$$

with $p + q = n$. Let $G \in \mathbf{M}_{n,m}$ be a matrix partitioned as in (8.2.3). Then

$$Sp(A_1) \cap Sp(A_2) = \{0\},\ G \text{ is a } g\text{-inverse of } A, \text{ and } \rho(A_1) = p$$

if and only if $G_1A_1 = I_p$, $G_1A_2 = 0$, $A_2G_2A_2 = A_2$, and $A_2G_2A_1 = 0$.

The implication of this result is that the matrices GA and I_n have the same first p rows. Corollary 8.2.13 has a mate.

COROLLARY 8.2.15. Let $A \in \mathbf{M}_{m,n}$ and $G \in \mathbf{M}_{n,m}$ be any g-inverse of A. Then a necessary and sufficient condition that the i-th row vector of GA is the same as the i-th row vector of I_n is that the i-th column vector of A is non-zero, and cannot be written as a linear combination of the remaining columns of A.

One of the implications of Corollary 8.2.15 is that if all the column vectors of A are linearly independent, i.e., $\rho(A) = n$, then $GA = I_n$. This means that G is a left inverse of A. The notions of generalized inverse and left inverse coincide in this special setting. We can rephrase these comments in another way. Suppose $A \in \mathbf{M}_{m,n}$ and $\rho(A) = m$. We have seen in Section 8.1 that A admits a right inverse G. From the very definition of right inverse, it is transparent that G is indeed a g-inverse of A. Corollary 8.2.13 implies that every g-inverse of A is indeed a right inverse of A. Similar remarks apply when $\rho(A) = n$.

COROLLARY 8.2.16. Let $A \in \mathbf{M}_{m,n}$ be partitioned as

$$A = \begin{bmatrix} \underset{p\times q}{A_{11}} & \underset{p\times r}{A_{12}} \\ \underset{s\times q}{A_{21}} & \underset{s\times r}{A_{22}} \end{bmatrix}$$

with $p + s = m$ and $q + r = n$. Let $G \in \mathbf{M}_{n,m}$ be any g-inverse of A. Partition G accordingly, i.e.,

$$G = \begin{bmatrix} \underset{q\times p}{G_{11}} & \underset{q\times s}{G_{12}} \\ \underset{r\times p}{G_{21}} & \underset{r\times s}{G_{22}} \end{bmatrix}.$$

Suppose each of the first q columns of A is non-zero and is not a linear combination of all other columns of A and each of the first p rows of A is non-zero and is not a linear combination of all other rows of A. Then G_{11} remains the same regardless of the choice of G.

PROOF. What the conditions of the proposition mean are

$$Sp((A_{11}|A_{12})') \cap Sp((A_{21}|A_{22})') = \{0\} \quad \text{and} \quad \rho(A_{11}|A_{12}) = p,$$

$$Sp\begin{bmatrix} A_{11} \\ A_{21} \end{bmatrix} \cap Sp\begin{bmatrix} A_{12} \\ A_{22} \end{bmatrix} = \{0\} \text{ and } \rho\begin{bmatrix} A_{11} \\ A_{21} \end{bmatrix} = q. \tag{8.2.4}$$

Under (8.2.4), by Corollary 8.2.12,

$$(A_{11}|A_{12})\begin{bmatrix} G_{11} \\ G_{21} \end{bmatrix} = I_p \text{ and } (A_{21}|A_{22})\begin{bmatrix} G_{11} \\ G_{21} \end{bmatrix} = 0.$$

These equations are equivalent to

$$A_{11}G_{11} + A_{12}G_{21} = I_p \quad \text{and} \quad A_{21}G_{11} + A_{22}G_{21} = 0. \tag{8.2.5}$$

Suppose F is another g-inverse of A partitioned in the same style as of G. Let F_{ij}'s stand for the blocks of F. Thus we must have

$$A_{11}F_{11} + A_{12}F_{21} = I_p \quad \text{and} \quad A_{21}F_{11} + A_{22}F_{21} = 0. \tag{8.2.6}$$

By subtracting (8.2.6) from (8.2.5), we observe that

$$A_{11}(G_{11}-F_{11})+A_{12}(G_{21}-F_{21}) = 0, \; A_{21}(G_{11}-F_{11})+A_{22}(G_{21}-F_{21}) = 0.$$

We can rewrite these equations as

$$\begin{bmatrix} A_{11} \\ A_{21} \end{bmatrix}(G_{11} - F_{11}) + \begin{bmatrix} A_{12} \\ A_{22} \end{bmatrix}(G_{21} - F_{21}) = 0.$$

This equation means that some linear combinations of the first q columns of A are the same as some linear combinations of the last $n-q$ columns of A. In view of (8.2.4), we must have

$$\begin{bmatrix} A_{11} \\ A_{21} \end{bmatrix} (G_{11} - F_{11}) = 0 \text{ and } \begin{bmatrix} A_{12} \\ A_{22} \end{bmatrix} (G_{21} - F_{21}) = 0.$$

Since

$$\rho \begin{bmatrix} A_{11} \\ A_{21} \end{bmatrix} = q,$$

$G_{11} - F_{11} = 0$. (Why?) This completes the proof.

The following is a special case of Corollary 8.2.16.

COROLLARY 8.2.17. If the i-th row of a matrix $A \in \mathbf{M}_{m,n}$ is non-zero and is not a linear combination of the remaining rows of A and the j-th column of A is non-zero and is not a linear combination of the remaining columns of A, then the (j,i)-th element of a generalized inverse G of A is a constant regardless of the choice of G.

We now focus on non-negative definite (nnd) matrices.

COROLLARY 8.2.18. Let $A \in \mathbf{M}_m$ be an nnd matrix partitioned as

$$A = \begin{bmatrix} \underset{p\times p}{A_{11}} & \underset{p\times q}{A_{12}} \\ \underset{q\times p}{A_{21}} & \underset{q\times q}{A_{22}} \end{bmatrix}$$

with $p + q = m$. Let $G \in \mathbf{M}_m$ be any g-inverse of A. Partition G accordingly, i.e.,

$$G = \begin{bmatrix} \underset{p\times p}{G_{11}} & \underset{p\times q}{G_{12}} \\ \underset{q\times p}{G_{21}} & \underset{q\times q}{G_{22}} \end{bmatrix}.$$

Suppose each of the first p rows of A is non-zero and is not a linear combination of the remaining rows of A. Then

$$G_{11} = (A_{11} - A_{12}A_{22}^{-}A_{21})^{-1},$$

for any g-inverse of A_{22}^{-} of A_{22}.

PROOF. The conditions of Corollary 8.2.12 are met. Consequently,

$$(A_{11}|A_{12})\begin{bmatrix} G_{11} \\ G_{21} \end{bmatrix} = A_{11}G_{11} + A_{12}G_{21} = I_p \tag{8.2.7}$$

and

$$(A_{21}|A_{22})\begin{bmatrix} G_{11} \\ G_{21} \end{bmatrix} = A_{21}G_{11} + A_{22}G_{21} = 0. \tag{8.2.8}$$

Choose and fix a g-inverse A_{22}^- of A_{22}. Pre-multiplying (8.2.8) by $A_{12}A_{22}^-$, we obtain

$$A_{12}A_{22}^-A_{21}G_{11} + A_{12}A_{22}^-A_{22}G_{21} = 0. \tag{8.2.9}$$

Since A is nnd, $A_{12}A_{22}^-A_{22} = A_{12}$. (Why?) (Since A is nnd, $Sp(A_{21}) \subset Sp(A_{22})$.) Thus we have from (8.2.9),

$$A_{12}A_{22}^-A_{21}G_{11} + A_{12}G_{21} = 0. \tag{8.2.10}$$

Subtracting (8.2.10) from (8.2.7), we obtain

$$(A_{11} - A_{12}A_{22}^-A_{21})G_{11} = I_p.$$

Hence $G_{11} = (A_{11} - A_{12}A_{22}^-A_{21})^{-1}$. (Note that G_{11} is unique. See Corollary 8.2.16.)

COROLLARY 8.2.19. Let $A \in \mathbf{M}_{m,n}$ be partitioned as

$$A = (\underset{m\times p}{A_1} \mid \underset{m\times q}{A_2})$$

with $p + q = n$. Let $G \in \mathbf{M}_{n,m}$ be any g-inverse of A. Let G be partitioned accordingly as in (8.2.3). If $Sp(A_1) \cap Sp(A_2) = \{0\}$ and $Sp(A_1) \oplus Sp(A_2) = \mathbf{C}^m$, then $p = A_1G_1$ is a projector from $\mathbf{C}^m$ onto $Sp(A_1)$ along $Sp(A_2)$.

Suppose $A = (A_1|A_2)$ is a partitioned matrix with A_1 being of order $m \times p$ and A_2 of order $m \times q$. Suppose we have g-inverses G_1 and G_2 of A_1 and A_2, respectively, available. We string G_1 and G_2 as in (8.2.3). Under what conditions G is a g-inverse of A? **P 8.2.11** provides an answer. In the following, we provide a sufficient condition.

COROLLARY 8.2.20. Let A_1 and A_2 be two matrices of orders $m \times p$ and $m \times q$, respectively. Let $A = (A_1|A_2)$ and G_1 and G_2 be g-inverses of A_1 and A_2, respectively. If $A_1G_1 + A_2G_2 = AF$ for some g-inverse F of A, then

$$G = \begin{bmatrix} G_1 \\ G_2 \end{bmatrix} \text{ is a } g\text{-inverse of } A.$$

PROOF. It suffices to show that $A_1G_1A_2 = 0$, and $A_2G_2A_1 = 0$. Post-multiply both sides of the equation $A_1G_1 + A_2G_2 = AF$ by $A = (A|A_2)$. This operation leads to

$$\begin{aligned}(A_1G_1 + A_2G_2)(A_1|A_2) &= AFA = A = (A_1|A_2) \\ &= (A_1G_1A_1 + A_2G_2A_1|A_1G_1A_2 + A_2G_2A_2),\end{aligned}$$

which gives $A_1G_1A_1 = 0$ and $A_2G_2A_1 = 0$, and the result is proved.

In **P 8.2.9**, we have seen how a single g-inverse of a matrix A generates all solutions of a consistent system $Ax = y$ of linear equations. In the following result, we demonstrate how a single generalized inverse of A generates all g-inverses of A.

P 8.2.21 Let G be a fixed g-inverse of a given matrix A. Any g-inverse G_1 of A has one of the following forms.

(1) $G_1 = G + U - GAUAG$ for some matrix U.
(2) $G_1 = G + V(I - AG) + (I - GA)W$ for some matrices V and W.

PROOF. The first thing we can check is that $AG_1A = A$ if G_1 has any one of the forms (1) and (2). Conversely, let G_1 be a given g-inverse of A. If we wish to write G_1 in the form (1), take $U = G_1 - G$. If we wish to write G_1 in the form (2), take $V = G_1 - G$ and $W = G_1AG$.

We now introduce a special notation. For any matrix A, let $\{A^-\}$ denote the collection of all g-inverses of A. In the following result we demonstrate that A is essentially determined by the class $\{A^-\}$ of all its g-inverses.

P 8.2.22 Let A and B be two matrices of the same order $m \times n$ such that $\{A^-\} = \{B^-\}$. Then $A = B$.

PROOF. What the condition of the theorem means is that if G is a g-inverse of A then it is also a g-inverse of B and vice versa. Let G be a g-inverse of A and

$$G_1 = G + (I_n \dot{-} A^*(AA^*)^- A)B^*.$$

We note that G_1 is also a g-inverse of A. For,

$$\begin{aligned} AG_1A &= AGA + A(I_n - A^*(AA^*)^-A)B^*A \\ &= A + (A - (AA^*)(AA^*)^-A)B^*A = A \end{aligned}$$

as $A = AA^*(AA^*)^-A$ by Corollary 8.2.7 using A^* in place of A. By the hypothesis of the proposition, G_1 is also a g-inverse of B. Thus

$$\begin{aligned} B = BG_1B &= BGB + B(I_n - A^*(AA^*)^-A)B^*B \\ &= B + B(I_n - A^*(AA^*)^-A)(I_n - A^*(AA^*)^-A)^*B^*B, \end{aligned}$$

since $(I_n - A^*(AA^*)^-A)$ is Hermitian and idempotent. See **P 8.2.8.** This implies that

$$B(I_n - A^*(AA^*)^-A)(I_n - A^*(AA^*)^-A)^*B^*B = 0.$$

Pre-multiplying the above by B^*, we have

$$B^*B(I_n - A^*(AA^*)^-A)(I_n - A^*(AA^*)^-A)^*B^*B = 0.$$

Consequently, $B^*B(I_n - A^*(AA^*)^-A) = 0$. (Why?) From this, we have $B(I_n - A^*(AA^*)^-A) = 0$, or equivalently, $B = BA^*(AA^*)^-A$. (Why?) Thus we are able to write $B = CA$ for some matrix C. Following the same line of reasoning, we can write $A = DB$. By focusing now on a variation of G, given above by

$$G_2 = G + B^*(I_m - A(A^*A)^-A^*),$$

one can show that $B = AE$ for some matrix E. In a similar vein, one can show that $A = BF$ for some matrix F. (Try.) Now,

$$B = BGB = CAGAE = CAE = BE,$$

which implies $DB = DBE = AE = B$. Hence $B = DB = A$. This completes the proof.

Suppose G is a g-inverse of A, i.e., $AGA = A$. It need not imply that A is a g-inverse of G, i.e., $GAG = G$. We will now introduce a special terminology.

DEFINITION 8.2.23. Let A be a matrix of order $m \times n$. A matrix G of order $n \times m$ is said to be a reflexive g-inverse of A if $AGA = A$ and $GAG = G$.

We use the notation A_r^- for a reflexive g-inverse of A. We now demonstrate the existence of reflexive g-inverses.

P 8.2.24 For any matrix $A \in \mathbf{M}_{m,n}$, a reflexive g-inverse of A exists.

PROOF. Let $\rho(A) = a$. By the Rank Factorization Theorem, write $A = RF$ with R of order $m \times a$ with rank a and F of order $a \times n$ with rank a. Let C and D be the right and left inverses of F and R, respectively, i.e., $FC = I_a$ and $DR = I_a$. Choose $G = CD$. Note that $AGA = RFCDRF = RI_aI_aF = RF = A$. On the other hand, $GAG = CDRFCD = CI_aI_aD = CD = G$.

If G is a g-inverse of A, one can show that $\rho(A) \leq \rho(G)$. The equality $\rho(A) = \rho(G)$ does indeed characterize reflexive g-inverses of A. We demonstrate this in the following proposition.

P 8.2.25 The following statements are equivalent for any two matrices $A \in \mathbf{M}_{m,n}$ and $G \in \mathbf{M}_{n,m}$.

(1) $AGA = A$ and $GAG = G$.

(2) $AGA = A$ and $\rho(A) = \rho(G)$.

PROOF. (1) $\Rightarrow$ (2). The statement $AGA = A$ implies that $\rho(A) \leq \rho(G)$ and the statement $GAG = G$ implies that $\rho(G) \leq \rho(A)$. Then we must have $\rho(A) = \rho(G)$.

To prove (2)$\Rightarrow$(1), note that

$$\begin{aligned}\rho(G) &= \rho(A) \quad \text{(by (2))}\\ &\leq \rho(GA) \leq \rho(G),\end{aligned}$$

which implies that $\rho(GA) = \rho(G)$. The matrix GA is idempotent. Then by **P 8.2.3**, A is a g-inverse of G.

The computation of a reflexive g-inverse of A can be done in several ways. Making use of the rank factorization of A is one possibility. Suppose we already have a g-inverse G of A. Let $G_1 = GAG$. One can check that G_1 is a reflexive g-inverse of A. Every reflexive g-inverse of A can be written in the form GAG for some g-inverse of G of A. Try.

It is clear that if G is a g-inverse of A, then $\rho(G) \geq \rho(A)$. For a given integer s satisfying $\rho(A) \leq s \leq \min\{m, n\}$, is it possible to find a

g-inverse G of A such that $\rho(G) = s$? In the following result we answer this question.

P 8.2.26 Let $A \in \mathbf{M}_{m,n}$ have the decomposition given by

$$A = P \begin{bmatrix} \Delta & 0 \\ 0 & 0 \end{bmatrix} Q,$$

where P and Q are nonsingular matrices of orders $m \times m$ and $n \times n$, respectively, Δ is a diagonal matrix of order $a \times a$ with rank a, where $a = \rho(A)$. Then:

(1) For any three matrices E_1, E_2, and E_3 of appropriate orders,

$$G = Q^{-1} \begin{bmatrix} \Delta^{-1} & E_1 \\ E_2 & E_3 \end{bmatrix} P^{-1}$$

is a g-inverse of A;

(2) For any two matrices E_1 and E_2 of appropriate orders,

$$G_r = Q^{-1} \begin{bmatrix} \Delta^{-1} & E_1 \\ E_2 & E_2 \Delta E_1 \end{bmatrix} P^{-1}$$

is a reflexive generalized inverse of A.

PROOF. One simply verifies that they are doing their intended jobs. Note that the matrix G of (1) has the property that

$$\rho(G) = \rho\left(\begin{bmatrix} \Delta^{-1} & E_1 \\ E_2 & E_3 \end{bmatrix}\right).$$

Given any integer s such that $\rho(A) \leq s \leq \min\{m, n\}$, one can choose E_1, E_2, and E_3 carefully so that $\rho(G) = s$. You can experiment with these matrices. One can directly verify that $\rho(G_r) = a = \rho(A)$.

Complements

8.2.1 Cancellation Laws. Prove the following.

(1) If $AB = AC$, $Sp(B) \subset Sp(A^*)$, and $Sp(C) \subset Sp(A^*)$, then $B = C$.

(*Hint:* The matrices B and C can be written as $B = A^*D$ and $C = A^*E$ for some matrices D and E. Therefore, $AA^*D =$

AA^*E or $AA^*(D-E)=0$ or $(D-E)^*AA^*(D-E)=0$ or $A^*(D-E)=0$.) (Thus A can be cancelled in $AB=AC$.)

(2) If $A \in \mathbf{M}_{m,n}$, $\rho(A)=n$, and $AB=AC$, then $B=C$.
(3) If $A \in \mathbf{M}_{m,n}$, $\rho(A)=m$, and $BA=CA$, then $B=C$.
(4) If $ABC=ABD$ and $\rho(AB)=\rho(B)$, then $BC=BD$.
(5) If $CAB=DAB$ and $\rho(AB)=\rho(A)$, then $CA=DA$.

8.2.2 If $A=0$, determine the structure of g-inverses of A.

8.2.3 Let J_n be a matrix of order $n\times n$ in which every entry is equal to unity. Let $a\neq b$ and $a-b+n=0$. Show that $(a-b)^{-1}I_n$ is a g-inverse of $(a-b)I_n+J_n$.

8.2.4 Let A be a matrix of order $m\times n$ and α and β column vectors of orders $m\times 1$ and $n\times 1$, respectively. Let G be any g-inverse of A. If either $\alpha\in Sp(A)$ or $\beta\in Sp(A')$, show that

$$G_1 = G - \frac{((G\alpha)(\beta' G))}{(1+\beta' G\alpha)}$$

is a g-inverse of $A+\alpha\beta'$, provided $1+\beta' G\alpha\neq 0$.

8.2.5 Let A be a matrix of order $n\times m$ with rank r and B a matrix of order $s\times m$ with rank $m-r$. Suppose $Sp(A^*)\cap Sp(B^*)=\{0\}$. Then

(1) A^*A+B^*B is nonsingular;
(2) $(A^*A+B^*B)^{-1}$ is a g-inverse of A^*A;
(3) $\begin{bmatrix} A^*A & B^* \\ B & 0 \end{bmatrix}$ is nonsingular provided that $s=m-r$.

8.2.6 Show that a Hermitian matrix has a Hermitian g-inverse.

8.2.7 Show that a non-negative definite matrix has a non-negative g-inverse.

8.2.8 Show that a Hermitian matrix A has a non-negative definite g-inverse if and only if A is non-negative definite.

8.2.9 If G_1 and G_2 are two g-inverses of A, show that $\alpha G_1+(1-\alpha)G_2$ is a g-inverse of A for any α.

8.2.10 If G is a g-inverse of a square matrix A, is G^2 a g-inverse of A^2? Explain fully.

8.2.11 Let $T=A+XUX'$ where A is nnd, U is symmetric such that $Sp(A)\subset Sp(T)$ and $Sp(X)\subset Sp(T)$. Then $Sp(X'T^-X)=Sp(X')$ for any g-inverse T^-.

8.2.12 Let $AX = B$ and $XC = D$ be two consistent system of matrix equations in the unknown matrix X. Show that they have a common solution in X if and only if $AD = BC$. If this condition holds, show that

$$X = A^{-}B + DC^{-} - ADC^{-}$$

is a common solution.

8.3. Geometric Approach: LMN-inverse

In defining a g-inverse G of $A \in \mathbf{M}_{m,n}$ in Section 8.2, emphasis has been laid on its use in providing a solution to a consistent system $Ax = y$ of linear equations in the form $x = Gy$. A necessary and sufficient condition for this is that the operator AG is an identity on the subspace $Sp(A) \subset \mathbf{C}^m$. Nothing has been specified about the values of AGy or Gy for $y \notin Sp(A)$. The basic theme of this section is that we want to determine a g-inverse $G \in \mathbf{M}_{n,m}$ such that $Sp(G)$ is a specified subspace in $\mathbf{C}^n$ and the kernel of G is a specified subspace in $\mathbf{C}^m$.

The concept of g-inverse introduced in Section 8.2 makes perfect sense in the realm of general vector spaces and transformations. Let $\mathbf{V}$ and $\mathbf{W}$ be two vector spaces and A a transformation from $\mathbf{V}$ and $\mathbf{W}$. (We retain the matrix symbol A for transformation too.) A transformation G from $\mathbf{W}$ to $\mathbf{V}$ is said to be a generalized inverse of A if $AGA = A$. The entire body of results of Section 8.2 carries over to this general framework. In this section, it is in this general framework we work on.

P 8.3.1 Let $G : \mathbf{W} \to \mathbf{V}$ be a g-inverse of a transformation $A : \mathbf{V} \to \mathbf{W}$, i.e., satisfies the equation $AGA = A$. Define

$$N = G - GAG, \ \mathbf{M} = \mathbf{R}(G - N), \ \mathbf{L} = \mathbf{K}(G - N), \tag{8.3.1}$$

where $\mathbf{R}$ denotes the range and $\mathbf{K}$ the kernel of a transformation. Then:

(1) $AN = 0, NA = 0$.
(2) $\mathbf{M} \cap \mathbf{K}(A) = \{0\}$ and $\mathbf{M} \oplus \mathbf{K}(A) = \mathbf{V}$.
(3) $\mathbf{L} \cap \mathbf{R}(A) = \{0\}$ and $\mathbf{L} \oplus \mathbf{R}(A) = \mathbf{W}$.

PROOF. (1) is clearly satisfied. To prove (2) let $x \in \mathbf{V}$, belong to both $\mathbf{M}$ and $\mathbf{K}(A)$. Then, obviously $Ax = 0$, and $x = (G - N)y$ for some y. Then $A(G - N)y = AGAGy = AGy = 0$. This implies that $0 = GAGy = (G - N)y = x$. Hence $\mathbf{M} \cap \mathbf{K}(A) = \{0\}$. Let

$x \in \mathbf{V}$. Obviously $x = GAx + (I - GA)x$, where $(I - GA)x \in \mathbf{K}(A)$ since $A(I - GA)x = 0$, and $GAx = (G - N)Ax \in \mathbf{R}(G - N)$. Thus $\mathbf{M} \oplus \mathbf{K}(A) = \mathbf{V}$. (3) is proved in an analogous manner.

Now we raise the following question. Let $\mathbf{M} \subset \mathbf{V}$ be any given complement of $\mathbf{K}(A), \mathbf{L} \subset \mathbf{W}$ be any given complement of $\mathbf{R}(A)$, and N be a given transformation such that $NA = 0, AN = 0$. Does there exist a g-inverse G of A such that the conditions (8.3.1), i.e., $N = G - GAG, \mathbf{M} = \mathbf{R}(G - N), \mathbf{L} = \mathbf{K}(G - N)$ hold? The answer is yes and the following propositions give a construction of such an inverse and also establish its uniqueness. We call this inverse the **LM**N-inverse, which was proposed in Rao and Yanai (1985).

P 8.3.2 Let A be a transformation from a vector space $\mathbf{V}$ into a vector space $\mathbf{W}$. Let $\mathbf{L}$ be a subspace of $\mathbf{W}$ such that $\mathbf{L} \cap \mathbf{R}(A) = \{0\}$ and $\mathbf{L} \oplus \mathbf{R}(A) = \mathbf{W}$. Let $\mathbf{M}$ be a subspace of $\mathbf{V}$ such that $\mathbf{M} \cap \mathbf{K}(A) = \{0\}$ and $\mathbf{M} \oplus \mathbf{K}(A) = \mathbf{V}$. Let N be a transformation from $\mathbf{W}$ to $\mathbf{V}$ such that $AN = 0$ and $NA = 0$. Then there exists an **LM**N-inverse.

PROOF. Let H be any g-inverse of A. Then we show that

$$G = \mathbf{P}_{\mathbf{M}\cdot\mathbf{K}(A)} H \mathbf{P}_{\mathbf{R}(A)\cdot\mathbf{L}} + N \tag{8.3.2}$$

is the desired **LM**N-inverse, where $\mathbf{P}_{\mathbf{M}\cdot\mathbf{K}(A)}$ (abbreviated as $\mathbf{P}_{\mathbf{M}}$ in the proof) is the projection operator on $\mathbf{M}$ along $\mathbf{K}(A)$ and $\mathbf{P}_{\mathbf{R}(A)\cdot\mathbf{L}}$ (abbreviated as $\mathbf{P}_A$) is the projection operator on $\mathbf{R}(A)$ along $\mathbf{L}$.

Consider $x(\in \mathbf{V}) = x_1(\in \mathbf{M}) + x_2(\in \mathbf{K}(A))$. Then

$$A\mathbf{P}_{\mathbf{M}}x = Ax_1 = Ax \text{ for every } x \in \mathbf{V} \Rightarrow A\mathbf{P}_{\mathbf{M}} = A. \tag{8.3.3}$$

Obviously,

$$\mathbf{P}_A A = A. \tag{8.3.4}$$

It is easy to verify using (8.3.3) and (8.3.4) that $AGA = A$, where G is as in (8.3.2), i.e., G is a g-inverse of A.

Let $x \in \mathbf{M}$ and write $HAx = x + x_0$. Then $AHAx = Ax + Ax_0 \Rightarrow Ax_0 = 0$ since $AHAx = Ax$, i.e., $x_0 \in \mathbf{K}$, and $\mathbf{P}_{\mathbf{M}}HAx = \mathbf{P}_{\mathbf{M}}x + \mathbf{P}_{\mathbf{M}}x_0 = x$. Then using (8.3.4)

$$\mathbf{P}_{\mathbf{M}}HAx = \mathbf{P}_{\mathbf{M}}H\mathbf{P}_A Ax = x \tag{8.3.5}$$

so that any $x \in \mathbf{M}$ belongs to $\mathbf{R}(\mathbf{P_M} H \mathbf{P}_A) = \mathbf{R}(G - N)$. Obviously any vector in $\mathbf{R}(\mathbf{P_M} H \mathbf{P}_A)$ belongs to $\mathbf{M}$, which shows that

$$\mathbf{R}(G - N) = \mathbf{M}. \tag{8.3.6}$$

Let $\ell(\in \mathbf{W}) = \ell_1(\in \mathbf{R}(A)) + \ell_2(\in \mathbf{L}) = Am + \ell_2$ for some m. Then $(G - N)\ell = \mathbf{P_M} H \mathbf{P}_A(Am + \ell_2) = \mathbf{P_M} HAm = 0 \Rightarrow A\mathbf{P_M} HAm = 0$. From (8.3.3), $A\mathbf{P_M} = A$ so that $A\mathbf{P_M} HAm = AHAm = Am$ which is zero. Hence $\ell = \ell_2 \in \mathbf{L}$. Further, for any $\ell \in \mathbf{L}, (G - N)\ell = 0$. Hence

$$\mathbf{K}(G - N) = \mathbf{L}. \tag{8.3.7}$$

It is easy to verify that

$$N = G - GAG \tag{8.3.8}$$

so that all the conditions for an $\mathbf{LM}N$-inverse are satisfied, and **P 8.3.2** is proved.

Now we settle the uniqueness problem.

P 8.3.3 Let A be a transformation from $\mathbf{V}$ to $\mathbf{W}$. The $\mathbf{LM}N$-inverse of A, which is a transformation from $\mathbf{W}$ to $\mathbf{V}$ is unique.

Let G_1 and G_2 be two transformations satisfying the conditions (8.3.1).

Note that

$$\mathbf{R}(G_i - N) = \mathbf{M} \Rightarrow (G_i - N)y \in \mathbf{M}$$

for all $y \in \mathbf{W}$ and $i = 1, 2$. Taking the difference for $i = 1$ and 2,

$$(G_1 - G_2)y \in \mathbf{M}. \tag{8.3.9}$$

Similarly $\mathbf{K}(G_i - N) = \mathbf{L} \Rightarrow (G_i - N)y = (G_i - N)\mathbf{P}_A y$ for $y \in \mathbf{W}$. Taking the difference for $i = 1$ and 2

$$\begin{aligned} (G_1 - G_2)y &= (G_1 - G_2)\mathbf{P}_A y \\ A(G_1 - G_2)y &= A(G_1 - G_2)\mathbf{P}_A y = 0 \text{ since } \mathbf{P}_A y = Az \text{ for some } z \\ &\Rightarrow (G_1 - G_2)y \in \mathbf{K}(A). \end{aligned} \tag{8.3.10}$$

Since $\mathbf{M} \cap \mathbf{K}(A) = 0$, (8.3.9) and (8.3.10) $\Rightarrow (G_1 - G_2)y = 0$ for all $y \in \mathbf{W}$. Hence $G_1 = G_2$.

Complements

8.3.1 Let $\mathbf{L}, \mathbf{M}, N$ be as defined in **P 8.3.2**. Then the $\mathbf{LM}N$-inverse G has the alternative definitions:

$$(1)\ AG = \mathbf{P}_{\mathbf{R}(A)\cdot\mathbf{L}},\ GA = \mathbf{P}_{\mathbf{M}\cdot\mathbf{K}(A)},\ G - GAG = N. \quad (8.3.11)$$

$$(2)\ AGA = A,\ \mathbf{R}(G - N) = \mathbf{M},\ \mathbf{K}(G - N) = \mathbf{L}. \quad (8.3.12)$$

8.3.2 Let G be any g-inverse of A and $\mathbf{M}$ is any complement of $\mathbf{K}(A)$ in $\mathbf{V}$. Then the transformation $A|\mathbf{M} : \mathbf{M} \to \mathbf{R}(A)$ is bijective, and the true inverse $(A|\mathbf{M})^{-1} : \mathbf{R}(A) \to \mathbf{M}$ exists.

FIGURE. **LM**N-inverse

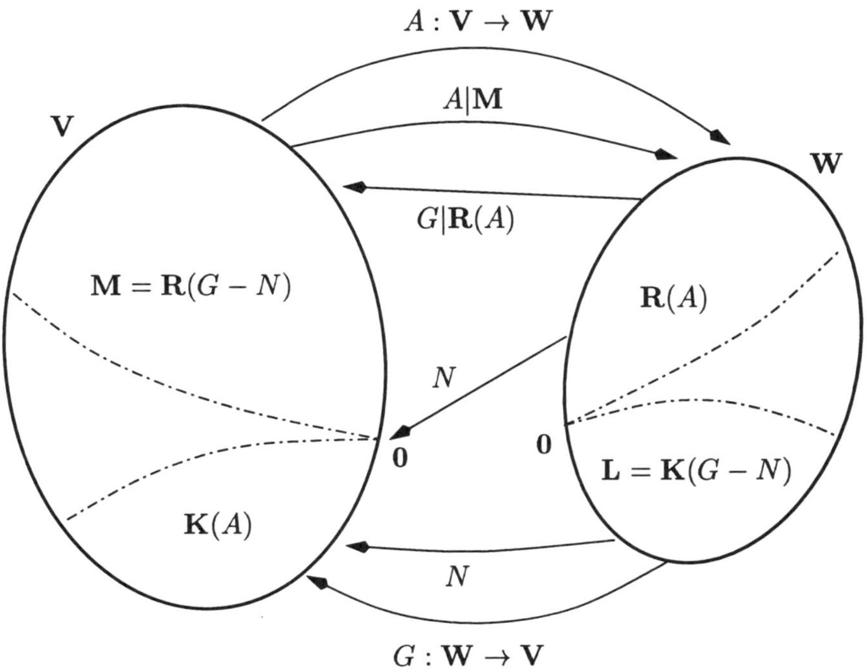

The figure given above shows the range and kernel spaces of the given transformation A and any associated g-inverse G for given subspaces $\mathbf{L}, \mathbf{M}$ and a transformation N. **P 8.3.4** shows that all generalized inverses can be generated by different choices of $\mathbf{L}, \mathbf{M}$ and N.

P 8.3.4 Let A be a transformation from a vector space $\mathbf{V}$ to a vector space $\mathbf{W}$. Let $\mathcal{F}$ be the collection of all entities $(\mathbf{L}, \mathbf{M}, N)$ satisfying the following:

(1) $\mathbf{L}$ is a subspace of $\mathbf{W}$ satisfying $\mathbf{L} \cap \mathbf{R}(A) = \{0\}$ and $\mathbf{L} \oplus \mathbf{R}(A) = \mathbf{W}$.
(2) $\mathbf{M}$ is a subspace of $\mathbf{V}$ satisfying $\mathbf{M} \cap \mathbf{K}(A) = \{0\}$ and $\mathbf{M} \oplus \mathbf{K}(A) = \mathbf{V}$.
(3) N is a transformation from $\mathbf{W}$ to $\mathbf{V}$ such that $NA = 0$ and $AN = 0$.

Let G be the collection of all g-inverses of A. Then there exists a bijection from $\mathcal{F}$ to G.

PROOF. Given $\mathbf{L}, \mathbf{M}$, and N satisfying (1), (2), and (3), let $G_{\mathbf{LM}N}$ be the unique $\mathbf{LM}N$-inverse of A. The map $f(\mathbf{L}, \mathbf{M}, N) = G_{\mathbf{LM}N}$ is a bijection from $\mathcal{F}$ to G. This proves **P 8.3.4**.

The case $N = 0$ is of special interest. The conditions (8.3.11) and (8.3.12) reduce to

$$AG = P_{\mathbf{R}(A) \cdot \mathbf{L}}, \; GA = P_{\mathbf{M} \cdot \mathbf{K}(A)}, \; G = GAG \tag{8.3.13}$$

and

$$AGA = A, \; \mathbf{R}(G) = \mathbf{M}, \; \mathbf{K}(G) = \mathbf{L}. \tag{8.3.14}$$

In other words, when $N = 0$, G is a reflexive g-inverse of A with range $\mathbf{M}$ and kernel $\mathbf{L}$. This special case deserves a mention.

COROLLARY 8.3.5. Let A be a transformation from a vector space $\mathbf{V}$ to a vector space $\mathbf{W}$. Let $\mathcal{L}_r$ be the collection of all entities $(\mathbf{L}, \mathbf{M})$ with the following properties:

(1) $\mathbf{L}$ is a subspace of $\mathbf{W}$ such that $\mathbf{L} \cap \mathbf{R}(A) = \{0\}$ and $\mathbf{L} \oplus \mathbf{R}(A) = \mathbf{W}$.
(2) $\mathbf{M}$ is a subspace of $\mathbf{V}$ such that $\mathbf{M} \cap \mathbf{K}(A) = \{0\}$ and $\mathbf{M} \oplus \mathbf{K}(A) = \mathbf{V}$.

Let G_r be the collection of all reflexive g-inverses of A. Then there exists a bijection from $\mathcal{L}_r$ to G_r.

A special case of this is of interest. Assume that the vector spaces $\mathbf{V}$ and $\mathbf{W}$ are equipped with inner products. We take $\mathbf{L}$ to be the orthogonal complement of $\mathbf{R}(A)$ and $\mathbf{M}$ to be the orthogonal complement of $\mathbf{K}(A)$. Let $P_{\mathbf{M}}$ be the orthogonal projection from $\mathbf{V}$ onto $\mathbf{M}$ and

$P_{\mathbf{R}(A)}$ the orthogonal projection from $\mathbf{W}$ onto $\mathbf{R}(A)$. Take $N = 0$. The conditions (8.3.13) and (8.3.14) reduce to

$$AG = P_{\mathbf{R}(A)},\ GA = P_{\mathbf{M}},\ GAG = G \tag{8.3.15}$$

and

$$AGA = A,\ \mathbf{R}(G) = \mathbf{M},\ \mathbf{K}(G) = \mathbf{L}, \tag{8.3.16}$$

respectively.

If the projection operators are self-adjoint, the equivalent sets (8.3.15) and (8.3.16) of conditions reduce to

$$AGA = A,\ GAG = G,\ AG = (AG)^*,\ GA = (GA)^*, \tag{8.3.17}$$

where $(AG)^*$ is the adjoint of the transformation AG in the inner product space $\mathbf{V}$ and $(GA)^*$ is the adjoint of the transformation GA in the inner product space $\mathbf{W}$. We now introduce a definition.

DEFINITION 8.3.6. Let A be a transformation from an inner product space $\mathbf{V}$ to an inner product space $\mathbf{W}$. A transformation G from $\mathbf{W}$ to $\mathbf{V}$ is said to be a Moore-Penrose inverse of A if (8.3.17) holds.

The existence of Moore-Penrose inverse is guaranteed and it is unique. It is after all an $\mathbf{LM}N$-inverse for some suitable choice of $\mathbf{L}, \mathbf{M}$, and N. Thus the concept of $\mathbf{LM}N$-inverse is more general than the Moore-Penrose inverse.

Another special case of interest is when we work with matrices. Let $A \in \mathbf{M}_{m,n}$ and $N \in \mathbf{M}_{n,m}$ be such that $AN = 0$ and $NA = 0$. Take $\mathbf{V} = \mathbf{C}^n$ and $\mathbf{W} = \mathbf{C}^m$. We now specify some subspaces of $\mathbf{V}$ and $\mathbf{W}$. Let B be a matrix of order $m \times p$ such that $Sp(B) \cap Sp(A) = \{0\}$ and $Sp(B) \oplus Sp(A) = \mathbf{W}$. Let C be a matrix of order $q \times n$ such that $\mathbf{K}(C) \cap \mathbf{K}(A) = \{0\}$ and $\mathbf{K}(C) \oplus \mathbf{K}(A) = \mathbf{V}$. The subspaces under focus are $Sp(B)$ and $\mathbf{K}(C)$. In the terminology of $\mathbf{L}$ and $\mathbf{M}, \mathbf{M} = \mathbf{K}(C)$ and $\mathbf{L} = Sp(B)$. We now seek a matrix $G \in \mathbf{M}_{n,m}$ such that

$$AGA = A,\ Sp(G - N) = \mathbf{K}(C),\ \mathbf{K}(G - N) = Sp(B).$$

An explicit form of G is given by

$$G = (A^*A + C^*C)^{-1}A^*AA^*(AA^* + BB^*)^{-1}.$$

8.4. Minimum Norm Solution

In this section, we take up again the problem of solving a system $Ax = y$ of consistent linear equations in the unknown x. The entity A is a transformation from an inner product space $\mathbf{V}$ to a vector space $\mathbf{W}$. For the equation to be consistent, we ought to have $y \in \mathbf{R}(A)$. Let $< \cdot, \cdot >$ be the inner product and $\| \cdot \|$ the associated norm on $\mathbf{V}$. The problem we focus on is:

$$\text{minimize } \|x\| \text{ over the set } \{x \in \mathbf{V} : \ Ax = y\}. \tag{8.4.1}$$

The following proposition solves this problem.

P 8.4.1 Let $P_{\mathbf{K}(A)}$ be the orthogonal projection from $\mathbf{V}$ onto $\mathbf{K}(A)$, the null space of A. Let G be a transformation from $\mathbf{W}$ to $\mathbf{V}$ such that

$$GA = I - P_{\mathbf{K}(A)}. \tag{8.4.2}$$

(We will settle the existence of G a little later.) Then $x^{(n)} = Gy$ solves the minimization problem (8.4.1). (We use the superscript (n) to indicate that the vector $x^{(n)}$ is a minimum norm solution.)

PROOF. Since $y \in \mathbf{R}(A)$, $y = Ab$ for some $b \in \mathbf{V}$. Then

$$x^{(n)} = Gy = GAb = (I - P_{\mathbf{K}(A)})b, \tag{8.4.3}$$

which means that $x^{(n)} \in [\mathbf{K}(A)]^{\perp}$, the orthogonal complement of $\mathbf{K}(A)$. Note that G is a g-inverse of A, i.e., $AGA = A$. Let $x \in \mathbf{V}$, and write $x = x_1 + x_2$ with $x_1 \in \mathbf{K}(A)$ and $x_2 \in [\mathbf{K}(A)]^{\perp}$. Then $Ax = Ax_1 + Ax_2 = Ax_2$. Further, $AGAx = A(I - P_{\mathbf{K}(A)})x = A(x - P_{\mathbf{K}(A)}x) = A(x - x_1) = Ax_2 = Ax$. Hence $AGA = A$. Since G is a g-inverse of A, $Gy = x^{(n)}$ is a solution to the system of equations. Now let x be any solution to the system of equations. It is clear that $x - x^{(n)} \in \mathbf{K}(A)$. Consequently,

$$< x - x^{(n)}, x^{(n)} > = 0. \tag{8.4.4}$$

Observe that

$$\begin{aligned} \|x\|^2 = < x, x > &= < x - x^{(n)} + x^{(n)}, x - x^{(n)} + x^{(n)} > \\ &= < x - x^{(n)}, x - x^{(n)} > + < x - x^{(n)}, x^{(n)} > \\ &\quad + < x^{(n)}, x - x^{(n)} > + < x^{(n)}, x^{(n)} > \\ &= \|x - x^{(n)}\|^2 + \|x^{(n)}\|^2 \geq \|x^{(n)}\|^2 \end{aligned}$$

using (8.4.4) in the last but one step. This completes the proof.

There is another way to look at the transformation $I - P_{\mathbf{K}(A)}$. Let $A^{\#}$ be the adjoint of A. Then $I - P_{\mathbf{K}(A)}$ is the orthogonal projection from $\mathbf{V}$ onto $\mathbf{R}(A^{\#})$. In our suggestive notation, $I - P_{\mathbf{K}(A)} = P_{\mathbf{R}(A^{\#})}$.

The existence of G is not a problem. Look back at (8.3.5). The operator $P_{\mathbf{M}}$ is precisely our $I - P_{\mathbf{K}(A)}$ here. There is a G satisfying (8.3.5). Put any inner product on the vector space $\mathbf{W}$. We could call G a minimum norm inverse.

A minimum norm inverse could be defined in general vector spaces not necessarily equipped with inner products. Let A be a transformation from a vector space $\mathbf{V}$ to a vector space $\mathbf{W}$. Let $\mathbf{M}$ be any subspace of $\mathbf{V}$ which is a complement of $\mathbf{K}(A)$, i.e., $\mathbf{M} \cap \mathbf{K}(A) = \{0\}$ and $\mathbf{M} \oplus \mathbf{K}(A) = \mathbf{V}$. Let $P_{\mathbf{M} \cdot \mathbf{K}(A)}$ be a projector from $\mathbf{V}$ onto $\mathbf{M}$ along $\mathbf{K}(A)$. One could define a transformation G from $\mathbf{W}$ to $\mathbf{V}$ as a minimum norm inverse if $GA = P_{\mathbf{M} \cdot \mathbf{K}(A)}$. Such a G is necessarily a g-inverse of A. (Prove this.)

8.5. Least Squares Solution

Let A be a transformation from a vector space $\mathbf{V}$ into an inner product space $\mathbf{W}$. Let $< \cdot, \cdot >$ be the inner product on $\mathbf{W}$ and $\| \cdot \|$ the corresponding norm. Let $y \in \mathbf{W}$ and $Ax = y$ be a system of equations possibly inconsistent. We focus on the problem:

$$\text{minimize } \|y - Ax\| \text{ over all } x \in \mathbf{V}. \tag{8.5.1}$$

If the system $Ax = y$ is consistent, any solution of the system, obviously, solves the problem (8.5.1). It is inconsistency that poses a challenge.

P 8.5.1 Let $P_{\mathbf{R}(A)}$ be the orthogonal projection from $\mathbf{W}$ onto $\mathbf{R}(A)$. Let G be any transformation from $\mathbf{W}$ to $\mathbf{V}$ such that

$$AG = P_{\mathbf{R}(A)}. \tag{8.5.2}$$

(Is the existence of G a problem?) Then $x^{(\ell)} = Gy$ solves the problem (8.5.1). (We use the suggestive superscript ℓ to indicate that $x^{(\ell)}$ is a least squares solution.)

PROOF. Note that

$$y - Ax^{(\ell)} = y - AGy = (I - P_{\mathbf{R}(A)})y \in [\mathbf{R}(A)]^{\perp}.$$

Since $A(x^{(\ell)}-x) \in \mathbf{R}(A)$ for any $x \in \mathbf{V}$, we have

$$< A(x^{(\ell)}-x), y - Ax^{(\ell)} >= 0. \tag{8.5.3}$$

For any $x \in \mathbf{V}$,

$$\begin{aligned}\|y-Ax\|^2 &=< y-Ax, y-Ax > \\ &=< (y-Ax^{(\ell)}) + A(x^{(\ell)}-x), (y-Ax^{(\ell)}) + A(x^{(\ell)}-x) > \\ &=< y-Ax^{(\ell)}, y-Ax^{(\ell)} > + < y-Ax^{(\ell)}, A(x^{(\ell)}-x) > \\ &\quad + < (x^{(\ell)}-x), y-Ax^{(\ell)} > + < A(x^{(\ell)}-x), A(x^{(\ell)}-x) > \\ &= \|y-Ax^{(\ell)}\|^2 + \|A(x^{(\ell)}-x)\|^2 \geq \|y-Ax^{(\ell)}\|^2.\end{aligned}$$

This completes the proof.

The label that $x^{(\ell)}$ is a least squares solution stems legitimately. Suppose $A = (a_{ij})$ is a matrix of order $m \times n$ complex entries. Let $y' = (y_1, y_2, \cdots, y_m) \in \mathbf{C}^m$. Let the vector space $\mathbf{C}^m$ be equipped with the standard inner product. Then for $x' = (x_1, x_2, \cdots, x_m) \in \mathbf{C}^m$,

$$\|y-Ax\|^2 = \sum_{i=1}^{m} |y_i - (a_{i1}x_1 + a_{i2}x_2 + \ldots + a_{in}x_n)|^2.$$

This expression appears routinely in many optimization problems. In the context of Linear Models in Statistics, y_i is a realization of a random variable Y_i in the model and $(a_{i1}x_1 + a_{i2}x_2 + \ldots + a_{in}x_n)$ its expected value. The difference $(y_i - (a_{i1}x_1 + a_{i2}x_2 + \ldots + a_{in}x_n))$ is labelled as the error. The objective is to minimize the sum of squared errors with respect to x_i's. Any solution to this problem is called a least squares solution.

The transformation G satisfying $AG = P_{\mathbf{R}(A)}$ could be labelled as a least squares solution. Such a G is necessarily a g-inverse of A. This notion can be defined in general vector spaces without any reference to inner products. Let A be a transformation from a vector space $\mathbf{V}$ to a vector space $\mathbf{W}$. Let $\mathbf{L}$ be any subspace of $\mathbf{W}$ which is a complement of $\mathbf{R}(A)$, i.e.,

$$\mathbf{R}(A) \cap \mathbf{L} = \{0\} \text{ and } \mathbf{R}(A) \oplus \mathbf{L} = \mathbf{W}.$$

Let $P_{\mathbf{R}(A)\cdot\mathbf{L}}$ be a projector from $\mathbf{W}$ onto $\mathbf{R}(A)$ along $\mathbf{L}$. We say that a transformation G from $\mathbf{W}$ to $\mathbf{V}$ a least squares inverse if $AG = P_{\mathbf{R}(A)\cdot\mathbf{L}}$.

8.6. Minimum Norm Least Squares Solution

Let A be a transformation from an inner product space $(\mathbf{V}, <\cdot,\cdot>_1)$ to an inner product space $(\mathbf{W}, <\cdot,\cdot>_2)$ with the corresponding norms $\|\cdot\|_1$ and $\|\cdot\|_2$. Let $y \in \mathbf{W}$. We focus on the problem:

$$\text{minimize } \|x^{(\ell)}\|_1 \text{ over the set}$$
$$\{x^{(\ell)} \in \mathbf{V} : \|y - Ax^{(\ell)}\|_2 = \min_{x \in \mathbf{V}} \|y - Ax\|_2\}. \tag{8.6.1}$$

First, we gather all the least squares solutions $x^{(\ell)}$ each of which minimizes $\|y - Ax\|_2$ over $\mathbf{V}$. Among these solutions, we seek one $x^{(\ell)}$ for which $\|x^{(\ell)}\|_1$ is minimum.

P 8.6.1 Let $P_{\mathbf{R}(A)}$ be the orthogonal projection from $\mathbf{W}$ onto $\mathbf{R}(A)$ and $P_{\mathbf{K}(A)}$ the orthogonal projection from $\mathbf{V}$ onto $\mathbf{K}(A)$. Let G be a transformation from $\mathbf{W}$ to $\mathbf{V}$ such that

$$AG = P_{\mathbf{R}(A)},\ GA = (I - P_{\mathbf{K}(A)}),\ GAG = G. \tag{8.6.2}$$

(Is the existence of G a problem? Look up (8.3.6).) Then $x^{(n\ell)} = Gy$ solves the problem (8.6.1).

PROOF. Let $x^{(\ell)}$ be any solution to the problem:

$$\text{minimize } \|y - Ax\|_1 \text{ over all } x \in \mathbf{V}. \tag{8.6.3}$$

From the proof of **P 8.5.1**, it is clear that $x^{(\ell)}$ satisfies the equation

$$Ax^{(\ell)} = P_{\mathbf{R}(A)}y. \tag{8.6.4}$$

The conditions (8.6.2) subsume the conditions (8.5.2). Consequently, $x^{(n\ell)}$ is also a solution to the minimization problem (8.6.3). By (8.6.4),

$$Ax^{(\ell)} - Ax^{(n\ell)} = P_{\mathbf{R}(A)}y - P_{\mathbf{R}(A)}y = 0$$

from which we have $x^{(\ell)} - x^{(n\ell)} \in \mathbf{K}(A)$. Observe that

$$x^{(n\ell)} = Gy = GAGy = GA(Gy) \in \mathbf{R}(GA) = \mathbf{R}(I - P_{\mathbf{K}(A)}).$$

Since $I - P_{\mathbf{K}(A)}$ is an orthogonal projection from $\mathbf{V}$ onto $[\mathbf{K}(A)]^{\perp}$, we have $x^{(n\ell)} \in [\mathbf{K}(A)]^{\perp}$, from which it follows that

$$< x^{(\ell)} - x^{(n\ell)}, x^{(n\ell)} >= 0.$$

Thus for any solution $x^{(\ell)}$ of the minimization problem (8.6.3),

$$\begin{aligned}\|x^{(\ell)}\|_1^2 &=< x^{(\ell)}, x^{(\ell)} >=< x^{(n\ell)} + (x^{(\ell)} - x^{(n\ell)}), x^{(n\ell)} + (x^{(\ell)} - x^{(n\ell)}) > \\ &= \|x^{(n\ell)}\|_1^2 + < x^{(n\ell)}, (x^{(\ell)} - x^{(n\ell)}) > + < (x^{(\ell)} - x^{(n\ell)}), x^{(n\ell)} > \\ &\quad + \|x^{(\ell)} - x^{(n\ell)}\|_1^2 = \|x^{(n\ell)}\|_1^2 + \|x^{(\ell)} - x^{(n\ell)}\|_1^2 \geq \|x^{(n\ell)}\|_1^2.\end{aligned}$$

This completes the proof.

The transformation G satisfying (8.6.2) can be called as a minimum norm least squares solution. This notion can be introduced in general vector spaces without involving inner products. Let A be a transformation from a vector space $\mathbf{V}$ to $\mathbf{W}$. Let $\mathbf{M}$ be a complement of the subspace $\mathbf{K}(A)$ in $\mathbf{V}$ and $\mathbf{L}$ a complement of $\mathbf{R}(A)$ in $\mathbf{W}$. Let $P_{\mathbf{M}\cdot\mathbf{K}(A)}$ and $P_{\mathbf{R}(A)\cdot\mathbf{L}}$ stand for the projectors in the usual connotation. A transformation G from $\mathbf{W}$ to $\mathbf{V}$ is said to be a minimum norm least squares inverse if

$$AG = P_{\mathbf{R}(A)\cdot\mathbf{L}},\ AG = P_{\mathbf{M}\cdot\mathbf{K}(A)},\ GAG = G.$$

8.7. Various Types of g-inverses

We have come across a variety of g-inverses in our sojourn. It is time to take stock of what has been achieved and then provide a summary of the deliberations. In the following $\mathbf{V}$ and $\mathbf{W}$ stand for vector spaces. We look at three possible scenarios.

Scenario 1 (1) $\mathbf{V}$ and $\mathbf{W}$ are general vector spaces.

(2) A stands for a transformation from $\mathbf{V}$ to $\mathbf{W}$.

(3) G stands for a transformation from $\mathbf{W}$ to $\mathbf{V}$.

(4) $\mathbf{R}(A)$ and $\mathbf{K}(A)$ stand for the range and kernel of the transformation A, respectively.

(5) $\mathbf{M}$ stands for any subspace of $\mathbf{V}$ which is a complement of the subspace $\mathbf{K}(A)$ of $\mathbf{V}$, i.e., $\mathbf{M} \cap \mathbf{K}(A) = \{0\}$ and $\mathbf{M} \oplus \mathbf{K}(A) = \mathbf{V}$.

(6) $P_{\mathbf{M}\cdot\mathbf{K}(A)}$ stands for the projector from $\mathbf{V}$ onto $\mathbf{M}$ along $\mathbf{K}(A)$.

(7) $\mathbf{L}$ stands for any subspace of $\mathbf{W}$ which is a complement of the subspace $\mathbf{R}(A)$.

(8) $P_{\mathbf{R}(A)\cdot\mathbf{L}}$ stands for the projector from $\mathbf{W}$ onto $\mathbf{R}(A)$ along $\mathbf{L}$.

Scenario 2 (1) $\mathbf{V}$ and $\mathbf{W}$ stand for general inner product spaces.

(2), (3) and (4) are the same as those under Scenario 1.

(5) $A^\#$ stands for the adjoint of the transformation A. (The map $A^\#$ is a transformation from $\mathbf{W}$ to $\mathbf{V}$.)

(6) $P_{\mathbf{R}(A)}$ stands for the orthogonal projection from $\mathbf{W}$ onto $\mathbf{R}(A)$.

(7) $P_{[\mathbf{K}(A)]^\perp}$ stands for the orthogonal projection from $\mathbf{V}$ onto $[\mathbf{K}(A)]^\perp$.

(8) $P_{\mathbf{R}(A^\#)}$ stands for the orthogonal projection from $\mathbf{V}$ onto $\mathbf{R}(A^\#)$.

Scenario 3 (1) $\mathbf{V}$ and $\mathbf{W}$ are unitary spaces, i.e., $\mathbf{V} = \mathbf{C}^n$ and $\mathbf{W} = \mathbf{C}^m$, equipped with their usual inner products.

(2) A is a matrix of order $m \times n$.

(3) G is a matrix of order $n \times m$.

(4) $\mathbf{R}(A)$ is the vector space spanned by the columns of A and $\mathbf{K}(A)$ the null space of A.

(5) A^* is the conjugate of A.

(6) $P_{A^*} = A^* M_1 A$, where M_1 is any matrix satisfying $(AA^*)M_1(AA^*) = AA^*$.

(7) $P_A = AM_2A^*$, where M_2 is any matrix satisfying $(A^*A)M_2(A^*A) = A^*A$.

Complements

8.7.1 Under scenario 1, we have defined the $\mathbf{LM}N$-inverse in Section 8.3.

(1) Show that the $\mathbf{LM}N$-inverse G can also be characterized as follows

$$G|\mathbf{R}(A) = (A|\mathbf{M})^{-1} \text{ and } G|\mathbf{L} = N|\mathbf{L}.$$

[Note that $A|\mathbf{M} : \mathbf{M} \to \mathbf{R}(A)$ is bijective so that $(A|\mathbf{M})^{-1} : \mathbf{R}(A) \to \mathbf{M}$ is well defined and unique.]

(2) The $\mathbf{LM}N$-inverse can also be defined as

$$Gy = (A|M)^{-1} y_1 + N y_2$$

where $y(\in \mathbf{W}) = y_1(\in \mathbf{R}(A)) + y_2(\in \mathbf{L})$.

(3) If instead of N, suppose we are given $\mathbf{R}(G - GAG)$, an $\mathbf{M}$ and an $\mathbf{L}$. Show that G is not unique in such a case.

In the following table, we provide a summary of the properties that various types of g-inverses should satisfy under each of the scenarios.

TABLE. A catalogue of different types of g-inverses

Description of G	Scenario 1	Scenario 2	Scenario 3
g-inverse	$AGA = A$	$AGA = A$	$AGA = A$
r-inverse	$GAG = G$	$GAG = G$	$GAG = G$
r, g-inverse	$AGA = A$ $GAG = G$	$AGA = A$ $GAG = G$	$AGA = A$ $GAG = G$
min norm g-inverse	$GA = P_{M \cdot K(A)}$	$GA = P_{R(A^{\#})}$	$GA = P_{A^*}$
min norm r, g-inverse	$GA = P_{M \cdot K(A)}$ $GAG = G$	$GA = P_{R(A^{\#})}$ $GAG = G$	$GA = P_{A^*}$ $GAG = G$
least squares g-inverse	$AG = P_{R(A) \cdot L}$	$AG = P_A$	$AG = P_A$
least squares r, g-inverse	$AG = P_{R(A) \cdot L}$ $GAG = G$	$AG = P_A$ $GAG = G$	$AG = P_A$ $GAG = G$
pre min norm least squares g-inverse	$AG = P_{R(A) \cdot L}$ $GA = P_{M \cdot K(A)}$	$AG = P_A$ $GA = P_{A^{\#}}$	$AG = P_A$ $GA = P_{A^*}$
min norm least squares g-inverse	$AG = P_{R(A) \cdot L}$ $GA = P_{M \cdot K(A)}$ $GAG = G$	$AG = P_A$ $GA = P_G$ (1)	$AG = P_A$ $GA = PG$ (2)

g-inverse=generalized inverse, r-inverse=reflexive inverse.
Equivalent conditions for (1) are

$$AGA = A, \; GAG = G, \; (AG)^{\#} = AG, \; (GA)^{\#} = GA$$

Equivalent conditions for (2) are

$$AGA = A, \; GAG = G, \; (AG)^{*} = AG, \; (GA)^{*} = GA$$

Some comments are in order on the above table.

(1) If G is a minimum norm inverse of A, then it is also a g-inverse of A.

(2) If G is a least squares inverse of A, then it is also a g-inverse of A.

(3) Suppose A is a matrix of order $m \times n$ and rank a. In Section 8.2, we presented the structure of a g-inverse G of A. Let

$$A = P \begin{bmatrix} \Delta & 0 \\ 0 & 0 \end{bmatrix} Q^*$$

be singular value decomposition of A, where P and Q are unitary matrices of orders $m \times m$ and $n \times n$, respectively, and Δ is a diagonal matrix of order $a \times a$ with $\rho(A) = a$. Then a g-inverse G of A is of the form

$$G = Q \begin{bmatrix} \Delta^{-1} & E_1 \\ E_2 & E_3 \end{bmatrix} P^*$$

for any arbitrary matrices E_1, E_2, and E_3. A reflexive g-inverse G of A is of the form

$$G = Q \begin{bmatrix} \Delta^{-1} & E_1 \\ E_2 & E_2\Delta E_1 \end{bmatrix} P^*$$

for any matrices E_1 and E_2. A matrix G of order $n \times m$ is a minimum norm g-inverse of A if G is of the form

$$G = Q \begin{bmatrix} \Delta^{-1} & E_1 \\ 0 & E_3 \end{bmatrix} P^*$$

for any matrices E_1 and E_3. A matrix G of order $n \times m$ is a minimum norm and reflexive g-inverse of A if G is of the form

$$G = Q \begin{bmatrix} \Delta^{-1} & E_1 \\ 0 & 0 \end{bmatrix} P^*$$

for any matrix E_1. A matrix G of order $n \times m$ is a least squares g-inverse of A if G is of the form

$$G = Q \begin{bmatrix} \Delta^{-1} & 0 \\ E_2 & E_3 \end{bmatrix} P^*$$

for any matrices E_2 and E_3. A matrix G of order $n \times m$ is a least squares and reflexive inverse of A if G is of the form

$$G = Q \begin{bmatrix} \Delta^{-1} & 0 \\ E_2 & 0 \end{bmatrix} P^*$$

for any matrix E_2. A matrix G of order $n \times m$ is a pre-minimum-norm-least-squares-inverse of A if G is of the form

$$G = Q \begin{bmatrix} \Delta^{-1} & 0 \\ 0 & E_3 \end{bmatrix} P^*$$

for any matrix E_3. A matrix G of order $n \times m$ is a minimum norm least squares inverse of A, i.e., Moore-Penrose inverse of A if G is of the form

$$G = Q \begin{bmatrix} \Delta^{-1} & 0 \\ 0 & 0 \end{bmatrix} P^*.$$

8.8. *G*-inverses Through Matrix Approximations

Given a matrix A of order $m \times n$ with complex entries, there may not exist a matrix G of order $n \times m$ such that $GA = I_n$. In such an event, we may try to find a matrix G such that AG and GA are close to I_m and I_n, respectively. Such a G may be called an approximate inverse of A. The underlying theme of this section is to pursue this idea of determining approximate inverses of A. It turns out that the g-inverses introduced and examined above are after all approximate inverses in some sense. (See Rao (1980).)

In order to develop the theory of approximate inverses, we need a general criterion to decide which one of the two given matrices is closer to the null matrix.

DEFINITION 8.8.1. Let S and R be two matrices of the same order $m \times n$. Assume that R is closer to a null matrix than S if $SS^* \geq RR^*$ or $S^*S \geq R^*R$. Assume that R is strongly closer to a null matrix than S if $SS^* \geq RR^*$ and $S^*S \geq R^*R$.

Some comments are in order on the definition.

(1) The notation $C \geq_L D$ (or $C \geq D$ for convenience of notation) for two matrices C and D means that $C - D$ is non-negative definite. (The subscript stands for Löwner ordering.)

(2) If $SS^* \geq RR^*$, it does not imply that $S^*S \geq R^*R$.

(3) Let $\sigma_1(S) \geq \sigma_2(S) \geq \ldots \geq \sigma_t(S) \geq 0$ and $\sigma_1(R) \geq \sigma_2(R) \geq \ldots \geq \sigma_t(R) \geq 0$ be the singular values of S and R, respectively, where $t = \min\{m, n\}$. If $\sigma_i(S) \geq \sigma_i(R)$ for all i, it does not follow that $SS^* \geq RR^*$ or $S^*S \geq R^*R$. But the reverse is true.

(4) If $SS^* \geq RR^*$ or $S^*S \geq R^*R$, then it follows that $\|S\| \geq \|R\|$ for any unitarily invariant norm $\|\cdot\|$ on the space of all matrices of order $m \times n$. We are jumping the gun again. We will not come across unitarily invariant norms until Chapter 11. The converse is not true. Thus the concept of closeness in the sense of Definition 8.6.1 is stronger than having a smaller norm.

We are now ready to establish results on matrix approximations via g-inverses.

P 8.8.2 Let A be a matrix of order $m \times n$ and G be any g-inverse of A, and L a least squares inverse of A. Then

$$(I_m - AG)^*(I_m - AG) \geq (I_m - AL)^*(I_m - AL).$$

PROOF. Let P_A be the orthogonal projection from $\mathbf{W} = \mathbf{C}^m$ onto $\mathbf{R}(A)$. We have seen that $AL = P_A$. This means that AL is idempotent and Hermitian. Further, $A^*AL = A^*$. (Why?) We check that

$$\begin{aligned}(AL - AG)^*(I_m - AL) &= AL(I_m - AL) - G^*A^*(I_m - AL)\\ &= 0 - G^*(A^* - A^*AL) = 0.\end{aligned}$$

Finally,

$$\begin{aligned}&(I_m - AG)^*(I_m - AG)\\ &= (I_m - AL + AL - AG)^*(I_m - AL + AL - AG)\\ &= (I_m - AL)^*(I_m - AL) + (AL - AG)^*(I_m - AL)\\ &\quad + (I_m - AL)^*(AL - AG) + (AL - AG)^*(AL - AG)\\ &= (I_m - AL)^*(I_m - AL) + 0 + 0 + (AL - AG)^*(AL - AG)\\ &\geq (I_m - AL)^*(I_m - AL). \quad \text{Why?}\end{aligned}$$

This completes the proof.

We discuss some implications of this result. The matrix $I_m - AL$ is closest to the null matrix among all matrices $I_m - AG$, with G being a g-inverse of A. Further, for any unitarily invariant norm $\|\cdot\|$ on the space of all $m \times m$ matrices,

$$\min_G \|I_m - AG\| = \|I_m - AL\|,$$

where the minimum is taken over all g-inverses G of A. In particular, specializing in the Euclidean norm on the space of all matrices of order $m \times m$, we have

$$\begin{aligned}\min_G [\operatorname{Tr}(I_m - AG)^*(I_m - AG)] &= \operatorname{Tr}[(I_m - AL)^*(I_m - AL)] \\ &= \operatorname{Tr}(I_m - AL). \quad \text{(Why?)}\end{aligned}$$

Note that for any matrix C of order $m \times m$ its Euclidean norm is defined by $[\operatorname{Tr}(C^*C)]^{1/2}$.

The following results can be established analogously.

P 8.8.3 For a given matrix A of order $m \times n$, let M be any minimum norm inverse of A. Then for any g-inverse G of A,

$$(I_n - GA)^*(I_n - GA) \geq (I_n - MA)^*(I_n - MA),$$

so that $I_n - MA$ is the closest to the null matrix among all matrices $I_n - GA$ with G being a g-inverse of A.

P 8.8.4 For a given matrix A of order $m \times n$, let Q be any pre-minimum-norm-least-squares-inverse of A. Then for any g-inverse G of A,

$$(I_m - AG)^*(I_m - AG) \geq (I_m - AQ)^*(I_m - AQ)$$

and

$$(I_n - GA)^*(I_n - GA) \geq (I_n - QA)^*(I_n - QA).$$

We now focus on the stronger notion of closeness. Recall that a matrix C is strongly closer to a null matrix than a matrix D if $CC^* \geq DD^*$ and $C^*C \geq D^*D$.

P 8.8.5 Let A be a matrix of order $m \times n$. The following hold.

(1) Let L be any least squares inverse as well as a reflexive inverse of A. Then for any g-inverse G of A,

$$(I_m - AG)^*(I_m - AG) \geq (I_m - AL)^*(I_m - AL),$$
$$(I_m - AG)(I_m - AG)^* \geq (I_m - AL)(I_m - AL)^*,$$

i.e., $I_m - AL$ is strongly closest to a null matrix among all matrices $I_m - AG$ with G being a g-inverse of A.

(2) Let M be any minimum norm inverse as well as a reflexive inverse of A. Then for any g-inverse G of A,

$$(I_n - GA)^*(I_n - GA) \geq (I_n - MA)^*(I_n - MA),$$
$$(I_n - GA)(I_n - GA)^* \geq (I_n - MA)(I_n - MA)^*,$$

i.e., $I_n - MA$ is strongly closest to a null matrix among all matrices $I_n - GA$ with G being a g-inverse of A.

(3) Let Q be a Moore-Penrose inverse of A. Then for any g-inverse G of A,

$$(I_m - AG)^*(I_m - AG) \geq (I_m - AQ)^*(I_m - AQ),$$
$$(I_m - AG)(I_m - AG)^* \geq (I_m - AQ)(I_m - AQ)^*,$$
$$(I_n - GA)^*(I_n - GA) \geq (I_n - QA)^*(I_n - QA),$$
$$(I_n - GA)(I_n - GA)^* \geq (I_n - QA)(I_n - QA)^*,$$

i.e., both $I_m - AQ$ and $I_n - QA$ are strongly closest to a null matrix among all matrices $I_m - AG$ and $I_n - GA$, respectively, with G being a g-inverse of A.

PROOF. A proof can be hammered out by imitating the theme in the proof of **P 8.8.2**.

All these results can be restated in the framework of unitarily invariant norms in the manner presented right after the proof of **P 8.8.2**.

Complements

8.8.1 Let

$$A = \begin{bmatrix} 1 & 0 & 1 \\ 0 & 1 & 0 \end{bmatrix} \text{ and } B = \begin{bmatrix} 1 & 0 & 0 \\ 0 & 1 & 0 \end{bmatrix}.$$

Show that $AA' \geq BB'$ but $A'A \geq B'B$ is not true. Compare the singular values of A and B.

8.8.2 Construct two matrices A and B such that the i-th singular value of A is greater than or equal to the i-th singular value of B for every i, but neither $AA^* \geq BB^*$ nor $A^*A \geq B^*B$.

8.9. Gauss-Markov Theorem

The focus in this section is on a random vector Y consisting of m real components with some joint distribution which, among other things, depends on a parameter vector $\beta \in \mathbf{R}^n$ and a scalar $\sigma \in \mathbf{R}^+ = (0, \infty)$ in the following way,

$$\begin{aligned} &E_{\beta,\sigma}(Y) = X\beta, \\ &D_{\beta,\sigma}(Y) = E_{\beta,\sigma}(Y - X\beta)(Y - X\beta)' = \sigma^2 I_m, \end{aligned} \tag{8.9.1}$$

where E stands for expectation and D for dispersion matrix, X is a matrix of order $m \times n$ with known entries. The model (8.9.1) specifying the mean, $E(Y)$ and dispersion matrix, $D(Y)$ (variance-covariance) matrix is called the Gauss-Markov model and is represented by the triplet $(Y, X\beta, \sigma^2 I_m)$. This model is widely used in statistics and the associated methodology is called the regression theory. The problems usually considered are the estimation of the unknown parameters β and σ and tests of hypotheses concerning them. We have touched upon this model earlier but now we show how the concepts of projection and g-inverses are useful in solving these problems. We begin the proceedings with a definition.

DEFINITION 8.9.1. Let $f(\beta) = Q\beta$, $\beta \in \mathbf{R}^n$, where Q is a given matrix of order $k \times n$. A statistic CY with C of order $k \times m$ is said to be a Linear Unbiased Minimum Dispersion (LUMD) estimator of $f(\cdot)$ if

$$E_{\beta,\sigma}(CY) = f(\beta), \; D_{\beta,\sigma}(FY) \geq D_{\beta,\sigma}(CY)$$

for all $\beta \in \mathbf{R}^n$ and $\sigma \in \mathbf{R}^+$, where FY is any statistic satisfying $E_{\beta,\sigma}(FY) = f(\beta)$ for all $\beta \in \mathbf{R}^n$ and $\sigma \in \mathbf{R}^+$.

The parametric function $Q\beta$ is a set of k linear functions of the components of β. The statistic CY is a set of k linear functions of the components of Y. The estimator CY is LUMD if it is an unbiased estimator of the parametric function and has the least dispersion matrix among all linear unbiased estimators of the parametric function, i.e., that $D_{\beta,\sigma}(CY)$ is closest to a null matrix among all matrices $D_{\beta,\sigma}(FY)$ for all $\beta \in \mathbf{R}^n$ and $\sigma \in \mathbf{R}^+$, where FY is an unbiased estimator of $f(\cdot)$. There is no guarantee that we will have at least one statistic FY which is an unbiased estimator of $f(\cdot)$. The following result answers some of the questions one confronts in the realm of Gauss-Markov models.

P 8.9.2 Let P be the orthogonal projector from $\mathbf{R}^m$ onto $\mathbf{R}(X)$ under the inner product $< u, v > = u'v$ for $u.v \in \mathbf{R}^m$. The following statements hold.

(1) There exists an unbiased estimator of $f(\cdot)$, i.e., $f(\cdot)$ is estimable, if and only if $Sp(Q') \subset Sp(X')$, i.e., $Q = AX$ for some matrix A.

(2) If $f(\cdot)$ is estimable, i.e., $Q = AX$ for some matrix A, APY is the $LUMD$ of $f(\cdot)$. Further,

$$D_{\beta,\sigma}(APY) = \sigma^2 APA' \quad \text{for all } \beta \in \mathbf{R}^n \text{ and } \sigma \in \mathbf{R}^+. \tag{8.9.2}$$

(3) Let $g(\sigma) = \sigma^2, \sigma \in \mathbf{R}^+$. If $\rho(X) = r$, an unbiased estimator of $g(\cdot)$ is given by

$$[1/(m-r)]Y'(I-P)Y. \tag{8.9.3}$$

PROOF. (1) Suppose AY is an unbiased estimator of $f(\cdot)$. This means that for every $\beta \in \mathbf{R}^n$ and $\sigma \in \mathbf{R}^+$,

$$Q\beta = E_{\beta,\sigma}(AY) = AX\beta,$$

which implies that $Q = AX$. Conversely, if $Q = AX$ for some matrix A, the AY is an unbiased estimator of $f(\cdot)$.

(2) Suppose $f(\cdot)$ is estimable. Then $Q = AX$ for some matrix A. Further,

$$E_{\beta,\sigma}(APY) = APX\beta = AX\beta = Q\beta$$

for all β and σ. This demonstrates that APY is an unbiased estimator of $f(\cdot)$. Let FY be an alternative unbiased estimator of $f(\cdot)$. This means that $FX = Q = AX$ which implies that $FP = AP$. Now for all β and σ,

$$\begin{aligned} D_{\beta,\sigma}(FY) - D_{\beta,\sigma}(APY) &= \sigma^2 FF' - \sigma^2 APP'A' = \sigma^2(FF' - FPP'F') \\ &= \sigma^2(FF' - FPF'), \quad \text{since } PP' = P, \\ &= \sigma^2 F(I-P)F' = \sigma^2 F(I-P)(I-P)'F', \end{aligned}$$

which is clearly non-negative definite; $D_{\beta,\sigma}(APY) = \sigma^2 APA$.

(3) Note that for all β and σ,

$$\begin{aligned} E_{\beta,\sigma}[Y'(I-P)Y] &= E_{\beta,\sigma}[(Y-X\beta)'(I-P)(Y-X\beta)] \\ &= E_{\beta,\sigma}[\text{Tr}(I-P)(Y-X\beta)(Y-X\beta)'] \\ &= \text{Tr}[(I-P)E_{\beta,\sigma}(Y-X\beta)(Y-X\beta)'] \\ &= \sigma^2\text{Tr}(I-P) = \sigma^2(m-\text{Tr}(P)) \\ &= \sigma^2(m-\rho(P)) = \sigma^2(m-r), \end{aligned}$$

from which the result (3) follows. This completes the proof. (What about uniqueness?) We state some of the consequences of the main result.

COROLLARY 8.9.3. If $Q = X$, then PY is the $LUMD$ of $X\beta$.

COROLLARY 8.9.4. If $Q = X$, the least squares estimator of $X\beta$ is PY.

COROLLARY 8.9.5. Suppose $f(\cdot)$ is estimable. The $LUMD$ estimator of $f(\cdot)$ is given by $Q\hat{\beta}$, where $\hat{\beta} = GX'Y$ and G is a g-inverse of $X'X$. Further, $D_{\beta,\sigma}(Q\hat{\beta}) = \sigma^2 QGQ'$ for all β and σ.

PROOF. To establish the result, we make use of the explicit representation of the projection operator P, namely, $P = XGX'$, where G is any g-inverse of $X'X$. Now the $LUMD$ estimator of $f(\cdot)$ is given by

$$APY = AXGX'Y = QGX'Y = Q\hat{\beta}$$

in our new terminology. Further, for all β and σ,

$$D_{\beta,\sigma}(Q\hat{\beta}) = \sigma^2 APA' = \sigma^2 AXGX'A' = \sigma^2 QGQ'.$$

The expression $\hat{\beta}$ is not unique when $X'X$ is a singular and depends on the particular choice of the g-inverse G used. However, $Q\hat{\beta}$ and QGQ' are unique for any choice of g-inverse G provided $f(\cdot)$ is estimable.

COROLLARY 8.9.6. The unbiased estimator of $g(\cdot)$ given in **P 8.9.2** (3) can be rewritten as

$$(m-r)^{-1}Y'(I-P)Y = (m-r)^{-1}(Y'Y - \hat{\beta}'X'Y).$$

Note: The material for this Chapter is drawn from the references: Rao (1945a, 1945b, 1946b, 1951, 1955, 1968, 1971, 1972b, 1973a, 1973b, 1973c, 1975, 1976a, 1976b, 1978a, 1978b, 1979a, 1980, 1981, 1985b), Rao and Mitra (1971a, 1971b, 1973, 1975), Rao, Bhimasankaram and Mitra (1972) and Rao and Yanai (1985a, 1985b).

CHAPTER 9

MAJORIZATION

In this chapter we introduce the notion of majorization and examine some of its ramifications. We also focus on the role of doubly stochastic matrices in characterizing majorization. The chain of ideas presented here will be made use of in understanding the relationship between the eigenvalues and singular values of a square matrix.

9.1. Majorization

For any n given numbers $x_1, x_2, \ldots, x_n$, let $x_{(1)} \geq x_{(2)} \geq \ldots \geq x_{(n)}$ be the arrangement of $x_1, x_2, \ldots, x_n$ in decreasing order of magnitude. Let $x = (x_1, x_2, \ldots, x_n)'$ and $y = (y_1, y_2, \ldots, y_n)'$ be two vectors in $\mathbf{R}^n$. Keeping up with the statistical tradition, we identify members of $\mathbf{R}^n$ as column vectors.

DEFINITION 9.1.1. We say that the vector x is majorized by the vector y (or y majorizes x), and use the symbol $x \ll y$, if

$$x_{(1)} + \ldots + x_{(i)} \leq y_{(1)} + \ldots + y_{(i)}, \; i = 1, \ldots, n-1, \quad (9.1.1)$$

$$x_1 + x_2 + \ldots + x_n = y_1 + y_2 + \ldots + y_n. \quad (9.1.2)$$

REMARKS.

A. The relationship $\ll$ defined above has the following properties.

(1) $x \ll x$ for every x in $\mathbf{R}^n$.
(2) If $x \ll y$ and $y \ll z$, then $x \ll z$.
(3) The properties stated in (1) and (2) above make the relation $\ll$ only a pre-order. The crucial element missing for the relation $\ll$ to be a partial order on $\mathbf{R}^n$ is that it does not have the property that $x \ll y$ and $y \ll x$ do imply that $x = y$. However, one can show

that the components of y is a permutation of the components of x. See (4) below for a precise statement.

(4) Let π be a permutation on $\{1, 2, \ldots, n\}$, i.e., π is a one-to-one map from $\{1, 2, \ldots, n\}$ onto $\{1, 2, \ldots, n\}$. For x in $\mathbf{R}^n$, let $x_\pi = (x_{\pi(1)}, x_{\pi(2)}, \ldots, x_{\pi(n)})'$. The components of x_π are just a permutation of the components of x. For any x in $\mathbf{R}^n$, $x \ll x_\pi$ for every permutation π of $\{1, 2, \ldots, n\}$. The above property can be rephrased as follows. With every permutation π of $\{1, 2, \ldots, n\}$, one can associate a permutation matrix $P_\pi = (p_{\pi ij})$ of order $n \times n$ defined as follows for $1 \leq i, j \leq n$:

$$p_{\pi ij} = \begin{cases} 1 & \text{if } \pi(i) = j, \\ 0 & \text{otherwise.} \end{cases}$$

Only one entry of every row and column of P_π is unity and the remaining $(n-1)$ entries consist of zeros. One can verify that for any x in $\mathbf{R}^n$ and any permutation π of $\{1, 2, \ldots, n\}$, $x_\pi = P_\pi x$. Thus we have $x \ll P_\pi x$ for every permutation π of $\{1, 2, \ldots, n\}$. We will elaborate the significance of this observation in the next section. Suppose $x \ll y$ and $y \ll x$. Then one can verify that $y = x_\pi$ for some permutation π of $\{1, 2, \ldots, n\}$.

B. The notion of majorization introduced above can be described in a slightly different fashion. For any $x = (x_1, x_2, \ldots, x_n)'$ in $\mathbf{R}^n$, let $x_{[1]} \leq x_{[2]} \leq \ldots \leq x_{[n]}$ be the arrangement of $x_1, x_2, \ldots, x_n$ in increasing order of magnitude. One can show that $x \ll y$ if and only if

$$x_{[1]} + \ldots + x_{[i]} \geq y_{[1]} + \ldots + y_{[i]}, \; i = 1, \ldots, n-1,$$
$$x_1 + x_2 + \ldots + x_n = y_1 + y_2 + \ldots + y_n.$$

To prove this assertion, one could use the fact that $x_{[i]} = x_{(n-i+1)}$ for $i = 1, 2, \ldots, n$.

C. The notion of majorization can be described in yet another form. For x and y in $\mathbf{R}^n$, let $< x, y > = \sum_{i=1}^{n} x_i y_i$. For any subset I of $\{1, 2, \ldots, n\}$, let $\mathcal{E}_I$ be the column vector in $\mathbf{R}^n$ whose i-th entry is given by

$$(\mathcal{E}_I)_i = \begin{cases} 1 & \text{if } i \in I, \\ 0 & \text{otherwise.} \end{cases}$$

For example, if $I = \{1,2\}$, then $\mathcal{E}_I = (1,1,0,0,\ldots,0)'$. Let $\#I$ denote the cardinality of I. For any x in $\mathbf{R}^n$, it can be verified that

$$\sum_{i=1}^{k} x_{(i)} = \max\{< x, \mathcal{E}_I >: \#I = k\}, k = 1,2,\ldots,n.$$

The following characterization of majorization is valid. For x and y in $\mathbf{R}^n, x \ll y$ if and only if the following two conditions hold.

(1) For any $I \subset \{1,2,\ldots,n\}$ with $\#I \leq n-1$, there exists $J \subset \{1,2,\ldots,n\}$ such that $\#I = \#J$ and $< x, \mathcal{E}_I > \leq < y, \mathcal{E}_J >$.

(2) $x_1 + \ldots + x_n = y_1 + \ldots + y_n$.

D. Another equivalent of majorization can be described as follows. A vector x is majorized by a vector y if and only if

$$\sum_{i=1}^{n}(x_i - a)^+ \leq \sum_{i=1}^{n}(y_i - a)^+ \text{ for every } a \in \mathbf{R} \text{ and } \sum_{i=1}^{n} x_i = \sum_{i=1}^{n} y_i.$$

EXAMPLES.

1. Let $X_1,\ldots,X_n$ be n random variables with some joint distribution function. Let $X_{(1)} \geq \ldots \geq X_{(n)}$ be the order statistics of $X_1,\ldots,X_n$ arranged in decreasing order of magnitude. Assume that $E|X_i| < \infty$ for every i. Let $X = (X_1,X_2,\ldots,X_n)'$, $x = (EX_1,EX_2,\ldots,EX_n)' = (x_1,x_2,\ldots,x_n)'$, say, and $y = (EX_{(1)},EX_{(2)},\ldots,EX_{(n)})'$. Then $x \ll y$. This can be proved as follows: for any $1 \leq k \leq n$,

$$\begin{aligned}\sum_{i=1}^{k} x_{(i)} &= \max\{< x, \mathcal{E}_I >: \#I = k\} \\ &= \max\{E(< X, \mathcal{E}_I >: \#I = k)\} \\ &\leq E(\max\{< X, \mathcal{E}_I >: \#I = k\}) \\ &= E(\sum_{i=1}^{k} X_{(i)}) = \sum_{i=1}^{k} EX_{(i)}.\end{aligned}$$

In view of Remark C, this essentially completes the proof of the above assertion.

2. Let $A_1, A_2, \ldots, A_n$ be n events in a probability space with $\Pr(A_i) = a_i, i = 1, 2, \ldots, n$. For each $1 \le j \le n$, let B_j denote the event that at least j of $A_1, A_2, \ldots, A_n$ occur. Equivalently,

$$B_j = \cup(A_{i_1} \cap A_{i_2} \cap \ldots \cap A_{i_j}),$$

where the union is taken over all $1 \le i_1 < i_2 < \ldots < i_j \le n$. Let $\Pr(B_j) = b_j, j = 1, 2, \ldots, n$. Let $x = (a_1, a_2, \ldots, a_n)$ and $y = (b_1, b_2, \ldots, b_n)$. Then $x \ll y$. This can be shown as follows. Let $X_i = I(A_i), i = 1, 2, \ldots, n$, where $I(A)$ is the indicator function of the event A. It is clear that $EX_i = a_i$ for every i. Let $X_{(j)}$ be the j-th order statistic of X_i's. Observe that for each $1 \le j \le n, X_{(j)} = 1$ if and only if at least j of $X_1, \ldots, X_n$ are each equal to unity, and otherwise equal to 0. Consequently, $I(B_j) = X_{(j)}$. The required assertion follows from Example 1 above.

3. Let S_{n-1} be the simplex in $\mathbf{R}^n$, i.e., $S_{n-1} = \{(p_1, p_2, \ldots, p_n) :$ each $p_i \ge 0$ and $\sum_{i=1}^{n} p_i = 1\}$. There is a unique smallest element in S_{n-1} according to the pre-order $\ll$ on S_{n-1} given by $(1/n, 1/n, \ldots, 1/n)$, i.e., $(1/n, 1/n, \ldots, 1/n) \ll \mathcal{P}$ for every $\mathcal{P}$ in S_{n-1}. There are n largest elements in S_{n-1} in the pre-order $\ll$ on S_{n-1}. One of them is given by $(1, 0, 0, \ldots, 0)$ which majorizes $\mathcal{P}$ for every $\mathcal{P}$ in S_{n-1}.

In the following definition we introduce an idea whose defining condition is slightly weaker than that of majorization. This notion plays a useful role in the formulation of some inequalities for eigenvalues.

DEFINITION 9.1.2. We say that a vector $x = (x_1, x_2, \ldots, x_n)'$ is weakly majorized by a vector $y = (y_1, y_2, \ldots, y_n)'$ and denote the relationship by $x \ll_w y$, if

$$\sum_{j=1}^{i} x_{(j)} \le \sum_{j=1}^{i} y_{(j)}, i = 1, 2, \ldots, n.$$

Complements

9.1.1 If x and y are any two vectors of the same order and z any vector, show that

$$\begin{pmatrix} x \\ z \end{pmatrix} \ll \begin{pmatrix} y \\ z \end{pmatrix} \text{ if and only if } x \ll y.$$

9.1.2 If $x \ll z$, $y \ll z$, and $0 \le \theta \le 1$, show that $\theta x + (1-\theta)y \ll z$.
9.1.3 If $x \ll y$, $x \ll z$, and $0 \le \theta \le 1$, show that $x \ll \theta y + (1-\theta)z$.

9.2. A Gallery of Functions

The notion of a function f from $\mathbf{R}^m$ to $\mathbf{R}^n$ preserving the pre-order $\ll$ is part of a natural progression of ideas which advance the majorization concept. In this section, we will introduce this notion more formally and study some examples.

DEFINITION 9.2.1. Let f be a map from $\mathbf{R}^m$ to $\mathbf{R}^n$. It is said to be Schur-convex if $x, y \in \mathbf{R}^m$ and $x \ll y \;\Rightarrow\; f(x) \ll_w f(y)$.

There are variations of Schur-convexity worth reporting. The function f is said to be strongly Schur-convex if

$$x, y \in \mathbf{R}^m \text{ and } x \ll_w y \;\Rightarrow\; f(x) \ll_w f(y).$$

The function f is said to be strictly Schur-convex if

$$x, y \in \mathbf{R}^m \text{ and } x \ll y \;\Rightarrow\; f(x) \ll f(y).$$

There are a host of other ways one could classify functions. We will now review some of these and examine their connection to Schur-convexity.

Let $\le$ be the usual partial order on $\mathbf{R}^n$. More precisely, let $x = (x_1, x_2, \ldots, x_n)'$ and $y = (y_1, y_2, \ldots, y_n)'$ be any two members of $\mathbf{R}^n$. Say that $x \le_e y$ iff $x_i \le y_i$ for all i. [suffix e stands for entrywise.]

DEFINITION 9.2.2. Let f be a map from $\mathbf{R}^m$ to $\mathbf{R}^n$. We define f to be monotonically increasing if

$$x, y \in \mathbf{R}^m \text{ and } x \le_e y \;\Rightarrow\; f(x) \le_e f(y),$$

f to be monotonically decreasing if $-f$ is monotonically increasing, and f to be monotone if f is either monotonically increasing or monotonically decreasing.

The notion that f is monotonically increasing is equivalent to the idea that f is coordinatewise increasing, i.e., if ξ and ξ' are real numbers such that $\xi \le \xi'$, then $f(x_1, x_2, \ldots, x_{i-1}, \xi, x_{i+1}, \ldots, x_n) \le f(x_1, x_2, \ldots, x_{i-1}, \xi', x_{i+1}, \ldots, x_n)$ for $1 \le i \le n$ and $x_1, x_2, \ldots, x_{i-1}, x_{i+1}, \ldots, x_n$ in $\mathbf{R}$.

Another notion of the same type is "convexity." This notion also uses the usual partial order $\le$ on $\mathbf{R}^n$.

DEFINITION 9.2.3. Let f be a map from $\mathbf{R}^m$ to $\mathbf{R}^n$. We define f to be convex if $f(px+(1-p)y) \leq_e pf(x)+(1-p)f(y)$, for every $0 \leq p \leq 1$ and x, y in $\mathbf{R}^m$, and f to be concave if $-f$ is convex.

The notion of "symmetry" can also be introduced for functions. The notion of a real-valued symmetric function is easy to guess. It is simply a function which is symmetric in its arguments. With some innovation, this notion can be extended to multi-valued functions.

DEFINITION 9.2.4. Let f be a map from $\mathbf{R}^m$ to $\mathbf{R}^n$. Say that f is symmetric if for every permutation π of $\{1, 2, \ldots, m\}$ there exists a permutation π' of $\{1, 2, \ldots, n\}$ such that

$$f(x_\pi) = (f(x)_{\pi'}), \text{ for all } x \text{ in } \mathbf{R}^m.$$

(See Remark A(4) above for the definition of x_π.)

In the case when $n = 1$, f is symmetric if and only if $f(x_\pi) = f(x)$ for all permutations π of $\{1, 2, \ldots, m\}$ and $x \in \mathbf{R}^m$. A simple example of a real valued symmetric function is given by $f(x_1, x_2, \ldots, x_m) = \sum_{i=1}^{m} x_i$, $x = (x_1, x_2, \ldots, x_m)' \in \mathbf{R}^m$.

One of the pressing needs now is to identify Schur-convex functions among other classes of functions introduced above. We need some more machinery to move in this direction. We will take up this issue in the next section.

9.3. Basic Results

We will now discuss inter-relations between various entities introduced in Section 9.2. At the core of the discussion lies doubly stochastic matrices and permutation matrices. We start with doubly stochastic matrices.

DEFINITION 9.3.1. A matrix $P = (p_{ij})$ of order $n \times n$ is said to be a doubly stochastic matrix if

(1) $p_{ij} \geq 0$ for all i and j,

(2) $\sum_{i=1}^{n} p_{ij} = 1$ for all j, and (3) $\sum_{j=1}^{n} p_{ij} = 1$ for all i.

Every permutation matrix of order $n \times n$ is a doubly stochastic matrix. See Remark A(4) of Section 9.2.

We will now discuss the structure of doubly stochastic matrices. Let $\mathbf{D}_n$ be the collection of all doubly stochastic matrices of order $n \times n$. The set $\mathbf{D}_n$ can be viewed as a subset of $\mathbf{R}^{n^2}$ and shown to be a compact convex subset of $\mathbf{R}^{n^2}$. The convexity of the set is obvious. If P and Q are members of $\mathbf{D}_n$ and $0 \leq p \leq 1$, then it is clear that $pP+(1-p)Q \in \mathbf{D}_n$. In the context of convex sets, the notion of an extreme point plays a useful role. If $\mathbf{A}$ is a convex subset of some vector space, then a member x of $\mathbf{A}$ is an extreme point of $\mathbf{A}$ if x cannot be written as a strict convex combination of two distinct members of $\mathbf{A}$, i.e., if $x = py + (1-p)z$ for some $0 < p < 1$ and $y, z \in A$, then $y = z$. It is natural to enquire about the extreme points of $\mathbf{D}_n$. Let $\mathbf{P}_n$ be the collection of all permutation matrices of order $n \times n$. See Remark A(4) above. It is obvious that $\mathbf{P}_n \subset \mathbf{D}_n$. The following result characterizes the extreme points of $\mathbf{D}_n$.

P 9.3.2 The set of all extreme points of $\mathbf{D}_n$ is precisely $\mathbf{P}_n$.

PROOF. Let P be a permutation matrix from $\mathbf{P}_n$ and suppose that $P = pD_1 + (1-p)D_2$ for some $0 < p < 1$ and $D_1, D_2 \in \mathbf{D}_n$. Let $P = (p_{ij}), D_1 = (d_{ij(1)})$ and $D_2 = (d_{ij(2)})$. Look at the first row of P. Note that $p_{1j} = 1$ for exactly one $j \in \{1, 2, \ldots, n\}$ and the rest of the entries in the first row are all equal to zero. Let $p_{1j_1} = 1$ for some $j_1 \in \{1, 2, \ldots, n\}$. Since $p_{1j_1} = pd_{1j_1(1)} + (1-p)d_{1j_1(2)}$ and $0 < p < 1$, it follows that $d_{1j_1(1)} = d_{1j_1(2)} = 1$. For $j \neq j_1, 0 = p_{1j} = pd_{1j(1)} + (1-p)d_{1j(2)}$, which implies that $d_{1j(1)} = d_{1j(2)} = 0$. Consequently, the first rows of P, D_1 and D_2 are identical. A similar argument can be used to show that the i-th rows of P, D_1 and D_2 are identical for any $i = 2, 3, \ldots, n$. Hence $P = D_1 = D_2$. Thus P is an extreme point of $\mathbf{D}_n$.

Conversely, let $D = (d_{ij})$ be an extreme point of $\mathbf{D}_n$. Suppose D is not equal to any of the permutation matrices in $\mathbf{P}_n$. Then there are some rows in D such that in each of the rows there are at least two positive entries. Start with one such row, i_1-th, say. Then $0 < d_{i_1j_1} < 1$ and $0 < d_{i_1j_2} < 1$ for some $j_1 < j_2$. Look at the j_2-th column. There must be another entry $d_{i_2j_2}$ such that $0 < d_{i_2j_2} < 1$. There must be an entry $d_{i_2j_3}$ in the i_2-th row such that $0 < d_{i_2j_3} < 1$. If we continue this way, we will obtain an infinite sequence (this process never ends!)

$$d_{i_1j_1}, d_{i_1j_2}; d_{i_2j_2}, d_{i_2j_3}; \ldots$$

in D such that each entry in the sequence is positive and less than unity. But we have only a finite number of subscripts (i,j)'s for d's with $i, j \in \{1, 2, \ldots, n\}$. Consequently, a subscript either of the form (i_r, j_r) or of the form (i_r, j_{r+1}) must repeat itself. Assume, without loss of generality, that (i_1, j_1) repeats itself. We now look at the following segment of the above sequence.

$$d_{i_1j_1}, d_{i_1j_2}; d_{i_2j_2}, d_{i_2j_3}; \ldots ; d_{i_sj_s}, d_{i_sj_{s+1}}; d_{i_1j_1} \tag{9.3.1}$$

with $j_{s+1} = j_1$. A possible scenario with $s = 4$ could look like as one depicted below with the entries forming a loop.

Entries of the matrix D

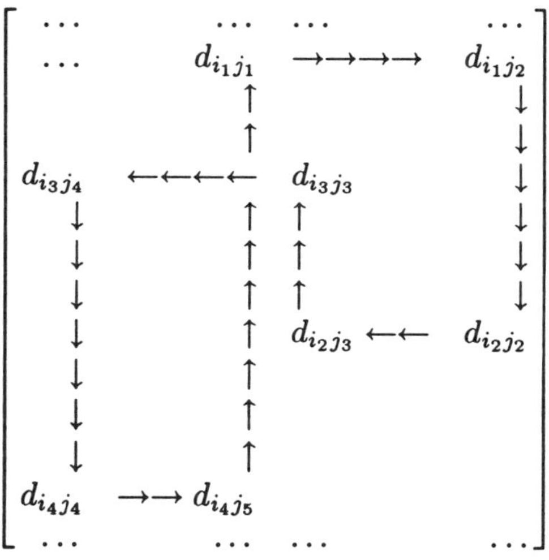

We now form two distinct matrices. Let δ be the minimum of all the entries in the loop (9.3.1). One matrix C_1 is created from D by retaining all the entries of D in C_1 with the exception of those in the loop (9.3.1). The entries in the loop are replaced by

$$d_{i_1j_1} - \delta, d_{i_1j_2} + \delta; d_{i_2j_2} - \delta, d_{i_2j_3} + \delta; \ldots ; d_{i_sj_s} - \delta, d_{i_sj_1} + \delta,$$

respectively and inserted into C_1. The other matrix C_2 is created in a similar way the only change being the replacement of δ by $-\delta$. We can now verify that C_1 and C_2 are distinct members of $\mathbf{D}_n$ and

$D = (1/2)C_1 + (1/2)C_2$. This contradicts the supposition that D is an extreme point of $\mathbf{D}_n$. This completes the proof.

A complete knowledge of extreme points of a convex set is helpful in recreating the set. Every member of the set can be written as a mixture of extreme points. This result is true in a more general framework but, for us, proving it in the environment of doubly stochastic matrices is rewarding and fulfilling. Before we establish this celebrated result, due to G.Birkhoff, we explore the notion of permanent of a matrix.

DEFINITION 9.3.3. Let $A = (a_{ij})$ be a square matrix of order $n \times n$ with real entries. The permanent of A, per(A), is defined by

$$\text{per}(A) = \sum_{\pi} \prod_{i=1}^{n} a_{i\pi(i)} \tag{9.3.2}$$

where the summation is taken over all permutations π of $\{1, 2, \ldots, n\}$.

There is one essential difference between per(A) and det(A). In the definition of the determinant, every product of the type $\prod_{i=1}^{n} a_{i\pi(i)}$ is multiplied by either $+1$ or -1 depending on whether π is an even or odd permutation. As an example, let

$$A = \begin{bmatrix} 1 & 0 & -1 \\ 2 & -2 & 1 \\ -2 & -3 & 4 \end{bmatrix}.$$

Check that per(A) $= 1$ and det(A) $= -1$. Another critical property of the permanent is that it is permutation-invariant. If we permute the rows and/or columns of A, the value of the permanent remains the same.

We will introduce a special notation for submatrices. Let $A = (a_{ij})$ be a square matrix of order $n \times n$. Let $I, J \subset \{1, 2, \ldots, n\}$. The submatrix $A(I, J)$ of A is a matrix of order $\#I \times \#J$ whose rows are indexed by the members of I and columns by J with entries plucked from A correspondingly. (Recall that $\#I$ stands for the cardinality of the set I.) Symbolically, one writes the submatrix as $A(I, J) = (a_{ij})_{i \in I, j \in J}$.

For example,

$$A = \begin{bmatrix} 1 & 0 & -1 \\ 2 & -2 & 1 \\ -2 & -3 & 4 \end{bmatrix}, \quad I = \{2,1\}, \text{ and } J = \{2,3\},$$

$$\Rightarrow A(I,J) = \begin{bmatrix} -2 & 1 \\ 0 & -1 \end{bmatrix}.$$

The submatrix $A(I,J)$ is technically different from the one given by $I' = \{1,2\}$ and $J' = \{2,3\}$, i.e.,

$$A(I',J') = \begin{bmatrix} 0 & -1 \\ -2 & 1 \end{bmatrix}.$$

The permanents of these two submatrices are the same. Since we are predominantly dealing with permanents in this sojourn, we are not concerned about how the members of I and J are arranged. If I and J do not have the same cardinality, per$A(I,J)$ does not make sense. Let us record a simple result on permanents which follows directly from the definition of the permanent.

P 9.3.4 If all the entries of a square matrix A of order $n \times n$ are non-negative, and I, J are subsets of $\{1,2,\ldots,n\}$ having the property that $\#I + \#J = n$, then

$$\text{per}(A) \geq \text{per}(A(I,J^c)) \times \text{per}(A(I^c,J)). \tag{9.3.3}$$

It is worth noting that, in **P 9.3.4**, I and J^c, the complement of set J, have the same cardinality. The following result plays a crucial role in many of the subsequent results.

P 9.3.5 Let $A = (a_{ij})$ be a square matrix of order $n \times n$ with non-negative entries. Then per$(A) = 0$ if and only if there are subsets I, J of $\{1,2,\ldots,n\}$ such that

$$\#I + \#J \geq n+1 \quad \text{and} \quad A(I,J) = 0. \tag{9.3.4}$$

PROOF. Suppose (9.3.4) holds. Since the permanent is permutation-invariant, we can take $I = \{1,2,\ldots,k\}$ and $J = \{k, k+1, \ldots, n\}$ for some k. Note that $\#I + \#I^c = n$ and $A(I, I^c) = 0$. Further, the last column of $A(I, I)$ is zero. Consequently,

$$\text{per}(A) = \text{per}(A(I, I)) \times \text{per}(A(I^c, I^c)) = 0.$$

See Complement 9.3.4. The converse can be established using induction. If $n = 1$, the validity of the converse is transparent. Suppose that the converse is true for all non-negative matrices of order $m \times m$ with $m \leq n$, n fixed. Let $A = (a_{ij})$ be a matrix of order $(n+1) \times (n+1)$ with non-negative entries and per(A)= 0. If all the entries of A are zeros, the proof ends right there. Assume that a_{ij} is positive for some i and j. Without loss of generality, one can take $a_{n+1,n+1} > 0$. (Why? Remember that we are dealing with permanents.) Take $I_0 = \{1, 2, \ldots, n\}$ and $J_0 = \{n+1\}$. By (9.3.3),

$$\begin{aligned} 0 = \text{per}(A) &\geq \text{per}(A(I_0, (J_0)^c)) \times \text{per}(A((I_0)^c, J_0)) \\ &= \text{per}(A(I_0, (J_0)^c)) \times a_{n+1,n+1} \geq 0. \end{aligned}$$

This implies that $\text{per}(A(I_0, (J_0)^c)) = 0$. Let $B = A(I_0, (J_0)^c)$. Thus B is a matrix of order $n \times n$ with permanent equal to zero. The rows and columns of B are indexed by the same set. By the induction hypothesis, there are subsets I_1 and J_1 of $\{1, 2, \ldots, n\}$ such that

$$\#I_1 + \#J_1 \geq n + 1 \text{ and } B(I_1, J_1) = A(I_1, J_1) = 0.$$

If $\#I_1 + \#J_1 \geq n + 2$, the proof ends. Take $I = I_1$ and $J = J_1$ in (9.3.3). Suppose $\#I_1 + \#J_1 = n + 1$. By (9.3.3), in the realm of $(n+1) \times (n+1)$-order matrices,

$$0 = \text{per}(A) \geq \text{per}(A(I_1, (J_1)^c)) \times \text{per}(A((I_1)^c, J_1)) \geq 0,$$

which implies that either $\text{per}(A(I_1, (J_1)^c))= 0$ or $\text{per}(A((I_1)^c, J_1)) = 0$. Assume, without loss of generality, that $A(I_1, (J_1)^c) = 0$. By the induction hypothesis, there are sets $I_2 \subset I_1$ and $J_2 \subset (J_1)^c$ such that $\#I_2 + \#J_2 \geq \#I_1 + 1$ and $A(I_2, J_2) = 0$. Note that

$$\begin{aligned} \#I_2 + \#(J_1 \cup J_2) &= \#I_2 + \#J_1 + \#J_2 \\ &\geq (\#I_1 + 1) + ((n+1) - \#I_1) = n + 2. \end{aligned}$$

Further, the fact that $A(I_2, J_1 \cup J_2) = 0$ follows from $A(I_2, J_2) = 0$ and $A(I_2, J_1)$ is a submatrix of $A(I_1, J_1)$ which is equal to zero. Take $I = I_2$ and $J = J_1 \cup J_2$ in (9.3.4).

A useful consequence of this result is the following.

P 9.3.6 If $A = (A_{ij})$ is a doubly stochastic matrix of order $n \times n$, then $\text{per}(A) > 0$.

PROOF. Suppose $\text{per}(A) = 0$. By **P 9.3.5**, there are subsets I and J of $\{1, 2, \ldots, n\}$ such that $\#I + \#J \geq n+1$ and $A(I, J) = 0$. After a suitable permutation of its rows and columns, the matrix A can be brought to the form

$$A* = \begin{bmatrix} 0 & B \\ C & D \end{bmatrix},$$

where 0 is of order $\#I \times \#J$ and the transformed matrix is also doubly stochastic. Note that

$$\begin{aligned} n &= \text{sum of all entries in } A* \\ &\geq \text{sum of all entries in } C + \text{sum of all entries in } B \\ &= \#I + \#J \geq n+1, \end{aligned}$$

which is a contradiction.

We are now in a position to describe the structure of the set of all doubly stochastic matrices of the same order.

P 9.3.7 Every matrix in $\mathbf{D}_n$ is a convex combination of members of $\mathbf{P}_n$, i.e., given any D in $\mathbf{D}_n$, there exist non-negative numbers $\lambda_P, P \in \mathbf{P}_n$ such that

$$\sum_{P \in \mathbf{P}_n} \lambda_P = 1 \text{ and } D = \sum_{P \in \mathbf{P}_n} \lambda_P P.$$

PROOF. We furnish a proof using the induction argument. For any matrix D, let $n(D)$ be the total number of non-zero entries in D. It is a well-defined map on $\mathbf{D}_n$. Obviously, $n(D) \geq n$. If $n(D) = n$, it is clear that D ought to be a permutation matrix. The induction method is used on the number $n(D)$. If $n(D) = n$, the proposition is evidently true. Assume that the proposition is true for all doubly stochastic matrices B for which $n(B) < n(D)$. We will show that the result is also valid for the doubly stochastic matrix $D = (d_{ij})$. Since per(D) is positive, there exists a permutation π on $\{1, 2, \ldots, n\}$ such that $d_{i\pi(i)} > 0$ for all i. (Why?) Let P_0 be the permutation matrix in $\mathbf{P}_n$ which corresponds to π. Let $\theta = \min_{1 \leq i \leq n} d_{i\pi(i)}$. Two cases arise. If $\theta = 1$, the matrix D is precisely equal to the permutation matrix P_0. The result is true. If $\theta < 1$, let

$$B = (\frac{1}{1-\theta})(D - \theta P_0).$$

The following features emerge from the definition of the matrix B.

(1) B is doubly stochastic.
(2) $n(B) < n(D)$.
(3) $D = (1-\theta)B + \theta P_0$.

By the induction hypothesis, B is a convex combination of permutation matrices. Hence D is a convex combination of permutation matrices. This completes the proof.

REMARK. It is clear that the cardinality of the set $\mathbf{P}_n$ of all permutation matrices of order $n \times n$ is $n!$ Given D in $\mathbf{D}_n$, a natural question one can ask is whether one requires all members of $\mathbf{P}_n$ in the representation of D in terms of members of $\mathbf{P}_n$ in the form given above. The answer is no. In fact, a crude upper bound for the number of members of $\mathbf{P}_n$ required in the representation of D above is n^2. (Why? Each number of $\mathbf{P}_n$ can be viewed as a member of $\mathbf{R}^{n^2}$.) But a good upper bound is given by $n^2 - 2n + 2$.

Majorization and doubly stochastic matrices are intimately related. One interesting feature of majorization is that if a vector y is a permutation of a vector x, then one can traverse from x to y through a series of intermediate vectors

$$x = v^{(0)} \ll v^{(1)} \ll \ldots \ll v^{(m)} = y,$$

where any two consecutive vectors differ only in two coordinates. One merely flips two coordinates of a vector to move on to the next vector in moving from x to y. In the following result we demonstrate that this can be achieved in the environment of majorization.

P 9.3.8 Let $x, y \in \mathbf{R}^n$. The following are equivalent.

(1) $x \ll y$.

(2) There is a finite number of vectors $u^{(0)}, u^{(1)}, \ldots, u^{(m)}$ in $\mathbf{R}^n$ such that $x = u^{(m)} \ll u^{(m-1)} \ll \ldots \ll u^{(0)} = y$, and for all k, $u^{(k)}$ and $u^{(k+1)}$ differ only in two coordinates.
(3) $x = Dy$ for some doubly stochastic matrix.

PROOF. We show that (1) $\Rightarrow$ (2). We use the induction argument on the dimension n. If $n = 1$, there is nothing to show. Assume that the implication is true for all vectors of order $m \times 1$ with m less than n. Let $x, y \in \mathbf{R}^n$. We can assume, without loss of generality, that the components of each of the vectors $x = (x_1, x_2, \ldots, x_n)'$ and $y = (y_1, y_2, \ldots, y_n)'$ are arranged in the decreasing order of magnitude. (Why?) We move from y to x successively the way outlined in (2). We produce, first, $u^{(1)}$. Note that $y_n \leq x_1 \leq y_1$. One can find k such that $y_{k-1} \leq x_1 \leq y_k$. Write $x_1 = ty_1 + (1-t)y_k$ for some $0 \leq t \leq 1$. Let

$$u^{(1)} = (x_1, y_2, \ldots, y_{k-1}, (1-t)y_1 + ty_k, y_{k+1}, \ldots, y_n)'.$$

The vectors $u^{(1)}$ and $u^{(0)}$ could possibly differ in the first and k-th coordinates. We will show that $u^{(1)} \ll u^{(0)}$. Towards this end, we will show, first, that $(x_1, (1-t)y_1 + ty_k)'$ is equal to

$$(ty_1 + (1-t)y_k, (1-t)y_1 + ty_k)' \ll (y_1, y_k)'. \tag{9.3.5}$$

It is clear that $x_1 + (1-t)y_1 + ty_k = y_1 + y_k$ and both x_1 and $(1-t)y_1 + ty_k$ are less than or equal to y_1. Consequently, (9.3.5) follows. (Why?) If we adjoin the vector $(x_1, (1-t)y_1 + ty_k)$ to $(y_2, y_3, \ldots, y_{k-1}, y_{k+1}, \ldots, y_n)$ and the vector (y_1, y_k) to $(y_2, y_3, \ldots, y_{k-1}, y_{k+1}, \ldots, y_n)$, it now follows that $u^{(1)} \ll u^{(0)}$. See Complement 9.1.1.

The next critical step is to demonstrate that

$$p = (x_2, x_3, \ldots, x_n)'$$
$$\ll (y_2, y_3, \ldots, y_{k-1}, (1-t)y_1 + ty_k, y_{k+1}, \ldots, y_n)' = q. \tag{9.3.6}$$

By design, we have

(1) $y_2 \geq y_3 \geq \ldots \geq y_{k-1} \geq x_1 \geq x_2 \geq \ldots,$
(2) $(1-t)y_1 + ty_k \geq y_{k+1} \geq y_{k+2} \ldots \geq y_n.$

The number $(1-t)y_1 + ty_k$ is intermediate to the numbers y_2 and y_{k+1}. The following inequalities are clear.

(1) $x_2 + x_3 + \ldots + x_r \leq y_2 + y_3 + \ldots + y_r$ for any $2 \leq r \leq k-1$.

(2) $(y_2 + y_3 + \ldots + y_{k-1}) + (1-t)y_1 + ty_k + (y_{k+1} + y_{k+2} + \ldots + y_r)$
$= (y_1 + y_2 + \ldots + y_r) - x_1 \quad \geq (x_1 + x_2 + \ldots + x_r) - x_1 =$
$x_2 + x_3 + \ldots + x_r$ for any $k \leq r \leq n$.

In (2), equality occurs when $r = n$. No matter where the number $(1-t)y_1 + ty_k$ occurs in the chain $y_2 \geq y_3 \geq \ldots y_{k-1} \geq y_{k+1}$, (9.3.6) follows. (Why?) Let us invoke the induction hypothesis. There are vectors $w^{(1)}, w^{(2)}, \ldots, w^{(m)}$ in $\mathbf{R}^{n-1}$ such that

$$p = w^{(m)} \ll w^{(m-1)} \ll \ldots \ll w^{(2)} \ll w^{(1)} = q.$$

Let $u^{(k)} = \binom{x_1}{w^{(k)}}$ for $k = 1, 2, \ldots, m$ and $u^{(0)} = y$. These vectors meet all the requirements stipulated in (2).

We now proceed with the implication (2) $\Rightarrow$ (3). In the chain of vectors, we note that $u^{(k+1)}$ is obtainable from $u^{(k)}$ by pre-multiplying $u^{(k)}$ by a suitable doubly stochastic matrix. More precisely, $u^{(k+1)} = (tI + (1-t)P_k)u^{(k)}$, where I is the identity matrix and P_k is some suitable permutation matrix which swaps two coordinates only. Stringing all these doubly stochastic matrices in a multiplicative way, we obtain a doubly stochastic matrix D such that $x = Dy$.

Finally, we tackle the implication (3) $\Rightarrow$ (1). Let $x = Dy$ for some doubly stochastic matrix $D = (d_{ij})$. Assume, without loss of generality, that the components of $x = (x_1, x_2, \ldots, x_n)'$ are in decreasing order of magnitude, i.e., $x_1 \geq x_2 \geq \ldots \geq x_n$. Likewise, assume that the components of $y = (y_1, y_2, \ldots, y_n)$ are in decreasing order of magnitude. (If not, let P_1 and P_2 be the permutation matrices so that the components of each of the vectors P_1x and P_2y are in decreasing order of magnitude. The equation $x = Dy$ can be rewritten as $P_1x = (P_1)D(P_2)^{-1}(P_2y)$. Note that $(P_1)D(P_2)^{-1}$ is doubly stochastic.) Note that for any $1 \leq k \leq n$,

$$\begin{aligned}
\sum_{i=1}^{k} x_i - \sum_{i=1}^{k} y_i &= \sum_{i=1}^{k}\sum_{j=1}^{n} d_{ij}y_j - \sum_{i=1}^{k} y_i \\
&= \sum_{j=1}^{n} t_jy_j - \sum_{i=1}^{k} y_i + y_k(k - \sum_{j=1}^{n} t_j) \\
&= \sum_{j=1}^{k}(y_j - y_k)(t_j - 1) + \sum_{j=k+1}^{n} t_j(y_j - y_k) \leq 0,
\end{aligned}$$

where $0 \leq t_j = \sum_{i=1}^{k} d_{ij} \leq 1$. These numbers are non-negative and sum up to k. This completes the proof.

We would like to identify Schur-convex functions among convex functions. The following result flows in this direction.

P 9.3.9 If a function f from $\mathbf{R}^m$ to $\mathbf{R}^n$ is convex and symmetric, then it is Schur-convex. If, in addition, f is monotonically increasing, then f is strongly Schur-convex.

PROOF. Let $x, y \in \mathbf{R}^m$ be such that $x \ll y$. By **P 9.3.8**, there exists a doubly stochastic matrix D such that $x = Dy$. By **P 9.3.7**, we can write $D = \sum_{P \in \mathbf{P}_m} \lambda_P P$, a convex combination of permutation matrices with $0 \leq \lambda_P$ and $\sum_{P \in \mathbf{P}_m} \lambda_P = 1$. Since f is convex,

$$f(x) = f(\sum_{P \in \mathbf{P}_m} \lambda_P P y) \leq \sum_{P \in \mathbf{P}_m} \lambda_P f(Py).$$

Since f is symmetric, for each $P \in \mathbf{P}_m$, there exists $Q_P \in \mathbf{P}_n$ such that $f(Py) = Q_P f(y)$ for all $y \in \mathbf{R}_m$. Consequently,

$$f(x) \leq \sum_{P \in \mathbf{P}_m} \lambda_P Q_P f(y).$$

Let $D_0 = \sum_{P \in \mathbf{P}_m} \lambda_P Q_P$. Clearly, D_0 is a doubly stochastic matrix of order $n \times n$. From the proof of **P 9.3.8**, it is clear that $f(x) \leq_e D_0 f(y)$ implying that $f(x) \ll_w f(y)$. Suppose, in addition, f is monotonically increasing. Let $x, y \in \mathbf{R}^m$ be such that $x \ll_w y$. There exists z in $\mathbf{R}^m$ such that $x \leq z \ll y$. See Complement 9.2.2. By what we have assumed about f, we have

$$f(x) \leq_e f(z) \ll_w f(y),$$
$$\Rightarrow f(x) \ll_w f(z) \ll_w f(y) \text{ and } f(x) \ll_w f(y).$$

This shows that f is strongly Schur-convex. One can jot down a number of consequences of **P 9.3.9**.

COROLLARY 9.3.10. Let f be a real-valued function of a real variable and

$$f_s(x_1, x_2, \ldots, x_n) = f_s(x) = (f(x_1), f(x_2), \ldots, f(x_n)), \; x \in \mathbf{R}^n.$$

If f is convex, then f_s is Schur-convex. If, in addition, f is monotonically increasing, then f_s is strongly Schur-convex.

PROOF. It is clear that f_s is convex. As for the symmetry of f_s, note that for any permutation Π of $\{1, 2, \ldots, n\}$, $f(x_\Pi) = (f(x))_\Pi$ for all $x \in \mathbf{R}^n$.

Let us introduce some additional operations on vectors. For any vector $x \in \mathbf{R}^n$, let u define $x^+ = (\max\{x_1, 0\}, \max\{x_2, 0\}, \ldots, \max\{x_n, 0\})'$, and $|x| = (|x_1|, |x_2|, \ldots, |x_n|)$.

COROLLARY 9.3.11.
(a) If $x \ll y$, then $|x| \ll_w |y|$.
(b) If $x \ll y$, then $x^+ \ll_w y^+$.
(c) If $x \ll y$, then $(x_1^2, x_2^2, \ldots, x_n^2) \ll_w (y_1^2, y_2^2, \ldots, y_n^2)$.

PROOF. For (a), take $f(t) = |t|$, $t \in \mathbf{R}$, in Corollary 9.3.10. For (b), take $f(t) = \max\{t, 0\}$, $t \in \mathbf{R}$, in Corollary 9.3.10. For (c), take $f(t) = t^2$, $t \in \mathbf{R}$.

Complements

9.3.1 Let $A = (a_{ij})$ be an orthogonal matrix with real entries, i.e., $A'A = I$. Show that $D = (a_{ij}^2)$ is a doubly stochastic matrix.
9.3.2 Let $x \ll y$. Show that there exists an orthogonal matrix $A = (a_{ij})$ such that $x = Dy$, where $D = (a_{ij}^2)$.
9.3.3 Let $A = (a_{ij})$ be a Hermitian matrix, $x = (\lambda_1, \lambda_2, \ldots, \lambda_n)'$, the vector of eigenvalues of A, $y = (a_{11}, a_{22}, \ldots, a_{nn})$, and $A = U \operatorname{Diag}\{\lambda_1, \lambda_2, \ldots, \lambda_n\} U^*$, the spectral decomposition of A, i.e., $U = (u_{ij})$ is a unitary matrix. Let $D = (|u_{ij}|^2)$. Show that $y = Dx$. Hence, or otherwise, show that $y \ll x$.
9.3.4 Let $f(x_1, x_2, \ldots, x_n) = \sum_{i=1}^{n} x_i \log x_i$; $x_1, x_2, \ldots, x_n > 0$. Show that f is Schur-convex on the set $(\mathbf{R}^+)^n$.
9.3.5 Let A be a square matrix of order $n \times n$ and I, J subsets of $\{1, 2, \ldots, n\}$. If $\#I + \#J = n$ and $A(I, J) = 0$, show that

$$\operatorname{per}(A) = \operatorname{per}(A(I, J^c)) \times \operatorname{per}(A(I^c, J)).$$

9.3.6 Let $f(x_1, x_2, \ldots, x_n) = \frac{1}{n}\sum_{i=1}^{n}(x_i - \bar{x})^2$ = variance of the numbers $x_1, x_2, \ldots, x_n$, $x = (x_1, x_2, \ldots, x_n) \in \mathbf{R}^n$, where $\bar{x} = \frac{1}{n}\sum_{i=1}^{n} x_i$. Show that f is Schur-convex.

9.3.7 Let f be a real-valued function defined on $\mathbf{R}^n$ whose first-order partial derivatives exist. Let $f_{(i)} = \frac{\partial}{\partial x_i} f$, $i = 1, 2, \ldots, n$. Show that f is Schur-convex if and only if the following hold.

(1) f is symmetric.

(2) $(x_i - x_j)(f_{(i)}(x) - f_{(j)}(x)) \geq 0$ for all x and i, j.

9.3.8 The k-th elementary symmetric polynomial, for any fixed $1 \leq k \leq n$, is defined by

$$\begin{aligned} S_k(x) &= S_k(x_1, x_2, \ldots, x_n) \\ &= \sum x_{i_1} x_{i_2} \ldots x_{i_k};\ x_1, x_2, \ldots, x_n > 0, \end{aligned}$$

where the summation is taken over all $1 \leq i_1 < i_2 < \ldots < i_k \leq n$. Show that $-S_k$ is Schur-convex over $(\mathbf{R}^+)^n$.

9.3.9 Let the components of the vectors $x' = (x_1, x_2, \ldots, x_n)$ and $y' = (y_1, y_2, \ldots, y_n)$ be non-negative. If $x \ll y$, show that $\prod_{i=1}^{n} x_i \geq \prod_{i=1}^{n} y_i$.

9.3.10 Let $A = (a_{ij})$ be a positive definite matrix of order $n \times n$. Show that $\det(A) \leq \prod_{i=1}^{n} a_{ii}$. (Use Complements 9.3.3 and 9.3.9.)

Note: For a historical and comprehensive treatment of majorization, Marshall and Olkin (1979) is a good source. A substantial portion of this chapter is inspired by the work of Ando (1982).

CHAPTER 10

INEQUALITIES FOR EIGENVALUES

Eigenvalues and singular values of a matrix play a dominant role in a variety of problems in Statistics, especially in Multivariate Analysis. In this chapter we will present a nucleus of results on the eigenvalues of a square matrix. Corresponding results for the singular values will also be presented. To begin with, we make some general remarks.

Let A be a matrix of order $n \times n$ with complex numbers as entries. Let $\lambda_1, \lambda_2, \ldots, \lambda_n$ be the eigenvalues of A. In order to indicate the dependence of the eigenvalues on the underlying matrix A, the eigenvalues of A are usually denoted by $\lambda_1(A), \lambda_2(A), \ldots, \lambda_n(A)$. The eigenvalues of A can be obtained as the roots of the determinantal polynomial equation,

$$|A - \lambda I_n| = 0,$$

in λ of degree n. Let $\sigma_1^2 \geq \sigma_2^2 \geq \ldots \geq \sigma_n^2$ (≥ 0) be the eigenvalues of the Hermitian matrix A^*A, or equivalently of AA^*. The numbers $\sigma_1 \geq \sigma_2 \geq \ldots \geq \sigma_n (\geq 0)$ are called the singular values of A. More precisely, the singular values are denoted by $\sigma_1(A) \geq \sigma_2(A) \geq \ldots \geq \sigma_n(A)$. Now, let A be a matrix of order $m \times n$. If $m \neq n$, the concept of eigenvalues of A makes no sense. However, singular values of A can be defined in a legitimate manner as the square roots of the eigenvalues of AA^* or of A^*A. We usually write the singular values of a matrix in decreasing order of magnitude. For a square matrix, some of the eigenvalues could be complex. But if the matrix is Hermitian, the eigenvalues are always real. In such a case, the eigenvalues are written in decreasing order of magnitude.

Thus two sets of numbers, namely the set of eigenvalues and the set of singular values, can be defined for a square matrix. The relationship between these two sets will be pinpointed later in this chapter.

All the results in this chapter are cast in the environment of the vector space $\mathbf{C}^n$ with its usual inner product $< \cdot, \cdot >$ given by

$$< x, y >= \sum_{i=1}^{n} x_i \bar{y}_i = y^* x,$$

for $x' = (x_1, x_2, \ldots, x_n)$ and $y' = (y_1, y_2, \ldots, y_n) \in \mathbf{C}^n$. Two vectors x and y are orthogonal ($x \perp y$ in notation) if $< x, y > = 0$ or $y^* x = 0$.

10.1. Monotonicity Theorem

Let A and B be two Hermitian matrices of the same order $n \times n$, and $C = A + B$. In this section, some inequalities connecting the eigenvalues of the three matrices A, B, and C will be presented.

P 10.1.1 (Monotonicity Theorem for Eigenvalues) Let

$$\alpha_1 \geq \ldots \geq \alpha_n; \ \beta_1 \geq \ldots \geq \beta_n; \ \gamma_1 \geq \ldots \geq \gamma_n$$

be the eigenvalues of the matrices A, B, and C, respectively. Then

$$(1) \qquad \alpha_1 + \beta_1 \quad \geq \quad \gamma_1 \quad \geq \begin{cases} \alpha_1 + \beta_n \\ \alpha_2 + \beta_{n-1} \\ \cdots \\ \alpha_n + \beta_1 \end{cases}$$

$$(2) \qquad \left. \begin{array}{l} \alpha_1 + \beta_2 \\ \alpha_2 + \beta_1 \end{array} \right\} \quad \geq \quad \gamma_2 \quad \geq \begin{cases} \alpha_2 + \beta_n \\ \alpha_3 + \beta_{n-1} \\ \cdots \\ \alpha_n + \beta_2 \end{cases}$$

$$(3) \qquad \left. \begin{array}{l} \alpha_1 + \beta_3 \\ \alpha_2 + \beta_2 \\ \alpha_3 + \beta_1 \end{array} \right\} \quad \geq \quad \gamma_3 \quad \geq \begin{cases} \alpha_3 + \beta_n \\ \alpha_4 + \beta_{n-1} \\ \cdots \\ \alpha_n + \beta_3 \end{cases}$$

$$\cdots \qquad\qquad\qquad\qquad \cdots$$

$$(n) \qquad \left. \begin{array}{l} \alpha_1 + \beta_n \\ \alpha_2 + \beta_{n-1} \\ \cdots \\ \alpha_n + \beta_1 \end{array} \right\} \quad \geq \quad \gamma_n \quad \geq \quad \alpha_n + \beta_n.$$

Proof. The inequalities on the left quoted above can be written succinctly as follows: for each $i = 1, 2, \ldots, n$,

$$\alpha_j + \beta_{i-j+1} \geq \gamma_i \text{ for } j = 1, 2, \ldots, i; \; i = 1, 2, \ldots, n. \tag{10.1.1}$$

The inequalities quoted above on the right can be written as

$$\gamma_i \geq \alpha_j + \beta_{n-j+i}, \; j = i, i+1, \ldots, n; \; i = 1, 2, \ldots, n. \tag{10.1.2}$$

Using a very simple argument, we will show that the inequalities (10.1.2) follow from (10.1.1). Let

$$u_1, u_2, \ldots, u_n; \; v_1, v_2, \ldots, v_n; \; w_1, w_2, \ldots, w_n$$

be the corresponding orthonormal eigenvectors of A, B, and C, respectively. Fix $1 \leq i \leq n$ and $1 \leq j \leq i$. Let

$$\begin{aligned}
\mathbf{S}_1 &= \operatorname{span}\{u_j, u_{j+1}, \ldots, u_n\}, \\
\mathbf{S}_2 &= \operatorname{span}\{v_{i-j+1}, v_{i-j+2}, \ldots, v_n\}, \\
\mathbf{S}_3 &= \operatorname{span}\{w_1, w_2, \ldots, w_i\}.
\end{aligned}$$

Note that $\dim(\mathbf{S}_1) = n - j + 1$, $\dim(\mathbf{S}_2) = n - i + j$, and $\dim(\mathbf{S}_3) = i$. Using the dimensional identity (see **P 1.5.7**),

$$\begin{aligned}
\dim(\mathbf{S}_1 \cap \mathbf{S}_2 \cap \mathbf{S}_3) &= \dim(\mathbf{S}_1) + \dim(\mathbf{S}_2 \cap \mathbf{S}_3) - \dim(\mathbf{S}_1 + (\mathbf{S}_2 \cap \mathbf{S}_3)) \\
&= \dim(\mathbf{S}_1) + \dim(\mathbf{S}_2) + \dim(\mathbf{S}_3) - \dim(\mathbf{S}_2 + \mathbf{S}_3) \\
&\quad - \dim(\mathbf{S}_1 + (\mathbf{S}_2 \cap \mathbf{S}_3)) \\
&\geq \dim(\mathbf{S}_1) + \dim(\mathbf{S}_2) + \dim(\mathbf{S}_3) - n - n \\
&= n - j + 1 + n - i + j + i - n - n = 1.
\end{aligned}$$

Consequently, there exists a vector x in $\mathbf{S}_1 \cap \mathbf{S}_2 \cap \mathbf{S}_3$ such that $x^* x = 1$. Since $x \in \mathbf{S}_1$, we can write $x = a_j u_j + \ldots + a_n u_n$ for some complex numbers $a_j, a_{j+1}, \ldots, a_n$. The property that $x^* x = 1$ implies that $\sum_{r=j}^{n} |a_r|^2 = 1$. Further,

$$x^* A x = (\sum_{r=j}^{n} \bar{a}_r u_r^*) A (\sum_{r=j}^{n} a_r u_r) = (\sum_{r=j}^{n} \bar{a}_r u_r^*)(\sum_{r=j}^{n} a_r \alpha_r u_r) = \sum_{r=j}^{n} |a_r|^2 \alpha_r.$$

We have used the fact that u_r's are orthonormal eigenvectors. Since the eigenvalues α_i's are written in decreasing order of magnitude, it now follows that $\alpha_j \geq x^*Ax \geq \alpha_n$. In our argument we need only the inequality $\alpha_j \geq x^*Ax$. In a similar vein, one can show that

$$\beta_{i-j+1} \geq x^*Bx \geq \beta_n, \ \gamma_1 \geq x^*Cx \geq \gamma_i,$$
$$\Rightarrow \alpha_j + \beta_{i-j+1} \geq x^*Ax + x^*Bx = x^*Cx \geq \gamma_i.$$

Thus (10.1.1) follows. The validity of the inequalities on the right hand side of the Monotonicity Theorem can be effected as follows. Note that

$$(-A) + (-B) = -C,$$

and the eigenvalues of $(-A), (-B)$ and $(-C)$ are respectively

$$-\alpha_n \geq \ldots \geq -\alpha_1; \ -\beta_n \geq \ldots \geq -\beta_1, \text{ and } -\gamma_n \geq \ldots \geq -\gamma_1.$$

By what we have proved earlier, it follows that for every $i = 1, 2, \ldots, n$ and $j = i, i+1, \ldots, n$,

$$(-\alpha_j) + (-\beta_{n-(j-i)}) \geq -\gamma_i,$$

from which we have

$$\gamma_i \geq \alpha_j + \beta_{n-(j-i)}.$$

(The map $T : \{1, 2, \ldots, n\} \rightarrow \{n, n-1, \ldots, 1\}$ involved here is given by $T(r) = n - r + 1$, $r \in \{1, 2, \ldots, n\}$.) This completes the proof.

The plethora of inequalities stated in **P 10.1.1** can be recast utilizing the majorization concept. Let

$$\lambda(C) = (\gamma_1, , \ldots, \gamma_n); \lambda(A) = (\alpha_1, \ldots, \alpha_n), \text{ and } \lambda(B) = (\beta_1, \ldots, \beta_n).$$

It follows from **P 10.1.1** that $\lambda(C) \ll (\lambda(A) + \lambda(B))$.

We would like to venture into the realm of singular values. The question we pop is what is the relationship between the singular values of A, B, and the sum $A + B$. We need a trick. Let $A \in \mathbf{M}_{m,n}$ and $q = \min\{m, n\}$. Let $\sigma_1 \geq \sigma_2 \geq \ldots \geq \sigma_q (\geq 0)$ be the singular values of A. Construct a new matrix by

$$\tilde{A} = \begin{bmatrix} 0 & A^* \\ A & 0 \end{bmatrix}.$$

Note that the matrix $\tilde{A}$ is of order $(m+n) \times (m+n)$. Further, $\tilde{A}$ is Hermitian.

P 10.1.2 The eigenvalues of the Hermitian matrix $\tilde{A}$ are:

$$\sigma_1 \geq \sigma_2 \geq \ldots \geq \sigma_q \geq 0 = \ldots = 0 \geq -\sigma_q \geq -\sigma_{q-1} \geq \ldots \geq -\sigma_1$$

(with $|m-n|$ zeros in the middle).

PROOF. Using a singular value decomposition of A, we will obtain a spectral decomposition of $\tilde{A}$.

Assume, for simplicity, that $m \leq n$. Let $A = P\Delta Q$ be a singular value decomposition of A, where P and Q are unitary matrices of appropriate order and $\Delta = (D|0)$ with $D = \text{diag}\{\sigma_1, \sigma_2, \ldots, \sigma_m\}$ and 0 is the zero matrix of order $m \times (n-m)$. Partition Q as: $Q = \binom{Q_1}{Q_2}$, where Q_1 is of order $m \times n$ and Q_2 of order $(n-m) \times n$. The fact that $Q^*Q = I_n$ implies that $Q_1^*Q_1 + Q_2^*Q_2 = I_n$. The singular value decomposition of A can be rewritten as: $A = PDQ_1$. Construct a matrix

$$U = \begin{bmatrix} Q_1/\sqrt{2} & P^*/\sqrt{2} \\ -Q_1/\sqrt{2} & P^*/\sqrt{2} \\ Q_2 & 0 \end{bmatrix}.$$

Note that U is a matrix of order $(m+n) \times (m+n)$. More interestingly, U is indeed a unitary matrix. Further,

$$\tilde{A} = U^* \begin{bmatrix} D & 0 & 0 \\ 0 & -D & 0 \\ 0 & 0 & 0 \end{bmatrix} U.$$

Thus we have a spectral decomposition of the Hermitian matrix $\tilde{A}$. The diagonal entries of the matrix in the middle of the representation above are indeed the eigenvalues of $\tilde{A}$. This completes the proof.

The above result will be instrumental in establishing a monotonicity theorem for singular values.

P 10.1.3 (Monotonicity Theorem for Singular Values). Let

$$\alpha_1 \geq \alpha_2 \geq \ldots \geq \alpha_q; \ \beta_1 \geq \beta_2 \geq \ldots \geq \beta_q; \text{ and } \gamma_1 \geq \gamma_2 \geq \ldots \geq \gamma_q$$

be the singular values of the matrices A, B, and C, respectively, where A and B are each of order $m \times n$, $C = A+B$, and $q = \min\{m, n\}$. Then we have the following results:

$$(1) \qquad \gamma_q \leq \begin{cases} \alpha_1 + \beta_q \\ \alpha_2 + \beta_{q-1} \\ \ldots \\ \alpha_q + \beta_1 \end{cases}$$

$$(2) \qquad \gamma_{q-1} \leq \begin{cases} \alpha_1 + \beta_{q-1} \\ \alpha_2 + \beta_{q-2} \\ \ldots \\ \alpha_{q-1} + \beta_1 \end{cases}$$

$$(3) \qquad \gamma_{q-2} \leq \begin{cases} \alpha_1 + \beta_{q-2} \\ \alpha_2 + \beta_{q-3} \\ \ldots \\ \alpha_{q-2} + \beta_1 \end{cases}$$

$$\ldots$$

$$(q) \qquad \gamma_1 \leq \alpha_1 + \beta_1$$

PROOF. A proof of this result can be obtained by a combination of **P 10.1.1** and **P 10.1.2**. For each of the matrices A, B and C, construct $\tilde{A}, \tilde{B}$ and $\tilde{C}$ as outlined in **P 10.1.2**. Observe that $\tilde{C} = \tilde{A} + \tilde{B}$. The eigenvalues of $\tilde{A}, \tilde{B}$, and $\tilde{C}$ written in decreasing order of magnitude are:

$$\alpha_1 \geq \ldots \geq \alpha_q \geq 0 = \ldots = 0 \geq -\alpha_q \geq \ldots \geq -\alpha_1,$$
$$\beta_1 \geq \ldots \geq \beta_q \geq 0 = \ldots = 0 \geq -\beta_q \geq \ldots \geq -\beta_1,$$
$$\gamma_1 \geq \ldots \geq \gamma_q \geq 0 = \ldots = 0 \geq -\gamma_q \geq \ldots \geq -\gamma_1,$$

respectively. Working carefully with the first q eigenvalues of $\tilde{C}$ and using **P 10.1.1**, one can establish the inequalities

$$\gamma_i \leq \alpha_j + \beta_{i-j+1},\ j = 1, 2, \ldots, i;\ i = 1, 2, \ldots, q,$$

on the right-hand side of the list presented in the theorem.

The contents of **P 10.1.3** can be recast in terms of weak majorization idea. By **P 10.1.2**, we have, working with the matrices $\tilde{A}, \tilde{B}$, and $\tilde{C}$,

$$\begin{aligned}&(\gamma_1, \gamma_2, \ldots, \gamma_q, 0, 0, \ldots, 0, -\gamma_q, -\gamma_{q-1}, \ldots, -\gamma_1)\\ &\ll (\alpha_1, \alpha_2, \ldots, \alpha_q, 0, 0, \ldots, 0, -\alpha_q, -\alpha_{q-1} \ldots, -\alpha_1)\\ &+(\beta_1, \beta_2, \ldots, \beta_q, 0, 0, \ldots, 0, -\beta_q, -\beta_{q-1}, \ldots, -\beta_1).\end{aligned}$$

From this, we obtain

$$(\gamma_1, \gamma_2, \ldots, \gamma_q) \ll_w (\alpha_1, \alpha_2, \ldots, \alpha_q) + (\beta_1, \beta_2, \ldots, \beta_q).$$

The proof that we have presented for the Monotonicity Theorem for the eigenvalues of Hermitian matrices is due to Ikebe, Inagaki, and Miyamoto (1987).

Complements

10.1.1 Let A, B and C be three matrices of the same order $m \times n$ with $C = A + B$. Note that $\sigma_i(A+B) \leq \sigma_i(A) + \sigma_i(B)$ for each i is not true.

Hint: Look at the example:

$$A = \begin{bmatrix} 1 & 0 \\ 0 & 0 \end{bmatrix} \quad \text{and} \quad B = \begin{bmatrix} 0 & 0 \\ 0 & 1 \end{bmatrix}.$$

10.1.2 Let A and B be Hermitian matrices of the same order $n \times n$ of which B is non-negative definite. Show that $\lambda_i(A) \leq \lambda_i(A+B)$, $i = 1, 2, \ldots, n$. If A and B are Hermitian matrices, then we have **Weyl's Theorem**

$$\begin{aligned}\lambda_j(A+B) &\leq \lambda_i(A) + \lambda_{j-i+1}(B) \quad \text{for } i \leq j,\\ \lambda_j(A+B) &\geq \lambda_i(A) + \lambda_{j-i+n}(B) \quad \text{for } i \geq j\end{aligned}$$

and for each $j = 1, \ldots, n$

$$\lambda_j(A) + \lambda_n(B) \leq \lambda_j(A+B) \leq \lambda_j(A) + \lambda_1(B).$$

10.2. Interlace Theorems

The focus in this section is to spell out the connection between the eigenvalues of a matrix and those of its submatrices. Let A be a Hermitian matrix of order $m \times m$ partitioned as

$$A = \begin{bmatrix} B & C \\ C^* & D \end{bmatrix},$$

where B is of order $n \times n$ for some $n < m$. It is clear that B is also a Hermitian matrix.

P 10.2.1 **(Interlace Theorem for Eigenvalues)** Let

$$\alpha_1 \geq \alpha_2 \geq \ldots \geq \alpha_m \text{ and } \beta_1 \geq \beta_2 \geq \ldots \geq \beta_n,$$

be the eigenvalues of A and B, respectively. Then

$$\alpha_1 \geq \beta_1 \geq \alpha_{m-n+1}; \alpha_2 \geq \beta_2 \geq \alpha_{m-n+2}; \cdots, \alpha_n \geq \beta_n \geq \alpha_m. \quad (10.2.1)$$

In particular, if $m = n + 1$, then

$$\alpha_1 \geq \beta_1 \geq \alpha_2; \;\; \alpha_2 \geq \beta_2 \geq \alpha_3; \;\; \ldots, \alpha_n \geq \beta_n \geq \alpha_{n+1}.$$

PROOF. Let $u_1, u_2, \ldots, u_m$ and $v_1, v_2, \ldots, v_n$, be the corresponding orthonormal eigenvectors of A and B, respectively. First, we show that $\alpha_k \geq \beta_k$ for each $k = 1, 2, \ldots, n$. For each $i = 1, 2, \ldots, n$, introduce the augmented vector

$$w_i = \begin{pmatrix} v_i \\ 0 \end{pmatrix}$$

of order $m \times 1$. Fix $1 \leq k \leq n$. Let $\mathbf{S}_1 = \text{span}\{u_k, u_{k+1}, \ldots, u_m\}$ and $\mathbf{S}_2 = \text{span}\{w_1, w_2, \ldots, w_k\}$. Both are subspaces of $\mathbf{R}^m$. Note that $\dim(\mathbf{S}_1) = m - k + 1$ and $\dim(\mathbf{S}_2) = k$. Further,

$$\begin{aligned} \dim(\mathbf{S}_1 \cap \mathbf{S}_2) &= \dim(\mathbf{S}_1) + \dim(\mathbf{S}_2) - \dim(\mathbf{S}_1 + \mathbf{S}_2) \\ &\geq (m - k + 1) + k - m = 1. \end{aligned}$$

Consequently, there exists a vector x in $\mathbf{S}_1 \cap \mathbf{S}_2$ such that $x^* x = 1$. Since $x \in \mathbf{S}_1$, it follows that $\alpha_k \geq x^* A x$. We have used this kind of argument in Section 10.1. Since $x \in \mathbf{S}_2$, we can write $x = a_1 w_1 + \ldots + a_k w_k$, for some unique scalars $a_1, a_2, \ldots, a_k$. Further, we have

$$1 = x^* x = \sum_{i=1}^{k} |a_i|^2 v_i^* v_i = \sum_{i=1}^{k} |a_i|^2,$$

$$\alpha_k \geq x^* A x = \sum_{i=1}^{k} |a_i|^2 (v_i^* | 0) \begin{pmatrix} B & C \\ C^* & D \end{pmatrix} \begin{pmatrix} v_i \\ 0 \end{pmatrix} = \sum_{i=1}^{k} |a_i|^2 v_i^* B v_i$$

$$\geq \beta_k \sum_{i=1}^{k} |a_i|^2 = \beta_k.$$

To establish the other bunch of inequalities in (10.2.1), look at the matrix

$$-A = \begin{bmatrix} -B & -C \\ -C^* & -D \end{bmatrix}.$$

The eigenvalues of $-A$ and $-B$ are

$$-\alpha_m \geq -\alpha_{m-1} \geq \ldots \geq -\alpha_1 \text{ and } -\beta_n \geq -\beta_{n-1} \geq \ldots \geq -\beta_1,$$

respectively. By what we have established above, we have now

$$-\alpha_m \geq -\beta_n; \ -\alpha_{m-1} \geq -\beta_{n-1}; \ \ldots, -\alpha_{m-n+1} \geq -\beta_1,$$

from which the desired inequalities follow. This completes the proof.

If one looks at the inequalities (10.2.1) a little closely, the following pattern emerges. The eigenvalues α's that figure on the left-hand side of the inequalities are the first n eigenvalues of A and the eigenvalues α's that figure on the right-hand side of the inequalities are the last n eigenvalues of A.

Our next goal is to obtain an analogue of the Interlace Theorem for the singular values of a matrix. The trick employed in **P 10.1.2** is also useful here.

P 10.2.2 (Interlace Theorem for Singular Values) Let $A \in \mathbf{M}_{m,n}$ be partitioned as

$$A = \begin{bmatrix} B & C \\ D & E \end{bmatrix},$$

where B is a matrix of order $p \times q$ for some $p \leq m$ and $q \leq n$. Let

$$\sigma_1 \geq \sigma_2 \geq \ldots \geq \sigma_r (\geq 0) \text{ and } \tau_1 \geq \tau_2 \geq \ldots \geq \tau_s (\geq 0)$$

be the singular values of A and B, respectively, where $r = \min\{m, n\}$ and $s = \min\{p, q\}$. Then

$$\tau_i \leq \sigma_i, \; i = 1, 2, \ldots, s.$$

PROOF. Observe that $s \leq r$. Let

$$\tilde{A} = \begin{bmatrix} 0 & A^* \\ A & 0 \end{bmatrix} \text{ and } \tilde{B} = \begin{bmatrix} 0 & B^* \\ B & 0 \end{bmatrix}.$$

Note that $\tilde{B}$ is a Hermitian submatrix of $\tilde{A}$. (Why?) As has been noted in **P 10.1.2**, the eigenvalues of $\tilde{A}$ are:

$$\sigma_1 \geq \ldots \geq \sigma_r \geq 0 = 0 = \ldots = 0 \geq -\sigma_r \geq \ldots \geq -\sigma_1$$

and that of $\tilde{B}$ are

$$\tau_1 \geq \ldots \geq \tau_s \geq 0 = \ldots = 0 \geq -\tau_s \geq \ldots \geq -\tau_1.$$

By **P 10.2.1**, by comparing the first s eigenvalues of $\tilde{A}$ and $\tilde{B}$, we obtain the desired inequalities.

The contents of **P 10.2.2** are disappointing. The result does not have as much pep as the one presented in **P 10.2.1**. It is the fault of the technique employed in the proof of **P 10.2.2**. We will take up the problem of obtaining a good analogue of Theorem 10.2.1 a little later in the complements for singular values.

Complements

10.2.1 Let $A \in \mathbf{M}_{m,n}$. Assume $m \geq n$. Let B be the matrix obtained from A by deleting one of the columns of A. Let

$$\sigma_1 \geq \sigma_2 \geq \ldots \geq \sigma_n \; (\geq 0) \text{ and } \tau_1 \geq \tau_2 \geq \ldots \geq \tau_{n-1} \; (\geq 0)$$

be the singular values of A and B, respectively. Show that

$$\sigma_1 \geq \tau_1 \geq \sigma_2 \geq \tau_2 \geq \sigma_3 \geq \ldots \geq \sigma_{n-1} \geq \tau_{n-1} \geq \sigma_n.$$

What happens if B is obtained from A by deleting a row of A?

Hint: Assume, without loss of generality, that it is the last column that is deleted from A. Note that $A = (B|b)$, where b is the last column of A. Look at A^*A and B^*B.

10.2.2 Let $A \in \mathbf{M}_{m,n}$. Assume $m < n$. Let B be the matrix obtained from A by deleting one column of A. Let

$$\sigma_1 \geq \sigma_2 \geq \ldots \geq \sigma_m \ (\geq 0) \text{ and } \tau_1 \geq \tau_2 \geq \ldots \geq \tau_m \ (\geq 0)$$

be the singular values of A and B, respectively. Show that

$$\sigma_1 \geq \tau_1 \geq \sigma_2 \geq \tau_2 \geq \ldots \geq \sigma_m \geq \tau_m.$$

A similar result holds if a row of A is deleted.

Some useful pointers emerge from Complements 10.2.1 and 10.2.2 for further generalizations. Let us recast the conclusions of the complements in a uniform manner. Before this, let us adopt the following terminology and conventions. For any matrix $A \in \mathbf{M}_{m,n}$, let

$$\sigma_1(A) \geq \sigma_2(A) \geq \ldots \geq \sigma_r(A) \ (\geq 0), \ r = \min\{m, n\}$$

be the singular values of A. Let us extend the range of the subscripts of the singular values of A by defining

$$\sigma_j(A) = 0 \quad \text{for } j > r.$$

The recasting is carried out as follows. Let $A \in \mathbf{M}_{m,n}$ be given. Let A_1 be the submatrix obtained from A by deleting either one row of A or one column of A. Then

$$\sigma_i(A) \geq \sigma_i(A_1) \geq \sigma_{i+1}(A), \ i = 1, 2, \ldots, \min\{m, n\}.$$

Suppose A_2 is a matrix obtained from A by deleting either two rows of A or two columns of A or one row and one column of A. We would like to explore the connection between the singular values of A and A_2. Suppose A_2 is obtained from A by deleting two columns, say j_1-th and

j_2-th, of A. Let A_1 be the matrix obtained from A by deleting the j_1-th column of A. It is clear that the matrix A_2 can be obtained from A_1 by deleting an appropriate column of A_1. The relationship between the singular values of A_1 and those of A_2 is clear. More precisely,

$$\sigma_i(A_1) \geq \sigma_i(A_2) \geq \sigma_{i+1}(A_1),\ i = 1, 2, \dots, \min\{m, n-1\}.$$

These inequalities can be stretched a little further. We can say that

$$\sigma_i(A_1) \geq \sigma_i(A_2) \geq \sigma_{i+1}(A_1),\ i = 1, 2, \dots, \{m, n\}.$$

We do have a good grip on the connection between the singular values of A_1 and those of A. Finally, we can jot down that

$$\sigma_i(A) \geq \sigma_i(A_2) \geq \sigma_{i+2}(A),\ i = 1, 2, \dots, \min\{m, n\}.$$

The argument can be pushed further. We have now an analogue of Theorem 10.2.1 for singular values.

10.2.3 Let $A \in \mathbf{M}_{m,n}$. Let A_r be any matrix obtained from A by deleting a total of r rows and columns. Show that

$$\sigma_i(A) \geq \sigma_i(A_r) \geq \sigma_{i+r}(A),\ \ i = 1, 2, \dots, \min\{m, n\}.$$

10.3. Courant-Fischer Theorem

The Monotonicity Theorem examines relationship between the singular values of a sum of two matrices and those of its constituents. The Interlace Theorem explores the connection between the singular values of a matrix and those of a submatrix. It is time to be a little introspective. The Courant-Fischer Theorem characterizes the singular values via optimization.

P 10.3.1 (Courant-Fischer Theorem for Eigenvalues) Let $A \in \mathbf{M}_n$ be a Hermitian matrix with eigenvalues

$$\alpha_1 \geq \alpha_2 \geq \ldots \geq \alpha_n.$$

Then for each $1 \leq k \leq n$,

$$\alpha_k = \max_{\mathbf{S} \subset \mathbf{C}^n, \dim(\mathbf{S}) = k} \ \min_{x \in \mathbf{S}, x^* x = 1} (x^* A x) \tag{10.3.1}$$

$$= \min_{\mathbf{T} \subset \mathbf{C}^n, \dim(\mathbf{T}) = n-k+1} \ \max_{x \in \mathbf{T}, x^* x = 1} (x^* A x). \tag{10.3.2}$$

In particular,

$$\alpha_1 = \max_{x \in \mathbf{C}^n, x^* x = 1} x^* A x, \; \alpha_n = \min_{x \in \mathbf{C}^n, x^* x = 1} x^* A x.$$

(The entities S and T are subspaces of $\mathbf{C}^n$.)

PROOF. Let $u_1, u_2, \ldots, u_n$ be the orthonormal eigenvectors corresponding to the eigenvalues, $\alpha_1, \ldots, \alpha_n$, of A. Fix $1 \leq k \leq n$. Let

$$\mathbf{S}_1 = \operatorname{span}\{u_k, u_{k+1}, \ldots, u_n\}.$$

Clearly, $\dim(\mathbf{S}_1) = n - k + 1$. Let $\mathbf{S}$ be any k-dimensional subspace of $\mathbf{C}^n$. Note that

$$\begin{aligned} \dim(\mathbf{S} \cap \mathbf{S}_1) &= \dim(\mathbf{S}) + \dim(\mathbf{S}_1) - \dim(\mathbf{S} + \mathbf{S}_1) \\ &\geq k + n - k + 1 - n = 1. \end{aligned}$$

Choose any y in $\mathbf{S} \cap \mathbf{S}_1$ such that $y^* y = 1$. Following by now the familiar argument, since $y \in \mathbf{S}_1$, we have $\alpha_k \geq y^* A y$. Consequently, since $y \in \mathbf{S}$,

$$\alpha_k \geq \min_{x \in \mathbf{S}, x^* x = 1} x^* A x.$$

This inequality is valid for any k-dimensional subspace $\mathbf{S}$ of $\mathbf{C}^n$. Therefore,

$$\alpha_k \geq \max_{\mathbf{S} \subset \mathbf{C}^n, \dim(\mathbf{S}) = k} \; \min_{x \in \mathbf{S}, x^* x = 1} (x^* A x).$$

We need to show now the reverse inequality. Let

$$\tilde{\mathbf{S}} = \operatorname{span}\{u_1, u_2, \ldots, u_k\}.$$

Note that $\tilde{\mathbf{S}}$ is a k-dimensional subspace of $\mathbf{C}^n$. Further, if $x \in \tilde{\mathbf{S}}$ and $x^* x = 1$, then

$$x^* A x \geq \alpha_k.$$

(Why?) It is clear that

$$\max_{\mathbf{S} \subset \mathbf{C}^n, \dim(\mathbf{S}) = k} \; \min_{x \in \mathbf{S}, x^* x = 1} (x^* A x) \;\geq\; \min_{x \in \tilde{\mathbf{S}}, x^* x = 1} x^* A x \;\geq\; \alpha_k.$$

Thus the identity (10.3.1) follows. As for the second identity, let

$$\mathbf{S}_1 = Sp\{u_1, u_2, \ldots, u_k\}.$$

It is clear that $\dim(\mathbf{S}_1) = k$. Let $\mathbf{T}$ be any $(n-k+1)$-dimensional subspace of $\mathbf{C}^n$. It is clear that $\dim(\mathbf{S}_1 \cap \mathbf{T}) \geq 1$. Choose any vector y in $\mathbf{S}_1 \cap \mathbf{T}$ such that $y^*y = 1$. Since $y \in \mathbf{S}_1, y^*Ay \geq \alpha_k$. Since $y \in \mathbf{T}$,

$$\max_{x \in \mathbf{T}, x^*x=1} (x^*Ax) \geq \alpha_k. \tag{10.3.3}$$

Let us recast this observation in the following way: for every $(n-k+1)$-dimensional subspace $\mathbf{T}$ of $\mathbf{C}^n$, (10.3.3) holds. Consequently,

$$\min_{\mathbf{T} \subset \mathbf{C}^n, \dim(\mathbf{T})=n-k+1} \quad \max_{x \in \mathbf{T}, x^*x=1} \quad (x^*Ax) \geq \alpha_k.$$

To establish the reverse inequality, take

$$\tilde{\mathbf{T}} = \operatorname{span}\{u_k, u_{k+1}, \ldots, u_n\}.$$

Further, if $x \in \tilde{\mathbf{T}}$, and $x^*x = 1$, then $x^*Ax \leq \alpha_k$. Hence

$$\alpha_k \geq \max_{x \in \tilde{\mathbf{T}}, x^*x=1} (x^*Ax) \geq \min_{\mathbf{T} \subset \mathbf{C}^n, \dim(\mathbf{T})=n-k+1} \max_{x \in \mathbf{T}, x^*x=1} (x^*Ax).$$

This establishes the identity (10.3.2). This completes the proof.

There is another angle at which one can look at the Fischer-Courant theorem. Let $A \in \mathbf{M}_n$ be a Hermitian matrix. Let $\alpha_1 \geq \alpha_2 \geq \ldots \geq \alpha_n$ be the eigenvalues of A with the corresponding orthogonal eigenvectors $u_1, u_2, \ldots, u_n$. Define a function $\rho : \mathbf{C}^n - \{0\} \to \mathbf{R}$ by

$$\rho(u) = \frac{u^*Au}{u^*u}, \quad u \in \mathbf{C}^n, u \neq 0.$$

The function $\rho(\cdot)$ is called the Raleigh quotient and has the following properties.

(1) $\alpha_1 \geq \rho(u) \geq \alpha_n, u \in \mathbf{C}^n, u \neq 0.$

(2) $\min_{u \in \mathbf{C}^n, u \neq 0} \rho(u) = \alpha_n$ with the minimum attaining at $u = u_n$.

(3) $\max_{u \in \mathbf{C}^n, u \neq 0} \rho(u) = \alpha_1$ with the maximum attaining at $u = u_1$.

(4) The function $\rho(\cdot)$ is stationary at, and only at, the eigenvectors of A, i.e., for example,

$$\frac{\delta \rho(u)}{\delta u}|_{u=u_i} = 0 \quad \text{for each } i.$$

(5) Let $1 \leq k \leq n$ be fixed. Let $\mathbf{S}_k = \text{span}\{u_k, u_{k+1}, \ldots, u_n\}$. Then

$$\max_{u \in \mathbf{S}_k, u \neq 0} \rho(u) = \alpha_k$$

and the maximum is attained at $u = u_k$.

(6) Let $T_k = \text{span}\{u_1, u_2, \ldots, u_k\}$. Then

$$\min_{u \in T_k, u \neq 0} \rho(u) = \alpha_k$$

and the minimum is attained at $u = u_k$.

Look at the Courant-Fischer theorem after going through the above properties of the Raleigh quotient. The properties of the Raleigh quotient involve both the eigenvalues and the chosen eigenvectors. The Courant-Fischer theorem characterizes the eigenvalues of A without involving eigenvectors.

An analogue of **P 10.3.1** for the singular values of a matrix is not hard to obtain. In the formulation of the relevant result, an expression like x^*Ax does not make sense for non-square matrices. We need to find an appropriate analogue for such an expression.

P 10.3.2 (Courant-Fischer Theorem for Singular Values) Let $A \in \mathbf{M}_{m,n}$ and

$$\sigma_1(A) \geq \sigma_2(A) \geq \ldots$$

be its singular values. Then for each $k \geq 1$,

$$\sigma_k(A) = \min_{\mathbf{S} \subset \mathbf{C}^n, \dim(\mathbf{S}) = n-k+1} \; \max_{x \in \mathbf{S}, x^*x=1} (x^*(A^*A)x)^{1/2} \tag{10.3.4}$$

$$= \max_{\mathbf{T} \subset \mathbf{C}^n, \dim(T) = k} \; \min_{x \in \mathbf{T}, x^*x=1} (x^*(A^*A)x)^{1/2}. \tag{10.3.5}$$

PROOF. If a subspace is not available with a specified dimension, we take the min-max and max-min numbers of (10.3.4) and (10.3.5) to be equal to zero. The result follows from the corresponding Courant-Fischer theorem for the eigenvalues of a matrix if we keep in mind the fact that the eigenvalues of A^*A are:

$$\sigma_1^2(A) \geq \sigma_2^2(A) \geq \dots$$

One can obtain the Interlace Theorem from the Courant-Fischer Theorem. First, we deal with the eigenvalues of a square matrix.

COROLLARY 10.3.3. **(Interlace Theorem)** Let $A \in \mathbf{M}_n$ be Hermitian and $\alpha_1 \geq \alpha_2 \geq \dots \geq \alpha_n$ its eigenvalues. Let B be a submatrix of A obtained by deleting some $n-r$ rows of A and the corresponding columns of A. Let $\beta_1 \geq \beta_2 \geq \dots \geq \beta_r$ be the eigenvalues of B. Then

$$\alpha_k \geq \beta_k \geq \alpha_{k+n-r}, \quad k = 1, 2, \dots, r.$$

PROOF. Note that B is a Hermitian matrix of order $r \times r$. Assume, without loss of generality, that B was obtained from A by deleting the last $n-r$ rows and last $n-r$ columns of A. For $1 \leq k \leq r$, by the Courant-Fischer theorem,

$$\begin{aligned}
\alpha_k &= \min_{\mathbf{T} \subset \mathbf{C}^n, \dim(\mathbf{T}) = n-k+1} \; \max_{x \in \mathbf{T}, x^*x=1} (x^*Ax) \\
&\geq \min_{\mathbf{T} \subset \mathbf{C}^n, \dim(\mathbf{T}) = n-k+1} \; \max_{z \in \mathbf{T}, z^*z=1} (z^*Az) \\
&= \min_{\mathbf{T} \subset \mathbf{C}^r, \dim(\mathbf{T}) = r-k+1} \; \max_{y \in \mathbf{T}, y^*y=1} (y^*By) = \beta_k
\end{aligned}$$

where z in the middle equation is a vector with the last $(n-r)$ components as zero. The last step requires some deft handling. It is left as an exercise to the reader. On the other hand, for each $1 \leq k \leq r$, by the Courant-Fischer theorem,

$$\begin{aligned}
\alpha_{k+n-r} &= \max_{\mathbf{S} \subset \mathbf{C}^n, \dim(\mathbf{S}) = k+n-r} \; \min_{x \in \mathbf{S}, x^*x=1} (x^*Ax) \\
&\leq \max_{\mathbf{S} \subset \mathbf{C}^n, \dim(\mathbf{S}) = k+n-r} \; \min_{z \in \mathbf{S}, z^*z=1} (z^*Az) \\
&= \max_{\mathbf{S} \subset \mathbf{C}^r, \dim(\mathbf{S}) = k} \; \min_{y \in \mathbf{S}, y^*y=1} (y^*By) = \beta_k.
\end{aligned}$$

This completes the proof.

Complements

10.3.1 Let $A, B \in \mathbf{M}_n$ be such that A is Hermitian and B is non-negative definite. Let

$$\alpha_1 \geq \alpha_2 \geq \ldots \geq \alpha_n \text{ and } \beta_1 \geq \beta_2 \geq \ldots \geq \beta_n$$

be the eigenvalues of A and $A + B$, respectively. Using the Courant-Fischer theorem, show that $\alpha_k \leq \beta_k$, $k = 1, 2, \ldots, n$. (These inequalities can also be obtained using the Monotonicity Theorem.)
10.3.2 Obtain the Interlace Theorem for the singular values of a matrix from the corresponding Courant-Fischer theorem.
10.3.3 **(Sturmian Separation Theorem)** Let A_r be the submatrix obtained by deleting the last $n - r$ rows and columns of a Hermitian matrix A, $r = 1, \ldots, n$. Then $\lambda_{k+1}(A_{i+1}) \leq \lambda_k(A_i) \leq \lambda_k(A_{i+1})$.

10.4. Poincaré Separation Theorem

The Monotonicity Theorem, Interlace Theorem and Courant-Fischer Theorem form a triumvirate of results on the eigenvalues and the singular values of matrices. The monotonicity theorem compares the eigenvalues of two Hermitian matrices A and B with their sum. The Interlace Theorem compares the eigenvalues of a Hermitian matrix and its principal submatrices. The Courant-Fischer Theorem characterizes the eigenvalues of a Hermitian matrix. Any one of these results is deducible from any one of the other results. Some of these implications have already been alluded earlier. The Poincaré Separation Theorem, which is the subject of discussion in this section, also falls into the same genre.

P 10.4.1 (Poincaré Separation Theorem for Eigenvalues) Let $A \in \mathbf{M}_n$ be a Hermitian matrix with eigenvalues $\alpha_1 \geq \alpha_2 \geq \ldots \geq \alpha_n$. Let B be any matrix of order $n \times k$ such that $B^*B = I_k$, i.e., the columns of B constitute a set of orthonormal vectors. Let $\beta_1 \geq \beta_2 \geq \ldots \geq \beta_k$ be the eigenvalues of the matrix B^*AB. Then

$$\alpha_i \geq \beta_i \geq \alpha_{i+n-k}, \; i = 1, 2, \ldots, k.$$

PROOF. Note that B^*AB is Hermitian. This result can be deduced from the Interlace Theorem. Let $B = (u_1, u_2, \ldots, u_k)$, where u_i is the

i-th column of B. Determine orthonormal vectors $u_{k+1}, u_{k+2}, \ldots, u_n$ such that $U = (u_1, u_2, \ldots, u_n)$ is unitary. Observe that the matrices U^*AU and A have the same set of eigenvalues. Further, B^*AB is a principal submatrix of U^*AU obtained by deleting the last $n-k$ rows and columns. Now the Interlace Theorem takes over. The result follows.

Now the question arises as to when $\alpha_i = \beta_i$ for $i = 1, \ldots, k$ in the Poincaré Separation Theorem. Let $u_1, u_2, \ldots, u_n$ be the corresponding orthonormal eigenvectors of A. Let $U = (u_1, \ldots, u_k)$ and take $B = UT$ for some unitary matrix T of order $k \times k$. Then $\alpha_i = \beta_i$, $i = 1, 2, \ldots, k$. Let us see what happens to B^*AB. Since $U^*AU = \text{Diag}(\alpha_1, \ldots, \alpha_k)$, $B^*AB = T^*U^*AUT = T^*\text{Diag}\{\alpha_1, \alpha_2, \ldots, \alpha_k\}T$.

Thus the eigenvalues of B^*AB are precisely $\alpha_1, \alpha_2, \ldots, \alpha_k$. This establishes that $\alpha_i = \beta_i$ for $i = 1, 2, \ldots, k$. One might ask in a similar vein as to when $\beta_i = \alpha_{i+n-k}$ holds for $i = 1, 2, \ldots, k$. This is left to the reader as an exercise.

We now need to launch an analogue of the Poincaré separation theorem for the singular values of a matrix. The following result which is easy to prove covers such a contingency.

P 10.4.2 (Poincaré Separation Theorem for Singular Values) Let $A \in \mathbf{M}_{m,n}$ with singular values

$$\sigma_1(A) \geq \sigma_2(A) \geq \ldots.$$

Let U and V be two matrices of order $m \times p$ and $n \times q$, respectively, such that $U^*U = I_p$ and $V^*V = I_q$. Let $B = U^*AV$ with singular values

$$\sigma_1(B) \geq \sigma_2(B) \geq \ldots.$$

Then, with $r = (m-p) + (n-q)$,

$$\sigma_i(A) \geq \sigma_i(B) \geq \sigma_{i+r}(A), \; i = 1, 2, \ldots, \min\{m, n\}.$$

Complements

10.4.1 Let $A, B \in \mathbf{M}_{m,n}$, $E = B - A$ and $q = \min\{m, n\}$. If $\sigma_1 \geq \ldots \geq \sigma_q$ are the singular values of A, $\tau_1 \geq \ldots \geq \tau_q$ are the singular values of B, and δ is the spectral norm of E, then

(1) $|\sigma_i - \tau_i| \leq \delta$ for all $i = 1, \ldots, q$; and
(2) $(\sigma_1 - \tau_1)^2 + \ldots + (\sigma_q - \tau_q)^2 \leq \|E\|_2^2$.

10.5. Singular Values and Eigenvalues

For a square matrix, one can determine its singular values as well as its eigenvalues. If the matrix is Hermitian, a precise relationship exists between its singular values and eigenvalues. In this section, we establish a set of inequalities connecting the singular values and eigenvalues of a general matrix.

P 10.5.1 Let $A \in \mathbf{M}_n$. Let $\alpha_1, \alpha_2, \ldots, \alpha_n$ be the eigenvalues of A arranged in such a way that

$$|\alpha_1| \geq |\alpha_2| \geq \ldots \geq |\alpha_n|.$$

Let $\sigma_1 \geq \sigma_2 \geq \ldots \geq \sigma_n$ be the singular values of A. Then

$$|\alpha_1 \alpha_2 \ldots \alpha_i| \leq \sigma_1 \sigma_2 \ldots \sigma_i, \; i = 1, 2, \ldots, n$$

with equality for $i = n$.

PROOF. By the Schur triangularization theorem, there exist unitary matrices U and V each of order $n \times n$ such that $U^*AV = \Delta$, where Δ is an upper triangular matrix with diagonal entries $\alpha_1, \alpha_2, \ldots,$ α_n. Fix $1 \leq i \leq n$. Write $U = (U_1|U_2)$ and $V = (V_1|V_2)$, where U_1 and V_1 are of order $n \times i$. Note that

$$U^*AV = \begin{bmatrix} U_1^* \\ U_2^* \end{bmatrix} A(V_1|V_2) = \begin{bmatrix} U_1^*AV_1 & U_1^*AV_2 \\ U_2^*AV_1 & U_2^*AU_2 \end{bmatrix}$$
$$= \Delta = \begin{bmatrix} \Delta_i & C \\ 0 & D \end{bmatrix}, \text{ say.}$$

It is now clear that $U_1^*AV_1$ is upper triangular with entries in the diagonal being $\alpha_1, \alpha_2, \ldots, \alpha_i$. It is indeed a submatrix of U^*AV. By the Interlace Theorem,

$$\sigma_j(U_1^*AV_1) \leq \sigma_j(U^*AV), \; j = 1, 2, \ldots, i.$$

Let us compute the singular values of U^*AV. For this we need to find the eigenvalues of $(U^*AV)^*(U^*AV) = V^*A^*AV$. The eigenvalues of A^*A and V^*A^*AV are identical. Consequently, $\sigma_j(U^*AV) = \sigma_j(A) =$ σ_j, $j = 1, 2, \ldots, n$. Finally,

$$|\det(U_1^* A V_1)| = |\det(\Delta_i)| = |\Pi_{j=1}^i \alpha_j| = \Pi_{j=1}^i \sigma_j(U_1^* A V_1^*) \leq \Pi_{j=1}^i \sigma_j.$$

When $i = n$, equality holds. (See Complement 10.5.1.) This completes the proof.

The multiplicative inequalities presented in **P 10.6.2** for the eigenvalues and singular values of a square matrix have an additive analogue. More precisely, if none of the eigenvalues and singular values is zero, then

$$(\ell n|\alpha_1|, \ell n|\alpha_2|, \dots, \ell n|\alpha_n|) \ll (\ell n\sigma_1, \ell n\sigma_2, \dots, \ell n\sigma_n),$$

where ℓn denotes the natural logarithm.

Complements

10.5.1 Let $A \in \mathbf{M}_n$ and $\sigma_1 \geq \sigma_2 \geq \dots \geq \sigma_n$ (≥ 0) its singular values. Show that $|\det(A)| = \sigma_1\sigma_2\dots\sigma_n$.

10.6. Products of Matrices, Singular Values and Horn's Theorem

One important operation on matrices is the operation of multiplication. In this section, we will present some results on a connection between the singular values of the product of two matrices and the singular values of its constituent matrices.

P 10.6.1 **(Horn's Theorem)** Let $A \in \mathbf{M}_{m,n}$ and $B \in \mathbf{M}_{n,p}$. Let $q = \min\{m,n,p\}, r = \min\{m,n\}, s = \min\{n,p\}$ and $t = \min\{m,p\}$. Let

$$\sigma_1(A) \geq \sigma_2(A) \geq \dots \geq \sigma_r(A) \ (\geq 0),$$
$$\sigma_1(B) \geq \sigma_2(B) \geq \dots \geq \sigma_s(B) \ (\geq 0),$$
$$\sigma_1(AB) \geq \sigma_2(AB) \geq \dots \geq \sigma_t(AB) \ (\geq 0),$$

be the singular values of A, B, and AB respectively. Then

$$\Pi_{j=1}^i \sigma_j(AB) \leq \Pi_{j=1}^i \sigma_j(A)\sigma_j(B), \ i = 1, 2, \dots, q. \quad (10.6.1)$$

If A and B are square matrices of the same order, i.e., $m = n = p$, then equality holds in (10.6.1) for $i = n$.

PROOF. The proof is outlined in a series of steps.

1. Let $AB = P\Delta Q^*$ be a singular value decomposition of AB for some unitary matrices P and Q of orders $m \times m$ and $p \times p$, respectively. The (k, k)-th entry of the matrix Δ is equal to $\sigma_k(AB)$, $k = 1, 2, \ldots, t$, and the rest of the entries of Δ are zeros.

2. Fix $1 \leq i \leq q$. Write $P = (P_1|P_2)$ and $Q = (Q_1|Q_2)$, where P_1 and Q_1 are of orders $m \times i$ and $p \times i$, respectively. Observe that $P_1^*(AB)Q_1$ is an $i \times i$ principal submatrix of $P^*(AB)Q$. Therefore, $P_1^*(AB)Q_1 = \text{diag}\{\sigma_1(AB), \sigma_2(AB), \ldots, \sigma_i(AB)\}$, and

$$\det(P_1^* ABQ_1) = \Pi_{j=1}^i \sigma_j(AB).$$

3. We want to focus on the matrix BQ_1, which is of order $n \times i$. By the polar decomposition theorem, we can find two matrices X of order $n \times i$ and W of order $i \times i$ such that X has orthonormal columns, W is non-negative definite, and $BQ_1 = XW$. Note that $W^2 = (BQ_1)^*(BQ_1) = Q_1^* B^* BQ_1$. Hence,

$$\begin{aligned}\det(W^2) &= \text{the product of the squares of the eigenvalues of } W \\ &= \text{the product of the eigenvalues of } Q_1^* B^* BQ_1 \\ &= \Pi_{j=1}^i \sigma_j(Q_1^* B^* BQ_1).\end{aligned}$$

(The singular values and eigenvalues are the same for a positive semi-definite matrix.) By the Poincaré separation theorem,

$$\Pi_{j=1}^i \sigma_j(Q_1^* B^* BQ_1) \leq \Pi_{j=1}^i \sigma_j(B^* B) = \Pi_{j=1}^i \sigma_j^2(B).$$

4. Let us focus on the matrix $P_1^* AX$. This is a square matrix of order $i \times i$. By Complement 10.5.1 and Poincaré Separation Theorem,

$$|\det(P_1^* AX)| = \Pi_{j=1}^i \sigma_j(P_1^* AX) \leq \Pi_{j=1}^i \sigma_j(A).$$

5. Combining all the steps we have carried out so far, we have

$$\begin{aligned}\Pi_{j=1}^i \sigma_j(AB) &= \det(P_1^* ABQ_1) = |\det(P_1^* ABQ_1)| = |\det((P_1^* AX)(W))| \\ &= |\det(P_1^* AX)| \; |\det(W)| \leq \Pi_{j=1}^i \sigma_j(A)\sigma_j(B).\end{aligned}$$

6. If $m = n = p$, then

$$\Pi_{j=1}^n \sigma_j(AB) = |\det(AB)| = |\det(A)| \quad |\det(B)| = \Pi_{j=1}^n \sigma_j(A)\sigma_j(B).$$

This completes the proof.

The multiplicative inequalities presented in **P 10.6.1** have an additive analogue too. Within the same scenario, we have

$$\sum_{j=1}^{i} \ell n \sigma_i(AB) \leq \sum_{j=1}^{i} \ell n(\sigma_i(A)\sigma_j(B)),\ i = 1, 2, \ldots, q.$$

If A and B are square matrices of the same order $n \times n$, then

$$\begin{aligned}&(\ell n\sigma_1(AB), \ell n\sigma_2(AB), \ldots, \ell n\sigma_n(AB))\\ &\ll (\ell n(\sigma_1(A)\sigma_2(B)), \ell n(\sigma_2(A)\sigma_2(B)), \ldots, \ell n(\sigma_n(A)\sigma_n(B))).\end{aligned}$$

10.7. Von Neumann's Theorem

For a square matrix, the trace of the matrix and the sum of all eigenvalues of the matrix are the same. Suppose we have two matrices A and B such that AB is a square matrix. The trace of the matrix AB is easy to compute. How is the trace related to the singular values of the individual matrices A and B? The von Neumann's Theorem is an answer to this question. But first, we need to prepare the reader for the von Neumann's Theorem (von Neumann (1937)).

P 10.7.1 Let $A \in \mathbf{M}_{m,n}$. Then $\text{tr}(AX) = 0$ for every matrix X of order $n \times m$ if and only if $A = 0$.

A stronger result than **P 10.7.1** is as follows.

P 10.7.2 Let $A \in \mathbf{M}_n$. Then $\text{tr}(AX) = 0$ for all Hermitian matrices X if and only if $A = 0$.

PROOF. Note that any matrix X of order $m \times m$ can be written as

$$X = X_1 + iX_2,$$

with both X_1 and X_2 Hermitian. More precisely, one can take

$$X_1 = (1/2)(X + X^*) \quad \text{and} \quad X_2 = (1/2i)(X - X^*).$$

P 10.7.3 Let $A \in \mathbf{M}_n$. Then $\text{tr}(AX)$ is real for all Hermitian matrices X if and only if A is Hermitian.

PROOF. Suppose $A = (a_{ij})$ and $X = (x_{ij})$ are Hermitian. Observe that

$$\operatorname{tr}(AX) = \sum_{i=1}^{m}\sum_{j=1}^{m} a_{ij}x_{ji}.$$

Since A and X are Hermitian, a_{ii} and x_{ii} are real for each i. Consequently, $a_{ii}x_{ii}$ is real for each i. Let $i \neq j$. Write $a_{ij} = a + ib$ and $x_{ji} = c + id$, where a, b, c and d are real numbers. Then

$$\begin{aligned} a_{ij}x_{ji} + a_{ji}x_{ij} &= a_{ij}x_{ji} + \bar{a}_{ij}\bar{x}_{ji} = (a+ib)(c+id) + (a-ib)(c-id) \\ &= 2(ac-bd) + i(ad+bc) - i(ad+bc) = 2(ac-bd) \end{aligned}$$

which is clearly real. Consequently, $\operatorname{tr}(AX)$ is real. Conversely, suppose, $\operatorname{tr}(AX)$ is real for all Hermitian matrices X. Then

$$\begin{aligned} \operatorname{tr}(AX) = \overline{\operatorname{tr}(AX)} &= \operatorname{tr}((AX)^*) = \operatorname{tr}(X^*A^*) = \operatorname{tr}(XA^*) = \operatorname{tr}(A^*X) \\ &\Rightarrow \operatorname{tr}((A - A^*)X) = 0 \end{aligned}$$

for all Hermitian matrices X. By **P 10.7.2**, we have $A - A^* = 0$, i.e., A is Hermitian.

P 10.7.4 Let $A \in \mathbf{M}_m$ be Hermitian. If $\operatorname{tr}(A) \geq Re\operatorname{tr}(AU)$ for all unitary matrices U, then A is non-negative definite. (The symbol Re stands as an abbreviation for "the real of part of.")

PROOF. Let us perform a spectral decomposition on A. Write

$$A = \lambda_1 u_1 u_1^* + \lambda_2 u_2 u_2^* + \ldots + \lambda_m u_m u_m^*,$$

where $\lambda_1, \lambda_2, \ldots, \lambda_m$ are the eigenvalues of A and $u_1, u_2, \ldots, u_m$ the corresponding orthonormal eigenvectors of A. Our goal is to prove that each $\lambda_i \geq 0$. Suppose not. Some eigenvalues are negative. Assume, without loss of generality, that

$$\lambda_1, \lambda_2, \ldots, \lambda_r \geq 0 \text{ and } \lambda_{r+1}, \lambda_{r+2}, \ldots, \lambda_m < 0,$$

for some $1 \leq r \leq n$. Let

$$\begin{aligned} B &= \lambda_1 u_1 u_1^* + \lambda_2 u_2 u_2^* + \ldots + \lambda_r u_r u_r^*, \\ C &= -\lambda_{r+1} u_{r+1} u_{r+1}^* - \lambda_{r+2} u_{r+2} u_{r+2}^* - \ldots - \lambda_m u_m u_m^*, \end{aligned}$$

$$U = u_1u_1^* + u_2u_2^* + \ldots + u_ru_r^* - u_{r+1}u_{r+1}^* - u_{r+2}u_{r+2}^* - \ldots - u_mu_m^*.$$

Check that B and C are non-negative definite, $C \neq 0$, $A = B - C$, U is Hermitian, and U is unitary. Further, $AU = B + C$. Observe that, by hypothesis,

$$\text{tr}(A) = \text{tr}(B) - \text{tr}(C) \geq \text{tr}(AU) = \text{tr}(B) + \text{tr}(C).$$

This inequality is possible only if $\text{tr}(C) = 0$. Since C is non-negative definite, this is possible only if $C = 0$. This is a contradiction. This completes the proof.

Now we take up a certain optimization problem. This has an important bearing on the proof of von Neumann's Theorem. Before we present the optimization result, let us take a detour.

Let $B \in \mathbf{M}_m$. The problem is to investigate under what condition the series

$$I_m + B + B^2 + \ldots$$

converges absolutely, and if it converges, identify the limit. A full discussion was undertaken in Chapter 11. But we are looking for a simple, sufficient, and easily verifiable condition. Let $M = \max_{1 \leq i,j \leq m} |b_{ij}|$. Observe that

(1) every entry in B is $\leq M$ in absolute value;
(2) every entry in B^2 is $\leq mM^2$ in absolute value;
(3) every entry in B^3 is $\leq m^2M^3$ in absolute value, and so on. Consequently, the series $I_m + B + B^2 + \ldots$ converges absolutely if $\sum_{k \geq 1} m^{k-1}M^k$ converges, or equivalently, if $\sum_{k \geq 1}(mM)^k$ converges. The geometric series converges if $mM < 1$. To answer the second question, if the series converges then the sum is equal to $(I_m - B)^{-1}$.

Let us make use of the above discussion. Let $X = (x_{ij})$ be an arbitrary but fixed Hermitian matrix of order $m \times m$. Let $M = \max_{1 \leq i,j \leq m} |x_{ij}|$. Then we can find $\epsilon_0 > 0$ such that both

$$I_m + i\epsilon X \quad \text{and} \quad I_m - i\epsilon X$$

are invertible for every $-\epsilon_0 < \epsilon < \epsilon_0$. To see this, take $B = i\epsilon X = (b_{ij})$ with ϵ real. Note that $\max_{1 \leq i,j \leq m} |b_{ij}| = |\epsilon|M$. The series

$$I_m + B + B^2 + \ldots$$

converges absolutely to $(I_m - B)^{-1} = (I_m - i\epsilon X)^{-1}$ if $|\epsilon| mM < 1$. One can take $\epsilon_0 = 1/mM$. In a similar vein, one can show that $(I_m + i\epsilon X)$ is invertible if $|\epsilon| < \epsilon_0$.

Fix $-\epsilon_0 < \epsilon < \epsilon_0$. Then it transpires that

$$V = (I_m + i\epsilon X)(I_m - i\epsilon X)^{-1} = (I_m - i\epsilon X)^{-1}(I_m + i\epsilon X),$$

and further, that V is unitary. Let us calculate

$$\begin{aligned}(I_m + i\epsilon X)(I_m - i\epsilon X)^{-1} &= (I_m + i\epsilon X)(I_m + i\epsilon X + (i\epsilon X)^2 + \ldots) \\ &= I_m + 2i\epsilon X + 2(i\epsilon X)^2 + 2(i\epsilon X)^3 + \ldots \\ &= 2(I_m - i\epsilon X)^{-1} - I_m. \qquad (10.7.1)\end{aligned}$$

Likewise,

$$(I_m - i\epsilon X)^{-1}(I_m + i\epsilon X) = 2(I_m - i\epsilon X)^{-1} - I_m.$$

Observe that

$$V^* = (I_m + i\epsilon X)^{-1}(I_m - i\epsilon X), \text{ and } V^*V = I_m,$$

which shows that V is unitary. We are now ready to establish the desired optimization result.

P 10.7.5 Let $A \in \mathbf{M}_m$. Let $\mathbf{U}_m$ be the collection of all unitary matrices of order $m \times m$. Then

$$\sup_{U \in \mathbf{U}_m} \text{Re tr}(AU)$$

is attained at some matrix $U_0 \in \mathbf{U}_m$. Further, AU_0 turns out to be non-negative definite.

PROOF. The fact that the supremum is attained at some matrix in $\mathbf{U}_m$ is not difficult to prove. This is a topological result. But the battle to show that AU_0 is non-negative definite is hard. We will present the proof in a series of steps.

1. Observe that the set $\mathbf{U}_m$ is compact when it is viewed as a subset of an appropriate unitary space with the usual topology. Further, the real valued maps

$$\text{tr}(AU) \quad \text{and Re tr}(AU)$$

as functions of $U \in \mathbf{U}_m$ are continuous. By a standard result in topology that the supremum of a continuous function on a compact is attained, it follows that the desired supremum is attained at some $U_0 \in \mathbf{U}_m$.

2. The next objective is to show that AU_0 is indeed Hermitian. By **P 10.7.3**, it suffices to show that $\text{tr}(AU_0X)$ is real for every Hermitian matrix X. Start with any Hermitian matrix X. Look up the discussion that followed **P 10.7.4**. There exists $\epsilon_0 > 0$, such that

$$I_m + i\epsilon X \quad \text{and} \quad I_m - i\epsilon X$$

are both nonsingular for every $-\epsilon_0 < \epsilon < \epsilon_0$. Further, the matrix

$$V = (I_m + i\epsilon X)^{-1}(I_m - i\epsilon X) = I_m + 2\epsilon i X + \epsilon^2 f(\epsilon, X)$$

is unitary, where $f(\epsilon, X)$ is a series of matrices involving ϵ and X. See (10.7.1). Note that

$$\begin{aligned} \text{Re tr}(AU_0V) &= \text{Re tr}(AU_0) + 2\epsilon\text{Re}[i\,\text{tr}(AU_0X)] + \epsilon^2\,\text{Re tr}(AU_0f(\epsilon, X)) \\ &\leq \text{Re tr}(AU_0), \end{aligned}$$

by the very definition of U_0, and V is unitary. This implies that

$$2\epsilon\text{Re}[i\,\text{tr}(AU_0X)] + \epsilon^2\,\text{Re tr}(AU_0f(\epsilon, X)) \leq 0$$

for every $-\epsilon_0 < \epsilon < \epsilon_0$. If $0 < \epsilon < \epsilon_0$, we indeed have the inequality

$$2\text{Re}[i\,\text{tr}(AU_0X)] + \epsilon\,\text{Re tr}(AU_0f(\epsilon, X)) \leq 0.$$

Taking the limit as $\epsilon \downarrow 0$, we observe that $\text{Re}[i\ \text{tr}(AU_0X)] \leq 0$. Arguing in a similar vein for $-\epsilon_0 < \epsilon < 0$, we can conclude that $\text{Re}[i\ \text{tr}(AU_0X)] \geq 0$. Hence $\text{Re}[i\ \text{tr}(AU_0X)] = 0$, from which it follows that $\text{tr}(AU_0X)$ is real. (Why?) This shows that AU_0 is Hermitian.

3. The final step is to show that AU_0 is non-negative definite. Note that by the very definition of U_0, $\text{Re tr}(AU_0) \geq \text{Re tr}(AU_0V)$ for every

Hermitian matrix V. By **P 10.7.4**, AU_0 is non-negative definite. This completes the proof.

This result is a little unsatisfactory. It uses a topological result to show the existence of the optimal unitary matrix U_0. But we do not know what exactly it is. Secondly, we do not know what the supremum of

$$\text{Re tr}(AU), \ U \in \mathbf{U}_m$$

actually is. We will ameliorate the deficiencies now.

P 10.7.6 Let $A \in \mathbf{M}_m$ with the singular values

$$\sigma_1(A) \geq \sigma_2(A) \geq \ldots \geq \sigma_m(A),$$

and singular value decomposition $A = P\Delta Q$, where P and Q are unitary matrices and $\Delta = \text{diag}\{\sigma_1(A), \sigma_2(A), \ldots, \sigma_m(A)\}$. Let $\mathbf{U}_m$ be the collection of all unitary matrices of order $m \times m$. Then

$$\max_{U \in \mathbf{U}_m} \text{Re tr}(AU) = \sum_{i=1}^{m} \sigma_i(A),$$

and the maximum is attained at $U_0 = Q^* P^*$. (U_0 need not be unique.)

PROOF. Let $U \in \mathbf{U}_m$. Let us compute

$$\text{Re tr}(AU) = \text{Re tr}(P\Delta QU) = \text{Re tr}(\Delta QUP) = \text{Re} \sum_{i=1}^{m} \sigma_i(A)[QUP]_{ii},$$

where $[QUP]_{ii}$ is the i-th diagonal entry of the unitary matrix QUP. Being a unitary matrix, $|[QUP]_{ii}| \leq 1$ for each i. Consequently,

$$|\text{Re tr}(AU)| \leq \sum_{i=1}^{m} \sigma_i(A)|[QUP]_{ii}| \leq \sum_{i=1}^{m} \sigma_i(A).$$

Let us compute specifically

$$\begin{aligned} \text{Re tr}(AU_0) &= \text{Re tr}(P\Delta QQ^* P^*) = \text{Re tr}(\Delta QQ^* P^* P) \\ &= \text{Re tr}(\Delta) = \sum_{i=1}^{m} \sigma_i(A). \end{aligned}$$

Incidentally, $AU_0 = P\Delta P^*$, which is clearly non-negative definite. The proof is complete.

P 10.7.5 and **P 10.7.6** solve the same optimization problem. The proof of **P 10.7.5** is originally due to von Neumann (1937). His proof is non-constructive but the methods used are fascinating. The proof provided under **P 10.7.6** is constructive.

Another nugget emerges from **P 10.7.5** and **P 10.7.6**. Suppose U_1 is a unitary matrix such that AU_1 is non-negative definite. Then U_1 maximizes Re tr(AU) over all $U \in \mathbf{U}_m$! This can be seen as follows. Since AU_1 is non-negative definite, its eigenvalues and singular values are identical. Consequently,

$$\operatorname{tr}(AU_1) = \sum_{i=1}^{m} \sigma_i(AU_1).$$

The singular values of AU_1 are precisely the positive square roots of the eigenvalues of $(AU_1)^*(AU_1) = U_1^* A^* AU_1$. The eigenvalues of $U_1^* A^* AU_1$ are identical to the eigenvalues of A^*A. The singular vales of A are the positive square roots of the eigenvalues of A^*A. Hence

$$\operatorname{tr}(AU_1) = \sum_{i=1}^{m} \sigma_i(AU_1) = \sum_{i=1}^{m} \sigma_i(A).$$

The avowed assertion now follows from **P 10.7.6**. We now come to the main result of this section.

P 10.7.7 **(Von Neumann's Theorem)** Let $A \in \mathbf{M}_{m,n}$ and $B \in \mathbf{M}_{n,m}$ be such that AB and BA are non-negative definite. Let $p = \min\{m, n\}$, $q = \max\{m, n\}$, and

$$\sigma_1(A) \geq \sigma_2(A) \geq \ldots \geq \sigma_p(A) \text{ and } \sigma_1(B) \geq \sigma_2(B) \geq \ldots \geq \sigma_p(B),$$

be the singular values of A and B, respectively. Set

$$\sigma_{p+1}(A) = \sigma_{p+2}(A) = \ldots = \sigma_q(A) = 0$$
$$\sigma_{p+1}(B) = \sigma_{p+2}(B) = \ldots = \sigma_q(B) = 0.$$

Then there exists a permutation τ of $\{1, 2, \ldots, q\}$ such that

$$\operatorname{tr}(AB) = \operatorname{tr}(BA) = \sum_{i=1}^{q} \sigma_i(A)\sigma_{\tau(i)}(B).$$

PROOF. The proof is carried out in several steps.

1. We work out a special case. Assume that $m = n$, A and B are non-negative definite, and $AB = BA$. This implies that AB is non-negative definite. (Why?) We can diagonalize both A and B simultaneously. There exists a unitary matrix U such that

$$A = U\Delta_1 U^* \text{ and } B = U\Delta_2 U^*,$$
$$\Delta_1 = \operatorname{diag}\{\alpha_1, \alpha_2, \ldots, \alpha_n\},\ \Delta_2 = \operatorname{diag}\{\beta_1, \beta_2, \ldots, \beta_n\},$$

where α_i's and β_i's are the eigenvalues of A and B respectively. Note that

$$\begin{aligned}\operatorname{tr}(AB) &= \operatorname{tr}(U\Delta_1 U^* U\Delta_2 U^*) = \operatorname{tr}(U\Delta_1\Delta_2 U^*) = \operatorname{tr}(\Delta_1\Delta_2 U^* U)\\ &= \operatorname{tr}(\Delta_1\Delta_2) = \sum_{i=1}^{n} \alpha_i\beta_i.\end{aligned}$$

Since the eigenvalues and singular values are identical for a non-negative definite matrix, we can write

$$\sum_{i=1}^{n} \alpha_i\beta_i = \sum_{i=1}^{n} \sigma_i(A)\sigma_{\tau(i)}(B),$$

for some permutation τ of $\{1, 2, \ldots, n\}$. The statement of the theorem is valid for this special case.

2. The strategy for the general case is as follows. Assume, without loss of generality, that $m \leq n$. Let A and B be given as stipulated in the theorem. We will construct two matrices A_0 and B_0 each of order $n \times n$ with the following properties.

a. A_0 and B_0 are non-negative definite.
b. A_0 and B_0 commute.
c. The eigenvalues of A_0 are $\sigma_1(A), \sigma_2(A), \ldots, \sigma_m(A), 0, 0, \ldots, 0$ ($n - m$ zeros) and the eigenvalues of B_0 are $\sigma_1(B), \sigma_2(B), \ldots, \sigma_m(B), 0, 0, \ldots, 0$ ($n - m$ zeros).
d. $\operatorname{tr}(AB) = \operatorname{tr}(A_0 B_0)$.

Now the conclusion of the theorem will follow from Step 1.

3. We want to simplify the hypothesis of the theorem. We would like to assume that $m = n$ and that A and B commute. Let us see how this can be done. Let $\lambda_1, \lambda_2, \ldots, \lambda_m$ be the eigenvalues of the $m \times m$ positive definite matrix AB. We claim that the eigenvalues of the $n \times n$ matrix BA are: $\lambda_1, \lambda_2, \ldots, \lambda_m, 0, 0, \ldots, 0$ ($n - m$ zeros). Let us prove a general result. Let λ be a non-zero eigenvalue of the non-negative definite matrix AB with multiplicity t with the corresponding linearly independent eigenvectors $u_1, u_2, \ldots, u_t$. Then λ is also an eigenvalue of BA with multiplicity at least t and $Bu_1, Bu_2, \ldots, Bu_t$ are the corresponding linearly independent eigenvectors. In fact, it will come out later that the multiplicity is exactly equal to t. First, check that

$$(BA)(Bu_i) = B(ABu_i) = B(\lambda u_i) = \lambda(Bu_i).$$

We now check that $Bu_1, Bu_2, \ldots, Bu_t$ are linearly independent. Suppose that $\theta_1(Bu_1) + \ldots + \theta_t(Bu_t) = 0$ for some scalars $\theta_1, \theta_2, \ldots, \theta_t$. Then we have

$$\begin{aligned} 0 &= A(\sum_{i=1}^{t} \theta_i B u_i) \\ &= \sum_{i=1}^{t} \theta_i (ABu_i) = \sum_{i=1}^{t} \theta_i(\lambda u_i) = \lambda \sum_{i=1}^{t} \theta_i u_i, \end{aligned}$$

which implies

$$\sum_{i=1}^{n} \theta_i u_i = 0,$$

as $\lambda \neq 0$. Since $u_1, u_2, \ldots, u_t$ are linearly independent, it follows that $\theta_1 = \theta_2 = \ldots = \theta_t = 0$, from which we have the desired linear independence of $Bu_1, Bu_2, \ldots, Bu_t$. In order to establish the claim, let $\mu_1, \mu_2, \ldots, \mu_r$ be the distinct eigenvalues of AB among $\lambda_1, \lambda_2, \ldots, \lambda_m$. If no $\mu_i = 0$, the argument will be a little easier. We will take up the more difficult case: one of the eigenvalues of AB is zero. Assume that $\mu_r = 0$. We will produce two tables. The first table summarizes all the information about the eigenvalues of AB and the second table deals with the matrix BA, where $m = t_1 + \ldots + t_r$.

Eigenvalues of AB	Multiplicity	Linearly independent eigenvectors
μ_1	t_1	$u_1^{(1)}, u_2^{(1)}, \ldots, u_{t_1}^{(1)}$
μ_2	t_2	$u_1^{(2)}, u_2^{(2)}, \ldots, u_{t_2}^{(2)}$
.	.	...
μ_{r-1}	t_{r-1}	$u_1^{(r-1)}, u_2^{(r-1)}, \ldots, u_{t_{r-1}}^{(r-1)}$
$\mu_r = 0$	t_r	Immaterial

Eigenvalues of BA	Multiplicity	Linearly independent eigenvectors
μ_1	t_1	$Bu_1^{(1)}, Bu_2^{(1)}, \ldots, Bu_{t_1}^{(1)}$
μ_2	t_2	$Bu_1^{(2)}, Bu_2^{(2)}, \ldots, Bu_{t_2}^{(2)}$
.	.	...
μ_{r-1}	t_{r-1}	$Bu_1^{(r-1)}, Bu_2^{(r-1)}, \ldots, Bu_{t_{r-1}}^{(r-1)}$
$\mu_r = 0$	$t_r + (n-m)$	Immaterial

Since Rank(AB)=Rank(BA)=number of non-zero eigenvalues of AB= number of non-zero eigenvalues of BA, the second table follows. Consequently, the eigenvalues of BA are $\lambda_1, \lambda_2, \ldots, \lambda_m, 0, 0, \ldots, 0$ ($n-m$ zeros). There may be more zeros among $\lambda_1, \lambda_2, \ldots, \lambda_m$.

Since AB is Hermitian, $AB = U\Delta U^*$, where $\Delta = \text{diag}\{\lambda_1, \lambda_2, \ldots, \lambda_m\}$ and U is unitary. Also, there exists a unitary matrix V such that

$$BA = V \begin{bmatrix} \Delta & 0 \\ 0 & 0 \end{bmatrix} V^*.$$

Partition $V = (V_1|V_2)$ where V_1 is of order $n \times m$. Note that V_1 has orthonormal columns, i.e., $V_1^* V_1 = I_m$.

Now $\Delta = U^*ABU$, which gives

$$BA = V_1 \Delta V_1^* = V_1 U^* ABUV_1^*.$$

Now define $A_1 = Y^*A$ and $A_2 = BY$ with $Y = UV_1^*$.

Let us list the properties of A_1 and A_2.

a. Each of A_1 and A_2 is of order $n \times n$.

b. A_1 and A_2 commute. As a matter of fact,

$$A_1A_2 = Y^*ABY = BA \quad \text{and} \quad A_2A_1 = BYY^*A = BA.$$

c. A_1A_2 and A_2A_1 are non-negative definite.

d. The singular values of A_1 are $\sigma_1(A), \sigma_2(A), \ldots, \sigma_m(A), 0, 0, \ldots, 0$ ($n - m$ zeros), since the singular values of A_1 are the positive square roots of the eigenvalues of $A_1^*A_1 = A^*YY^*A = A^*A$.

e. The singular values of A_2 are $\sigma_1(B), \sigma_2(B), \ldots, \sigma_m(B), 0, 0, \ldots, 0$ ($n - m$ zeros).

f. $\text{Tr}(A_1A_2) = \text{tr}(AB)$.

Thus we are justified in assuming that $m = n$ and that A and B commute in the statement of the theorem.

Let us restate the conditions of the theorem unequivocally.

Let A and $B \in \mathbf{M}_m$. Assume that A and B commute and that AB is non-negative definite.

Step 2 needs to be reasserted. We want to find two matrices A_0 and B_0 each of order $m \times m$ with the following properties.

a. A_0 and B_0 are non-negative definite.

b. A_0 and B_0 commute.

c. The singular values of A_0 are $\sigma_1(A), \sigma_2(A), \ldots, \sigma_m(A)$. The singular values of B_0 are $\sigma_1(B), \sigma_2(B), \ldots, \sigma_m(B)$.

d. $\text{Tr}(AB) = \text{tr}(A_0B_0)$.

4. We are now working under the simplified scenario of Step 3. Let $\mu_1, \ldots, \mu_r$ be the distinct eigenvalues of AB with multiplicities $t_1, \ldots, t_r$ adding up to m. The information is summarized in the following table.

Eigenvalues of AB	Multiplicity	Orthonormal eigenvectors	Space spanned
μ_1	t_1	$u_1^{(1)}, u_2^{(1)}, \ldots, u_{t_1}^{(1)}$	$\mathbf{M}_1$
μ_2	t_2	$u_1^{(2)}, u_2^{(2)}, \ldots, u_{t_2}^{(2)}$	$\mathbf{M}_2$
.	.	$\ldots \quad \ldots$	.
μ_r	t_r	$u_1^{(r)}, u_2^{(r)}, \ldots, u_{t_r}^{(r)}$	$\mathbf{M}_r$

The subspaces spanned have the following properties.

a. $\mathbf{M}_1 \oplus \mathbf{M}_2 \oplus \ldots \oplus \mathbf{M}_r = \mathbf{C}^m$.

b. $\mathbf{M}_i \perp \mathbf{M}_j$ for every $i \neq j$.

c. If $u \in \mathbf{M}_k$ for some $1 \leq k \leq r$, then $Au \in \mathbf{M}_k$. For this, it suffices to show that $Au_i^{(k)} \in \mathbf{M}_k$ for every $1 \leq i \leq t_k$. Note that $(AB)(Au_i^{(k)}) = A(ABu_i^{(k)}) = A(\mu_k u_i^{(k)}) = \mu_k(Au_i^{(k)})$, implying that $Au_i^{(k)}$ is an eigenvector of AB corresponding to the eigenvalue μ_k. Thus $Au_i^{(k)} \in \mathbf{M}_k$.

d. If $u \in \mathbf{M}_k$ for some $1 \leq k \leq r$, then $Bu \in \mathbf{M}_k$.

Let U consist of all the eigenvectors compiled in the table above:

$$U = (u_1^{(1)}, \ldots, u_{t_1}^{(1)}, u_1^{(2)}, \ldots, u_{t_2}^{(2)}, \ldots, u_1^{(r)}, \ldots, u_{t_r}^{(r)}).$$

Note that U is a unitary matrix. Let $\hat{A} = U^*AU$ and $\hat{B} = U^*BU$. Observe that:

α. The matrix $\hat{A}$ is block-diagonal with i-th diagonal block matrix A_i being of order $t_i \times t_i$.

β. The matrix $\hat{B}$ is block-diagonal with i-th diagonal block matrix B_i being of order $t_i \times t_i$.

γ. $\hat{A}\hat{B} = U^*AUU^*BU = U^*ABU$ is block diagonal with the i-th diagonal matrix $\mu_i I_{t_i}$, where I_{t_i} is a unit matrix of order t_i.

The properties α, β, and γ follow from the properties enunciated under a, b, c, and d above.

δ. $\hat{A}$ and $\hat{B}$ commute.

ϵ. $\hat{A}\hat{B}$ is non-negative definite.

φ. A_i and B_i commute for each i. More precisely, $A_iB_i = \mu_i I_{t_i}$ for each i.

ξ. A_iB_i is non-negative definite.

η. The singular values of A and $\hat{A}$ are identical. The singular values of $\hat{B}$ and B are identical.

ζ. The singular values of $\hat{A}$ are the same as the singular values of $A_1, A_2, \ldots, A_r$ put together. The singular values of $\hat{B}$ are the same as the singular values of $B_1, B_2, \ldots, B_r$ put together.

For each $1 \leq i \leq r$, if only we can find a permutation τ_i of $\{1, 2, \ldots, t_i\}$ such that

$$\operatorname{tr}(A_iB_i) = \sum_{j=1}^{t_i} \sigma_j(A_i)\sigma_{\tau_i(j)}(B_i),$$

our goal will be accomplished. In line with Steps 1 and 2, if we can find two matrices A_{i0} and B_{i0} with the following properties,

a. A_{i0} and B_{i0} are non-negative definite,
b. A_{i0} and B_{i0} commute,
c. the singular values of A_i and A_{i0} are identical, the singular values of B_i and B_{i0} are identical,
d. $\operatorname{tr}(A_iB_i) = \operatorname{tr}(A_{i0}B_{i0})$,

our mission will be accomplished.

5. In view of Steps 1,2,3 and 4, the assumptions of the theorem can be simplified as follows.

a. $m = n$.
b. A and B commute.
c. $AB = \lambda I_m$ for some $\lambda \geq 0$.

The goal is now to find two matrices A_0 and B_0 with the following properties.

α. A_0 and B_0 are non-negative definite.

β. A_0 and B_0 commute.

γ. The singular values of A and A_0 are identical. The singular values of B and B_0 are identical.

δ. $\operatorname{tr}(AB) = \operatorname{tr}(A_0B_0)$.

Case 1. $\lambda > 0$. By **P 10.7.5**, there exists a unitary matrix U such that AU is non-negative definite. Define $A_0 = AU$ and $B_0 = U^*B$. The matrices A_0 and B_0 have the following properties.

α. A_0 and B_0 are non-negative definite. The case of A_0 is clear.

Note that $B_0 = U^*B = U^*(\lambda A^{-1}) = \lambda U^*A^{-1} = \lambda(AU)^{-1}$, from which it follows that B_0 is non-negative definite.

β. A_0 and B_0 commute. Note that $A_0B_0 = AB = \lambda I_m = B_0A_0$.

γ. The singular values of A and A_0 are identical. The singular values of B and B_0 are identical.

δ. $\text{tr}(AB) = \text{tr}(A_0B_0)$. The goal is achieved.

Case 2. $\lambda = 0$. We can proceed in exactly the same way as in Case 1. We will have the matrices A_0 and B_0 with the properties α, β, γ, and δ but with one exception. We do not know whether or not B_0 is non-negative definite. We need to do some more work.

Let $\mu_1, \mu_2, \ldots, \mu_r$ be the distinct eigenvalues of A_0 with multiplicities $t_1, t_2, \ldots, t_r$. If none of the eigenvalues of A_0 is zero, then A_0 is nonsingular and $B_0 = 0$, which is non-negative definite. The desired objective is achieved. Assume that one of the eigenvalues of A_0 is zero. Take $\mu_r = 0$, for example. Let us summarize some of this information in the following table, where $m = t_1 + \ldots + t_r$.

Eigenvalues of A	Multiplicity	Orthonormal eigenvectors	Space spanned
μ_1	t_1	$u_1^{(1)}, u_2^{(1)}, \ldots, u_{t_1}^{(1)}$	$\mathbf{M}_1$
.	.	... ...	.
μ_r	t_r	$u_1^{(r)}, u_2^{(r)}, \ldots, u_{t_r}^{(r)}$	$\mathbf{M}_r$

Let u be any eigenvector of A that corresponds to a non-zero eigenvalue μ of A. We claim that $Bu = 0$. Note that

$$0 = ABu = B(Au) = B(\mu u) = \mu(Bu).$$

Since $\mu \neq 0$, it follows that $Bu = 0$. Let W be the matrix composed of the orthonormal eigenvectors of A listed above in the order they were written down as columns. It is clear that W is a unitary matrix. Note that by what we have noted down about Bu for eigenvectors u of A that correspond to non-zero eigenvalues of A, W^*A_0W is a block diagonal matrix with $\mu_i I_{t_i}$ in the i-th diagonal block, $i = 1, \ldots, r$, and

$$W^*B_0W = \begin{bmatrix} 0 & 0 \\ 0 & \tilde{B}_0 \end{bmatrix},$$

for some matrix $\tilde{B}_0$ of order $t_r \times t_r$ and appropriate dimensions for 0 matrices.

The matrices W^*A_0W and W^*B_0W would have met the requirements of α, β, γ, and δ under Case 1 but for the matrix $\tilde{B}_0$, which need not be non-negative definite. By **P 10.7.5**, there exists a unitary matrix W_0 of order $t_r \times t_r$ such that $\tilde{B}_0 W_0$ is non-negative definite. Define

$$A_{00} = W^*A_0W,$$

and

$$V = \begin{bmatrix} I_{t_1} & 0 & \ldots & 0 & 0 \\ 0 & I_{t_2} & \ldots & 0 & 0 \\ \cdot & \cdot & \ldots & \cdot & \cdot \\ 0 & 0 & \ldots & I_{t_{r-1}} & 0 \\ 0 & 0 & \ldots & 0 & W_0 \end{bmatrix}.$$

Note that V is unitary. Let

$$A_{00} = W^*A_0WV \text{ and } B_{00} = W^*B_0WV.$$

The following properties of A_{00} and B_{00} flow easily now.

α. A_{00} and B_{00} are non-negative definite.

β. A_{00} and B_{00} commute. In fact, $A_{00}B_{00} = 0$.

γ. The singular values of A_{00} and A_0 are identical. The singular values of A_0 and A are identical. The singular values of B_{00} and B_0 are identical. The singular values of B_0 and B are identical.

Our mission is successful.

As a consequence of von Neumann's theorem, one can solve another optimization problem. The details follow.

P 10.7.8 Let $A \in \mathbf{M}_{m,n}$ and $B \in \mathbf{M}_{n,m}$.

(1) Then

$$\sup_{U \in \mathbf{U}_n, V \in \mathbf{U}_m} \text{Re tr}(AUBV)$$

is attained at some unitary matrices $U = U_0 \in \mathbf{U}_n$ and $V = V_0 \in \mathbf{U}_m$.

(2) Further,

$$\max_{U \in \mathbf{U}_n, V \in \mathbf{U}_m} \text{Re tr}(AUBV) = \sum_{i=1}^{p} \sigma_i(A)\sigma_i(B),$$

where $p = \min\{m, n\}$.

PROOF. One can offer a topological proof of (1). One needs to observe that the spaces of unitary matrices are all compact. We will prove (2). Let U_0 and V_0 be a solution to the problem in (1). Since

$$\text{Re tr}((AU_0B)V_0) \geq \text{Re tr}((AU_0B)V)$$

for all unitary matrices V, it follows that AU_0BV_0 is non-negative definite. See **P 10.7.5**. In a similar vein, since

$$\text{Re tr}(AU_0BV_0) = \text{Re tr}((BV_0A)U_0) \geq \text{Re tr}((BV_0A)U)$$

for all unitary matrices U, it follows that BV_0AU_0 is non-negative definite. Let us focus on the matrices AU_0 and BV_0. The two matrices $(AU_0)(BV_0)$ and $(BV_0)(AU_0)$ are non-negative definite. By von Neumann's theorem, there exists a permutation τ of $\{1, 2, \ldots, p\}$ such that

$$\begin{aligned}\max_{U \in \mathbf{U}_n, V \in \mathbf{U}_m} \text{Re tr}(AUBV) &= \text{Re tr}(AU_0BV_0) = \text{tr}\,(AU_0BV_0) \\ &= \sum_{i=1}^{p} \sigma_i(AU_0)\sigma_{\tau(i)}(BV_0) \\ &= \sum_{i=1}^{p} \sigma_i(A)\sigma_{\tau(i)}(B).\end{aligned}$$

The proof will be complete if we can show that τ can be taken to be the identity permutation.

Assume, without loss of generality, that $m = n$. In Step 2 of the proof of von Neumann's theorem, for the matrices AU_0 and BU_0 we can find two matrices A_0 and B_0 each of order $m \times m$ with the following properties.

(1) A_0 and B_0 are non-negative definite.

(2) A_0 and B_0 commute.

(3) The singular values of A and A_0 are identical. The singular values of B and B_0 are identical.

Since A_0 and B_0 commute, we can find a unitary matrix W such that

$$\begin{aligned} W^* A_0 W &= \text{diag}\{\sigma_1(A), \sigma_2(A), \ldots, \sigma_m(A)\}, \\ W^* B_0 W &= \text{diag}\{\sigma_{\tau(1)}(B), \sigma_{\tau(2)}(B), \ldots, \sigma_{\tau(m)}(B)\}, \end{aligned}$$

for some permutation matrix τ of $\{1, 2, \ldots, m\}$. Let w_i be the i-th column vector of W. Each w_i is a common eigenvector for both A_0 and B_0 for the eigenvalues $\sigma_i(A_0)$ and $\sigma_{\tau(i)}(B_0)$, respectively. Let W_0 be a unitary matrix with the following properties.

$$W_0 w_1 = w_{\tau^{-1}(1)},\ W_0 w_2 = w_{\tau^{-1}(2)},\ \ldots,\ W_0 w_m = w_{\tau^{-1}(m)}.$$

(Exercise: Construct W_0 explicitly.) Let us determine the eigenvalues of $A_0 W_0^* B_0 W_0$. Observe that

$$\begin{aligned} A_0 W_0^* B_0 W_0 w_1 &= A_0 W_0^* B_0 w_{\tau^{-1}(1)} = A_0 W_0^* (\sigma_1(B_0) w_{\tau^{-1}(1)}) \\ &= \sigma_1(B_0) A_0 w_1 = \sigma_1(B_0) \sigma_1(A_0) w_1. \end{aligned}$$

Thus $\sigma_1(A_0)\sigma_1(B_0)$ is an eigenvalue of $A_0 B_0$ with an eigenvector w_1. Likewise, it follows that the eigenvalues of $A_0 B_0$ are: $\sigma_i(A_0)\sigma_i(B_0), i = 1, 2, \ldots, m$. Thus

$$\begin{aligned} \sum_{i=1}^{m} \sigma_i(A_0)\sigma_i(B_0) &= \text{tr}(A_0 W_0^* B_0 W_0) \\ &\leq \max_{U \in \mathbf{U}_m, V \in \mathbf{U}_m} \text{Re tr}(A_0 U B_0 V) \\ &= \sum_{i=1}^{m} \sigma_i(A_0)\sigma_{\tau(i)}(B_0) \leq \sum_{i=1}^{m} \sigma_i(A_0)\sigma_i(B_0). \end{aligned}$$

Consequently,

$$\max_{U\in\mathbf{U}_m, V\in\mathbf{U}_m} \operatorname{tr}(A_0UB_0V) = \sum_{i=1}^{m} \sigma_i(A_0)\sigma_i(B_0) = \sum_{i=1}^{m} \sigma_i(A)\sigma_i(B).$$

This completes the proof.

Complements

10.7.1 Let $A \in \mathbf{M}_n$ have singular values $\sigma_n \le \ldots \le \sigma_1$. Denote by r_i the Euclidean norm of the i-th row of $A, i = 1,\ldots,n$, and by $R_1 \le R_2 \le \ldots \le R_n$ the ordered values of r_i. Show that

$$\sum_{i=1}^{k} \sigma_{n-i+1}^2 \le \sum_{i=1}^{k} R_i^2 \text{ for } k = 1,2,\ldots,n.$$

Write down a similar upper bound for norms of columns.

10.7.2 (Majorization Theorem) Let $A \in M_n$ be Hermitian. The vector of diagonal entries of A majorizes the vector of eigenvalues of A. [*Hint:* The theorem can be proved by induction by assuming that the result is valid for Hermitian matrices of dimension k for all $k \le n-1$ and using the Interlace Theorem (**P 10.2.1**).

10.7.3 Let A and $B \in \mathbf{M}_n$ be Hermitian matrices with ordered eigenvalues $\lambda_1 \ge \ldots \ge \lambda_n$ and $\mu_1 \ge \ldots \ge \mu_n$. Further let $\nu_1 \ge \ldots \ge \nu_n$ be the eigenvalues of $A - B$. Show that the vector $(\lambda_1 - \mu_1, \ldots, \lambda_n - \mu_n)$ is majorized by the vector $(\nu_1, \ldots, \nu_n)$.

10.7.4 (Weyl's Theorem) Let $A, B \in \mathbf{M}_n$ be Hermitian and $\rho(B)$, the rank of B, be less than or equal to r. Let the eigenvalues $\lambda_i(A), \lambda_i(B)$ and $\lambda_i(A+B)$ be arranged in decreasing order. Then

(1) $\lambda_i(A+B) \le \lambda_{i-r}(A) \le \lambda_{i-2r}(A+B),\ i = 2r+1,\ldots,n.$

(2) $\lambda_i(A) \le \lambda_{i-r}(A+B) \le \lambda_{i-2r}(A),\ i = 2r+1,\ldots,n.$

(3) If $A = U\Lambda U^*$ with $U = (u_1|\ldots|u_n) \in \mathbf{M}_n$ unitary and $\Lambda = \operatorname{diag}(\lambda_1,\ldots,\lambda_n)$ with $\lambda_1 \ge \ldots \ge \lambda_n$ and if

$$B = \lambda_1 u_1 u_1^* + \ldots + \lambda_r u_r u_r^*$$

then $\lambda_{\max}(A - B) = \lambda_{r+1}(A)$.

10.7.5 (Analogue of Weyl's result of Complement 10.1.2 for singular values). Let $A, B \in \mathbf{M}_{m,n}$, $q = \min\{m, n\}$ and σ_i denote singular values. Show that, using 10.7.4,

$$\sigma_{r+s-1}(A+B) \leq \sigma_r(A) + \sigma_s(B).$$

In particular,

$$\sigma_1(A+B) \leq \sigma_1(A) + \sigma_1(B)$$
$$\sigma_q(A+B) \leq \min\{\sigma_q(A) + \sigma_1(B),\ \sigma_1(A) + \sigma_q(B)\}.$$

10.7.6 Let $A, B \in \mathbf{M}_{m,n}$ and $q = \min\{m, n\}$. Show that

$$\sigma_{r+s-1}(AB^*) \leq \sigma_r(A)\sigma_s(B).$$

10.7.7 Let $A \in \mathbf{M}_n$ be a Hermitian matrix with non-negative eigenvalues $\lambda_1 \geq \ldots \geq \lambda_n \geq 0$. Show that for each $r = 1, \ldots, n$, the product $\lambda_n \lambda_{n-1} \ldots \lambda_{n-r+1}$ is less than or equal to the product of the r smallest main diagonal entries of A.

Note: References to material in this Chapter are: Horn and Johnson (1985, 1991), Rao (1973c) and von Neumann (1937).

CHAPTER 11

MATRIX APPROXIMATIONS

The basic line of inquiry in this chapter proceeds along the following lines. Let **A** be a given collection of matrices each of order $m \times n$. Let B be a given matrix of order $m \times n$. Determine a matrix A in **A** such that A is closest to B. The notion of one matrix being too close to another can be sanctified with the introduction of the notion of a norm on the spaces of matrices. In this chapter, we will introduce the notion of a norm on a general vector space and establish a variety of matrix approximation theorems.

11.1. Norm on a Vector Space

Let **V** be a vector space either over the field **C** of complex numbers or the field **R** of real numbers. The notion of a norm on the vector space **V** is at the heart of many a development in this chapter. In what follows, we will assume that the underlying field is the field **C** of complex numbers. If **R** is the underlying field of the vector space, we simply replace **C** by **R** in all the deliberations with obvious modifications.

In Chapter 2, we have already introduced the concept of a norm arising out of an inner product. The definition of norm introduced here is more general which need not be generated by an inner product.

DEFINITION 11.1.1. A map $\|\cdot\|$ from **V** to **R** is said to be a vector norm on **V** if it has the following properties.

(1) $\|x\| \geq 0$ for all x in **V** and $\|x\| = 0$ if and only if $x = 0$.

(2) $\|\alpha x\| = |\alpha| \, \|x\|$ for all x in **V** and $\alpha \in$ **C**.

(3) $\|x + y\| \leq \|x\| + \|y\|$ for all x and y in **V**.

The pair $(\mathbf{V}, \|\cdot\|)$ is called a normed vector space. Using the norm $\|\cdot\|$ on the vector space **V**, one can define a distance function $d(\cdot,\cdot)$

between any two vectors of **V**. More precisely,

$$d(x, y) = \|x - y\|, \; x, y \in \mathbf{V}.$$

In view of the definition of the distance $d(\cdot, \cdot)$, one can interpret $\|x\|$ as the distance between the vectors x and 0, or simply, the length of x. Property (3) is often called the triangle inequality satisfied by the norm and it implies that $d(x, y) \leq d(x, z) + d(z, y)$ for any three vectors $x, y,$ and z in the vector space **V**. Some examples are in order.

EXAMPLE 11.1.2. Take $\mathbf{V} = \mathbf{C}^n$. Let $x = (x_1, \ldots, x_n) \in \mathbf{V}$. The following maps are all norms on the vector space **V**:

(1) $\|x\|_\infty = \max_{1 \leq i \leq n} |x_i|$ (L_∞ – norm).
(2) $\|x\|_1 = \sum_{i=1}^n |x_i|$ (L_1 – norm).
(3) $\|x\|_2 = (\sum_{i=1}^n |x_i|^2)^{1/2}$ (L_2 – norm or Euclidean norm).
(4) $\|x\|_p = (\sum_{i=1}^n |x_i|^p)^{1/p}$ (L_p – norm with $p \geq 1$ fixed).

The norms presented above are all geometrically different. Let us amplify this statement. Let $\|\cdot\|$ be any given norm on a vector space **V** over the field **C** of complex numbers. The unit ball **O** in **V** is defined by

$$\mathbf{O} = \{x \in \mathbf{V} : \|x\| \leq 1\}.$$

The following exercise is designed to make one realize how the norms presented above are all geometrically different.

EXERCISE 11.1.3. Take $\mathbf{V} = \mathbf{R}^2$. View **V** as a vector space over **R**. Graph unit balls in each of the normed vector spaces: $(\mathbf{V}, \|\cdot\|_\infty)$, $(\mathbf{V}, \|\cdot\|_1)$, $(\mathbf{V}, \|\cdot\|_2)$, $(\mathbf{V}, \|\cdot\|_{1.5})$, and $(\mathbf{V}, \|\cdot\|_3)$.

There are many more norms one can introduce on the vector space $\mathbf{V} = \mathbf{C}^n$. For example, let $A = (a_{ij})$ be any positive definite matrix of order $n \times n$ with complex entries. Let $x' = (x_1, \ldots, x_n) \in \mathbf{V}$. Define

$$\|x\|_A = (\sum_{i=1}^n \sum_{j=1}^n a_{ij} x_i \bar{x}_j)^{1/2} = (x^* A x)^{1/2}.$$

Then $\|\cdot\|_A$ is a norm on the vector space **V**. (Recall that $\bar{x}$ is the conjugate of the complex number x.)

There are other ways to generate new norms. One can add two norms to get a new norm. One can multiply a norm by a fixed positive number to get a new norm.

The examples of norms given for the vector space $\mathbf{C}^n$ can be transferred to any vector space $\mathbf{V}$ over the field $\mathbf{C}$ of complex numbers. Let $\{v_1, v_2, \dots, v_n\}$ be a basis of the vector space $\mathbf{V}$. Let $v \in \mathbf{V}$. One can write $v = x_1 v_1 + x_2 v_2 + \dots + x_n v_n$, with $x_1, x_2, \dots, x_n$ in $\mathbf{C}$. Further, the above representation is unique. Let $\|\cdot\|$ be any norm on $\mathbf{C}^n$. Define (abuse of notation) $\|v\| = \|x\|$, where $x' = (x_1, x_2, \dots, x_n) \in \mathbf{C}^n$.

Complements

11.1.1 Take $\mathbf{V} = \mathbf{C}^n$. Fix $0 < p < 1$. Let $x' = (x_1, x_2, \dots, x_n) \in \mathbf{V}$. Why is the map $\|\cdot\|_p$ defined below not a norm on $\mathbf{V}$

$$\|x\|_p = (\sum_{i=1}^{n} |x_i|^p)^{1/p}.$$

11.1.2 Let $\mathbf{V} = \mathbf{R}^2$. Graph the unit ball in the real normed space $(\mathbf{V}, \|\cdot\|_\infty + \|\cdot\|_1)$.

11.2. Norm on Spaces of Matrices

Let m and n be two fixed positive integers. Let $\mathbf{M}_{m,n}$ be the collection of all matrices $A = (a_{ij})$ of order $m \times n$ with complex entries a_{ij}'s. It is clear that $\mathbf{M}_{m,n}$ is a vector space with the usual operations of addition of matrices and scalar multiplication of matrices over the field $\mathbf{C}$ of complex numbers. As a matter of fact, the space $\mathbf{M}_{m,n}$ can be identified with the vector space $\mathbf{C}^{mn}$ by arranging the entries of each matrix A in $\mathbf{M}_{m,n}$ as an mn-tuple in some order. The following are some of the natural norms one can introduce on the vector space $\mathbf{M}_{m,n}$ analogous to those introduced in Example 11.1.2.

EXAMPLE 11.2.1. Let $A = (a_{ij}) \in \mathbf{M}_{m,n}$.

(1) $\|A\|_\infty = \max_{1 \le i \le m, 1 \le j \le n} |a_{ij}|$ (L_∞ – norm);
(2) $\|A\|_F = (\sum_{i=1}^{m} \sum_{j=1}^{n} |a_{ij}|^2)^{1/2}$ (Frobenius norm);
(3) $\|A\|_p = (\sum_{i=1}^{m} \sum_{j=1}^{n} |a_{ij}|^p)^{1/p}$ (L_p – norm with $p \ge 1$ fixed).

Let us now concentrate on the vector space $\mathbf{M}_n$. (We take $m = n$.) (For simplicity, we abbreviate $\mathbf{M}_{n,n}$ as $\mathbf{M}_n$.) The vector space $\mathbf{M}_n$

has an additional algebraic property: if $A, B \in \mathbf{M}_n$, then the product $AB \in \mathbf{M}_n$. It is only natural to expect a norm on $\mathbf{M}_n$ to respect the product operation of matrices.

DEFINITION 11.2.2. A norm $\|\cdot\|$ on $\mathbf{M}_n$ is said to be a *matrix norm* on $\mathbf{M}_n$ if $\|AB\| \leq \|A\|\,\|B\|$ for all A and B in $\mathbf{M}_n$.

From the definition of a matrix norm, the following are transparent:

(1) $\|I_n\| \geq 1$ (I_n is the identity matrix of order $n \times n$).
(2) $\|A^k\| \leq (\|A\|)^k$ for any A in $\mathbf{M}_n$ and positive integer k.
(3) $\|A^{-1}\| \geq (\|A\|)^{-1}$ for any nonsingular matrix A in $\mathbf{M}_n$.

Why do we need to introduce the matrix norm? Is the plain norm not good enough to measure the magnitude of matrices? The clue lies in Property (2) spelled out above. The limiting behavior of powers of a matrix, say A, is the subject of many an inquiry in applied mathematics and statistics. Suppose for some suitable matrix norm $\|\cdot\|$ we discover that $\|A\| < 1$. By property (2), it transpires that A^k converges to zero as $k \to \infty$.

Are there not other ways of measuring the magnitude of matrices? Surely, eigenvalues must have a say on the magnitude of matrices. Let $\lambda_1, \lambda_2, \ldots, \lambda_n$ be the eigenvalues of a matrix A.

DEFINITION 11.2.3. The spectral radius of A is defined to be the quantity

$$\rho_s(A) = \max_{1 \leq i \leq n} |\lambda_i|.$$

One could propose the spectral radius as a measure of the magnitude of matrices. But it does not satisfy any of the properties of a norm.

(1) $\rho_s(A)$ could be equal to zero without A being equal to zero. For example, look at the matrix

$$A = \begin{bmatrix} 0 & 1 \\ 0 & 0 \end{bmatrix}.$$

(2) It is quite possible that $\rho_s(A+B) > \rho_s(A) + \rho_s(B)$ (failure of the triangle inequality). For example, look at the matrices,

$$A = \begin{bmatrix} 0 & 1 \\ 0 & 0 \end{bmatrix} \text{ and } B = \begin{bmatrix} 0 & 0 \\ 1 & 0 \end{bmatrix}.$$

(3) It is quite possible that $\rho_s(AB) > \rho_s(A)\rho_s(B)$. For example, look at the matrices,

$$A = \begin{bmatrix} 0 & 1 \\ 1 & 0 \end{bmatrix} \text{ and } B = \begin{bmatrix} 1 & 1 \\ 0 & 1 \end{bmatrix}.$$

May be the eigenvalues are not a good choice to develop a norm. The singular values are the ones we need to use to define the norm or magnitude of matrices. We will pursue this task a little later. However, we will not abandon the spectral radius. It has some uses.

There is a wonderful relationship between the spectral radius of a matrix and its matrix norm. This is stated below.

P 11.2.4 Let $A \in \mathbf{M}_n$ and $\|\cdot\|$ any matrix norm on $\mathbf{M}_n$. Then

$$\rho_s(A) \leq \|A\|.$$

PROOF. By the very definition of the spectral radius, there exists a scalar λ and a non-zero column vector x such that $Ax = \lambda x$ and $\rho_s(A) = |\lambda|$. Let B be the matrix of order $n \times n$ such that every column of B is the same vector x. Observe that $AB = \lambda B$. Since $\|\cdot\|$ is a matrix norm,

$$\|A\| \, \|B\| \geq \|AB\| = \|\lambda B\| = |\lambda| \, \|B\|,$$

from which it follows that $\rho_s(A) \leq \|A\|$.

A matrix norm is useful in investigating the existence of an inverse of a matrix. First, we need to examine the precise relationship between the individual entries of a matrix and its matrix norm. Then we need a criterion for the convergence of a matrix series.

Let I_{ij} be the matrix of order $n \times n$ such that the (i,j)-th entry of I_{ij} is unity and the rest of the entries are all zeros. Let $\|\cdot\|$ be a matrix norm on $\mathbf{M}_n$. Let

$$\theta = \max_{1 \leq i \leq n, 1 \leq j \leq n} \|I_{ij}\|.$$

Let $A = (a_{ij}) \in \mathbf{M}_n$. Then one can verify that $I_{ij}AI_{ij} = a_{ij}I_{ij}$ for all i and j.

P 11.2.5 Let $A \in \mathbf{M}_n$ and $\|\cdot\|$ a matrix norm on $\mathbf{M}_n$. Then $|a_{ij}| \leq \theta\|A\|$ for all i and j.

PROOF. Note that

$$|a_{ij}|\,\|I_{ij}\| = \|a_{ij}I_{ij}\| = \|I_{ij}AI_{ij}\| \le \|I_{ij}\|^2\|A\|,$$
$$\Rightarrow |a_{ij}| \le \|I_{ij}\|\,\|A\| \le \theta\|A\|.$$

P 11.2.6 Let $A \in \mathbf{M}_n$, $\|\cdot\|$ a matrix norm on $\mathbf{M}_n$, and $\{a_k\}, k \ge 0$ a sequence of scalars. Then the series $\sum_{k\ge 0} a_k A^k$ converges if the series $\sum_{k\ge 0} |a_k|\,\|A\|^k$ of real numbers converges. (By convention, $A^0 = I_n$.)

PROOF. Let $A^k = (a_{ij}^{(k)})$ for each $k \ge 0$. By **P 11.2.5**,

$$|a_{ij}^{(k)}| \le \theta\|A^k\|$$

for all i and j, and $k \ge 1$. Since $\|A^k\| \le \|A\|^k$, the series $\sum_{k\ge 0}|a_k|\,|a_{ij}^{(k)}|$ of real numbers converges. Hence $\sum_{k\ge 0} a_k a_{ij}^{(k)}$ converges.

Now we can settle the existence of inverse of a matrix.

P 11.2.7 Let $A \in \mathbf{M}_n$ and $\|\cdot\|$ a matrix norm on $\mathbf{M}_n$. If $\|I_n - A\| < 1$, then A^{-1} exists and is given by the series

$$A^{-1} = \sum_{k\ge 0}(I_n - A)^k.$$

PROOF. From **P 11.2.6**, note that the series $\sum_{k\ge 0}(I_n - A)^k$ converges. Let N be any positive integer. Then

$$\begin{aligned} A\Big(\sum_{k=0}^{N}(I_n - A)^k\Big) &= (I_n - (I_n - A))\sum_{k=0}^{N}(I_n - A)^k \\ &= I_n - (I_n - A)^{N+1}. \end{aligned}$$

Since $\|I_n - A\| < 1$, $(I_n - A)^{N+1}$ converges to 0 as $N \to \infty$. Consequently,

$$A\Big(\sum_{k\ge 0}(I_n - A)^k\Big) = I_n,$$

from which the desired result follows.

Not all norms are matrix norms. It will be instructive to check how many of the norms introduced in Example 11.2.1 pass muster. See Complements 11.2.1 to 11.2.3.

We need more examples of matrix norms. Instead of plunking down some examples, it would be useful to develop a method of generating a variety of norms on $\mathbf{M}_n$ using norms on the vector space $\mathbf{C}^n$. Start with any norm $\|\cdot\|$ on $\mathbf{C}^n$. For each $A \in \mathbf{M}_n$, define

$$\|A\|_{in} = \sup_{x\in\mathbf{C}^n, x\neq 0} \frac{\|Ax\|}{\|x\|}. \tag{11.2.1}$$

As usual, the vectors $x \in \mathbf{C}^n$ are regarded as column vectors so that matrix multiplication Ax makes sense. If we can show that $\|A\|_{in}$ is finite, we can say that

$$\|Ax\| \leq \|A\|_{in}\|x\| \tag{11.2.2}$$

for every vector $x \in \mathbf{C}^n$. The eventual goal is to demonstrate that the map $\|\cdot\|_{in}$ is a matrix norm on $\mathbf{M}_n$. If this is the case, one can call $\|\cdot\|_{in}$, the *induced matrix norm* in $\mathbf{M}_n$ induced by the norm $\|\cdot\|$ on $\mathbf{C}^n$. The letters "*in*" in the norm are an abbreviation of the phrase "induced norm". It is clear that $\|A\|_{in}$ is non-negative. Observe that

$$\|A\|_{in} = \sup_{x\in\mathbf{C}^n, \|x\|=1} \|Ax\|.$$

Using topological arguments, one can show that there exists a vector $x_0 \in \mathbf{C}^n$ (depending on A) such that

$$\|x_0\| = 1 \quad \text{and} \quad \|A\|_{in} = \|Ax_0\|. \tag{11.2.3}$$

This demonstrates that $\|A\|_{in}$ is finite. (The topological arguments use the facts that the map $\|\cdot\|$ is a continuous function from $\mathbf{C}^n$ to $\mathbf{R}$ and that the set $\{x \in \mathbf{C}^n : \|x\| = 1\}$ is a compact subset of $\mathbf{C}^n$ in the usual topology of $\mathbf{C}^n$.) If $A = 0$, then $\|A\|_{in} = 0$. Conversely, suppose $\|A\|_{in} = 0$. This implies that $Ax = 0$ for every vector $x \in \mathbf{C}^n$ with $\|x\| = 1$. Hence $A = 0$. (Why?) If α is any complex number, then

$$\|\alpha A\|_{in} = \sup_{x\in\mathbf{C}^n, \|x\|=1} \|\alpha A_x\| = |\alpha| \sup_{x\in\mathbf{C}^n, \|x\|=1} \|Ax\| = |\alpha|\,\|A\|.$$

We now set upon the triangle inequality. Let A and $B \in \mathbf{M}_n$. Note that for each vector $x \in \mathbf{C}^n$, $\|(A+B)x\| \leq \|Ax\| + \|Bx\|$. From this inequality, it follows that $\|A+B\|_{in} \leq \|A\|_{in} + \|B\|_{in}$. Finally, we need to show that $\|AB\|_{in} \leq (\|A\|_{in})(\|B\|_{in})$. By what we have pointed out in (11.2.3), there exists a vector x_0 (of course, depending on AB) such that $\|x_0\| = 1$ and

$$\begin{aligned}\|AB\|_{in} &= \|ABx_0\| = \|A(Bx_0)\| \leq \|A\|_{in}\|Bx_0\| \\ &\leq \|A\|_{in}\|B\|_{in}\|x_0\| = \|A\|_{in}\|B\|_{in}.\end{aligned}$$

Thus we have shown that $\|\cdot\|_{in}$ is indeed a matrix norm.

The definition of the induced norm is something one introduces routinely in functional analysis. The matrix A from $\mathbf{M}_n$ can be viewed as a linear operator from the normed linear space $(\mathbf{C}^n, \|\cdot\|)$ to the normed linear space $(\mathbf{C}^n, \|\cdot\|)$. The definition of the induced norm of A is precisely the operator norm of A.

There is no dearth of matrix norms. Every norm on $\mathbf{C}^n$ induces a matrix norm on $\mathbf{M}_n$. Some examples are included below.

P 11.2.8 For $A = (a_{ij}) \in \mathbf{M}_n$, define

$$\|A\|_{\infty,in} = \max_{1\leq i\leq n} \sum_{j=1}^{n} |a_{ij}|.$$

(First we form the row sums of absolute values of entries of A and then take the maximum of the row sums to compute $\|A\|_{\infty,in}$.) Then $\|\cdot\|_{\infty,in}$ is a matrix norm on $\mathbf{M}_n$.

PROOF. The main idea of the proof is to show that $\|\cdot\|_{\infty,in}$ is the induced matrix norm on $\mathbf{M}_n$ induced by the L_∞-norm, $\|\cdot\|_\infty$ on $\mathbf{C}^n$. For any given matrix $A \in \mathbf{M}_n$, let us compute $\|Ax\|_\infty$ on $\mathbf{C}^n$. Let $x' = (x_1, x_2, \ldots, x_n) \in \mathbf{C}^n$ be such that $\|x\|_\infty = \max_{1\leq i\leq n}|x_i| = 1$. Then

$$\|Ax\|_\infty = \max_{1\leq i\leq n} |\sum_{j=1}^{n} a_{ij}x_j| \leq \max_{1\leq i\leq n} \sum_{j=1}^{n} |a_{ij}|\,|x_j|$$

$$\leq (\max_{1\leq i\leq n} \sum_{j=1}^{n} |a_{ij}|)\|x\|_\infty = \max_{1\leq i\leq n} \sum_{j=1}^{n} |a_{ij}|. \qquad (11.2.4)$$

Let

$$\max_{1\leq i\leq n} \sum_{j=1}^{n} |a_{ij}| = \sum_{j=1}^{n} |a_{kj}| \quad \text{for some } 1 \leq k \leq n.$$

If we can show that there exists a vector $\alpha' = (\alpha_1, \alpha_2, \ldots, \alpha_n)$ in $\mathbf{C}^n$ such that $\|\alpha\|_\infty = 1$ and $\|A\alpha\|_\infty \geq \max_{1\leq i\leq n} \sum_{j=1}^{n} |a_{ij}|$, it would then imply that $\|A\|_{\infty,in} \geq \max_{1\leq i\leq n} \sum_{j=1}^{n} |a_{ij}|$ and the assertion of the proposition follows by using (11.2.4). Take for each $j = 1, 2, \ldots, n$

$$\alpha_j = \begin{cases} |a_{kj}|/a_{kj} & \text{if } \ a_{kj} \neq 0, \\ 0 & \text{if } \ a_{kj} = 0. \end{cases}$$

Then the k-th entry in the column vector of $A\alpha$ is given by $\sum_{j=1}^{n} |a_{kj}|$. Consequently,

$$\|A\alpha\|_\infty = \max_{1\leq i\leq n} |\sum_{j=1}^{n} a_{ij}\alpha_j| \geq \sum_{j=1}^{n} |a_{kj}| = \max_{1\leq i\leq n} \sum_{j=1}^{n} |a_{ij}|.$$

The norm given in **P 11.2.8** is helpful in answering affirmatively the invertibility of diagonally dominant matrices. First, we need a definition.

DEFINITION 11.2.9. A matrix $A = (a_{ij})$ of order $n \times n$ is said to be diagonally dominant if

$$|a_{ii}| > \sum_{j=1, j\neq i}^{n} |a_{ij}|, \ i = 1, \ldots, n.$$

As an example, let a and b be two scalars such that $|a| > (n-1)|b|$. Take $a_{ij} = a$ if $i = j$ and $a_{ij} = b$ if $i \neq j$. Every diagonal entry of A is equal to a and every off-diagonal entry is equal to b. Then A is a diagonally dominant matrix.

COROLLARY 11.2.10. Every diagonally dominant matrix is invertible.

Proof. Let $A = (a_{ij})$ be diagonally dominant. It then follows that every diagonal entry of A is non-zero. Let $D = \text{diag}\{a_{11}, a_{22}, \ldots, a_{nn}\}$. The matrix D is invertible. Let $B = I_n - D^{-1}A = (b_{ij})$. Every diagonal entry of B is zero. Let us compute

$$\|B\|_{\infty,in} = \max_{1\leq i\leq n} \sum_{j=1}^{n} |b_{ij}| = \max_{1\leq i\leq n} \sum_{j=1, j\neq i}^{n} |a_{ij}/a_{ii}| < 1.$$

By **P 11.2.7**, $I_n - B$ is invertible. But $I_n - B = D^{-1}A$. Hence A is invertible.

Now we focus on the matrix norm $\|\cdot\|_{1,in}$ on $\mathbf{M}_n$ induced by the norm $\|\cdot\|_1$ on $\mathbf{C}^n$. The following result focuses on this particular induced norm. This norm is analogous to the norm $\|\cdot\|_{\infty,in}$. The only difference is that column sums are involved in the norm instead of row sums.

P 11.2.11 For each matrix $A = (a_{ij})$ in $\mathbf{M}_n$, we have

$$\|A\|_{1,in} = \max_{1\leq j\leq n} \sum_{i=1}^{n} |a_{ij}|.$$

(First, the column sums of the absolute values of the entries of A are formed and then the maximum among the column sums is determined.)

Proof. Let $A = (a_1, a_2, \ldots, a_n)$, where a_i is the i-th column of A. Let $x' = (x_1, x_2, \ldots, x_n) \in \mathbf{C}^n$ be such that $\|x\|_1 = 1$. Observe that

$$\begin{aligned}\|Ax\|_1 &= \|x_1a_1 + x_2a_2 + \ldots + x_na_n\|_1 \\ &\leq \sum_{j=1}^{n} |x_j|\, \|a_j\|_1 = \sum_{j=1}^{n} |x_j| \sum_{i=1}^{n} |a_{ij}| \\ &\leq (\max_{1\leq j\leq n} \sum_{j=1}^{n} |a_{ij}|) \sum_{s=1}^{n} |x_s| = \max_{1\leq j\leq n} \sum_{i=1}^{n} |a_{ij}|.\end{aligned}$$

Consequently, $\|A\|_{1,in} \leq \max_{1\leq j\leq n} \sum_{i=1}^{n} |a_{ij}|$. Let δ_j be the i-th unit vector, i.e., a column vector in which the j-th entry is unity and the rest of the entries are zeros, $j = 1, 2, \ldots, n$. Note that $\|\delta_j\|_1 = 1$ and

$$\|A\delta_j\|_1 = \sum_{i=1}^{n} |a_{ij}| \leq \|A\|_{1,in} \quad \text{for each } j.$$

Hence $\max_{1\leq j\leq n} \sum_{i=1}^{n} |a_{ij}| \leq \|A\|_{1,in}$. This completes the proof.

There is another important induced matrix norm induced by the Euclidean norm on $\mathbf{C}^n$. This is such an important norm, let us devote some space to it. This norm is called the spectral norm on $\mathbf{M}_n$.

DEFINITION 11.2.12. For each $A \in \mathbf{M}_n$, the spectral norm of A is defined by

$$\|A\|_S = \sup_{x\in\mathbf{C}^n, \|x\|_2=1} \|Ax\|_2.$$

We need to compute precisely the spectral norm for any given matrix. The singular values of A play a pivotal role in the computation.

P 11.2.13 Let $A \in \mathbf{M}_n$ and $\lambda_1 \geq \lambda_2 \geq \ldots \geq \lambda_n \geq 0$ be the singular values of A. Then

$$\|A\|_S = \lambda_1,$$

i.e., the spectral norm of A is the largest singular value of A.

PROOF. Note that $\lambda_1^2, \lambda_2^2, \ldots, \lambda_n^2$ are the eigenvalues of A^*A. Let $x_{(1)}, x_{(2)}, \ldots, x_{(n)}$ be a set of orthonormal eigenvectors associated with the eigenvalues $\lambda_1^2, \lambda_2^2, \ldots, \lambda_n^2$. Let $x \in \mathbf{C}^n$ be such that $\|x\|_2 = 1$. Write $x = \alpha_1 x_{(1)} + \alpha_2 x_{(2)} + \ldots + \alpha_n x_{(n)}$ for some scalars $\alpha_1, \alpha_2, \ldots, \alpha_n$ in $\mathbf{C}$. The condition that $\|x\|_2 = 1$ implies that $\sum_{i=1}^{n} |\alpha_i|^2 = 1$. Observe that

$$\|Ax\|_2^2 = \|A(\sum_{i=1}^{n} \alpha_i x_{(i)})\|_2^2 = \|\sum_{i=1}^{n} \alpha_i \lambda_i x_{(i)}\|_2^2 = \sum_{i=1}^{n} |\alpha_i|^2 \lambda_i^2 \leq \lambda_1^2.$$

Consequently, $\|A\|_S \leq \lambda_1$. But

$$\|Ax_{(1)}\|_2 = \|\lambda_1 x_{(1)}\|_2 = \lambda_1 \Rightarrow \|A\|_s = \lambda_1.$$

Earlier, we have introduced the notion of the spectral radius of a matrix. If A is a symmetric matrix, we can conclude that $\rho_s(A) = \|A\|_S$. For any matrix A, we can only assert that $\rho_s(A) \leq \|A\|_S$. We would like to spend some more time on the spectral radius. The spectral radius provides a nice criterion under which convergence of powers of a matrix is guaranteed. But first, we need the following results. The first result is easy to prove.

P 11.2.14 Let $\|\cdot\|$ be any matrix norm on $\mathbf{M}_n$ and T a nonsingular matrix in $\mathbf{M}_n$. Define the map $\|\cdot\|_T$ on $\mathbf{M}_n$ by

$$\|A\|_T = \|T^{-1}AT\|, \ A \in \mathbf{M}_n.$$

Then $\|\cdot\|_T$ is a matrix norm on $\mathbf{M}_n$.

P 11.2.15 Given $A \in \mathbf{M}_n$ and $\epsilon > 0$, there exists a matrix norm $\|\cdot\|$ such that

$$\rho_s(A) \leq \|A\| < \rho_s(A) + \epsilon.$$

PROOF. The inequality $\rho_s(A) \leq \|A\|$ was already given in **P 11.2.4**. Let $\lambda_1, \lambda_2, \ldots, \lambda_n$ be the eigenvalues of A. By Schur triangularization theorem (See Chapter 5), there exists a unitary matrix U and an upper triangular matrix $D = (d_{ij})$ such that $A = UDU^*$ and $d_{ii} = \lambda_i$ for all i. For $t > 0$, let $G_t = \text{diag}\{t, t^2, \ldots, t^n\}$. Observe that $G_tDG_t^{-1}$ is upper triangular and in fact, given by

$$G_tDG_t^{-1} = \begin{bmatrix} \lambda_1 & d_{12}/t & d_{13}/t^2 & \ldots & d_{1n}/t^{n-1} \\ 0 & \lambda_2 & d_{23}/t & \ldots & d_{2n}/t^{n-2} \\ 0 & 0 & \lambda_3 & \ldots & d_{3n}/t^{n-3} \\ \cdot & \cdot & \cdot & \ldots & \cdot \\ 0 & 0 & 0 & \ldots & \lambda_n \end{bmatrix}.$$

Choose $t > 0$ such that the sum of all off-diagonal entries of $G_tDG_t^{-1}$ is $< \epsilon$. Define a new matrix norm $\|\cdot\|$ on $\mathbf{M}_n$ by

$$\|B\| = \|(UG_t^{-1})^{-1}B(UG_t^{-1})\|_{1,in}, \ B \in \mathbf{M}_n.$$

Recall the structure of the norm $\|\cdot\|_{1,in}$ from **P 11.2.11**. By **P 11.2.14**, $\|\cdot\|$ is indeed a matrix norm on $\mathbf{M}_n$. Let us compute

$$\begin{aligned} \|A\| &= \|(UG_t^{-1})^{-1}A(UG_t^{-1})\|_{1,in} = \|G_tU^*AUG_t^{-1}\|_{1,in} \\ &= \|G_tDG_t^{-1}\|_{1,in} < \max_{1 \leq j \leq n} |\lambda_j| + \epsilon = \rho_s(A) + \epsilon. \end{aligned}$$

This completes the proof.

We are now ready to provide the connection between the spectral radius of a matrix and the limiting behavior of powers of the matrix.

P 11.2.16 Let $A \in \mathbf{M}_n$. Then $A^k \to 0$ as $k \to \infty$ if and only if $\rho_s(A) < 1$.

PROOF. Suppose $\rho_s(A) < 1$. By **P 11.2.15**, there exists a matrix norm $\|\cdot\|$ on $\mathbf{M}_n$ such that $\|A\| < 1$. (Why?) Consequently, A^k converges to zero as $k \to \infty$. See the discussion following Definition 11.2.2. Conversely, suppose A^k converges to 0 as $k \to \infty$. There is nothing to prove if $\rho_s(A) = 0$. Assume that $\rho_s(A) > 0$. Let λ be any non-zero eigenvalue of A. Then there exists a non-zero vector x in $\mathbf{C}^n$ such that $Ax = \lambda x$. Consequently, $A^k x = \lambda^k x$. Since $A^k x$ converges to 0 as $k \to \infty$, and $x \neq 0$, it follows that λ^k converges to 0 as $k \to \infty$. But this is possible only when $|\lambda| < 1$. Hence $\rho_s(A) < 1$.

The spectral radius of a matrix A has a close connection with the asymptotic behavior of $\|A^k\|$ for matrix norms $\|\cdot\|$. The following result spells out the precise connection.

P 11.2.17 Let $\|\cdot\|$ be any matrix norm on $\mathbf{M}_n$. Then for any $A \in \mathbf{M}_n$,

$$\rho_s(A) = \lim_{k\to\infty} (\|A^k\|)^{1/k}.$$

PROOF. If $\lambda_1, \lambda_2, \ldots, \lambda_n$ are the eigenvalues of A, then $\lambda_1^k, \lambda_2^k, \ldots, \lambda_n^k$ are the eigenvalues of A^k for any positive integer k. Consequently, $\rho_s(A^k) = [\rho_s(A)]^k$. By **P 11.2.4**, $\rho_s(A^k) \leq \|A^k\|$. Hence $\rho_s(A) \leq (\|A^k\|)^{1/k}$. Let $\varepsilon > 0$ and $B = (\rho_s(A) + \varepsilon)^{-1}A$. The matrix B has spectral radius $\rho_s(B) < 1$. By **P 11.2.16**, $\|B^k\| \to 0$ as $k \to \infty$. We can find $m \geq 1$ such that $\|B^k\| < 1$ for all $k \geq m$. Equivalently, $\|A^k\| < (\rho_s(A) + \varepsilon)^k$ for all $k \geq m$. This means that $(\|A^k\|)^{1/k} \leq \rho_s(A) + \varepsilon$ for all $k \geq m$. Thus we have $\rho_s(A) \leq (\|A^k\|)^{1/k} \leq \rho_s(A) + \varepsilon$ for all $k \geq m$. Since $\varepsilon > 0$ is arbitrary, the desired conclusion follows.

Complements

11.2.1 Show that the L_∞-norm on $\mathbf{M}_n$ is not a matrix norm ($n \geq 2$). However, show that the norm defined by

$$\|A\| = n(\max_{1\leq i\leq n,\, 1\leq j\leq n} |a_{ij}|)$$

for $A = (a_{ij}) \in \mathbf{M}_n$ is a matrix norm on $\mathbf{M}_n$. See Example 11.2.1.

11.2.2 Show that the Frobenius norm on $\mathbf{M}_n$ is a matrix norm on $\mathbf{M}_n$. See Example 11.2.1.

11.2.3 Show that the L_p-norm on $\mathbf{M}_n$ is a matrix norm on $\mathbf{M}_n$ if and only if $1 \leq p \leq 2$.

(An outline of the proof. The objective of this exercise is to examine which norms in Example 11.2.1 are matrix norms. If $p = 1$ or 2, it is easy to demonstrate that the L_p-norm is indeed a matrix norm. Let $1 < p < 2$. Determine q such that $\frac{1}{p} + \frac{1}{q} = 1$. Note that $0 < p - 1 = \frac{p}{q} < 1$. Let $A = (a_{ij})$ and $B = (b_{ij}) \in \mathbf{M}_n$. Then

$$\|AB\|_p^p = \sum_{i=1}^{n}\sum_{j=1}^{n}|\sum_{r=1}^{n} a_{ir}b_{rj}|^p$$

$$\leq \sum_{i=1}^{n}\sum_{j=1}^{n}\left[\left(\sum_{r=1}^{n}|a_{ir}|^p\right)^{\frac{1}{p}}\left(\sum_{s=1}^{n}|b_{sj}|^q\right)^{\frac{1}{q}}\right]^p$$

(by Hölder's inequality)

$$\leq \sum_{i=1}^{n}\sum_{r=1}^{n}|a_{ir}|^p \sum_{j=1}^{n}\left(\sum_{s=1}^{n}(|b_{sj}|^p)^{\frac{q}{p}}\right)^{\frac{p}{q}}$$

$$\leq \|A\|_p^p\left(\sum_{j=1}^{n}\sum_{s=1}^{n}|b_{sj}|^p\right) \leq \left(\|A\|_p^p\right)\left(\|B\|_p^p\right).$$

The inequality $(a^\theta + b^\theta) \leq (a+b)^\theta$ for any $a \geq 0,\ b \geq 0$, and $\theta \geq 1$ is at the heart of the last step above.)

11.2.4 For any norm $\|\cdot\|$ on $\mathbf{C}^n$, show that $\|I_n\|_{in} = 1$.

11.2.5 For any matrix $A = (a_{ij})$ in $\mathbf{M}_n$, show that $\|A\|_{1,in} = \|A^*\|_{\infty,in}$, where A^* is the adjoint of A.

11.2.6 If A is symmetric, show that A^k converges to 0 if and only if $\rho_s(A) < 1$ using the spectral decomposition of A.

11.3. Unitarily Invariant Norms

Let m and n be two positive integers. Let $\mathbf{M}_{m,n}$ be the collection of all matrices of order $m \times n$ with complex entries. In many statistical applications, we will be concerned with data matrices with m being the sample size and n the number of variables. Generally, m will be much larger than n. We do not have the convenience of matrix multiplication being operational in the space $\mathbf{M}_{m,n}$. Consequently, the idea of a matrix norm does not make sense in such a general space. In this section, we will look at a particular class of norms on the space $\mathbf{M}_{m,n}$ and find a way to determine the structure of such norms. First, we start with a definition.

DEFINITION 11.3.1. A real valued function $\|\cdot\|$ on the vector space $\mathbf{M}_{m,n}$ is said to be a unitarily invariant norm and denoted by $\|\cdot\|_{ui}$, if it has the following properties.

(1) $\|A\| \geq 0$ for all $A \in \mathbf{M}_{m,n}$.
(2) $\|A\| = 0$ if and only if $A = 0$.
(3) $\|\alpha A\| = |\alpha| \, \|A\|$ for every $\alpha \in \mathbf{C}$ and $A \in \mathbf{M}_{m,n}$.
(4) $\|A + B\| \leq \|A\| + \|B\|$ for all A and B in $\mathbf{M}_{m,n}$.
(5) $\|UAV\| = \|A\|$ for all $A \in \mathbf{M}_{m,n}$ and unitary matrices U and V of orders $m \times m$ and $n \times n$, respectively.

The first four properties are the usual properties of a norm. The fifth property is the one which adds spice to the theme of this section. If $\mathbf{M}_{m,n}$ is a real vector space we use the term orthogonally in the place of unitarily invariant norm. We will discuss in an informal way how such an invariant norm looks like. Let $A \in \mathbf{M}_{m,n}$. Let $\sigma_1(A) \geq \sigma_2(A) \geq \ldots \geq \sigma_r(A)$ be the singular values of A, where $r = \min\{m, n\}$. By the singular value decomposition theorem, there exist two unitary matrices P and Q of orders $m \times m$ and $n \times n$, respectively, such that $A = PDQ$, where

$$D = \begin{cases} (D_1|0) & \text{if } r = m, \\ \begin{bmatrix} D_1 \\ 0 \end{bmatrix} & \text{if } r = n, \end{cases}$$

with $D_1 = \operatorname{diag}\{\sigma_1(A), \sigma_2(A), \ldots, \sigma_r(A)\}$ and 0's are the appropriate zero matrices. Then

$$\|A\|_{ui} = \|PDQ\|_{ui} = \|P^*PDQQ^*\|_{ui} = \|D\|_{ui}.$$

Note that $\|D\|$ is purely a function of the singular values of A. Let us denote this function by φ, i.e.,

$$\|D\|_{ui} = \varphi(\sigma_1(A), \sigma_2(A), \ldots, \sigma_r(A)).$$

The question then arises as what kind of properties the function $\varphi(\cdot)$ should possess. One thing is pretty clear. The function $\varphi(\cdot)$ must be a symmetric function of its arguments. We will make these notions more precise shortly. Before determining the structure of unitarily invariant norms, let us look at some examples.

EXAMPLE 11.3.2. The Frobenius norm on the vector space $\mathbf{M}_{m,n}$ is unitarily invariant. Let $A = (a_{ij})$ and $\sigma_1 \geq \sigma_2 \geq \ldots \geq \sigma_r \geq 0$ be singular values of A, where $r = \min\{m, n\}$. Then

$$\|A\|_F = (\sum_{i=1}^{m}\sum_{j=1}^{n}|a_{ij}|^2)^{1/2} = (\mathrm{Tr}(A^*A))^{1/2}$$

$$= (\text{Sum of all the eigenvalues of } A^*A)^{1/2} = (\sum_{i=1}^{r}\sigma_i^2)^{1/2}.$$

If U and V are unitary matrices of order $m \times m$ and $n \times n$, respectively, then

$$\|UAV\|_F = [\mathrm{tr}((UAV)^*(UAV))]^{1/2} = [\mathrm{tr}(V^*A^*U^*UAV)]^{1/2}$$
$$= [\mathrm{tr}(V^*A^*AV)]^{1/2} = [\mathrm{tr}(A^*A)]^{1/2}.$$

The eigenvalues of A^*A and V^*A^*AV are the same for any unitary matrix V. (Why?) Thus the Frobenius norm on $\mathbf{M}_{m,n}$ is seen to be unitarily invariant.

EXAMPLE 11.3.3. The spectral norm on $\mathbf{M}_{m,n}$ is also unitarily invariant. In Section 11.2, we have defined the spectral norm on the vector space $\mathbf{M}_n$. It can be defined in two equivalent ways on the space $\mathbf{M}_{m,n}$ too. Let $A \in \mathbf{M}_{m,n}$. One way is:

$$\|A\|_S = \sup_{x \in \mathbf{C}^n, x \neq 0} \frac{\|Ax\|_2}{\|x\|_2}.$$

Another way is: $\|A\|_S = \sigma_1 = \max\{\sigma_1, \sigma_2, \ldots, \sigma_r\} = (\rho_s(A^*A))^{1/2}$. One can check that both approaches lead to the same answer. Note that $\rho_s(A^*A)$ is the spectral radius of the matrix A^*A. One can also check that the spectral norm is unitarily invariant.

Now we take up the case of determining the structure of unitarily invariant norms. Let $\mathbf{P}_n$ be the group of all permutations of the set $\{1, 2, \ldots, n\}$. Every member π of $\mathbf{P}_n$ is a one-to-one map from $\{1, 2, \ldots, n\}$ to $\{1, 2, \ldots, n\}$ and is called a permutation. For each $x' = (x_1, x_2, \ldots, x_n) \in \mathbf{R}^n$ and $\pi \in \mathbf{P}_n$, let $x'_\pi = (x_{\pi(1)}, x_{\pi(2)}, \ldots, x_{\pi(n)})$. The vector x_π is an n-tuple which permutes the components of x. We want to introduce another entity. Let $\mathbf{J}_n$ denote the collection of all $n \times n$ diagonal matrices with each diagonal entry being equal to either $+1$ or -1. We are ready to introduce a special class of functions.

DEFINITION 11.3.4. A real valued function φ from $\mathbf{R}^n$ to $\mathbf{R}$ is said to be a symmetric gauge function if it has the following properties.

(1) $\varphi(x) > 0$ for all $x \in \mathbf{R}^n$ with $x \neq 0$.
(2) $\varphi(\alpha x) = |\alpha|\varphi(x)$ for all $x \in \mathbf{R}^n$ and $\alpha \in \mathbf{R}$.
(3) $\varphi(x + y) \leq \varphi(x) + \varphi(y)$ for all x and y in $\mathbf{R}^n$.
(4) $\varphi(x_\pi) = \varphi(x)$ for all x in $\mathbf{R}^n$ and $\pi \in \mathbf{P}_n$.
(5) $\varphi(Jx) = \varphi(x)$ for all x in $\mathbf{R}^n$ and $J \in \mathbf{J}_n$.

The first three properties merely stipulate that the function φ be a norm on the real vector space $\mathbf{R}^n$. The fourth property stipulates that the function φ be symmetrical in its arguments (permutation invariance). The fifth property exhorts the function to remain the same if signs are changed at any number of arguments.

The usual L_p-norms on the vector space $\mathbf{R}^n$ are some good examples of symmetric gauge functions. The sum of any two symmetric gauge functions is one of the family. A positive multiple of a symmetric gauge function retains all the properties of a symmetric gauge function. Here are some more examples.

EXAMPLE 11.3.5. Let $1 \leq k \leq n$ be fixed. For $x' = (x_1, x_2, \ldots, x_n) \in \mathbf{R}^n$, let

$$\varphi_k(x) = \max_{1 \leq i_1 < i_2 < \ldots < i_k \leq n} (|x_{i_1}| + |x_{i_2}| + \ldots + |x_{i_k}|).$$

One can check that $\varphi_k(\cdot)$ is indeed a symmetric gauge function. If $k = 1$, then $\varphi_k(\cdot)$ is the usual l_∞-norm on $\mathbf{R}^n$. If $k = n$, then $\varphi_k(\cdot)$ is the usual l_1-norm on $\mathbf{R}^n$.

We establish some simple properties of symmetric gauge functions.

P 11.3.6 Let φ be a symmetric gauge function.

(1) If $x' = (x_1, x_2, \ldots, x_n) \in \mathbf{R}^n$ and $0 \leq p_1, p_2, \ldots, p_n \leq 1$, then

$$\varphi(p_1 x_1, p_2 x_2, \ldots, p_n x_n) \leq \varphi(x_1, x_2, \ldots, x_n).$$

(2) If $x_i \geq 0$ for $i = 1, \ldots, n$ and $y_1 \geq x_1, y_2 \geq x_2, \ldots, y_n \geq x_n$, then

$$\varphi(x_1, x_2, \ldots, x_n) \leq \varphi(y_1, y_2, \ldots, y_n).$$

(3) There exists a constant $k > 0$ such that for all $(x_1, x_2, \ldots, x_n)' \in \mathbf{R}^n$,

$$k(\max_{1 \leq i \leq n} |x_i|) \leq \varphi(x_1, x_2, \ldots, x_n) \leq k(\sum_{i=1}^{n} |x_i|).$$

As a matter of fact, $k = \varphi(1, 0, 0, \ldots, 0)$.

(4) The function φ is continuous.

PROOF. (1) Assume that $0 \leq p_i < 1$ for exactly one i and $p_j = 1$ for the rest of j's. For simplicity, take $i = 1$ and write $p_1 = p$. We will show that

$$\varphi(px_1, x_2, \ldots, x_n) \leq \varphi(x_1, x_2, \ldots, x_n).$$

The general result would then follow in an obvious way. Let

$$u = (\frac{1+p}{2})(x_1, x_2, x_3, \ldots, x_n)', \; v = (\frac{1-p}{2})(-x_1, x_2, x_3, \ldots, x_n)'.$$

Note that $u + v = (px_1, x_2, x_3, \ldots, x_n)$. By Properties (2), (3), and (5) of a symmetric gauge function,

$$\begin{aligned} \varphi(px_1, x_2, \ldots, x_n) &= \varphi(u+v) \leq \varphi(u) + \varphi(v) \\ &= (\frac{1+p}{2})\varphi(x) + (\frac{1-p}{2})\varphi(x) = \varphi(x). \end{aligned}$$

(2) Since $0 \leq x_i \leq y_i$, we can write $x_i = p_i y_i$ for some $0 \leq p_i \leq 1$. Now (2) follows from (1).

(3) Observe that

$$\begin{aligned} \varphi(x_1, x_2, \ldots, x_n) &= \varphi(x_1 + 0, 0 + x_2, 0 + x_3, \ldots, 0 + x_n) \\ &\leq \varphi(x_1, 0, 0, \ldots, 0) + \varphi(0, x_2, x_3, \ldots, x_n) \\ &\leq |x_1|\varphi(1, 0, 0, \ldots, 0) + \varphi(0, x_2, 0, \ldots, 0) \\ &\quad + \varphi(0, 0, x_3, x_4, \ldots, x_n) \end{aligned}$$

$$= k|x_1| + k|x_2| + \varphi(0, 0, x_3, x_4, \ldots, x_n)$$

$$\cdots \qquad \cdots \qquad \cdots$$

$$\leq k(\sum_{i=1}^{n} |x_i|).$$

Also, since for each i,

$$(0, 0, \ldots, 0, |x_i|, 0, \ldots, 0) \leq (|x_1|, |x_2|, \ldots, |x_n|)$$

coordinate-wise, by (2),

$$\begin{aligned} |x_i|\ \varphi(0, 0, \ldots, 0, 1, 0, \ldots, 0) &= \varphi(0, 0, \ldots, 0, |x_i|, 0, \ldots, 0) \\ &\leq \varphi(|x_1|, |x_2|, \ldots, |x_n|) = \varphi(x_1, x_2, \ldots, x_n), \end{aligned}$$

which implies that $\varphi(x_1, x_2, \ldots, x_n) \geq k|x_i|$ for each i. Hence

$$\varphi(x_1, x_2, \ldots, x_n) \geq k(\max_{1 \leq i \leq n} |x_i|).$$

(4) Since φ is a norm,

$$\begin{aligned} &|\varphi(x_1, x_2, \ldots, x_n) - \varphi(y_1, y_2, \ldots, y_n)| \\ &\quad \leq \varphi(x_1 - y_1, x_2 - y_2, \ldots, x_n - y_n) \leq k(\sum_{i=1}^{n} |x_i - y_i|). \end{aligned}$$

The continuity of φ now follows.

We have already indicated that a unitarily invariant norm of a matrix is basically a function of the singular values of the matrix. We will show that any such norm is generated by a symmetric gauge function.

P 11.3.7 Let $\|\cdot\|_{ui}$ be a unitarily invariant norm on the vector space $\mathbf{M}_{m,n}$. Assume, for simplicity, $m \leq n$. For each $(x_1, x_2, \ldots, x_m) \in \mathbf{R}^m$, let

$$X = (\underset{m \times m}{D} \ : \ \underset{m \times (n-m)}{0}), \text{ and } \varphi(x_1, x_2, \ldots, x_m) = \|X\|_{ui},$$

where $D = \operatorname{diag}\{x_1, x_2, \ldots, x_m\}$. Then φ is a symmetric gauge function on $\mathbf{R}^m$. Conversely, if φ is a symmetric gauge function on $\mathbf{R}^m$, then the map $\|\cdot\|$ defined by

$$\|A\| = \varphi(\sigma_1(A), \sigma_2(A), \ldots, \sigma_m(A)), \ A \in \mathbf{M}_{m,n}$$

is a unitarily invariant norm on the vector space $\mathbf{M}_{m,n}$.

PROOF. Note that the eigenvalues of XX^* are $|x_1|^2, |x_2|^2, \dots, |x_m|^2$. Consequently, the singular values of X are $|x_1|, |x_2|, \dots, |x_m|$. It is a routine job to check that φ is a symmetric gauge function. Conversely, let φ be a given symmetric gauge function. The singular values of a matrix can be written in any order. Since φ is symmetric, it is clear that the map $\|\cdot\|$ induced by φ is well defined and has the following properties.

(1) $\|A\| \geq 0$ for all $A \in \mathbf{M}_{m,n}$.
(2) $\|A\| = 0$ if and only if $A = 0$.
(3) $\|\alpha A\| = |\alpha| \, \|A\|$ for all $\alpha \in \mathbf{C}$ and $A \in \mathbf{M}_{m,n}$. (The singular values of αA are $|\alpha|\sigma_1(A), \dots, |\alpha|\sigma_m(A)$.)
(4) If U and V are unitary matrices of order $m \times m$ and $n \times n$, respectively, then $\|UAV\| = \|A\|$.

The critical step would be to show that the map $\|\cdot\|$ satisfies the triangle inequality. Let A and $B \in \mathbf{M}_{m,n}$ with singular values

$$\sigma_1(A) \geq \sigma_2(A) \geq \dots \geq \sigma_m(A), \text{ and } \sigma_1(B) \geq \sigma_2(B) \geq \dots \geq \sigma_m(B).$$

Define

$$\begin{aligned} \sigma(A) &= (\sigma_1(A), \sigma_2(A), \dots, \sigma_m(A))', \\ \sigma(B) &= (\sigma_1(B), \sigma_2(B), \dots, \sigma_m(B))', \\ \sigma(A+B) &= (\sigma_1(A+B), \sigma_2(A+B), \dots, \sigma_m(A+B)). \end{aligned}$$

By **P 10.1.3**, the vector $\sigma(A+B)$ is weakly majorized by the vector $\sigma(A)+\sigma(B)$. By **P 9.3.8** and Complement 9.3.2, there exists a doubly stochastic matrix S of order $m \times m$ such that

$$\sigma(A+B) \leq S(\sigma(A)+\sigma(B))$$

coordinate-wise. Every doubly stochastic matrix can be written as a convex combination of some permutation matrices, i.e.,

$$S = \theta_1 P_1 + \dots + \theta_r P_r$$

where θ_i's are non-negative, $\sum_{i=1}^r \theta_i = 1$, and P_i's are permutation matrices. Observe that

$$\|A+B\| = \varphi(\sigma(A+B)) \leq \varphi(S(\sigma(A)+\sigma(B))) \text{ (by } \mathbf{P\ 11.3.6}\ (2))$$

$$\leq \varphi(S\sigma(A)) + \varphi(S\sigma(B)) \text{ (by the triangle inequality)}$$
$$= \varphi(\sum_{i=1}^{r} \theta_i P_i \sigma(A)) + \varphi(\sum_{i=1}^{r} \theta_i P_i \sigma(B))$$
$$\leq \sum_{i=1}^{r} \theta_i \varphi(P_i \sigma(A)) + \sum_{i=1}^{r} \theta_i \varphi(P_i \sigma(B))$$
$$\text{(by the triangle inequality)}$$
$$= \sum_{i=1}^{r} \theta_i \varphi(\sigma(A)) + \sum_{i=1}^{r} \theta_i \varphi(\sigma(B))$$
$$\text{(by the permutation invariance property of } \varphi)$$
$$= \varphi(\sigma(A)) + \varphi(\sigma(B)) = \|A\| + \|B\|.$$

This completes the proof.

Complements

11.3.1 (Ky Fan (1951)) Let $\sigma_1 \geq \ldots \geq \sigma_r \geq 0$ and $\sigma'_1 \geq \ldots \geq \sigma'_r \geq 0$ be two sets of values. Show that for any gauge function φ on $\mathbf{R}^r$,

$$\varphi(\sigma_1, \ldots, \sigma_r) \geq \varphi(\sigma'_1, \ldots, \sigma'_r)$$

if and only if $\sigma_1 + \ldots + \sigma_i \geq \sigma'_1 + \ldots + \sigma'_i$, $i = 1, \ldots, r$. This result is useful in solving optimization problems as given in the next example.

11.3.2 Let $\sigma_i(A_1), \sigma_i(A_2)$, $i = 1, \ldots, r = \min(m, n)$ be the singular values of A_1 and $A_2 \in \mathbf{M}_{m,n}$. Show that $\|A_1\| \geq \|A_2\|$ for any unitarily invariant norm $\|\cdot\|$ on $\mathbf{M}_{m,n}$ if and only if

$$\sigma_1(A_1) + \ldots + \sigma_i(A_1) \geq \sigma_1(A_2) + \ldots + \sigma_i(A_2),\ i = 1, \ldots, r.$$

This result will be used in Section 11.5 in some optimization problems.

11.3.3 Let $G \in \mathbf{M}_m$, and $P \in \mathbf{M}_m$ be a symmetric idempotent matrix of rank, $\rho(P) = k \leq m$. Show that

$$\lambda_{m-k+i}(GG^*) \leq \lambda_i(G^*PG) \leq \lambda_i(GG^*),\ i = 1, \ldots, k.$$

11.3.4 (Ky Fan and Hoffman (1955)). Let $A \in \mathbf{M}_n$. Then $\lambda_i(A + A^*) \leq 2\sigma_i(A)$.

11.3.5 Let $A \in \mathbf{M}_{m,n}$ with $\rho(A) = r$ and $B \in \mathbf{M}_{m,n}$ with $\rho(B) = k$. Then

(1) $\sigma_i(A - B) \geq \sigma_{i+k}(A), \quad i + k \leq r$
$\qquad\qquad\quad \geq 0, \qquad\qquad i + k > r.$

(2) The equalities in (1) are attained if and only if $k \leq r$ and

$$B = \sigma_1(A)U_1U_1^* + \ldots + \sigma_k(A)U_kU_k^*$$

while the singular value decomposition of A is

$$A = \sigma_1(A)U_1U_1^* + \ldots + \sigma_r(A)U_rU_r^*.$$

PROOF. Since $\rho(B) = k$, it has a rank factorization $B = CD$, such that $C \in \mathbf{M}_{m,k}$, $C^*C = I_k$ and $D \in \mathbf{M}_{k,n}$. Then

$$(A - CD)^*(A - CD) \geq A^*(I - CC^*)A.$$

$$\begin{aligned} \Rightarrow \quad \sigma_i^2(A - B) &= \lambda_i(A - B)^*(A - B) \\ &\geq \lambda_i[A^*(I - CC^*)A] \\ &\geq \lambda_{k+i}(AA^*) = \sigma_{k+i}^2(A) \end{aligned}$$

by the result of Complement 11.3.3, noting that $(1-CC^*)$ is idempotent and has rank equal to $(m - k)$. Obviously $\sigma_i(A - B) \geq 0$ for $i + k > r$. This proves (1).

The sufficiency part of (2) is trivial. But the proof of necessity is a bit involved. For details the reader referred to Rao (1980).

11.3.6 If A, B and $A - B$ are Hermitian and non-negative definite matrices of order m and if B is utmost rank k, then

(1) $\lambda_i(A - B) \geq \lambda_{k+i}(A)$.

(2) A necessary and sufficient condition for equality in (1) to hold for all i is that $B = \lambda_1(A)V_1V_1' + \ldots + \lambda_k(A)V_kV_k'$ where $V_1, \ldots, V_k$ are the first k eigenvectors of A.

11.4. Some Matrix Optimization Problems

In this section we consider some matrix optimization problems which are useful in solving matrix approximations considered in the next section. For simplicity of notation, we will consider matrices with real entries only. The results readily extend to complex matrices if we replace the phrase "transpose" by "conjugate transpose", and the phrase "symmetric" by "Hermitian."

P 11.4.1 Let $A \in \mathbf{M}_n$ be a symmetric matrix with eigenvalues $\lambda_1 \geq \lambda_2 \geq \ldots \geq \lambda_n$ and $x_1, \ldots, x_n$ be some orthonormal vectors. (Note that we are writing the eigenvalues of A in decreasing order of magnitude.) Then

$$(1)\ \sum_{i=1}^{k} x_i' A x_i \leq \sum_{i=1}^{k} \lambda_i,\ k = 1, \ldots, n-1,$$

$$(2)\ \sum_{i=1}^{n} x_i' A x_i = \sum_{i=1}^{n} \lambda_i.$$

PROOF. Let $X_k = (x_1|\ldots|x_k)$ and $A = P \wedge P'$, be the spectral decomposition of A, where $\wedge = \operatorname{diag}(\lambda_1, \ldots, \lambda_n)$ and P an orthogonal matrix. Then for any $1 \leq k \leq n$,

$$\sum_{i=1}^{k} x_i' A x_i = \operatorname{tr}(X_k' A X_k) = \operatorname{tr}(X_k' P \wedge P' X_k)$$

$$= \operatorname{tr}(\wedge P' X_k X_k' P) = \operatorname{tr} \wedge (QQ') = \sum_{1}^{n} \lambda_i q_{ii}$$

where $Q = P' X_k$ so that $Q'Q = I$ (of order $k \times k$) and q_{ii} is the i-th diagonal element of the idempotent matrix QQ'. When $k = n$, $QQ' = I$ (of order $n \times n$), which proves (2). When $k \leq n-1$, we have

$$\sum_{i=1}^{n} \lambda_i q_{ii} \leq \sum_{i=1}^{k} \lambda_i q_{ii} + (k - \sum_{i=1}^{k} q_{ii}) \lambda_{k+1}$$

$$= \sum_{i=1}^{k} (\lambda_i - \lambda_{k+1}) q_{ii} + k \lambda_{k+1}$$

$$\leq \sum_{i=1}^{k}(\lambda_i - \lambda_{k+1}) + k\lambda_{k+1} = \sum_{1}^{k} \lambda_i,$$

since $\sum_{i=1}^{n} q_{ii} = 1$, tr $(QQ') = \text{tr}(Q'Q) = k$, and $q_{ii} \leq 1$.

In view of results (1) and (2) of **P 11.4.1**, we may say that the eigenvalues λ_i majorize $x_i'Ax_i$ and hence, in particular, that λ_i majorize the diagonal elements of the symmetric matrix A. More formally, let $x' = (x_1'Ax_1, x_2'Ax_2, \ldots, x_n'Ax_n)$, $y' = (\lambda_1, \lambda_2, \ldots, \lambda_n)$, and $z' = (a_{11}, a_{22}, \ldots, a_{nn})$, where a_{ii} is the i-th diagonal entry of A. Then

$$x \ll y \quad \text{and} \quad z \ll y.$$

The matrix A need not be non-negative definite.

The equality in (1) of **P 11.4.1** is attained for a given k, when $x_1, \ldots, x_k$ are the eigenvectors of A corresponding to the eigenvalues $\lambda_1, \ldots, \lambda_k$.

The result of **P 11.4.1** readily extends to the maximization of a bilinear form as stated in the next proposition.

P 11.4.2 Let $z_i' = (x_i' : y_i')$ be mutually orthonormal vectors, $i = 1, \ldots, p$, where x_i is an m-vector and y_i is an n-vector. Further, let $A \in \mathbf{M}_{m,n}$ have singular values $\alpha_1 \geq \alpha_2 \geq \ldots \geq \alpha_r > \alpha_{r+1} = \alpha_{r+2} = \ldots = \alpha_p = 0$, where $p = \min(m, n)$ and A has rank r. Then

$$\sum_{i=1}^{k} 2x_i'Ay_i \leq \sum_{i=1}^{k} \alpha_i, \; k = 1, \ldots, p.$$

The maximum is attained when x_i and y_i are the singular vectors of A associated with the singular value α_i, $i = 1, \ldots, k$.

PROOF. Let

$$M = \begin{bmatrix} 0 & A \\ A' & 0 \end{bmatrix}$$

which is symmetric of order $(m+n) \times (m+n)$. Note that $\alpha_1, \ldots, \alpha_p$ are the p largest eigenvalues of the symmetric matrix M. Then for any $1 \leq k \leq p$,

$$\sum_{i=1}^{k} 2x_i'Ay_i = \sum_{i=1}^{k} z_i'Mz_i \leq \sum_{i=1}^{k} \alpha_i,$$

by **P 11.4.1**, which proves the first part of the result. The second part is left to the reader as an exercise. (It will be instructive to write down all the eigenvalues of M.)

P 11.4.3 Let $A = (a_{ij}) \in \mathbf{M}_{m,n}$ with singular values $\alpha_1 \geq \alpha_2 \geq \ldots \geq \alpha_r > \alpha_{r+1} = \alpha_{r+2} = \ldots = \alpha_p = 0$, where $p = \min\{m, n\}$. Then

$$(1) \sum_{i=1}^{k} a_{ii}^2 \leq \sum_{i=1}^{k} \alpha_i^2, \quad k = 1, \ldots, p,$$

$$(2) \sum_{i=1}^{k} a_{ii} \leq \sum_{i=1}^{k} \alpha_i, \quad k = 1, \ldots, p.$$

Equality in (1) holds if and only if the leading $k \times k$ principal submatrix of A is diagonal and $|a_{ii}| = \alpha_i,\ i = 1, \ldots, k$. Equality in (2) holds when equality holds in (1) and $a_{ii} \geq 0,\ i = 1, \ldots, k$.

PROOF. We have

$$\sum_{i=1}^{k} a_{ii}^2 \leq \sum_{i=1}^{k} (\sum_{j=1}^{n} a_{ij}^2) = \sum_{i=1}^{k} e_i' A A' e_i \leq \sum_{i=1}^{k} \alpha_i^2,$$

where the second inequality follows from **P 11.4.1** and e_i is the i-th column of I_m. The inequality (1) is proved. The inequality (2) follows similarly from **P 11.4.2** setting $x_i = 2^{-1/2} e_i$ and y_i equal to $2^{-1/2}$ (i-th column of I_n).

P 11.4.4 (Abel's identity). Let $a_1, \ldots, a_n$ and $b_1, \ldots, b_n$ be two finite sequences of scalars. Then

$$\sum_{i=1}^{n} a_i b_i = \sum_{i=1}^{n-1} \Big((a_i - a_{i+1}) \sum_{j=1}^{i} b_j \Big) + a_n \sum_{j=1}^{n} b_j.$$

PROOF. The right-hand side of the above identity may be written as

$$\sum_{i=1}^{n-1} (a_i \sum_{j=1}^{i} b_j) - \sum_{i=1}^{n-1} (a_{i+1} \sum_{j=1}^{i} b_j) + a_n \sum_{j=1}^{n} b_j$$

$$= \sum_{i=1}^{n-1}(a_i \sum_{j=1}^{i} b_j) - \sum_{i=2}^{n}(a_i \sum_{j=1}^{i-1} b_j) + a_n \sum_{j=1}^{n} b_j$$
$$= \sum_{i=2}^{n-1} a_i(\sum_{j=1}^{i} b_j - \sum_{j=1}^{i-1} b_j) + a_1 b_1 - a_n \sum_{j=1}^{n-1} b_j + a_n \sum_{j=1}^{n} b_j$$
$$= \sum_{i=2}^{n-1} a_i b_i + a_1 b_1 + a_n b_n = \sum_{i=1}^{n} a_i b_i.$$

A more general version of the following result has been established in Chapter 10. In the special case reported here, the proof is simple and instructive.

P 11.4.5 (von Neumann (1937)). Let A and $B \in \mathbf{M}_n$ be both symmetric with eigenvalues $\lambda_1 \geq \ldots \geq \lambda_n$ and $\mu_1 \geq \ldots \geq \mu_n$. Then

$$\sum_{i=1}^{n} \lambda_i \mu_{n-i+1} \leq \operatorname{tr}(AB) \leq \sum_{i=1}^{n} \lambda_i \mu_i.$$

Equality holds on the right when $B = \sum_{i=1}^{n} \mu_i P_i P_i'$ and equality on the left when $B = \sum_{i=1}^{n} \mu_{n-i+1} P_i P_i'$, where P_i is an eigenvector of A for the eigenvalue λ_i, $i = 1, \ldots, n$.

PROOF. Let $A = P \wedge P'$ be the spectral decomposition of A with $\wedge = \operatorname{diag}(\lambda_1, \ldots, \lambda_n)$ and P orthogonal. Then

$$\operatorname{tr}(AB) = \operatorname{tr}(P \wedge P'B) = \operatorname{tr}(\wedge P'BP) = \Sigma \lambda_i b_{ii},$$

where b_{ii} is the i-th diagonal element of $P'BP$. Using Abel's identity **(P 11.4.4)**, we obtain

$$\operatorname{tr}(AB) = \sum_{i=1}^{n-1}\left((\lambda_i - \lambda_{i+1}) \sum_{j=1}^{i} b_{jj}\right) + \lambda_n \sum_{j=1}^{n} b_{jj}$$
$$\leq \sum_{i=1}^{n-1}\left((\lambda_i - \lambda_{i+1}) \sum_{j=1}^{i} \mu_j\right) + \lambda_n \sum_{j=1}^{n} \mu_j = \sum_{i=1}^{n} \lambda_i \mu_i,$$

where the inequality follows from **P 11.4.1**, since $P'BP$ and B have the same eigenvalues. The inequality on the left-hand side follows by replacing B by $-B$ in the inequality on the right-hand side.

P 11.4.6 Let $A \in \mathbf{M}_{m,n}$ and $B \in \mathbf{M}_{n,m}$, with singular values $\alpha_1 \geq \ldots \geq \alpha_p$ and $\beta_1 \geq \beta_2 \geq \ldots \geq \beta_p$, respectively, where $p = \min(m,n)$. Then

$$-\sum_{i=1}^{p} \alpha_i \beta_i \leq \operatorname{tr}(AB) \leq \sum_{i=1}^{p} \alpha_i \beta_i. \tag{11.4.1}$$

Equality holds on the right when $B = \sum_{i=1}^{p} \beta_i Q_i P_i'$ and equality on the left when $B = \sum_{i=1}^{p} \beta_i(-Q_i)P_i'$, where P_i and Q_i are singular vectors of A for the singular value $\alpha_i, i = 1, \ldots, p$.

PROOF. Note that the eigenvalues of the symmetric matrix

$$\begin{bmatrix} 0 & A \\ A' & 0 \end{bmatrix}$$

are $\alpha_1 \geq \ldots \geq \alpha_p \geq 0 = \ldots = 0 \geq -\alpha_p \geq \ldots \geq -\alpha_1$, where the number of 0's is $|m-n|$. By **P 11.4.5**, we have

$$\begin{aligned} 2\operatorname{tr}(AB) &= \operatorname{tr}\begin{bmatrix} 0 & A \\ A' & 0 \end{bmatrix}\begin{bmatrix} 0 & B' \\ B & 0 \end{bmatrix} \\ &\leq \sum_{1}^{p} \alpha_i\beta_i + \sum_{1}^{p}(-\alpha_i)(-\beta_i) = 2\sum_{1}^{p}\alpha_i\beta_i. \end{aligned} \tag{11.4.2}$$

The rest of the results are easily established.

Complements

11.4.1 Let $A \in \mathbf{M}_{m,n}$ and $B \in \mathbf{M}_{m,n}$ with singular value decompositions $P\Delta_1 Q'$ and $R\Delta_2 S'$, singular values $\alpha_1 \geq \ldots \geq \alpha_r$ and $\beta_1 \geq \ldots \geq \beta_r$, respectively, where $r = \min\{m,n\}$. Show that

$$|\operatorname{tr}(AUBV)| \leq \alpha_1\beta_1 + \ldots + \alpha_r\beta_r,$$

where U and V are any unitary matrices. Further, show that the upper bound is attained when $U = QR'$ and $V = SP'$. Show also that, if

$m = n$, $\text{tr}(AU) \leq \alpha_1 + \ldots + \alpha_r$ for any unitary matrix U with the upper bound attaining when $U = QP'$. (The results follow from von Neumann's propositions **P 11.4.5** and **P 11.4.6**.)

11.4.2 (Ky Fan and Hoffman (1955)) Let A be a square matrix. Show that $\lambda_i(A + A') \leq 2\sigma_i(A) = 2[\lambda_i(A'A)]^{1/2}$ where $\lambda_i(A + A')$ is the i-th largest eigenvalue of $(A + A')$ and $\sigma_i(A)$ is the i-th singular value of A.

11.4.3 Let $A \in \mathbf{M}_{m,n}$ and $q = \min\{m, n\}$, and $N_k(A) = \sigma_1(A) + \ldots + \sigma_k(A)$ denote the sum of the k largest singular values of A. Show that $N_k(\cdot)$ is a norm on $\mathbf{M}_{m,n}$ for $k = 1, \ldots, q$, and that when $m = n$, $N_k(\cdot)$ is a matrix norm on $\mathbf{M}_n$ for $k = 1, \ldots, n$.

The function $N_k(A)$ is called Ky Fan norm.

11.4.4 Let A_1 and $A_2 \in \mathbf{M}_n$ symmetric with $A_1 - A_2$ being non-negative definite. Show that $\lambda_i(A_1) \geq \lambda_i(A_2)$, $i = 1, \ldots, n$.

11.5. Matrix Approximations

In this section, we will look at some problems of approximating a given matrix by a matrix with specific structural properties. Let $A \in \mathbf{M}_n$ be a given matrix. First we consider the problem of finding a symmetric matrix $B \in \mathbf{M}_n$, which is closest to A with respect to a given norm. In the following we provide an explicit solution to this problem.

P 11.5.1 Let $A \in \mathbf{M}_n$ and $B = (A+A')/2$. Then B is a symmetric matrix closest to A with respect to any orthogonally invariant norm $\|\cdot\|$.

PROOF. Let C be any symmetric matrix in $\mathbf{M}_n$. Note that

$$\begin{aligned}
\|A - B\| &= \|(A - A')/2\| = (1/2)\|(A - C) - (A' - C)\| \\
&= (1/2)\|(A - C) - (A - C)'\| \\
&\leq (1/2)[\|(A - C)\| + \|(A - C)'\|] \\
&= (1/2)[\|(A - C)\| + \|(A - C)\|] = \|(A - C)\|.
\end{aligned}$$

A matrix B of order $n \times n$ with real entries is said to be skew-symmetric if $B = -B'$. We can find explicitly a skew-symmetric matrix closest to a given matrix.

P 11.5.2 Let $A \in \mathbf{M}_n$ and $B = (A - A')/2$. Then B is a skew-symmetric matrix closest to A with respect to any orthogonally invariant norm $\|\cdot\|$.

These results can be formulated for matrices with complex entries. If A is a given matrix of order $n \times n$ with complex entries, the objective is to find a hermitian matrix B closest to A. Likewise, one can find an anti-hermitian (A matrix B is anti-hermitian if $B = -B^*$.) matrix closest to A.

Our next objective is to find an orthogonal matrix closest to A. In Complements 11.5.1 and 11.5.3, we address this problem. In the following, we prove very general results useful in solving statistical problems.

P 11.5.3 Let A and $B \in \mathbf{M}_{m,n}$ with singular value decompositions $A = P\Delta_1 Q'$ and $B = R\ \Delta_2\ S'$, respectively. Then

$$\min_{U \in \mathbf{O}_m, T \in \mathbf{O}_n} \|UA - BT\|_F = \|U_* A - BT_*\|_F,$$

where $\mathbf{O}_k$ is the class of all $k \times k$ orthogonal matrices, $U_* = RP'$ and $T_* = SQ'$, and $\|\cdot\|_F$ is the Frobenius norm.

PROOF. Consider

$$\begin{aligned}\|UA - BT\|_F^2 &= \operatorname{tr}[(UA - BT)'(UA - BT)] \\ &= \operatorname{tr}(A'A) + \operatorname{tr}(B'B) - 2\operatorname{tr}(A'U'BT).\end{aligned}$$

We have to find the minimum of the last term in the above expression, when U and T roam over orthogonal matrices. Using von Neumann's result stated in the Complement 11.4.1, we obtain the desired result. The result may not hold for all unitarily invariant norms.

A number of results follow from **P 11.5.3**. These consequences are listed as Complements 11.5.1 to 11.5.6.

Next in line is the non-negative definiteness property. For a given matrix A of order $n \times n$, we seek a non-negative definite matrix B closest to A under the Frobenius norm.

P 11.5.4 Let A be a matrix of order $n \times n$ with real entries. Let $B = (A + A')/2$. Let $B = QH$ be a polar decomposition of B with Q being orthogonal and H non-negative definite. Then $C = (B + H)/2$ is non-negative definite and closest to A under the Frobenius norm $\|\cdot\|$. Moreover, C is unique.

PROOF. First, we show that any non-negative definite matrix closest to A is also closest to B, the symmetric part of A. For this, we note a special property of the Frobenius norm. Let $D = (d_{ij})$ and $E = (e_{ij})$ be symmetric and skew-symmetric matrices, i.e., $D = D'$ and $E = -E'$, respectively. Then

$$\begin{aligned}\|D+E\|_F^2 &= \sum_{i=1}^{n}\sum_{j=1}^{n}(d_{ij}+e_{ij})^2 = \sum_{i=1}^{n}\sum_{j=1}^{n}d_{ij}^2 + \sum_{i=1}^{n}e_{ij}^2 + 2\sum_{i=1}^{n}\sum_{j=1}^{n}d_{ij}e_{ij} \\ &= \sum_{i=1}^{n}\sum_{j=1}^{n}d_{ij}^2 + \sum_{i=1}^{n}\sum_{j=1}^{n}e_{ij}^2 + 0 = \|D\|_F^2 + \|E\|_F^2.\end{aligned}$$

Let X be any non-negative definite matrix of order $n \times n$. Then

$$\begin{aligned}\|A-X\|_F^2 &= \|(A+A')/2 - X + (A-A')/2\|_F^2 \\ &= \|B-X\|_F^2 + \|(A-A')/2\|_F^2,\end{aligned}$$

in view of the facts that $B - X$ is symmetric and $(A - A')/2$ is skew-symmetric. Thus minimizing $\|A - X\|_F$ over all non-negative definite matrices X is equivalent to minimizing $\|B - X\|_F$ over all non-negative definite matrices X. Let us work with the symmetric matrix B. Let $B = Z\Lambda Z'$ be the spectral decomposition of B, where Z is an orthogonal matrix, $\Lambda = \text{diag}\{\lambda_1, \lambda_2, \ldots, \lambda_n\}$, and $\lambda_1, \lambda_2, \ldots, \lambda_n$ are the eigenvalues of B. Let us replace each negative eigenvalue by zero in the spectral decomposition of B. More precisely, let for each $1 \le i \le n$,

$$d_i = \begin{cases} \lambda_i & \text{if } \lambda_i \ge 0, \\ 0 & \text{if } \lambda_i < 0, \end{cases}$$

and $C = ZDZ'$, where $D = \text{diag}\{d_1, d_2, \ldots, d_n\}$. It is clear that C is non-negative definite. Let us see whether C is closest to B. Let X be any non-negative definite matrix. Note that

$$\|B-X\|_F^2 = \|Z\Lambda Z' - ZZ'XZZ'\|_F^2 = \|\Lambda - Z'XZ\|_F^2.$$

The last equality follows from the fact that the Frobenius norm is orthogonally invariant. Let $Y = Z'XZ = (y_{ij})$. Then

$$\|B - X\|^2 = \sum_{i \neq j} y_{ij}^2 + \sum_{i=1}^{n} (\lambda_i - y_{ii})^2 \geq \sum_{i=1}^{n} (\lambda_i - y_{ii})^2 \geq \sum_{\lambda_i < 0} \lambda_i^2.$$

The last inequality follows from the fact that Y is non-negative definite and $y_{ii} \geq 0$. On the other hand,

$$\|B - C\|_F^2 = \|Z\Lambda Z' - ZDZ'\|_F^2 = \|\Lambda - D\|_F^2 = \sum_{(\lambda_i > 0)} \lambda_i^2.$$

The matrix C is the right choice. Two facts emerge from these deliberations. The first one is that the matrix C which minimizes $\|A - X\|^2$ over all non-negative definite matrices X is unique. The second one is that this minimum can be explicitly spelled out, i.e.,

$$\min \|A - X\|_F^2 = \sum_{\lambda_i < 0} \lambda_i^2 + \|(A - A')/2\|^2.$$

The final step consists of tying up the matrix C with the polar decomposition of B. The polar decomposition of B is derivable directly from the spectral decomposition of B. Let for each $1 \leq i \leq n$,

$$e_i = \begin{cases} +1 & \text{if } \lambda_i \geq 0, \\ -1 & \text{if } \lambda_i < 0, \end{cases}$$

and $E = \text{diag}\{e_1, e_2, \ldots, e_n\}$. Note that E is an orthogonal matrix and

$$B = Z\Lambda Z' = ZEZ'Z \,\text{diag}\{|\lambda_1|, |\lambda_2|, \ldots, |\lambda_n|\} Z' = QH,$$

where $Q = ZEZ'$ and $H = Z \,\text{diag}\{|\lambda_1|, |\lambda_2|, \ldots, |\lambda_n|\} Z'$. It is clear that Q is orthogonal and H is non-negative definite. This is the polar decomposition of B. Note that

$$(B + H)/2 = (Z\Lambda Z' + Z \,\text{diag}\{|\lambda_1|, |\lambda_2|, \ldots, |\lambda_n|\} Z')/2 = ZDZ' = C.$$

It is worthwhile to note that the result of **P 11.5.4** has been proved to be operational under the Frobenius norm. No concrete results are available for orthogonally invariant norms. However, for the spectral norm $\|\cdot\|_S$, Halmos (1972) has obtained the following result.

Let A be a matrix of order $n \times n$ with real entries. Let $B_1 = (A+A')/2$ and $B_2 = (A-A')/2$ be the symmetric and skew-symmetric parts of A, respectively. Let

$$C = B_1 + [\delta^2 I_n + B_2^2]^{1/2},$$

where $\delta = \min\{r \geq 0 : r^2 I_n + B_2^2$ is non-negative definite and $B_1 + (r^2 I_n + B_2^2)^{\frac{1}{2}}$ is non-negative definite$\}$. Then C is closest to A under the spectral norm.

Now we concentrate on the problem of approximating a given matrix $A \in \mathbf{M}_{m,n}$ of rank r by a matrix B of lower rank $k < r$ such that $\|A-B\|$ is a minimum under an orthogonality invariant norm $\|\cdot\|$. Such an optimization problem occurs in the representation of high dimensional data in lower dimensions. The symbol $\mathbf{M}_{m,n}(r)$ stands for the collection of all matrices of order $m \times n$ with rank $\leq r$.

P 11.5.5 (Mirsky (1960), Eckart and Young (1936).) Let $A \in \mathbf{M}_{m,n}(r)$ with the singular value decomposition $A = \sigma_1(A)P_1Q_1' + \ldots + \sigma_r(A)P_rQ_r'$, and define $B_* = \sigma_1(A)P_1Q_1' + \ldots + \sigma_k(A)P_kQ_k'$, $k \leq r$. Then

$$\min_{B \in \mathbf{M}_{m,n}(k)} \|A - B\| = \|A - B_*\|$$

for any orthogonally invariant norm $\|\cdot\|$.

PROOF. Note that the above result provides the best approximation of A of rank r by a matrix of lower rank k. The proof consists of showing that $\sigma_i(A - B_*) \leq \sigma_i(A - B)$, $i = 1, \ldots, p$, $p = \min\{m, n\}$. Then the result follows by using the Complement 11.3.2.

Let $B \in \mathbf{M}_{m,n}$ of rank k. Since $\rho(B) = k$, it has a rank factorization $B = CD$ where $C \in \mathbf{M}_{m,k}(k)$ such that $C'C = I$ and $D \in \mathbf{M}_{k,n}(k)$. Then

$$\begin{aligned}
&(A-B)'(A-B) \\
&= (A - CC'A + CC'A - CD)'(A - CC'A + CC'A - CD) \\
&= A'(I - CC')A + (C'A - D)'(C'A - D) \geq A'(I - CC')A
\end{aligned}$$

(This inequality can also be established by regressing A on C in the model $A = CD + \varepsilon$. The least squares solution in D is obtained by minimizing $(A - CD)'(A - CD)$ with respect to D. The solution is given by $D = (C'C)^{-1}C'A$.)

Then using the Complement 11.4.3, since $I - CC'$ is idempotent,

$$\begin{aligned}\sigma_i^2[(A - B)] &= \lambda_i[(A - B)'(A - B)] \\ &\geq \lambda_i A'[(I - CC')A] \geq \sigma_{i+k}^2(A),\ i + k \leq r,\end{aligned}$$

which gives that $\sigma_i(A - B) \geq \sigma_{i+k}(A)$. But

$$\sigma_i^2(A - B_*) = \sigma_i^2(\sigma_{k+1}(A)P_{k+1}Q'_{k+1} + \ldots + \sigma_r(A)P_rQ'_r) = \sigma_{i+k}^2(A),$$

so that $\sigma_i^2(A - B_*) \leq \sigma_i^2(A - B)$, as required.

What is the closest approximation by a normal matrix X to a given matrix $A \in \mathbf{M}_n$? For this we have to solve the optimization problems:

$$\begin{aligned}&\text{Minimize} \quad \varphi(X) \equiv \|A - X\|_F^2 \\ &\text{subject to} \quad g(X) \equiv X'X - XX' = 0.\end{aligned}$$

There is no closed form solution to the problem. A numerical algorithm is given by Bao and Rokne (1987).

Complements

In all the following problems, A and B stand as generic symbols for matrices in $\mathbf{M}_{m,n}$. The symbol $\mathbf{O}_k$ stands for the collection of all $k \times k$ orthogonal matrices.

11.5.1 Show that $\min\{\|UA - BT\|_F : T \in \mathbf{M}_{m,n} \text{ and } U \in \mathbf{O}_n\}$ is attained at $BT = P_B UA$ and $U = RP'$, where $AA' = P \wedge_1 P'$ and $P_B = R \wedge_2 R'$ are the spectral decompositions of AA' and P_B, the orthogonal projection operator on the space spanned by the columns of B, respectively.
11.5.2 Let $A \in \mathbf{M}_n$. Show that $\min\{\|A - T\|_F : T \in \mathbf{O}_n\} = \|A - T_*\|_F$ with $T_* = PQ'$, where $P\Delta Q'$ is the singular value decomposition of A.
11.5.3 Show that the result of Complement 11.5.2 holds for any unitarily invariant norm. (Ky Fan and Hoffman (1955).)
11.5.4 Extend the result of Complement 11.5.3 to the case where A and T are $m \times n$ matrices and the columns of T are orthogonal.

11.5.5 Let $A \in \mathbf{M}_{m,n}$. Show that the minimum of $\|A - BT\|$, under Frobenius norm, over $T \in \mathbf{O}_n$ is attained at $T = QP'$, where $P \wedge Q'$ is the s.v.d. of $A'B$. [Use the result that for any orthogonal matrix U, $|\mathrm{tr}\, CU| \le \mathrm{tr}\Delta$, where Δ is the diagonal matrix of the singular values of C.]

11.5.6 Let $\Sigma \in \mathbf{M}_n(r)$ be non-negative definite with the spectral decomposition $\Sigma = \lambda_1 P_1 P_1' + \ldots + \lambda_r P_r P_r'$, where $\mathbf{M}_n(r)$ is the class of all $n \times n$ matrices of rank $\le r$. Show that for any unitarily invariant norm $\|\cdot\|$, the minimum of $\|\Sigma - L\|$ when L runs through idempotent matrices of rank $k \le r$ is attained at

$$L_* = P_1 P_1' + \ldots + P_k P_k'.$$

11.6. M, N-invariant Norm and Matrix Approximations

Rao (1979b, 1980, 1985d) extended the concept of unitarily invariant norm to a more general M, N-invariant norm which is useful in what is known as dimensionality reduction problems in statistical multivariate analysis. We define an M, N-invariant norm, prove some results in matrix approximations and indicate some applications.

DEFINITION 11.6.1. A matrix norm in the space of $\mathbf{M}_{m\times n}$ (i.e., matrices of order $m \times n$) is said to be an M, N-invariant norm if in addition to conditions (1), (2) and (3) in Definition 11.1.1, the following condition is satisfied:

$$(4) \qquad \|VXU\| = \|X\|$$

for every $X \in \mathbf{M}_{m,n}$ and any $V \in \mathbf{M}_m$ and $U \in \mathbf{M}_n$ such that with respect to given positive definite matrices $M \in \mathbf{M}_m$ and $N \in \mathbf{M}_n$, $V^*MV = M$ and $U^*NU = N$. (Denote such a norm by $\|X\|_{MNi}$.)

The following proposition provides the connection between M, N-invariant and unitarily invariant norms.

P 11.6.2 Let $M \in \mathbf{M}_m$ and $N \in \mathbf{M}_n$ be positive definite matrices and $M^{1/2}$ and $N^{1/2}$ be Gramian square roots of M and N respectively, and $M^{-1/2}$ and $N^{-1/2}$ be those of M^{-1} and N^{-1}. The following results hold.

(1) If $\|X\|_1$ is a unitarily invariant norm of X, then $\|M^{1/2}XN^{1/2}\|_1$ is an M, N-invariant norm of X.

(2) If $\|X\|_2$ is an M, N-invariant norm of X, then $\|M^{-1/2}XN^{-1/2}\|_2$ is a unitarily invariant norm of X.

Proof. Let $U \in \mathbf{M}_m$ and $V \in \mathbf{M}_n$ be such that $U^*MU = M$ and $V^*NV = N$. Then

$$\begin{aligned}\|M^{1/2}UXVN^{1/2}\|_1 &= \|M^{1/2}UM^{-1/2}M^{1/2}XN^{1/2}N^{-1/2}VN^{1/2}\|_1\\ &= \|M^{1/2}XN^{1/2}\|_1\end{aligned}$$

since $M^{1/2}UM^{-1/2}$ and $N^{-1/2}VN^{1/2}$ are unitary matrices. This proves (1) of **P 11.6.2**. Part (2) is proved in a similar manner.

P 11.6.3 Let $A \in \mathbf{M}_{m,n}$, and $M \in \mathbf{M}_m$ and $N \in \mathbf{M}_n$ be positive definite matrices. Let $B \in \mathbf{M}_{m,r}$ and $C \in \mathbf{M}_{n,k}$ be such that $B^*MB = I_r$ and $C^*NC = I_k$. Then

$$\sigma_i(B^*AC) \leq \sigma_i(M^{-1/2}AN^{-1/2}),\ i = 1, 2, \dots, q, \tag{11.6.1}$$

where $q = \min\{r, k\}$. The upper bound in (11.6.1) is attained when B consists of the first r columns of $M^{-1/2}P$ and C consists of the first k columns of $N^{-1/2}Q$, where P and Q are unitary matrices in the s.v.d. of $M^{-1/2}AN^{-1/2}$.

Proof. The result (11.6.1) follows from Poincaré's Separation Theorem **P 10.4.2** for singular values.

P 11.6.4 Let $A \in \mathbf{M}_{m,n}$, $X \in \mathbf{M}_{m,a}$ and $Y \in \mathbf{M}_{n,b}$ be given matrices. Let $C \in \mathbf{M}_{a,n}$, $R \in \mathbf{M}_{b,n}$, and $G \in \mathbf{M}_{m,n}$ with rank $\rho(G) = k \leq r = \rho(A)$. Denote orthogonal projection matrices on the column spaces of X and Y by

$$P = X(X'M^{-1}X)^{-}X'M^{-1},\ Q = Y(Y'N^{-1}Y)^{-}Y'N^{-1}$$

where M and N are given positive definite matrices. Then:

(1) $$\min_C \|A - XC\| = \|A - PA\| \tag{11.6.2}$$

for any M, N-invariant norm. (Choose any C such that $XC = PA$.)

(2) $$\min_R \|A - RY'\| = \|A - AQ'\| \tag{11.6.3}$$

for any M, N-invariant norm. (Choose any R such that $RY' = AQ'$.)

(3) $$\min_{C,R} \|A - XC - RY'\| = \|A - PA - AQ' + PAQ'\| \quad (11.6.4)$$

for any M, N-invariant norm. (Choose $XC = PA$, $RY' = (I - P)AQ'$.)

(4) $$\min_{\rho(G)=k} \|A - G\| = \|A - G_0\| \quad (11.6.5)$$

for any M, N-invariant norm, where

$$M^{-1/2} G_0 N^{-1/2} = \sigma_1 U_1 V_1' + \ldots + \sigma_k U_k V_k', \quad (11.6.6)$$

and σ_i, U_i and V_i are as in the s.v.d.

$$\begin{aligned} M^{-1/2} A N^{-1/2} &= \sigma_1 U_1 V_1' + \ldots + \sigma_k U_k V_k' \\ &\quad + \sigma_{k+1} U_k V_k' + \ldots + \sigma_r U_r V_r' . \end{aligned} \quad (11.6.7)$$

PROOF. The results (11.6.2) and (11.6.3) follow from the inequalities on singular values

$$\sigma_i(M^{-1/2}(A - XC)N^{-1/2}) \geq \sigma_i(M^{-1/2}(A - PA)N^{-1/2}), \; i = 1, 2, \ldots$$
$$\sigma_i(M^{-1/2}(A - RY')N^{-1/2}) \geq \sigma_i(M^{-1/2}(A - AQ)N^{-1/2}), \; i = 1, 2, \ldots$$

by using Ky Fan's Theorem (Complements 11.3.1 and 11.3.2).

It is interesting to note that the result (11.6.2) is independent of N and (11.6.3) is independent of M. The solutions (11.6.2) and (11.6.3) are useful in studying regression problems in multivariate analysis.

Result (3) is proved as follows. For any given R

$$\begin{aligned} & N^{-1/2}(A - XC - RY')' M (A - XC - RY') N^{-1/2} \\ & \geq N^{-1/2}(A - RY')'(I - P)' M (I - P)(A - RY') N^{-1/2} \end{aligned} \quad (11.6.8)$$

from which it follows, using the notation $\sigma_i(\cdot)$ for the i-th singular value,

$$\begin{aligned} & \sigma_i[N^{-1/2}(A - XC - RY')M^{-1/2}] \\ & \quad \geq \sigma_i[N^{-1/2}(A - RY')'(I - P)'M^{-1/2}] \end{aligned}$$

$$= \sigma_i[M^{-1/2}(I-P)(A-RY')N^{-1/2}]$$
$$\geq \sigma_i[M^{-1/2}(I-P)(A-AQ')N^{-1/2}]$$
$$= \sigma_i[M^{-1/2}(A-PA-AQ'+PAQ')N^{-1/2}]. \qquad (11.6.9)$$

Result (3) follows from (11.6.9) by using Ky Fan's Theorem (Complement 11.3.2).

Result (4) is a direct consequence of the Complement 11.3.5.

P 11.6.5 Let A, X, Y, C, R and G be as defined in **P 11.6.4** with the additional conditions $\rho(X) = a$, $\rho(Y) = b$ and $\rho(G) = k \leq \min(m-a, n-b)$. Let P and Q be projection operators as in **P 11.6.4** and

$$\sigma_1 U_1 V_1' + \ldots + \sigma_r U_r V_r'$$

be the s.v.d. of

$$M^{-1/2}(I-P)A(I-Q')N^{-1/2}.$$

Then

$$\inf \|A - XC - RY' - G\|$$

taken over C, R and G, for any M, N-invariant norm, is attained at

$$XC = PA, \; RY' = (I-P)AQ' \qquad (11.6.10)$$
$$G = M^{1/2}[\sigma_1 U_1 V_1' + \ldots + \sigma_k U_k V_k']N^{1/2}. \qquad (11.6.11)$$

PROOF. Observe that

$$\sigma_i[N^{-1/2}(A-XC-RY'-G)M^{-1/2}]$$
$$\geq \sigma_i[N^{-1/2}(A-RY'-G)'(I-P)'M^{-1/2}]$$
$$= \sigma_i[M^{-1/2}(I-P)(A-RY'-G)N^{-1/2}]$$
$$\geq \sigma_i[M^{-1/2}(I-P)(A-G)(I-Q)'N^{-1/2}]$$
$$\geq \sigma_{i+k}[M^{-1/2}(I-P)A(I-Q)'N^{-1/2}], \; (i+k) \leq r$$

which shows, using Complement 11.3.5, that

$$\|A-XC-RY'-G\| \geq \|(I-P)A(I-Q')\|$$

for any M, N-invariant norm. It can be easily verified that the equality is attained when C, R and G are chosen as in (11.6.10) and (11.6.11).

Note that the solution for optimum C and R may not be unique.

11.7. Fitting a Hyperplane to a Set of Points

In discussing vector spaces, we introduced the concept of a subspace of a vector space $\mathbf{V}$, as a subset of vectors closed under the operations of addition and scalar multiplication. We now define subsets of vectors which are obtained by translating a subspace. They are called hyperplanes.

DEFINITION 11.7.1. Let $\mathbf{V}$ be a vector space of dimension m, and α a vector in $\mathbf{V}$, and $\mathbf{B}$ a subspace of dimension $k \le m$. A k dimensional hyperplane specified by α and $\mathbf{B}$ is defined to be the set of vectors

$$H_k(\alpha, \mathbf{B}) = \{\alpha + x : x \in \mathbf{B}\}. \tag{11.7.1}$$

In the special case when $\mathbf{V} = \mathbf{R}^m$, the elements of $\mathbf{V}$ are m-vectors and the set (11.7.1) can be represented as

$$H_k(b, A) = \{x : Ax = b\} \tag{11.7.2}$$

where $A \in \mathbf{M}_{m-k,m}$ with $\rho(A) = m - k$ and $b \in \mathbf{M}_{k,1}$.

A problem of great interest in statistics and econometrics is the following. We have a given set of n points (vectors), $x_1, \dots, x_n$, in a vector space $\mathbf{V}$ of dimension m. We would like to fit a k dimensional hyperplane to the n points, i.e., determine a k dimensional hyperplane to which the given set of points are closest in some sense. We will develop some criteria of closeness and discuss how to fit such hyperplanes. This problem was raised by Karl Pearson (1901) and solved in particular cases. Pearson's solution is the forerunner of principal component and related analyses which are currently used in statistical multivariate analysis. First we formulate the problem in general terms and consider some special cases.

Let us consider a $H_k(\alpha, \mathbf{B})$ for given k, α and $\mathbf{B}$, and let z_i be such that

$$\|x_i - z_i\|^2 = \inf_{z \in H_k(\alpha, \mathbf{B})} \|x_i - z\|^2 \tag{11.7.3}$$

where $\|\cdot\|$ is a chosen vector norm defined on $\mathbf{V}$. By (11.7.3) we have associated with the given point x_i a point z_i on $H_k(\alpha, \mathbf{B})$. We now define a compound measure of closeness by

$$L = w_1\|x_1 - z_1\|^2 + \dots + w_n\|x_n - z_n\|^2 \tag{11.7.4}$$

using suitable weights $w_1, \dots, w_n$. Then L is a function of α and $\mathbf{B}$ for given k. We now determine α and $\mathbf{B}$ by minimizing L with respect to α and $\mathbf{B}$. We have used only the concept of a vector norm in the above formulation of the problem and not any particular structure of $\mathbf{V}$.

Let us now consider $\mathbf{V} = \mathbf{R}^m$, in which case $x_1, \dots, x_n$ are column vectors which can be represented as a matrix $X = (x_1|\cdots|x_n) \in \mathbf{M}_{m,n}$. The subspace $\mathbf{B}$ is represented by a matrix $B = (b_1|\cdots|b_k)$ where $b_i \in \mathbf{R}^m$ and $b_1, \dots, b_k$ constitute a basis of $\mathbf{B}$. In such a case a point on $H_k(\alpha, \mathbf{B})$ can be written as $\alpha + By$, where $y \in \mathbf{M}_{k,1}$. Let us associate with x_i a point $\alpha + Bz_i$ in $H_k(\alpha, \mathbf{B})$. Then the set of points on $H_k(\alpha, \mathbf{B})$ associated with $x_1, \dots, x_n$ can be written in a matrix form

$$\alpha 1' + BZ, \ Z = (z_1|\cdots|z_n) \in \mathbf{M}_{k,n}.$$

Now the problem can be formulated as that of minimizing a matrix norm

$$\|X - \alpha 1' - BZ\| \tag{11.7.5}$$

with respect to α, B and Z. A general solution to this problem for a wide class of norms is given in the following proposition.

P 11.7.2 Let $Q = 1(1'N^{-1}1)^{-1}1'N^{-1}$, where 1 is the vector with all components as unity, and $F = M^{-1/2}X(1-Q')N^{-1/2}$ with the s.v.d.

$$F = \sigma_1 U_1 V_1' + \dots + \sigma_r U_r V_r'.$$

Then a set of α, B and Z which minimize (11.7.5) for any M, N-invariant norm is given by

$$\begin{aligned} \alpha &= AN^{-1}1(1'N^{-1}1)^{-1} \\ B &= M^{1/2}(U_1|\cdots|U_k) \\ Z' &= N^{1/2}(\sigma_1 V_1|\cdots|\sigma_k V_k). \end{aligned}$$

PROOF. The results of **P 11.7.2** follow from the general theorem proved in **P 11.6.5** by choosing the matrices involved appropriately.

COROLLARY 11.7.3. The solution for α, B and Z which minimize (11.7.5) for any unitarily invariant norm is

$$\begin{aligned} \alpha &= \frac{1}{n} X1 = \bar{x} \text{ (say)} \\ B &= (U_1|\cdots|U_k), \ Z' = (\sigma_1 V_1|\cdots|\sigma_k V_k) \end{aligned}$$

where σ_i, U_i and V_i are as in the s.v.d.

$$X - \bar{x}1' = \sigma_1 U_1 V_1' + \ldots + \sigma_r U_r V_r'.$$

Note 1. If the norm chosen is Frobenius norm, the solution is the same as in Corollary 11.7.3. We can then compute the minimum value

$$\min_{\alpha,B,Z} \|X - \alpha 1' - BZ\|_F^2 = \sigma_{k+1}^2 + \ldots + \sigma_r^2.$$

In statistics, the closeness of fit of k dimensional hyperplane to given points is measured by the index

$$\frac{\sigma_{k+1}^2 + \ldots + \sigma_r^2}{\sigma_1^2 + \ldots + \sigma_r^2}.$$

Note 2. Let $V_1' = (v_{i1}, \ldots, v_{in})$. Then from the Corollary 11.7.3, we find the best representation of x_i on H_k is

$$\bar{x} + \sigma_1 v_{1i} U_1 + \sigma_2 v_{2i} U_2 + \ldots + \sigma_k v_{ki} U_k$$

so that by choosing $U_1, \ldots, U_k$ as coordinate axes and $\bar{x}$ as the origin, the m vector x_i can be represented as the point $(\sigma_1 v_{1i}, \ldots, \sigma_k v_{ki})$ in a k dimensional space. The coordinates $\sigma_1 v_{1i}, \ldots, \sigma_k v_{ki}$ are called the first k principal components of x_i.

Note 3. Let us get back to the compound measure of closeness (11.7.4) and consider some special vector norms. The results are given in the following propositions.

P 11.7.4 Let $x \in \mathbf{M}_{m,1}$ and $z = \alpha + Bc$ where $\alpha \in \mathbf{M}_{m,1}$, $B \in \mathbf{M}_{m,k}$, $\rho(B) = k$, and $c \in \mathbf{M}_{k,1}$. Let Σ be a positive definite matrix. Then

$$\min_c (x - z)'\Sigma^{-1}(x - z) \tag{11.7.6}$$

is attained at

$$c = (B'\Sigma^{-1}B)^{-1}B'\Sigma^{-1}(x - \alpha)$$

and the minimum value is

$$(x - \alpha)'(\Sigma^{-1} - \Sigma^{-1}P_B)(x - \alpha) \tag{11.7.7}$$

where $P_B = B(B'\Sigma^{-1}B)^{-1}B'\Sigma^{-1}$.

PROOF. The results are easily established by differentiating the quadratic expression in (11.7.6) with respect to the variable c and solving for c (see Section 6.5 for vector and matrix derivatives).

Now we consider the problem of minimizing the expression (11.7.4) with norm as in (11.7.6)

$$w_1(x_1 - z_1)'\Sigma^{-1}(x_1 - z_1) + \ldots + w_n(x_n - z_n)'\Sigma^{-1}(x_n - z_n) \quad (11.7.8)$$

where $z_i = \alpha + Bc_i$. First we minimize each term of (11.7.8) with respect to its c value and compute the minimum value using **P 11.7.4**,

$$\sum_{i=1}^{n} w_i(x_i - \alpha)'(\Sigma^{-1} - \Sigma^{-1}P_B)(x_i - \alpha). \quad (11.7.9)$$

Minimizing (11.7.9) with respect to α, we find the minimum to be

$$\sum_{i=1}^{n} w_i(x_i - \bar{x})'(\Sigma^{-1} - \Sigma^{-1}P_B)(x_i - \bar{x})$$
$$= \sum_{i=1}^{n} w_i(x_i - \bar{x})'\Sigma^{-1}(x_i - \bar{x}) - \sum_{i=1}^{n} w_i(x_i - \bar{x})'\Sigma^{-1}P_B(x_i - \bar{x}). \quad (11.7.10)$$

To further minimize with respect to B, we have to maximize

$$\sum_{i=1}^{n} w_i(x_i - \bar{x})'\Sigma^{-1}P_B(x_i - \bar{x}) = \operatorname{tr}(\Sigma^{-1}P_B S)$$

where

$$S = \sum_{i=1}^{n} w_i(x_i - \bar{x})(x_i - \bar{x})'.$$

P 11.7.5 The maximum of $\operatorname{tr}(\Sigma^{-1}P_B S)$ over B is attained at $B = \Sigma^{1/2}Q_*$, where Q_* is the matrix of first k eigenvectors of the matrix $\Sigma^{-1/2}S\Sigma^{-1/2}$.

PROOF. Note that

$$\begin{aligned}\operatorname{tr}(\Sigma^{-1}P_BS) &= \operatorname{tr}(\Sigma^{-1/2}\Sigma^{-1/2}P_B\Sigma^{1/2}\Sigma^{-1/2}S) \\ &= \operatorname{tr}(\Sigma^{-1/2}P_B\Sigma^{1/2}\Sigma^{-1/2}S\Sigma^{-1/2}) \\ &= \operatorname{tr}(PT)\end{aligned}$$

where $T = \Sigma^{-1/2}S\Sigma^{-1/2}$ and $P = \Sigma^{-1/2}P_B\Sigma^{1/2}$ is an idempotent matrix of rank k. Then P can be expressed as

$$P = QQ', \ Q \in \mathbf{M}_{m,k} \text{ and } Q'Q = I_k.$$

Now

$$\operatorname{tr}(PT) = \operatorname{tr}(Q'TQ).$$

Using **P 11.4.1**, the optimum Q which maximizes $\operatorname{tr}(PT)$ is the matrix of the first k eigenvectors of T, which may be denoted by Q_*. Then

$$P\Sigma^{1/2}P_B\Sigma^{-1/2} = P = Q_*Q_*'. \tag{11.7.11}$$

It is easy to see that the choice $B = \Sigma^{-1/2}Q_*$ satisfies equation (11.7.11). This proves **P 11.7.5**.

Note: The vectors $c_1, \ldots, c_n$ associated with the optimum B provide the best representation in a k dimensional space of points $x_1, \ldots, x_n$ in the $m(> k)$ dimensional space. They are called canonical coordinates and used in graphical representation of multivariate data as discussed in Rao (1948).

CHAPTER 12

OPTIMIZATION PROBLEMS IN STATISTICS AND ECONOMETRICS

12.1. Linear Models

In this chapter, we consider some general optimization problems which are useful in the statistical analysis of linear models. A linear model is of the form

$$\underset{n\times 1}{Y} = \underset{n\times m}{X} \underset{m\times 1}{\beta} + \underset{n\times 1}{\epsilon} \tag{12.1.1}$$

where Y is an n-vector random variable, X is an $n \times m$ matrix of rank $\rho(X) = r \leq m$, β is an m-vector of unknown parameters and ϵ is an n-vector of error variables with $E(\epsilon) = 0$ and covariance matrix $D(\epsilon) = \sigma^2 V$ an $n \times n$ non-negative definite matrix. The problems of interest are the estimation of the fixed parameters β and σ^2 and the estimation or the prediction of the random component ϵ.

A mixed linear model is of the form

$$Y = X\beta + U_1\xi_1 + \ldots + U_k\xi_k + \epsilon \tag{12.1.2}$$

where Y, X, β and ϵ are as in (12.1.1), U_i is an $n \times p_i$ matrix, ξ_i is a p_i-vector random variable such that $E(\xi_i) = 0$ and

$$\text{Cov}(\xi_i) = \sigma_i^2 I_{p_i},\ \text{Cov}(\xi_i, \xi_j) = 0,\ i \neq j,\ \text{Cov}(\xi_i, \epsilon) = 0 \tag{12.1.3}$$

for $i = 1, \ldots, k$. The problems of interest are the estimation of $\beta, \sigma^2, \sigma_1^2, \ldots, \sigma_k^2$ and the prediction of the random components $\xi_1, \ldots, \xi_k$ and ϵ.

12.2. Some Useful Lemmas

In this section, we consider some general results used in optimization problems in statistics. The following notation and assumptions are used throughout this chapter.

(1) $Sp(A)$ and $\rho(A)$ stand for the space generated by the columns of the matrix A and the rank of A, respectively. $A^{\perp}$ is a matrix of maximum rank such that $A'A^{\perp} = 0$.
(2) V is $n \times n$ non-negative definite matrix, X is $n \times m$ matrix and $W = (V|X)$. Note that $\rho(W) = \rho(V + XX')$ and $Sp(V|X) = Sp(V + XX')$.

P 12.2.1 The linear equations in matrices L of order $n \times k$ and Λ of order $m \times k$,

$$\begin{aligned} VL + X\Lambda &= C_1 \\ X'L \qquad &= C_2 \end{aligned} \tag{12.2.1}$$

admit solutions for any C_1 of order $n \times k$ and C_2 of order $m \times k$ such that $Sp(C_1) \subset Sp(W) = Sp(V|X)$ and $Sp(C_2) \subset Sp(X')$, respectively.

PROOF. We only have to show that

$$Sp\begin{bmatrix} C_1 \\ C_2 \end{bmatrix} \subset Sp\begin{bmatrix} V & X \\ X' & 0 \end{bmatrix}$$

i.e., if there exists a vector $(a' : b')$ such that

$$[a'|b']\begin{bmatrix} V & X \\ X' & 0 \end{bmatrix} = 0, \tag{12.2.2}$$

then

$$[a'|b']\begin{bmatrix} C_1 \\ C_2 \end{bmatrix} = 0. \tag{12.2.3}$$

Equation (12.2.2) is equivalent to

$$a'V + b'X' = 0, \ a'X = 0 \Rightarrow a'Va = 0,$$
$$\Rightarrow \ Va = 0, \ Xb = 0, \ X'a = 0 \Rightarrow \ a'W = 0, \ b'X' = 0.$$

Then $a'C' = 0$ since $Sp(C_1) \subset Sp(W)$ and $b'C_2 = 0$ since $Sp(C_2) \subset Sp(X')$, which proves Lemma 12.2.1.

The next lemma due to Rao (1989) is concerned with the minimization of a matrix function in the sense of Löwner.

P 12.2.2 Let L_0 and Λ_0 be a solution of the matrix equations in L of order $n \times k$ and Λ of order $m \times k$ given by

$$\begin{aligned} VL + X\Lambda &= F, \\ X'L \quad &= P, \end{aligned} \tag{12.2.4}$$

where $Sp(F) \subset Sp(V|X)$ and $Sp(P) \subset Sp(X')$. Then

$$L'VL - L'F - F'L \geq_L -F'L_0 - P'\Lambda_0 = -L_0'F - \Lambda_0'P \tag{12.2.5}$$

for all L such that $X'L = P$.

PROOF. A general solution of $X'L = P$ is $L = L_0 + X^{\perp}A$, where A is an arbitrary matrix. Substituting in the expression on the left hand side of (12.2.5), we find (using the equation $VL_0 = F - X\Lambda_0$)

$$\begin{aligned} L'VL - L'F - F'L &= L_0'VL_0 - L_0'F - F'L_0 + (X^{\perp}A)'V(X^{\perp}A) \\ &\geq_L L_0'VL_0 - L_0'F - F'L_0 = -F'L_0 - P'\Lambda_0 \end{aligned}$$

since $(X^{\perp}A)'V(X^{\perp}A)$ is non-negative definite.

Complements

12.2.1 (A special case of Lemma 12.2.1) Consider the nonhomogeneous quadratic form $Q(\ell) = \ell'V\ell - 2\ell'w$ where ℓ is an n-vector and w is an n-vector. Let p be an m-vector such that $p \in R(X')$. Then

$$\min_{X'\ell=p} Q(\ell) = \begin{cases} -\infty & \text{if } w \notin R(V|X) \\ -\ell_0'w - p'\lambda_0 & \text{if } w \in R(V|X) \end{cases}$$

where ℓ_0, λ_0 is a solution of

$$V\ell + X\lambda = w, \quad X'\ell = p.$$

12.2.2 Consider the matrix A and a g-inverse A^- of A partitioned as

$$A = \begin{bmatrix} V & X \\ X' & 0 \end{bmatrix}, \quad A^- = \begin{bmatrix} C_1 & C_2 \\ C_3 & -C_4 \end{bmatrix},$$

where $V \in \mathbf{M}_n$ and nnd; and $X \in \mathbf{M}_{m,n}$. Show that:

1a. $\rho(A) \leq m+n$; $\rho(A) = m+n$ if $\rho(V) = n$ and $\rho(X) = m$.
1b. $\rho(A) = \rho(V|X) + \rho(X)$.
2a. $XC_2'X = X,\ XC_3X = X$.
2b. $XC_4X' = XC_4'X' = VC_3'X' = XC_3V = VC_2X' = XC_2'V$.
2c. $X'C_1X, X'C_1V, VC_1X$ are null matrices.
2d. $VC_1VC_1V = VC_1V = VC_1'VC_1V = VC_1'V$.
2e. $\text{Tr}VC_1 = \rho(V|X) - \rho(X) = \text{Tr}VC_1'$.
3a. If $Sp(X) \subset Sp(V)$, then one choice of C_1, C_2, C_3, C_4 is

$$C_1 = V^- - V^-XC_3, C_2 = C_3', C_3 = (X'V^-X)^-X'V^-, C_4 = (X'V^-X)^-.$$

3b. In general

$$C_1 = W - WXC_3,\ C_2 = C_3',\ C_3 = TX'W,\ C_4 = -U + T$$

where $T = (X'(V + X'UX)^-X)^-$, $W = (V + X'UX)^-$ and U is any matrix such that $Sp(X) \subset Sp(V + XUX')$ and $Sp(V) \subset Sp(V + XUX')$.

12.3. Estimation in a Linear Model

We apply the result of **P 12.2.1** to estimate $X\beta, \epsilon$ and σ^2, the triplet characterizing the linear model

$$Y = X\beta + \epsilon,\ E(\epsilon) = 0,\ D(\epsilon) = \sigma^2 V, \tag{12.3.1}$$

where we use the symbol D to denote the dispersion (variance covariance) matrix. The reason for choosing $X\beta$ as a more natural parameter for estimation rather than β will be made clear once we define estimability of a linear parametric function.

DEFINITION 12.3.1. A set of k parametric functions $P'\beta$ where P is an $m \times k$ matrix, is said to be unbiasedly estimable by linear functions of Y if there exists a $k \times n$ matrix L such that $E(L'Y) = P'\beta$, a sufficient condition for which is $Sp(P) \subset Sp(X')$. Note that the entire m-vector parameter β is estimable only if $Sp(I) = Sp(X')$, i.e., when $\rho(X') = m$. The function $X\beta$ is always estimable.

Definition 12.3.2. A linear function $L_0'Y$ is said to be the minimum dispersion unbiased estimator (MDUE) of an estimable function $P'\beta$ if $L_0'X = P'$ and

$$D(L_0'Y) = \sigma^2 L_0'VL_0 \leq_L \sigma^2 L'VL = D(L'Y)$$

for all L such that $L'X = P'$.

The following theorem provides the main result for computing the MDUE of estimable parametric functions of β.

P 12.3.1 Let L_0 and Λ_0 be a solution of the equations

$$\begin{aligned} VL + X\Lambda &= 0 \\ X'L \qquad &= X'. \end{aligned} \tag{12.3.2}$$

Then $L_0'Y$ is the MDUE of $X\beta$ with the dispersion matrix, $D(L_0'Y) = -\sigma^2 X\Lambda_0$.

Proof. The covariance matrix of $L'Y$ is $\sigma^2 L'VL$. Using the result of **P 12.2.1**, the minimum of $L'VL$ with L subject to $X'L = X'$ is attained as stated in the theorem.

The following results are consequences of **P 12.3.1**.

(1) If $C_1, C_2, C_3, -C_4$ are the partitions of any g-inverse of the matrix of the equations in (12.3.2), then $L_0 = C_2X'$ and $\Lambda_0 = -C_4X'$ and the estimate of $X\beta$ is

$$X\hat{\beta} = L_0'Y = XC_2'Y, \tag{12.3.3}$$

$$\text{and} \quad D(X\hat{\beta}) = \sigma^2(-X\Lambda_0) = \sigma^2 XC_4X'. \tag{12.3.4}$$

(2) Let P be an $m \times k$ matrix such that $Sp(P) \subset Sp(X')$ in which case the function $P'\beta$ is estimable and the MDUE of $P'\beta$ is $P'\hat{\beta}$ where $\hat{\beta} = C_2'Y$ with the covariance matrix $\sigma^2 P'C_4P$.

To establish this result, we note the following.

(1) There exists an $n \times k$ matrix A such that $P = X'A$.
(2) If T is MDUE of a vector parameter θ, then $A'T$ is MDUE of $A'\theta$ (Why?).

Now we choose $T = X\hat{\beta} = XC_2'Y$ (i.e., defining $\hat{\beta} = C_2'Y$) which is the MDUE of $X\beta$. Then using (2), the MDUE of $A'X\beta(= P'\beta)$ is $A'X\hat{\beta} = A'XC_2'Y = P'C_2'Y = P'\hat{\beta}$. Further

$$\begin{aligned} D(P'\hat{\beta}) &= D(A'X\hat{\beta}) \\ &= A'D(X\hat{\beta})A = \sigma^2 A'XC_4X'A \quad \text{using (12.3.3)} \\ &= \sigma^2 P'C_4P. \end{aligned} \tag{12.3.5}$$

The above results show that we may consider $\hat{\beta}$ as formally an estimate of β and $\sigma^2 C_4$ as the dispersion of $\hat{\beta}$, in the sense that the MDUE of an estimable function $P'\beta$ is $P'\hat{\beta}$ and $D(P'\hat{\beta}) = \sigma^2 P'\text{Cov}(\hat{\beta})P = \sigma^2 P'C_4P$, although β as such may not be estimable.

As we have estimated $X\beta$ by $X\hat{\beta}$, it would appear natural to estimate ϵ, the error vector, by the residual $Y - X\hat{\beta}$. Is there a direct way of predicting ϵ by minimizing a suitable criterion function? We attempt to do this by considering linear estimators of the form $L'Y$, where L is an $n \times n$ matrix.

P 12.3.2 The linear predictor $L'Y$ of ϵ has a bounded mean dispersion error (MDE) for all β only if L is such that $X'L = 0$. Let (L_e, V_e) be any solution of the equation

$$\begin{aligned} VL + X\Lambda &= V, \\ X'L &= 0. \end{aligned} \tag{12.3.6}$$

Then $L_0'Y$ is the unbiased predictor of ϵ with the minimum mean dispersion error of $\sigma^2 X\Lambda_e$.

PROOF. The MDE of predicting ϵ by $L'Y$ is

$$\begin{aligned} E[(L'Y - \epsilon)(L'Y - \epsilon)'] &= E[(L'X\beta + (L' - I)\epsilon)(L'X\beta + (L' - I)\epsilon)'] \\ &= L'X\beta\beta'X'L + \sigma^2(L - I)'V(L - I). \end{aligned} \tag{12.3.7}$$

For any given L, the elements of (12.3.7) tend to infinity as the components of β tend to infinity. For boundedness, it is therefore necessary that $X'L = 0$. Then the MDE is

$$\begin{aligned} \sigma^2(L - I)'V(L - I) &= \sigma^2(L'VL - L'V - VL' + V) \\ &\geq_L E[(L_e'Y - \epsilon)(L_e'Y - \epsilon)'] \end{aligned}$$

using the result of **P 12.2.2**, establishing that MMDE of ϵ is $L_0'Y$ with the dispersion matrix (apart from the multiplier σ^2)

$$\begin{aligned}(L_e'VL_e - L_e'V - VL_e + V) &= (L_e'(V - X\Lambda_e) - L_e'V - V + X\Lambda_e + V)\\ &= (I - L_e)'X\Lambda_e = \sigma^2 X\Lambda_e,\end{aligned}$$

where L_e and Λ_e are any solutions of (12.3.6), and **P 12.3.2** is proved.

From **P 12.3.1** and **P 12.3.2**, the estimates of $X\beta$ and ϵ are $L_0'Y$ and $L_e'Y$, respectively. Do they add up to Y? Let us consider

$$\begin{aligned}&D(L_0'Y + L_e'Y - Y)\\ &= \sigma^2(L_0 + L_e - I)'V(L_0 + L_e - I)\\ &= \sigma^2(L_0 + L_e - I)'(VL_0 + VL_e - V) && (12.3.8)\\ &= \sigma^2(L_0 + L_e - I)'X(-\Lambda_e - \Lambda_0)\\ &= \sigma^2(L_0'X + L_e'X - X)(-\Lambda_e - \Lambda_0) = 0 && (12.3.9)\end{aligned}$$

using the equations (12.3.2) and (12.3.6) and substituting for VL_0 and VL_e in (12.3.8) and for $X'L_0$ and $X'L_e$ in (12.3.9). The result shows that $L_0'Y + L_e'Y = Y$ with probability one. This result can also be established by using the explicit solutions to L_0 and L_e in terms of the C_i matrices.

12.4. A Trace Minimization Problem

Let $\mathcal{C}$ be the collection of all matrices C of order $m \times n$ with real entries, X a given matrix of order $n \times r$, $V_1, V_2, \ldots, V_k$ given matrices each of order $m \times n$, and $p_1, p_2, \ldots, p_k$ given real scalars. Let

$$\mathcal{C}_1 = \{C \in \mathcal{C} : CX = 0,\ \mathrm{Tr}(CV_i') = p_i \text{ for each } i\}. \qquad (12.4.1)$$

The objective of this section is to minimize $\mathrm{Tr}(CC')$ over all $C \in \mathcal{C}_1$. If the collection $\mathcal{C}_1$ is empty, we have no case to answer. Assume that $\mathcal{C}_1$ is non-empty. Let P_X be the projection operator from $\mathbf{R}^n$ onto the subspace $Sp(X)$, which has an explicit representation

$$P_X = X(X'X)^-X', \qquad (12.4.2)$$

where $(X'X)^-$ is any g-inverse of $X'X$. Let $Q = I - P_X$ and note that $CX = 0 \Rightarrow C = DQ$ for some matrix D of order $m \times n$.

We are now ready to solve the problem. Consider the following system of linear equations

$$\sum_{i=1}^{k}[\text{tr}(V_iQV_j')]\lambda_i = p_j, \; j = 1, 2, \ldots, k \tag{12.4.3}$$

in unknown $\lambda_1, \lambda_2, \ldots, \lambda_k$ and check on their consistency. Let F be the symmetric matrix of order $k \times k$ given by

$$F = (\text{tr}(V_iQV_j')), \tag{12.4.4}$$

i.e., the (i,j)-th entry of F is given by $\text{tr}(V_iQV_j')$. The strategy is the following standard one. Let a be any vector of order $k \times 1$ orthogonal to the columns of F. Then show that it is also orthogonal to the vector p, where $p' = (p_1, p_2, \ldots, p_k)$. This would then imply that $p \in Sp(F)$ and hence the system (12.4.3) is solvable. Let $a' = (a_1, a_2, \ldots, a_k)$. If a is orthogonal to the columns of F, then we have

$$\sum_{i=1}^{k} a_i\text{tr}(V_iQV_j') = 0 \quad \text{for each} \quad j = 1, 2, \ldots, k. \tag{12.4.5}$$

The operation trace is linear, so that from (12.4.5), we obtain

$$\text{tr}((\sum_{i=1}^{k} a_iV_i)QV_j') = 0 \quad \text{for each} \quad j = 1, 2, \ldots, k. \tag{12.4.6}$$

Multiplying (12.4.6) by a_j and summing over j, we obtain

$$\text{tr}((\sum_{i=1}^{k} a_iV_i)Q(\sum_{i=1}^{k} a_iV_i)') = 0. \tag{12.4.7}$$

Since the matrix involved in the trace operation of (12.4.7) is non-negative definite, we have

$$((\sum_{i=1}^{k} a_iV_i)Q(\sum_{i=1}^{k} a_iV_i)') = 0.$$

Since Q is non-negative definite,

$$Q(\sum_{i=1}^{k} a_i V_i)' = 0 = \sum_{i=1}^{k} a_i Q V_i'. \tag{12.4.8}$$

We now show that $a'p = 0$. Let $C \in \mathcal{C}_1$. Observe that

$$\begin{aligned} a'p = \sum_{i=1}^{k} a_i p_i &= \sum_{i=1}^{k} a_i \text{tr}(CV_i') = \sum_{i=1}^{k} a_i \text{tr}(DQV_i') \\ &= \text{tr}(D(\sum_{i=1}^{k} a_i Q V_i')) = 0 \quad (\text{from } (12.4.8)). \end{aligned}$$

Thus the solvability of the system (12.4.3) of equations is assured. We are now ready to state the result which will solve the optimization problem.

P 12.4.1 The minimum of $\text{tr}(CC')$ over all $C \in \mathcal{C}_1$ is attained at any matrix C_* given by

$$C_* = \sum_{i=1}^{k} \lambda_i V_i Q, \tag{12.4.9}$$

where $\lambda_1, \lambda_2, \ldots, \lambda_k$ is a solution to the following system of linear equations:

$$\sum_{i=1}^{k} [\text{tr}(V_i Q V_j')] \lambda_i = p_j, \ j = 1, 2, \ldots, k.$$

PROOF. First, let us check whether C_* belongs to $\mathcal{C}_1$. We need to verify, first, that $C_* X = 0$. This follows from the fact that

$$QX = (I - P_X)X = X - X(X'X)^- X'X = X - X = 0.$$

Second, we need to verify that $\text{tr}(C_* V_j') = p_j$ for each j. From the stipulation that λ_i's satisfy the linear equations above, it follows that

$$\text{tr}(C_* V_j') = \sum_{i=1}^{k} [\text{tr}(V_i Q V_j')] \lambda_i = p_j.$$

Thus $C_* \in \mathcal{C}_1$. The next objective is to show that C_* is optimal. Let C be any matrix in $\mathcal{C}_1$. One can always write $C = C_* + G$, for a suitable matrix G. The fact that $C \in \mathcal{C}_1$ implies, at the outset, that

$$0 = CX = (C_* + G)X = C_*X + GX = GX,$$

or more relevantly, that

$$GX = 0. \tag{12.4.10}$$

This implies that $GP_X = 0$. (Why?) Further,

$$GQ = G(I - P_X) = G. \tag{12.4.11}$$

Additionally,

$$p_j = \text{tr}(CV_j') = \text{tr}((C_* + G)V_j') = \text{tr}(C_*V_j') + \text{tr}(GV_j') = p_j + \text{tr}(GV_j')$$

for each j implies that

$$\text{tr}(GV_j') = 0 \quad \text{for each } j. \tag{12.4.12}$$

Crucially, from (12.4.11) and (12.4.12), we observe that

$$\begin{aligned}\text{tr}(C_*G') &= \text{tr}((\sum_{i=1}^{k} \lambda_i V_i Q)G') = \text{tr}(\sum_{i=1}^{k} \lambda_i V_i Q G') \\ &= \text{tr}(\sum_{i=1}^{k} \lambda_i V_i G') = \sum_{i=1}^{k} \lambda_i \text{tr}(V_i G') = 0. \end{aligned} \tag{12.4.13}$$

Finally, we observe that

$$\begin{aligned}\text{tr}(CC') &= \text{tr}[(C_* + G)(C_* + G)'] \\ &= \text{tr}(C_*C_*') + \text{tr}(GG') + \text{tr}(C_*G') + \text{tr}(GC_*') \\ &= \text{tr}(C_*C_*') + \text{tr}(GG') \geq \text{tr}(C_*C_*').\end{aligned}$$

This completes the proof.

We now focus on another problem of the same type as the one proposed at the beginning of this section. We assume that $m = n$. Let $V_1, V_2, \ldots, V_k$ be a finite collection of known symmetric matrices of order $m \times m$. Let X be a given matrix of order $m \times r$. Let $p_1, p_2, \ldots, p_k$ be real scalars. Let

$$\mathcal{C}_2 = \{C : C = C' \text{ of order } m \times m, CX = 0, \text{ tr } (CV_j) = p_j, \forall j\}.$$

The objective is to minimize $\text{tr}(C^2)$ over all $C \in \mathcal{C}_2$. The following result provides an explicit solution to the problem.

P 12.4.2 The minimum of $\text{tr}(C^2)$ is attained at any matrix C_* given by

$$C_* = \sum_{i=1}^{k} \lambda_i Q V_i Q, \tag{12.4.14}$$

where $\lambda_1, \lambda_2, \ldots, \lambda_k$ is a solution to the system of linear equations:

$$\sum_{i=1}^{k} [\text{tr}(QV_iQV_j)]\lambda_i = p_j, \ j = 1, 2, \ldots, k. \tag{12.4.15}$$

We will not give a proof of this result. One can imitate the proof of **P 12.4.1** to establish this result. One can notice right away that C_* is symmetric and that C_* belongs to $\mathcal{C}_2$. The consistency of the system of equations (12.4.15) follows in a similar vein. (Try it.) The solution to the optimization problems posed above may not be unique. Nonuniqueness arises if the matrices involved in equations (12.4.3) and (12.4.15) are nonsingular. In practice, we are not unduly concerned with the uniqueness problem.

12.5. Estimation of Variance

Let us get back to the basic model: $Y = X\beta + \varepsilon$ with $E\varepsilon = 0$ and $D(\varepsilon) = \sigma^2 V$ where $\sigma^2 > 0$ is unknown. In this section, we assume that $V = I$. The main objective of this section is the estimation of σ^2. Let $Y'CY$ be a quadratic form in the data Y, for some symmetric matrix C, which we would like to propose as an estimator of σ^2. The basic requirement is that it be unbiased. Let us record a very general result on the expected value of a quadratic form.

P 12.5.1 Let Y be a random vector of order $n \times 1$ with mean vector μ and dispersion matrix $\sigma^2 V$. Let C be any symmetric matrix of order $n \times n$. Then

$$E(Y'CY) = \sigma^2 \text{tr}(CV) + \mu' C \mu. \tag{12.5.1}$$

PROOF. Observe that $V = E(Y - \mu)(Y - \mu)' = EYY' - \mu\mu'$. Also,

$$\begin{aligned} E(Y'CY) &= E\,\text{tr}(Y'CY) = E\,\text{tr}(CYY') = \text{tr}(E(CYY')) \\ &= \text{tr}(CE(YY')) = \text{tr}(C(\sigma^2 V + \mu\mu')) = \sigma^2 \text{tr}(CV) + \mu' C \mu. \end{aligned}$$

In our model, we identify that $V = I$ and $\mu = X\beta$. Consequently,

$$E(Y'CY) = \sigma^2\text{tr}(C) + \beta X'CX\beta.$$

Since we would like to have the quadratic form to be unbiased, we need to nullify the nuisance element present in its expected value, namely $\beta X'CX\beta$. We impose the following conditions on the matrix C:

$$\text{tr}C = 1 \quad \text{and} \quad CX = 0. \tag{12.5.2}$$

Next, we would like the unbiased estimator $Y'CY$ to have minimum variance. But we have to venture into the realm of fourth order moments of the random variables Y_i's. The variance of $Y'CY$ can be computed but it is complicated. One way to get out of this rigmarole is to assume that the random variables involved in the linear model have a multivariate normal distribution. Under this assumption

$$\text{Var}(Y'CY) = 2\sigma^2\text{tr}(C^2).$$

In order to find a minimum variance estimator $Y'CY$ of σ^2, we need to minimize $\text{tr}C^2$ subject to the conditions: $\text{tr}C = 1$ and $CX = 0$. This problem falls into the orbit of **P 12.4.2**. We identify that $k = 1$; $V_1 = I$; and $p_1 = 1$. The optimal matrix C_* is given by $C_* = \lambda_1 QV_1Q = \lambda_1 Q^2 = \lambda_1 Q$, where λ_1 satisfies the equation $\text{tr}(QV_1QV_1)\lambda_1 = p_1 = 1$, from which we obtain, $\lambda_1 = 1/\text{tr}(Q)$. Consequently, the minimum variance quadratic unbiased estimator of σ^2 is given by

$$(\text{tr}\, Q)^{-1}Y'QY = (\text{tr}\, Q)^{-1}Y'(I - X(X'X)^- X')Y. \tag{12.5.3}$$

Let us make a few comments on the above derivation. There is a standard approach available to solve this problem. One uses the method of least squares to estimate β. As a matter of fact, a solution to the resultant normal equations in $\hat{\beta}$ is given by $\hat{\beta} = (X'X)^- X'Y$. Then one shows that the residual sum of squares

$$(Y - X\hat{\beta})'(Y - X\hat{\beta}) = Y'(I - X(X'X)^- X')Y$$

has expectation $(n - \rho(X))\sigma^2$, from which one obtains an unbiased estimator of σ^2. This estimator is precisely what we have obtained in (12.5.3). Our approach has given us something more. We showed that the estimator (12.5.3) has minimum variance under the normality assumption. The next goal is to dispense away with the normality assumption. This is the objective in the next section.

12.6. The Method of MINQUE: A Prologue

The acronym in the title of this section stands for minimum norm quadratic unbiased estimation. Let us work in the realm of the linear model: $Y = X\beta + \varepsilon$ with $E\varepsilon = 0$ and $D(\varepsilon) = \sigma^2 I$. In the last section, under the pretext of multivariate normality, we obtained the minimum variance quadratic unbiased estimator of σ^2. In this section, we will make an attempt to obtain an optimal quadratic estimator of σ^2 without invoking the normality assumption.

If the error vector $\varepsilon' = (\varepsilon_1, \varepsilon_2, \dots, \varepsilon_n)$ is observable, then a natural estimator of σ^2 is

$$(1/n)\sum_{i=1}^{n} \varepsilon_i^2 = (1/n)\varepsilon'\varepsilon = \varepsilon'(\frac{1}{n}I)\varepsilon.$$

But only the data vector Y is observable. We need to build an estimator of σ^2 based on Y. As professed earlier, we will entertain only quadratic estimators of σ^2. Let C be a symmetric matrix of order $n \times n$. We have seen, in the last section, that $Y'CY$ is an unbiased estimator of σ^2 if $CX = 0$ and $\mathrm{tr}(C) = 1$. Under these restrictions, the quadratic form $Y'CY$ simplifies to

$$\begin{aligned} Y'CY &= (X\beta + \varepsilon)'C(X\beta + \varepsilon) \\ &= \beta'X'CX\beta + \varepsilon'C\varepsilon + \beta'X'C\varepsilon + \varepsilon'CX\beta = \varepsilon'C\varepsilon. \end{aligned}$$

Consequently, the difference between what is contrived and what is ideal works out to be:

$$Y'CY - \varepsilon'(n^{-1}I)\varepsilon = \varepsilon'(C - n^{-1}I)\varepsilon.$$

The critical idea behind the method of MINQUE is to choose C such that it is close to $(1/n)I$. Choosing the Frobenius norm ($\|B\|_F = \mathrm{tr}(BB')$), the objective is to minimize $\|C - n^{-1}I\|_F$ subject to unbiasedness conditions that $CX = 0$ and tr $(C) = 1$. Observe that

$$\|C - n^{-1}I\|_F = \mathrm{tr}((C - n^{-1}I)(C - n^{-1}I)) = \mathrm{tr}(C^2) - n^{-1}.$$

Consequently, the above problem reduces to minimizing $\mathrm{tr}(C^2)$ subject to the conditions that $CX = 0$ and $\mathrm{tr}(C) = 1$. This is precisely what we have done in Section 12.5. The solution was identified and we obtained the estimator of σ^2 the usual one based on the residuals. There are two aspects of this method worth recording. One is that the usual estimator of σ^2 is optimal based on the underlying criterion of the method of MINQUE. The other aspect is that the method is amenable to generalization. This is what we are proposing to carry out in subsequent sections.

12.7. Variance Components Models and Unbiased Estimation

Variance components models (random and mixed effects models) are routinely postulated for a variety of datasets. They also play a useful role in analyzing repeated measurements data. Any such model can be described by a linear model, $Y = X\beta + \varepsilon$, with $E\varepsilon = 0$ and

$$D(\varepsilon) = \theta_1 V_1 + \theta_2 V_2 + \ldots + \theta_k V_k = V(\theta), \quad \text{say}, \tag{12.7.1}$$

where $V_1, V_2, \ldots, V_k$ are known non-negative definite matrices, $\theta_1, \theta_2, \ldots, \theta_k$ are unknown real numbers and $\theta' = (\theta_1, \theta_2, \ldots, \theta_k)$. There may be additional conditions imposed on θ_i's. For example, θ_i's are bound by the restriction that $V(\theta)$ be non-negative definite. For the time being, let us not put too much emphasis on the range of values for the vector θ, except for the minor technical condition that the range of values for the vector θ has a non-empty interior, which is a topological condition. The quantities $\theta_1, \theta_2, \ldots, \theta_k$ are called the variance components of the model. There may be a certain element of ambiguity in the specification of the matrices $V_1, V_2, \ldots, V_k$. We can rewrite, for example,

$$\theta_1 V_1 + \ldots + \theta_k V_k = (2\theta_1)(2^{-1}V_1) + \ldots + (2\theta_k)(2^{-1}V_k).$$

The matrices $2^{-1}V_1, 2^{-1}V_2, \ldots, 2^{-1}V_k$ are known non-negative definite matrices. Shall we call $2\theta_1, 2\theta_2, \ldots, 2\theta_k$ as the variance components of the model? If we fix the matrices $V_1, V_2, \ldots, V_k$, variance components are probably uniquely determined. Another complication could arise. It may be possible that there are two distinct vectors $(\theta_1, \theta_2, \ldots, \theta_k)$ and $(\theta_1^*, \theta_2^*, \ldots, \theta_k^*)$ such that $\theta_1 V_1 + \ldots + \theta_k V_k = \theta_1^* V_1 + \ldots + \theta_k^* V_k$. In such a contingency, estimation of all the variance components is not possible.

We now assume that the matrices $V_1, V_2, \ldots, V_k$ are fixed and the variance components are unambiguously defined. The main objective of this section is to find some good estimators of the variance components based on the data Y. More generally, let

$$p_1\theta_1 + p_2\theta_2 + \ldots + p_k\theta_k \tag{12.7.2}$$

be a linear combination of the variance components, where $p_1, p_2, \ldots, p_k$ are known. As usual, we will entertain only quadratic estimators $Y'CY$ for the linear combination of the variance components. We need to impose conditions on the matrix of the quadratic form in order that it be unbiased. Note that

$$p_1\theta_1 + p_2\theta_2 + \ldots + p_k\theta_k = E(Y'CY) = \sum_{i=1}^{k} \theta_i \text{tr}(CV_i) + \beta' X'CX\beta.$$

A set of necessary and sufficient conditions for $Y'CY$ to be an unbiased estimator of the given linear combination of the variance components are:

$$X'CX = 0 \text{ and } \text{tr}(CV_i) = p_i, \; i = 1, 2, \ldots, k. \tag{12.7.3}$$

Will there be a symmetric matrix C at all satisfying (12.7.3)? We need a criterion for the existence of a solution to (12.7.3) in terms of the matrices $V_1, V_2, \ldots, V_k$ and X. Construct the following symmetric matrix of order $k \times k$,

$$H = (\text{tr}(V_iV_j - PV_iPV_j)), \tag{12.7.4}$$

where P denotes, as usual, the projection operator $X(X'X)^-X'$. The matrix H is computable once we know V_i's and X. We will show that there is a matrix C satisfying (12.7.3) if and only if the coefficient vector p consisting of the components $p_1, p_2, \ldots, p_k$ belongs to $Sp(H)$. We will now prove this assertion. A general solution to the equation $X'CX = 0$ in unknown C is given by $B - PBP$ for some symmetric matrix B. (How?) Suppose there exists a matrix C satisfying (12.7.3). Then $C = B - PBP$ for some symmetric matrix B. Note that for each i,

$$\begin{aligned} p_i = \text{tr}(CV_i) &= \text{tr}(BV_i - PBPV_i) = \text{tr}(BV_i - BPV_iP) \\ &= \text{tr}(B(V_i - PV_iP)) = (\text{vec}(V_i - PV_iP))'(\text{vec}(B)). \end{aligned}$$

Let us take a critical look at the last step. We use the vec operation to express the trace of a product of two symmetric matrices. Recall the vec operation from Section 6.2. The vec operation performed on a matrix stacks all the entries of the matrix column by column into a single column vector. If $E = (e_{ij})$ and $F = (f_{ij})$ are two symmetric matrices of the same order, then $\operatorname{tr}(EF) = \sum_i \sum_j e_{ij} f_{ij} = (\operatorname{vec}(E))'(\operatorname{vec}(F))$. (Try it.) Note that

$$p = \begin{bmatrix} p_1 \\ p_2 \\ \vdots \\ p_k \end{bmatrix} = \begin{bmatrix} [\operatorname{vec}(V_1 - PV_1P)]' \\ [\operatorname{vec}(V_2 - PV_2P)]' \\ \cdots \\ [\operatorname{vec}(V_k - PV_kP)]' \end{bmatrix} [\operatorname{vec}(B)]. \tag{12.7.5}$$

Let T be the matrix of order $k^2 \times k$ given by

$$T = (\operatorname{vec}(V_1 - PV_1P), \operatorname{vec}(V_2 - PV_2P), \dots, \operatorname{vec}(V_k - PV_kP)).$$

Then (12.7.5) can be rewritten as

$$p = T'\operatorname{vec}(B).$$

What this means is that the vector p is a linear combination of the columns of T'. Consequently, $p \in Sp(T')$, the range space of the matrix T'. But $Sp(T') = Sp(T'T)$. Let us compute $T'T$. The (i,j)-th entry of $T'T$ is given by, in view of what we have observed about the relationship between the vec operation and trace,

$$\begin{aligned} [\operatorname{vec}(V_i - PV_iP)]'[\operatorname{vec}(V_j - PV_jP)] &= \operatorname{tr}((V_i - PV_iP)(V_j - PV_jP)) \\ &= \operatorname{tr}(V_iV_j - V_iPV_jP - PV_iPV_j + PV_iPPV_jP) \\ &= \operatorname{tr}(V_iV_j - PV_iPV_j), \end{aligned}$$

since the projection operator is idempotent. This entry is precisely the (i,j)-th entry of H. Consequently, $T'T = H$, and hence $p \in Sp(H)$.

Conversely, suppose $p \in Sp(H) = Sp(T')$. Then there exists a column vector ℓ of order $k^2 \times 1$ such that $p = T'\ell$. Now undo the vec operation. Let B be the matrix of order $k \times k$ such that $\operatorname{vec}(B) = \ell$. (How?) The matrix B may not be symmetric. Let

$$C = (1/2)[B + B' - P(B + B')P]. \tag{12.7.6}$$

This matrix C satisfies all the constraints of (12.7.3). The symmetric matrix C defined in (12.7.6) satisfies $X'CX = 0$. In order to establish this, one merely uses the fact that $PX = X$ and $X'P = X'$. Using (12.7.6) and after some simplification, we get

$$\begin{aligned} \text{tr}(CV_i) &= (1/2)\text{tr}[(B+B')V_i - P(B+B')PV_i] \\ &= [\text{vec}(V_i - PV_iP)]'[\text{vec}(B)]. \end{aligned} \qquad (12.7.7)$$

A critical look at (12.7.7) yields that $\text{tr}(CV_i) = p_i$ for every i. The above presentation clearly spells out when one can estimate a given linear combination of the variance components unbiasedly using quadratic estimators. Let us enshrine this in the following proposition.

P 12.7.1 Let $p_1\theta_1 + \ldots + p_k\theta_k$ be a linear combination of the variance components. Then there exists a quadratic unbiased estimator of the combination if and only if the vector of coefficients $p \in Sp(H)$.

12.8. Normality Assumption and Invariant Estimators

Let us get back to the main track. In what follows, we assume that there are quadratic unbiased estimators available to estimate a given linear combination of the variance components. We need to venture beyond unbiasedness. One approach would be to seek a symmetric matrix C such that $Y'CY$ is unbiased and has minimum variance in the class of all unbiased quadratic estimators of the given linear combination of the variance components. But the computation of the variance of the quadratic form involves third and fourth order moments of the error random vector and it is complicated. However, the variance of an unbiased quadratic estimator has a simple form of expression provided we assume normality for the distribution of the error random vector. Under normality, $\text{var}(Y'CY) = 2\text{tr}((CV)^2) + 4\beta'X'CVCX\beta$. But the minimization of $\text{var}(Y'CY)$ over all quadratic unbiased estimators is fraught with difficulties. The variance involves not only the dispersion matrix V but also the parameter vector β. The supposition that $Y'CY$ is an unbiased estimator of $p'\theta$ imposes the condition that $X'CX = 0$. This will not make the term involving β to go away. One way to get rid of this term is to impose the stronger condition that $CX = 0$. We not only want our estimators to be unbiased but have a little extra to guarantee that $CX = 0$. Is there a statistical or mathematical interpretation of the condition $CX = 0$? We will provide one interpretation of the condition. Let β_0 be an arbitrary but fixed vector of order $m \times 1$. Consider the new data vector $Y_* = Y + X\beta_0$, which is observable as soon as Y is observed. Let $\beta_* = \beta + \beta_0$. Since β is an unknown parameter vector in our original linear model, β_* is also unknown. Consider the linear model: $Y_* = X\beta_* + \varepsilon$. Structurally, this model is no different from our original model. Either of them could be used to estimate the parameter vector β and the variance components. Any estimator one provides to estimate the variance components whether one uses the data Y or Y_* should lead to the same value. This is the idea behind in the following definition.

DEFINITION 12.8.1. A quadratic form $Y'CY$ in the data is said to be an invariant unbiased estimator of a linear combination $p'\theta$ of variance components if it is unbiased and $Y'CY = (Y + X\beta_0)'C(Y + \beta_0)$ for all $\beta_0 \in \mathbf{R}^m$.

There is nothing new in the idea of invariance. In a substantial number of decision theoretic procedures, invariance property plays a crucial role in deriving some optimal procedures. The invariance property introduced in Definition 12.8.1 is akin to shift invariance. The collection of shift transformations, $Y \to Y + X\beta_0$, is indexed by β_0, where β_0 roams freely over the Euclidean space $\mathbf{R}^m$.

The condition of unbiasedness imposes restrictions on the matrix C. We have already seen that we must have that $X'CX = 0$ and $\text{tr}(CV_i) = p_i$ for all i. The additional property of invariance imposes also a restriction on C. The following,

$$\begin{aligned} Y'CY &= (Y + X\beta_0)'C(Y + X\beta_0) \\ &= Y'CY + 2Y'CX\beta_0 + \beta_0'X'CX\beta_0, \end{aligned} \tag{12.8.1}$$

is valid for all β_0 if and only if $CX = 0$. Let us paraphrase the above in the following proposition.

P 12.8.2. A quadratic form $Y'CY$ in the data is an invariant unbiased estimator of a linear combination $p'\theta$ of the variance components if and only if

$$CX = 0 \quad \text{and } \text{tr}(CV_i) = p_i,\ i = 1, 2, \ldots, k. \tag{12.8.2}$$

Understandably, the conditions (12.8.2) are stronger than the conditions (12.7.3). The next task is to mount an inquiry under what conditions there exists a symmetric matrix C meeting the requirements of (12.8.2)?

Let us provide an outline how this can be done. The equations $\text{tr}(CV_i) = p_i$, $i = 1, 2, \ldots, k$, can be viewed as a bunch of linear equations in the entries of C using the vec operation. But C needs to satisfy an additional restriction that $CX = 0$. In order to accommodate this restriction, determine the general form of C satisfying $CX = 0$, and accommodate this in the other set of constraints which can be rewritten as

$$\text{tr}(CV_i) = [\text{vec}(V_i)]'[\text{vec}(C)] = p_i, \ i = 1, 2, \ldots, k.$$

This is the technique we followed in Section 12.7 towards the build-up of **P 12.7.1**. We will not pursue this natural approach. We will try to exploit **P 12.7.1** to determine precise conditions under which we can find an invariant unbiased quadratic estimator of $p'\theta$. The key idea is to build another linear model in the framework of **P 12.7.1**. Any symmetric matrix C satisfying $CX = 0$ must be of the form $C = BAB'$ for some symmetric matrix A and for some matrix B of maximal rank satisfying $B'B = I$, $B'X = 0$ and $BB' = Q = I - P$, where P is the projection operator $X(X'X)^-X'$. (Show this.) Introduce a transformation $Y_* = B'Y$. Observe that $EY_* = B'X\beta = 0$ and $D(Y_*) = \theta_1 V_{1*} + \ldots + \theta_k V_{k*}$, where $V_{i*} = B'V_iB$. Thus we have a linear model: $Y_* = \varepsilon_*$ with the design matrix $X_* = 0$, $E\varepsilon_* = 0$ and $D(\varepsilon_*) = D(Y_*)$. Further, the new quadratic form

$$Y_*'AY_* = Y'BAB'Y = Y'CY.$$

We use the new linear model to discuss unbiased estimation of $p'\theta$. By **P 12.7.1**, $Y_*'AY_*$ is an unbiased estimator of $p'\theta$ if and only if $p \in Sp(H_*)$, where

$$H_* = (\text{tr}(V_{i*}V_{j*} - P_*V_{i*}P_*V_{j*})),$$

with P_* as the projection operator associated with the design matrix X_* of the model. But $X_* = 0$. Consequently, $P_* = 0$. Let us simplify the matrix H_*. Observe that $\text{tr}(V_{i*}V_{j*} - P_*V_{i*}P_*V_{j*}) = \text{tr}(B'V_iBB'V_jB) = \text{tr}(B'V_iQV_jB) = \text{tr}(BB'V_iQV_j) = \text{tr}(QV_iQV_j)$. Consequently, $H_* = (\text{tr}(QV_iQV_j))$. Let us rewrite the matrix H_* involving the projection operator directly. Observe that

$$\begin{aligned}\operatorname{tr}(QV_iQV_j) &= \operatorname{tr}((I-P)V_i(I-P)V_j) \\ &= \operatorname{tr}(V_iV_j) + \operatorname{tr}(PV_iPV_j) - \operatorname{tr}(PV_iV_j) - \operatorname{tr}(PV_jV_i).\end{aligned}$$

One might get the impression that the invariance property is just introduced to accommodate the normality assumption, viz., to make the variance of the unbiased estimator $Y'CY$ of $p'\theta$ look elegant. This is partly true. But the concept of invariance is intuitively appealing and it is not unreasonable to demand that the estimator be invariant, in addition.

12.9. The Method of MINQUE

We have broached unbiased estimation and invariant unbiased estimation of variance components in Sections 12.7 and 12.8. From now on, we will focus on developing an optimality criterion to select an optimal invariant unbiased estimator and optimal unbiased estimator of a variance component.

In many variance components models, the error random vector has a structural representation, $\varepsilon = U_1\xi_1 + U_2\xi_2 + \ldots + U_k\xi_k$, where U_i is a known matrix of order $n_i \times n_i$, ξ_i is a random vector of order $n_i \times 1$ with mean vector 0 and dispersion matrix $\sigma_i^2 I_{n_i}$, $i = 1, 2, \ldots, k$, and I_{n_i} is the identity matrix of order $n_i \times n_i$. Further, $\xi_1, \xi_2, \ldots, \xi_k$ are assumed to be independent. Under this canopy of assumptions, we have

$$\mathrm{D}(\varepsilon) = \sigma_1^2 U_1U_1' + \sigma_2^2 U_2U_2' + \ldots + \sigma_k^2 U_kU_k'.$$

In conformity with (12.7.1), we identify V_i with U_iU_i' for each i. If ξ_i is observable, then a natural estimator of σ_i^2 is $(1/n_i)\xi_i'\xi_i$. Consequently, a natural estimator of $p_1\sigma_1^2 + p_2\sigma_2^2 + \ldots + p_k\sigma_k^2$ is

$$\begin{aligned}\sum_{i=1}^{k}(p_i/n_i)\xi_i'\xi_i &= \sum_{i=1}^{k}(p_i/n_i)\xi_i'I_{n_i}\xi_i \\ &= (\xi_1', \xi_2', \ldots, \xi_k')D(\xi_1', \xi_2', \ldots, \xi_k')', \qquad (12.9.1)\end{aligned}$$

where D is the block-diagonal matrix given by

$$D = \operatorname{diag}\{(p_1/n_1)I_{n_1}, (p_2/n_2)I_{n_2}, \ldots, (p_k/n_k)I_{n_k}\}.$$

There is a reason in rewriting the natural estimator in the way it was written in (12.9.1). We will see shortly. But we need to develop an estimator of the given linear combination of variance components in terms of the data Y. Let us focus on invariant unbiased estimators of $p'\theta$. We have already noted down the necessary and sufficient conditions for the quadratic estimator $Y'CY$ to be an invariant unbiased estimator of $p'\theta$. See **P 12.8.2**. Let us rewrite the estimator in terms of the structural representation of ε. Note that, by (12.9.11),

$$\begin{aligned} Y'CY &= \beta'X'CX\beta + +2(U_1\xi_1 + U_2\xi_2 + \ldots + U_k\xi_k)'CX\beta \\ &\quad + (U_1\xi_1 + U_2\xi_2 + \ldots + U_k\xi_k)'C(U_1\xi_1 + U_2\xi_2 + \ldots + U_k\xi_k) \\ &= (U_1\xi_1 + U_2\xi_2 + \ldots + U_k\xi_k)'C(U_1\xi_1 + U_2\xi_2 + \ldots + U_k\xi_k) \\ &= (\xi_1', \xi_2', \ldots, \xi_k')D_*(\xi_1', \xi_2', \ldots, \xi_k')', \end{aligned} \tag{12.9.2}$$

where D_* is the block matrix whose (i,j)-th block is given by $U_i'CU_j$. The critical idea behind the method of MINQUE is to get the proposed invariant unbiased estimator of $p'\theta$ involving data close to the natural estimator. This is tantamount to getting D_* close to D in some norm. The Frobenius norm of $(D_* - D)$ is given by

$$\begin{aligned} \|D_* - D\|_F &= \text{tr}((D_* - D)(D_* - D)') = \text{tr}((D_* - D)^2) \\ &= \sum_{i=1}^{k} \text{tr}((U_i'CU_i - (p_i/n_i)I_{n_i})^2) + \sum_{\substack{i=1 \\ i \neq j}}^{k} \sum_{j=1}^{k} \text{tr}(U_i'CU_jU_j'CU_i) \\ &= \sum_{i=1}^{k} \text{tr}(U_i'CU_iU_i'CU_i) + \sum_{i=1}^{k} (p_i/n_i)^2 \text{tr}(I_{n_i}) \\ &\quad - 2\sum_{i=1}^{k} (p_i/n_i)\text{tr}(U_i'CU_i) + \sum_{\substack{i=1 \\ i \neq j}}^{k} \sum_{j=1}^{k} \text{tr}(CU_jU_j'CU_iU_i') \\ &= \sum_{i=1}^{k} \text{tr}(CU_iU_i'CU_iU_i') + \sum_{i=1}^{k} p_i^2/n_i \\ &\quad - 2\sum_{i=1}^{k} (p_i/n_i)\text{tr}(CU_iU_i') + \sum_{\substack{i=1 \\ i \neq j}}^{k} \sum_{j=1}^{k} \text{tr}(CV_jCV_i) \end{aligned}$$

$$= \sum_{i=1}^{k} \mathrm{tr}(CV_iCV_i) + \sum_{i=1}^{k} p_i^2/n_i - 2\sum_{i=1}^{k} p_i^2/n_i$$

$$+ \sum_{i=1}^{k}\sum_{j=1}^{k} \mathrm{tr}(CV_jCV_i)$$

$$= \mathrm{tr}((C\sum_{i=1}^{k} V_i)^2) - \sum_{i=1}^{k} p_i^2/n_i = \mathrm{tr}((CV_0)^2) - \sum_{i=1}^{k} p_i^2/n_i,$$

where $V_0 = V_1 + \ldots + V_k$. The objective now is to determine a symmetric matrix C which satisfies $CX = 0$ and tr $(CV_i) = p_i$, $i = 1, 2, \ldots, k$ and minimizes

$$\mathrm{tr}((CV_0)^2). \tag{12.9.3}$$

The optimization problem seems to fall within the purview of **P 12.4.2.** Not quite. We need to work a little bit more to bring the above problem to the format of **P 12.4.2.** Since V_0 is non-negative definite, we can find a non-negative definite matrix B such that $B^2 = V_0$. The matrix B can be termed as a square root of B. For convenience, one may write B as $V_0^{1/2}$. Note that

$$\mathrm{tr}((CV_0)^2) = \mathrm{tr}((CV_0CV_0)) = \mathrm{tr}((CB^2CB^2)) = \mathrm{tr}((BCB)(BCB)).$$

Note that BCB is symmetric. In many applications, it will turn out that V_0 will be positive definite. We will now assume that V_0 is positive definite. Consequently, B will be positive definite. In such a case, knowing C is equivalent to knowing BCB. Let $A = BCB$. We need to transform the constraints into a set of constraints involving A. Note that $CX = 0$ is equivalent to $B^{-1}AB^{-1}X = 0$, which is equivalent to $A(B^{-1}X) = 0$. The condition that $\mathrm{tr}(CV_i) = p_i$ is equivalent to $\mathrm{tr}(B^{-1}AB^{-1}V_i) = p_i$, which is equivalent to $\mathrm{tr}(A(B^{-1}V_iB^{-1})) = p_i$. Let

$$X_* = B^{-1}X \quad \text{and} \quad V_{i*} = B^{-1}V_iB^{-1},\ i = 1, 2, \ldots, k. \tag{12.9.4}$$

The optimization problem translates into the following problem. Determine a symmetric matrix A such that

$$AX_* = 0 \quad \text{and } \mathrm{tr}(AV_{i*}) = p_i,\ i = 1, 2, \ldots, k \tag{12.9.5}$$

and it minimizes $\text{tr}(A^2)$. This problem fits nicely into the format of **P 12.4.2**. An optimal solution A_* is given by

$$A_* = \sum_{i=1}^{k} \lambda_i Q_* V_{i*} Q_*, \tag{12.9.6}$$

where $\lambda_1, \lambda_2, \ldots, \lambda_k$ satisfy the linear equations

$$\sum_{j=1}^{k} [\text{tr}(Q_* V_{i*} Q_* V_{j*})] \lambda_i = p_i, \; i = 1, 2, \ldots, k, \quad \text{and}$$

$$Q_* = I - X_*(X_*' X_*)^- X_*'.$$

Once we compute A_*, the optimal invariant unbiased estimator $Y'C_*Y$ can be computed with $C_* = B^{-1} A_* B^{-1}$. We label $Y'C_*Y$ as an optimal estimator according to the criterion of MINQUE.

12.10. Optimal Unbiased Estimation

We are still in the operational mode of a linear model with error random vector containing several variance components: $Y = X\beta + \varepsilon$, where $E\varepsilon = 0$, $D(\varepsilon) = \sigma_1^2 V_1 + \ldots + \sigma_k^2 V_k, V_1, \ldots, V_k$ are known non-negative definite matrices and $\sigma_1^2, \ldots, \sigma_k^2$ are unknown. Let $p_1\sigma_1^2 + \ldots + p_k\sigma_k^2 = p'\theta$ be a linear combination of the variance components $\sigma_1^2, \ldots, \sigma_k^2$, where $p' = (p_1, \ldots, p_k)$ is known and $\theta = (\sigma_1^2, \ldots, \sigma_k^2)$. We are seeking an optimal unbiased estimator $Y'CY$ of $p'\theta$. Unbiasedness means that we are seeking a symmetric matrix C satisfying

$$X'CX = 0 \quad \text{and } \text{tr}(CV_i) = p_i \quad \text{for all } i. \tag{12.10.1}$$

Optimality means that we are seeking a symmetric matrix C such that

$$\text{tr}((CV_0)^2), \; V_0 = V_1 + \ldots + V_k \tag{12.10.2}$$

is a minimum. The entire process boils down to a mathematical problem of minimizing (12.10.2) with respect to the symmetric matrix C subject to (12.10.1). We will use the following result in our optimization problem.

P 12.10.1 The minimum of $\text{tr}(A^2)$ over all symmetric matrices A subject to the conditions $X'AX = 0$ and $\text{tr}(AV_i) = p_i$ for all i, is attained at

$$A_* = \sum_{i=1}^{k} \lambda_i (V_i - PV_iP), \tag{12.10.3}$$

where $P = X(X'X)^- X'$ is the projection operator associated with the design matrix X, and $\lambda_1, \lambda_2, \ldots, \lambda_k$ satisfy the linear equations:

$$\sum_{i=1}^{k} [\text{tr}(V_iV_j - PV_iPV_j)]\lambda_i = p_j,\ j = 1, 2, \ldots, k. \tag{12.10.4}$$

PROOF. In order to prove that the given A_* in (12.10.3) is optimal, we must ascertain that A_* is admissible, i.e., A_* satisfies the given constraints. This poses no problem. Next, we must determine the general structure of a symmetric matrix A satisfying $X'AX = 0$. A symmetric matrix A satisfies $X'AX = 0$ if and only if $A = T - PTP$, for some symmetric matrix T. In a similar vein, there exists a symmetric matrix T_* such that $A_* = T_* - PT_*P$. We can rewrite any general solution A of the constraints as

$$\begin{aligned} A &= A_* + (A - A_*) = A_* + [(T - T_*) - P(T - T_*)P] \\ &= A_* + (S - PSP), \end{aligned}$$

where $S = T - T_*$. Since

$$\text{tr}((T - PTP)V_i) = p_i \quad \text{for all} \quad i,$$

$$\text{tr}((T_* - PT_*P)V_i) = p_i \quad \text{for all} \quad i,$$

it follows that we must have

$$\text{tr}((S - PSP)V_i) = 0 \quad \text{for all} \quad i.$$

If we can show that $\text{tr}(A^2) \geq \text{tr}((A_*)^2)$, the battle is won. Observe that

$$\begin{aligned} \text{tr}(A^2) &= \text{tr}((A_* + (S - PSP))(A_* + (S - PSP))) \\ &= \text{tr}((A_*)^2) + \text{tr}((S - PSP)(S - PSP)) \end{aligned}$$

$$+\,2\mathrm{tr}(A_*(S-PSP))$$
$$=\mathrm{tr}((A_*)^2)+\mathrm{tr}((S-PSP)(S-PSP))$$
$$+\,2\sum_{i=1}^{k}\lambda_i(V_i-PV_iP)(S-PSP)$$
$$=\mathrm{tr}((A_*)^2)+\mathrm{tr}((S-PSP)^2)$$
$$+\,2\sum_{i=1}^{k}\lambda_i(V_i-PV_iP)S \quad \text{(Why?)}$$
$$=\mathrm{tr}((A_*)^2)+\mathrm{tr}((S-PSP)^2)\geq \mathrm{tr}((A_*)^2).$$

This completes the proof.

This result is not applicable as it is to our optimization problem. We need to do some more work. In what follows we will assume that V_0 is non-singular. Let B be a positive definite square root of V_0, i.e., $B^2 = V_0$. Note that we can write

$$\mathrm{tr}((CV_0)^2)=\mathrm{tr}((CB^2CB^2))=\mathrm{tr}((BCB)^2).$$

Let $A = BCB$, or equivalently, $C = B^{-1}AB^{-1}$. Knowing A is equivalent to knowing C. Let $X_* = X'B^{-1}$, $V_{i*} = B^{-1}V_iB^{-1}$, $i = 1, 2, \ldots, k$. The constraints can be rewritten involving A. More precisely, $0 = X'CX = X'B^{-1}AN^{-1}X = X_*'AX_*$ and $p_i = \mathrm{tr}(CV_i) = \mathrm{tr}((B^{-1}V_iB^{-1}))$ $= \mathrm{tr}(AV_{i*})$. The optimization problem we originally started with is equivalent to maximizing $\mathrm{tr}(A^2)$ over all symmetric matrices A subject to the constraints, $X_*'AX_* = 0$ and $\mathrm{tr}(AV_{i*}) = p_i$ for all i.

This problem fits admirably into the framework of **P 12.10.1**. Let P_* be the projection operator associated with the matrix X_*, i.e., $P_* = X_*(X_*'X_*)^-X_*'$. An optimal solution is given by

$$A_*=\sum_{i=1}^{k}\lambda_i(V_{i*}-P_*V_{i*}P_*),$$

where $\lambda_1, \lambda_2, \ldots, \lambda_k$ satisfy the following linear equations:

$$\sum_{i=1}^{k}[\mathrm{tr}(V_{i*}V_{j*}-P_*V_{i*}P_*V_{j*})]\lambda_i=p_j,\ j=1,2,\ldots,k.$$

Once we obtain A_*, we can compute the required optimal matrix C_* by, $C_* = B^{-1}A_*B^{-1}$. As a final flourish, we can lay down an optimal unbiased estimator of $p'\theta$ as $Y'C_*Y$.

12.11. Total Least Squares

We have seen that a linear predictor $\beta'x$ of a random variable y, where $x \in \mathbf{R}^m$ is a vector of predictor variables, is estimated by $\hat{\beta}'x$, where $\hat{\beta}$ is the solution to the optimization problem

$$\min_{\beta \in \mathbf{R}^m} \sum_{i=1}^{n} (y_i - \beta' x_i)^2, \tag{12.11.1}$$

where $(y_i, x_i), i = 1, \ldots, n$ are observed data (see Sections 12.1-12.3). The expression (12.11.1) can be written as

$$\min_{\beta \in \mathbf{R}^m} (Y - X\beta)'(Y - X\beta), \tag{12.11.2}$$

where $Y' = (y_1, \ldots, y_n)$ and $X' = (x_1 | \ldots | x_n)$ is an $m \times n$ matrix. The optimum β is a solution of the equation

$$X\hat{\beta} = P_X Y \Rightarrow \hat{\beta} = (X'X)^{-1}X'Y, \tag{12.11.3}$$

where P_X is the orthogonal projection operator on $Sp(X)$.

We may also formulate the problem as one of finding the optimum $\hat{\theta}$ for

$$\min_{\theta \in Sp(X)} \|(Y|X) - (\theta|X)\|^2 \tag{12.11.4}$$

for a suitably chosen norm, and then solving the consistent equation $\hat{\theta} = X\hat{\beta}$ to obtain $\hat{\beta}$. We now seek a solution to another problem of estimating a structural relationship $\theta = \gamma'\beta$ between the latent variables $\theta \in \mathbf{R}$ and $\gamma \in \mathbf{R}^m$, which are not directly observable. However, surrogate variables y and x for θ and γ are available, which bear the stochastic relationships

$$\begin{aligned} y &= \theta + \epsilon_1, \;\; V(\epsilon_1) = \sigma^2, \\ x &= \gamma + \epsilon_2, \;\; D(\epsilon_2) = E(\epsilon_2\epsilon_2') = V, \end{aligned} \tag{12.11.5}$$

where $V(\cdot)$ stands for variance and $D(\cdot)$ for the dispersion (variance-covariance) matrix. We suggest a method of estimating β on the basis of independent observations $(y_i, x_i), i = 1, \ldots, n$.

The matrix of observations and the corresponding latent variables are respectively

$$\begin{bmatrix} y_1 & x_1' \\ \cdot & \cdot \\ y_n & x_n' \end{bmatrix} \text{ and } \begin{bmatrix} \theta_1 & \gamma_1' \\ \cdot & \cdot \\ \theta_n & \gamma_n' \end{bmatrix} = \begin{bmatrix} \gamma_1'\beta & \gamma_1' \\ \cdot & \cdot \\ \gamma_n'\beta & \gamma_n' \end{bmatrix} \tag{12.11.6}$$

using the structural relation $\theta = \gamma'\beta$. We determine β by minimizing

$$\left\| \begin{matrix} y_1 - \gamma_1'\beta & (x_1 - \gamma_1)' \\ \cdots & \cdots \\ y_n - \gamma_n'\beta & (x_n - \gamma_n)' \end{matrix} \right\| \tag{12.11.7}$$

for a suitably chosen norm. From statistical considerations, a suitable norm is

$$\alpha \sum_{i=1}^{n} (y_i - \gamma_i'\beta)^2 + \sum_{i=1}^{n} (x_i - \gamma_i)' V^{-1} (x_i - \gamma_i), \tag{12.11.8}$$

where $\alpha = 1/\sigma^2$. We minimize (12.11.8) with respect to β and $\gamma_1, \ldots, \gamma_n$ in two stages.

P 12.11.1 The minimum value of (12.11.8) with respect to $\gamma_1, \ldots, \gamma_n$ is

$$\frac{\alpha}{1 + \alpha\beta'V\beta} \sum_{i=1}^{n} (y_i - \beta'x_i)^2 = \frac{\alpha}{1 + \alpha\beta'V\beta} (y - X\beta)'(Y - X\beta), \tag{12.11.9}$$

where $Y' = (y_1, \ldots, y_n)$ and $X' = (x_1 | \ldots | x_n)$.

PROOF. Differentiating (12.11.8) with respect to γ_i, we have

$$[\alpha(y_i - \gamma_i'\beta)]\beta + V^{-1}(x_i - \gamma_i) = 0 \tag{12.11.10}$$

which gives

$$\begin{aligned} \gamma_i'\beta &= \frac{\alpha\beta'V\beta y_i + x_i'\beta}{1 + \alpha\beta'V\beta} \\ y_i - \gamma_i'\beta &= \frac{y_i - x_i'\beta}{1 + \alpha\beta'V\beta}. \end{aligned} \tag{12.11.11}$$

Also from (12.11.10)

$$x_i - \gamma_i = [\alpha(y_i - \gamma_i'\beta)]V\beta. \tag{12.11.12}$$

Substituting (12.11.11) and (12.11.12) in (12.11.8) we have the expression (12.11.9).

It is interesting to note that the expression (12.11.9) to be minimized is same as that for least squares except for the multiplying factor which is also a function of β. The method of estimating β by minimizing (12.11.9) is called **total least squares.**

P 12.11.2 The vector β which satisfies the equation

$$\begin{bmatrix} Y'Y & Y'X \\ X'Y & X'X \end{bmatrix} \begin{bmatrix} -1 \\ \beta \end{bmatrix} = \lambda \begin{bmatrix} \sigma^2 & 0 \\ 0 & V \end{bmatrix} \begin{bmatrix} -1 \\ \beta \end{bmatrix} \tag{12.11.13}$$

for the smallest value of λ minimizes the expression (12.11.9).

PROOF. Differentiating (12.11.9) with respect to β and equating to zero we get

$$\frac{\alpha^2 V\beta}{(1+\alpha\beta'V\beta)^2}(Y - X\beta)'(Y - X\beta) + \frac{\alpha}{1+\alpha\beta'V\beta}X'(Y - X\beta) = 0. \tag{12.11.14}$$

Denoting the expression (12.11.9) by λ, equation (12.11.14) reduces to

$$X'X\beta - X'Y = \lambda V\beta. \tag{12.11.15}$$

Now

$$\begin{aligned}\lambda &= \frac{\alpha}{1+\alpha\beta'V\beta}(Y - X\beta)'(Y - X\beta) \\ &= \frac{\alpha}{1+\alpha\beta'V\beta}[Y'(Y - X\beta) - \beta'X'(Y - X\beta)] \\ &= \frac{\alpha}{1+\alpha\beta'V\beta}(Y'Y - Y'X\beta + \lambda\beta'V\beta), \text{ using (12.11.15)}\end{aligned}$$

i.e.,

$$\sigma^2\lambda = Y'Y - Y'X\beta. \tag{12.11.16}$$

The two expressions (12.11.15) and (12.11.16) give equation (12.11.13).

The method of obtaining β is now clear. Let λ be the smallest eigenvalue of

$$\begin{bmatrix} Y'Y & Y'X \\ X'Y & X'X \end{bmatrix} \text{ with respect to } \begin{bmatrix} \sigma^2 & 0 \\ 0 & V \end{bmatrix} \tag{12.11.17}$$

for which the corresponding eigenvector has its first element not zero. Let λ_k be such an eigenvalue and be not repeated. Further let $p_{k1}, \ldots, p_{k,(m+1)}$ be the eigenvector corresponding to λ_k. Then the unique estimate of β is $\hat{\beta}' = -(p_{k1})^{-1}(p_{k2}, \ldots, p_{k,(m+1)})$. If λ_k is repeated, then any combination of the corresponding eigenvectors will provide a minimizing solution.

The solution we obtained is more general than the total least squares estimate of β computed on the assumption that V is a diagonal matrix with each diagonal element equal to the same σ^2. If we have some knowledge of σ^2 and V, then a better estimate can be obtained.

Now we consider another situation where some of the surrogate x variables are latent variables observed without error. In such a case we write the structural equation as $\theta = x_1'\beta_1 + \gamma_2'\beta_2$, where for θ and γ_2 we have surrogate variables y and x_2. Let us assume, for simplicity, that the errors in y and the components of x_2 are independent and have the same variance. In such a case, the total least squares estimate of β_1 and β_2 may be obtained by minimizing

$$\sum_{i=1}^{n}(y_i - x_{1i}'\beta_1 - \gamma_{2i}'\beta_2)^2 + \sum_{i=1}^{n}(x_{2i} - \gamma_{2i})'(x_{2i} - \gamma_{2i}) \tag{12.11.18}$$

with respect to $\gamma_{2i}, i = 1, \ldots, n; \beta_1$ and β_2. As in **P 12.11.1**, it can be shown that the minimum value of (12.11.18), when minimized with respect to $\gamma_{2i}, i = 1, \ldots, n$ is

$$\frac{1}{1+\beta_2'\beta_2}\sum_{i=1}^{n}(y_i - x_{1i}'\beta_1 - x_{2i}'\beta_2)^2$$
$$= \frac{1}{1+\beta_2'\beta_2}(Y - X_1\beta_1 - X_2\beta_2)'(Y - X_1\beta_1 - X_2\beta_2) \tag{12.11.19}$$

with the usual notation for Y, X_1 and X_2. A further minimization with respect to β_1 gives estimate of β_1 given β_2 as $\hat{\beta}_1 = (X_1'X_1)^{-1}X_1'(Y - X_2\beta_2)$.

Substituting this estimate in (12.11.18), the expression to be minimized with respect to β_2 comes out to be

$$\frac{1}{1+\beta_2'\beta_2}(Y - X_2\beta_2)'(I - P_{X1})(Y - X_2\beta_2), \tag{12.11.20}$$

where P_{X1} is the orthogonal operator on $Sp(X_1)$. The rest of minimization with respect to β_2 is done on the same lines as in **P 12.11.2.**

Complements

12.11.1 Consider the matrix (12.11.7), which may be written as,

$$(Y \mid X) - (\Gamma\beta \mid \Gamma) = (Y - \Gamma\beta \mid X - \Gamma)$$

and choose the squared norm

$$\text{tr}[(Y - \Gamma\beta \mid X - \Gamma)W^{-1}(Y - \Gamma\beta \mid X - \Gamma)'] \tag{12.11.24}$$

where W is a positive definite matrix of order $(m+1)\times(m+1)$. Try to minimize the above expression with respect to β and Γ. The problem so formulated is the most general version of total least squares.

Note: The material of the first 9 sections of this chapter is based on the papers by Rao (1959, 1965, 1972a, 1972c, 1976b, 1984, 1987, 1989), Rao and Mitra (1968b, 1969), Rao and Kleffe (1988) and Rao and Toutenburg (1995).

The model (12.11.5) is called errors-in-variables model in the statistical literature. Some discussion of these models is carried out in Fuller (1987), Anderson (1976, 1980, 1984) and Anderson and Sawa (1982). Ammann and Van Ness (1988, 1989) have conducted some simulation studies to assess some statistical properties of least squares and total least squares estimators in the context of a simple linear regression problem. The multivariate errors-in-variables regression model has been considered by Gleser (1981). The total least squares methods is also called orthogonal regression. Finally, one cannot fail to mention the timely contributions of Van Huffel and Vandewalle (1991) and Van Huffel and Zha (1993) to the total least squares criterion.

CHAPTER 13

QUADRATIC SUBSPACES

13.1. Basic Ideas

The basic object of investigation in this chapter is the collection **A** of all $n \times n$ symmetric matrices with real entries and certain subsets of **A**, where n is any positive fixed integer. It is clear that the collection **A** is a real vector space with the usual operations of addition and scalar multiplication of matrices. More precisely, let $A = (a_{ij})$ and $B = (b_{ij})$ be any two members of **A**. The sum of A and B is defined to be the $n \times n$ matrix $C = (c_{ij})$ with $c_{ij} = a_{ij} + b_{ij}$ for all i and $j \in \{1, 2, \ldots, n\}$. Symbolically, one can write $C = A + B$. Further, if α is any real number, the scalar multiple of the matrix A by the scalar α is the $n \times n$ matrix $D = (d_{ij})$ with $d_{ij} = \alpha a_{ij}$ for all i and j. Symbolically, one can write $D = \alpha A$. One can introduce, in a perfectly natural manner, an inner product $< \cdot, \cdot >$ on the vector space **A**. For any two matrices $A = (a_{ij})$ and $B = (b_{ij})$ in **A**, let

$$< A, B > = \sum_{i=1}^{n} \sum_{j=1}^{n} a_{ij} b_{ij} = \mathrm{tr(BA)}.$$

The following properties of the inner product are transparent.

(1) For any A in **A**, $< A, A > = \sum_{i=1}^{n} \sum_{j=1}^{n} a_{ij}^2 \geq 0$.

(2) For any A in **A**, $< A, A > = 0$ if and only if $A = 0$.

(3) For any A and B in **A**, $< A, B > = < B, A >$.

(4) The inner product $< \cdot, \cdot >$ on the product space $\mathbf{A} \times \mathbf{A}$ is bilinear, i.e., for any A, B and C in **A** and α and β real numbers,

$$\begin{aligned} < \alpha A + \beta B, C > &= \alpha < A, C > + \beta < B, C >, \\ < A, \alpha B + \beta C > &= \alpha < A, B > + \beta < A, C > . \end{aligned}$$

In view of the properties enunciated above, the phrase "inner product" is apt for the entity $< \cdot, \cdot >$. It is time to introduce a special subset of the vector space **A**.

DEFINITION 13.1.1. A subset **B** of **A** is called a quadratic subspace of **A** if the following hold.

(1) **B** is a subspace of A.
(2) If $B \in \mathbf{B}$, then $B^2 \in \mathbf{B}$.

Some examples are in order.

EXAMPLE 13.1.2.

(1) At one end of the spectrum, the whole collection **A** is a quadratic subspace of **A**. At the other end, $\mathbf{B} = \{0\}$, the set consisting of only the zero matrix, is a quadratic subspace of **A**.
(2) Let A be a fixed symmetric idempotent matrix of order $n \times n$. Let
$$\mathbf{B} = \{\alpha A : \alpha \text{ real}\}.$$
Then **B** is a quadratic subspace of **A**.
(3) The idea behind Example (2) can be extended. Let A and B be two symmetric idempotent matrices of order $n \times n$ satisfying $AB = 0$. (The condition $AB = 0$ implies that $BA = 0$. Why?) Let
$$\mathbf{B} = \{\alpha A + \beta B : \alpha \text{ and } \beta \text{ real}\}.$$
Then **B** is a quadratic subspace of **A**. It is clear that **B** is a subspace of **A**. Observe that
$$(\alpha A + \beta B)^2 = \alpha^2 A^2 + \beta^2 B^2 + \alpha\beta AB + \alpha\beta BA = \alpha^2 A + \beta^2 B.$$
(4) Example (3) can be further generalized to handle more than two idempotent matrices. This is one way of providing many examples of quadratic subspaces. These quadratic subspaces have a certain additional property which make them stand out as a special breed. This will be apparent when the notion of commutative quadratic subspaces is introduced.
(5) Let us look at the case $n = 2$. The vector space **A** is three-dimensional. Let A be any matrix in **A**. By the spectral decomposition theorem, one can write $A = CDC'$ for some orthogonal

matrix C of order 2×2 and a diagonal matrix $D = \text{diag}\{a, b\}$. Every orthogonal matrix C is of the form

$$C = \begin{bmatrix} \cos(\varphi) & -\sin(\varphi) \\ \sin(\varphi) & \cos(\varphi) \end{bmatrix}$$

for some angle φ. The matrix A can now be written informatively as

$$A = \begin{bmatrix} \cos(\varphi) & -\sin(\varphi) \\ \sin(\varphi) & \cos(\varphi) \end{bmatrix} \begin{bmatrix} a & 0 \\ 0 & b \end{bmatrix} \begin{bmatrix} \cos(\varphi) & \sin(\varphi) \\ -\sin(\varphi) & \cos(\varphi) \end{bmatrix}$$

for some scalars a and b, and angle φ. The matrix A is idempotent if and only if $a = 0$ or 1 and $b = 0$ or 1. The only three-dimensional quadratic subspace of **A** is **A** itself. The zero-dimensional quadratic subspace of **A** is $\{0\}$. A one-dimensional subspace of **A** is precisely the vector space spanned by a single idempotent matrix. There is a lot of flexibility in the construction of two-dimensional quadratic subspaces of **A**. Choose an angle φ. Let

$$A = \begin{bmatrix} \cos(\varphi) & -\sin(\varphi) \\ \sin(\varphi) & \cos(\varphi) \end{bmatrix} \begin{bmatrix} 1 & 0 \\ 0 & 0 \end{bmatrix} \begin{bmatrix} \cos(\varphi) & \sin(\varphi) \\ -\sin(\varphi) & \cos(\varphi) \end{bmatrix},$$

$$B = \begin{bmatrix} \cos(\varphi) & -\sin(\varphi) \\ \sin(\varphi) & \cos(\varphi) \end{bmatrix} \begin{bmatrix} 0 & 0 \\ 0 & 1 \end{bmatrix} \begin{bmatrix} \cos(\varphi) & \sin(\varphi) \\ -\sin(\varphi) & \cos(\varphi) \end{bmatrix}.$$

Let **B** be the vector space spanned by A and B. Then **B** is a two-dimensional quadratic subspace of **A**. Note that A and B are idempotent and $AB = 0$. There is another way of constructing a two-dimensional quadratic subspace of **A**. Let $A = I_2$, the identity matrix of order 2×2, and B is any other idempotent matrix such that A and B are linearly independent. (It will be instructive to construct an example of B.) Let **B** be the vector space spanned by A and B. Then **B** is a two-dimensional quadratic subspace of **A**. (Why? **P 13.1.3** might be helpful.) In this example, note that $AB \neq 0$.

It is time to jot down a few simple properties of quadratic subspaces.

P 13.1.3 Let **B** be a subspace of **A**. Then the following statements are equivalent.

(1) **B** is a quadratic subspace of **A**.

(2) If $A, B \in \mathbf{B}$, then $(A+B)^2 \in \mathbf{B}$.
(3) If $A, B \in \mathbf{B}$, then $AB + BA \in \mathbf{B}$.
(4) If $A \in \mathbf{B}$, then $A^k \in \mathbf{B}$, $k = 1, 2, \ldots$.

PROOF. It is clear that Statement (4) implies Statement (1). Suppose (1) is true. One thing is clear. If $A \in \mathbf{B}$, then $A^2, A^4, A^8, A^{16}, \ldots \in \mathbf{A}$. It is not immediately obvious that A^3 also belongs to $\mathbf{B}$. Since $\mathbf{B}$ is a vector space, $A + A^2 \in \mathbf{B}$. By the presumption of (1), $(A + A^2)^2 \in \mathbf{B}$. But $(A + A^2)^2 = A^2 + 2A^3 + A^4$. Since $\mathbf{B}$ is a vector space, $2A^3 \in \mathbf{B}$. Hence $A^3 \in \mathbf{B}$. In a similar vein, one can show that every integral power of A belongs to $\mathbf{B}$. Try it yourself. Thus (4) follows. It is clear that Statements (1) and (2) are equivalent. To show that (1)$\Rightarrow$(3), note that $(A+B)^2 = A^2 + B^2 + (AB + BA)$. It now follows that $AB + BA \in \mathbf{B}$. It is clear that (3)$\Rightarrow$(1).

P 13.1.4 Let $\mathbf{B}$ be a quadratic subspace of $\mathbf{A}$.

(1) If $A, B \in \mathbf{B}$, then $ABA \in \mathbf{B}$.
(2) Let $A \in \mathbf{B}$ be fixed. Let $\mathbf{C} = \{ABA : B \in \mathbf{B}\}$. Then $\mathbf{C}$ is a quadratic subspace of $\mathbf{B}$.
(3) If $A, B, C \in \mathbf{B}$, then $ABC + CBA \in \mathbf{B}$.

PROOF. (1) Let $A, B \in \mathbf{B}$. Observe that

$$\begin{aligned}(A + AB + BA)^2 &= A^2 + (AB + BA)^2 + A(AB + BA) + (AB + BA)A \\ &= A^2 + (AB + BA)^2 + (A^2B + BA^2) + 2ABA.\end{aligned}$$

Since $\mathbf{B}$ is a quadratic subspace of $\mathbf{A}$, $(A + AB + BA)^2, A^2, (AB + BA)^2, A^2B + BA^2 \in \mathbf{B}$. Hence $ABA \in \mathbf{B}$.

(2) Let B_1 and $B_2 \in \mathbf{B}$, and α and β real numbers. Then

$$\alpha AB_1A + \beta AB_2A = A(\alpha B_1 + \beta B_2)A \in \mathbf{B}.$$

This shows that $\mathbf{C}$ is a vector space. Let $B \in \mathbf{B}$. Then $(ABA)^2 = ABAABA = A(BA^2B)A$. Since $\mathbf{B}$ is a quadratic subspace, $A^2 \in \mathbf{B}$ and hence $BA^2B \in \mathbf{B}$, by (1). Hence $(ABA)^2 \in \mathbf{C}$.

(3) From the deliberations carried out so far, it is clear that

$$\begin{aligned}A(BC + CB) + (BC + CB)A &= ABC + ACB + BCA + CBA \in \mathbf{B}, \\ B(AC + CA) + (AC + CA)B &= BAC + BCA + ACB + CAB \in \mathbf{B}, \\ C(AB + BA) + (AB + BA)C &= CAB + CBA + ABC + BAC \in \mathbf{B}.\end{aligned}$$

A simple arithmetic,

$$[A(BC+CB)+(BC+CB)A]-[B(AC+CA)+(AC+CA)B]$$
$$+[C(AB+BA)+(AB+BA)C]=2(ABC+CBA),$$

yields that $ABC + CBA \in \mathbf{B}$. The next item on the agenda is how to recognize a quadratic subspace when one sees one. Checking whether or not a particular subset $\mathbf{B}$ of $\mathbf{A}$ is a subspace is basically a simple act. Assume that one has a subspace $\mathbf{B}$ of $\mathbf{A}$ in hand. Let $\mathbf{B}_1 = \{B_1, B_2, \ldots, B_k\}$ be a basis of the vector space $\mathbf{B}$. By checking a few things about $\mathbf{B}_1$, one can verify that $\mathbf{B}$ is indeed a quadratic subspace of $\mathbf{A}$. The following result is in this direction.

P 13.1.5 Let $\mathbf{B}$ be a subspace of $\mathbf{A}$. The following statements are equivalent.

(1) $A \in \mathbf{B}_1 \Rightarrow A^2 \in \mathbf{B}$. ($\mathbf{B}$ is a quadratic subspace of $\mathbf{A}$.)
(2) $A, B \in \mathbf{B}_1 \Rightarrow (A+B)^2 \in \mathbf{B}$.
(3) $A, B \in \mathbf{B}_1 \Rightarrow AB + BA \in \mathbf{B}$.

PROOF. It is clear that Statement (1) implies Statement (2). Consider the implication (2)$\Rightarrow$(3). Let $A, B \in \mathbf{B}_1$. Observe that $(A+B)^2 = A^2+AB+BA+B^2 \in \mathbf{B}$, by (2). Since $\mathbf{B}$ is a vector space, $AB+BA \in \mathbf{B}$. Finally, consider the implication (3)$\Rightarrow$(1). Let $A \in \mathbf{B}$. If $A \in \mathbf{B}_1$, by taking $B = A$, one can observe that $AB + BA = 2A^2 \in \mathbf{B}$. Hence $A^2 \in \mathbf{B}$. If $A \notin \mathbf{B}_1$, one can write $A = \alpha_1 \mathbf{B}_1 + \ldots + \alpha_k \mathbf{B}_k$ for some scalars $\alpha_1, \alpha_2, \ldots, \alpha_k$. Note that

$$A^2 = \sum_{i=1}^{k} \alpha_i^2 B_i^2 + \sum_{i<j} \alpha_i \alpha_j (B_i B_j + B_j B_i).$$

By just what was observed, each $B_i^2 \in \mathbf{B}$. By (3), $B_i B_j + B_j B_i \in \mathbf{B}$. Consequently, $A^2 \in \mathbf{B}$. This completes the proof.

Complements

13.1.1 Show that the idempotent matrices $A = I_2$ and

$$B = \begin{bmatrix} \cos^2\varphi & \cos\varphi\sin\varphi \\ \cos\varphi\sin\varphi & \sin^2\varphi \end{bmatrix}$$

for any angle φ are linearly independent.

13.1.2 For what values of φ_1 and φ_2 are the idempotent matrices

$$A = \begin{bmatrix} \cos^2\varphi_1 & \cos\varphi_1\sin\varphi_1 \\ \cos\varphi_1\sin\varphi_1 & \sin^2\varphi_1 \end{bmatrix}$$

$$B = \begin{bmatrix} \sin^2\varphi_2 & -\cos\varphi_2\sin\varphi_2 \\ -\cos\varphi_2\sin\varphi_2 & \cos^2\varphi_2 \end{bmatrix}$$

linearly independent?

13.1.3 For what values of φ_1 and φ_2 are the idempotent matrices

$$A = \begin{bmatrix} \cos^2\varphi_1 & \cos\varphi_1\sin\varphi_1 \\ \cos\varphi_1\sin\varphi_1 & \sin^2\varphi_1 \end{bmatrix}$$

$$B = \begin{bmatrix} \cos^2\varphi_2 & \cos\varphi_2\sin\varphi_2 \\ \cos\varphi_2\sin\varphi_2 & \sin^2\varphi_2 \end{bmatrix}$$

linearly independent?

13.1.4 For $n = 2$, show that the matrices

$$A = I_2,\ B = \begin{bmatrix} \frac{1}{2} & \frac{1}{2} \\ \frac{1}{2} & \frac{1}{2}, \end{bmatrix},\ \text{and } C = \begin{bmatrix} 1 & 0 \\ 0 & 0 \end{bmatrix}$$

form a basis for the vector space **A**.

13.1.5 Construct a basis of $n(n+1)/2$ idempotent matrices for the vector space **A**, the collection of all $n \times n$ symmetric matrices. Prove a similar result for complex matrices.

13.2. The Structure of Quadratic Subspaces

The spectral decomposition of a symmetric matrix is the focal point of this section. Recall that the spectral decomposition of a symmetric matrix provides a representation of the matrix as a linear combination of idempotent matrices. More precisely, let A be a symmetric matrix of order $n \times n$, and $\lambda_1, \lambda_2, \ldots, \lambda_n$ be its eigenvalues. Then we can write

$$A = \lambda_1 Q_1 + \lambda_2 Q_2 + \ldots + \lambda_n Q_n$$

for some idempotent matrices $Q_1, Q_2, \ldots, Q_n$ each of rank one. Further, these idempotent matrices have the property that $Q_i Q_j = 0$ for all $i \neq j$. Some of the eigenvalues could be zero. Some eigenvalues could be repetitive. Let us tighten the representation in the following way.

DEFINITION 13.2.1. A symmetric matrix A is said to have a sparse spectral representation if it can be written as

$$A = \mu_1 P_1 + \mu_2 P_2 + \ldots + \mu_k P_k$$

for some distinct non-zero scalars $\mu_1, \mu_2, \ldots, \mu_k$, and non-zero idempotent matrices $P_1, P_2, \ldots, P_k$ with the property that $P_i P_j = 0$ for all $i \neq j$.

A sparse spectral representation is always possible for any symmetric non-zero matrix. Start with the usual spectral decomposition of the matrix. Throw away all those idempotent matrices in the representation that correspond to zero eigenvalues. If a non-zero eigenvalue repeats, add up all the idempotent matrices that correspond to this eigenvalue. A sparse representation of the matrix entails.

The following matrix is useful in what follows. This matrix is a relative of the well-known Vandermonde matrix. Let $\mu_1, \mu_2, \ldots, \mu_k$ be non-zero distinct real numbers and

$$\Delta = \begin{bmatrix} \mu_1 & \mu_1^2 & \cdots & \mu_1^k \\ \mu_2 & \mu_2^2 & \cdots & \mu_2^k \\ \cdot & \cdot & \cdots & \cdot \\ \mu_k & \mu_k^2 & \cdots & \mu_k^k \end{bmatrix}.$$

The matrix Δ of order $k \times k$ is non-singular. In fact, the determinant of Δ is given by $|\Delta| = (\Pi_{i=1}^k \mu_i)[\Pi_{i>j}(\mu_i - \mu_j)]$.

If one looks at the development of ideas so far, the idempotent matrices seem to play a substantial role in the formation of quadratic subspaces. The trend continues with the following critical result from which something important emerges about the idempotent matrices in the sparse spectral decomposition of matrices in a quadratic subspace.

P 13.2.2 Let **B** be a quadratic subspace of **A** and $A \in \mathbf{B}$. Let $A = \mu_1 P_1 + \mu_2 P_2 + \ldots + \mu_k P_k$ be a sparse spectral decomposition of A. Then each $P_i \in \mathbf{B}$.

PROOF. The critical idea in the proof is to demonstrate that each P_i is a linear combination of $A, A^2, \ldots, A^k$. Since **B** is a quadratic subspace of **A**, it would then follow that each $P_i \in \mathbf{B}$. To accomplish this, note that

$$A = \mu_1 P_1 + \mu_2 P_2 + \ldots + \mu_k P_k,$$
$$A^2 = \mu_1^2 P_1 + \mu_2^2 P_2 + \ldots + \mu_k^2 P_k,$$
$$\cdot \quad \cdot \quad \cdot \quad \cdots \quad \cdot$$
$$A^k = \mu_1^k P_1 + \mu_2^k P_2 + \ldots + \mu_k^k P_k.$$

To begin with, let us show that P_1 is a linear combination of $A, A^2, \ldots, A^k$. Consider the system of linear equations $\Delta\beta = \gamma$ in unknown vector $\beta' = (\beta_1, \beta_2, \ldots, \beta_k)$, where $\gamma' = (1, 0, 0, \ldots, 0)$ is a vector of order $1 \times k$. Since Δ is non-singular, the system of equations has a unique solution β, i.e., satisfying the equations

$$\mu_1\beta_1 + \mu_1^2\beta_2 + \ldots + \mu_1^k\beta_k = 1,$$
$$\mu_2\beta_1 + \mu_2^2\beta_2 + \ldots + \mu_2^k\beta_k = 0,$$
$$\cdot \quad \cdot \quad \cdots \quad \cdot$$
$$\mu_k\beta_1 + \mu_k^2\beta_2 + \ldots + \mu_k^k\beta_k = 0.$$

For this solution, observe that

$$\sum_{i=1}^{k} \beta_i A^i = \sum_{i=1}^{k} \beta_i (\sum_{j=1}^{k} \mu_j^i P_j) = \sum_{j=1}^{k} (\sum_{i=1}^{k} \mu_j^i \beta_i) P_j = P_1.$$

The general argument should be clear by now. This completes the proof.

P 13.2.2 is helpful in many ways. Some of the benefits are chronicled in the following result. The existence of a basis for a quadratic subspace consisting of idempotent matrices is the centerpiece of this section.

COROLLARY 13.2.3. Let **B** be a quadratic subspace of **A**.

(1) If $A \in \mathbf{B}$, then the Moore-Penrose inverse $A^+ \in \mathbf{B}$.
(2) If $A \in \mathbf{B}$, then $AA^+ \in \mathbf{B}$.
(3) There exists a basis of **B** consisting of idempotent matrices.

PROOF. (1) Let $A \in \mathbf{B}$ and $A = \mu_1 P_1 + \mu_2 P_2 + \ldots + \mu_k P_k$ be a sparse spectral decomposition of A. Note that

$$A^+ = (1/\mu_1)P_1 + (1/\mu_2)P_2 + \ldots + (1/\mu_k)P_k.$$

By **P 13.2.2**, $A^+ \in \mathbf{B}$.

(2) Let $A \in \mathbf{B}$. Continuing the argument set out in (1), note that $AA^+ = P_1 + P_2 + \ldots + P_k$. Hence $AA^+ \in \mathbf{B}$.

(3) Let $\{B_1, B_2, \ldots, B_k\}$ be a basis of the vector space **B**. Let

$$\begin{aligned} B_1 &= \lambda_{11} P_{11} + \lambda_{12} P_{12} + \ldots + \lambda_{1m_1} P_{1m_1}, \\ B_2 &= \lambda_{21} P_{21} + \lambda_{22} P_{22} + \ldots + \lambda_{2m_2} P_{2m_2}, \\ &\cdot \quad \cdot \quad \cdot \quad \cdots \quad \cdot \\ B_k &= \lambda_{k1} P_{k1} + \lambda_{k2} P_{k2} + \ldots + \lambda_{km_k} P_{km_k}, \end{aligned}$$

be sparse spectral decompositions of $B_1, B_2, \ldots, B_k$, respectively. Since the collection $\{B_1, B_2, \ldots, B_k\}$ spans the vector space **B**, the collection $\{P_{ij} : 1 \leq i \leq k,\ 1 \leq j \leq m_i\}$ of idempotent matrices span the vector space **B**. Consequently, one can find k elements from the collection $\{P_{ij} : 1 \leq i \leq k,\ 1 \leq j \leq m_i\}$ which form a basis for **B**.

We will now introduce some special quadratic subspaces. All the examples presented in Section 13.1 have one particular feature in common. Every pair of members of the quadratic subspaces commute. We will formalize this particular feature in the following.

DEFINITION 13.2.4. A quadratic subspace **B** of **A** is said to be commutative if $AB = BA$ for every pair A and B of elements in **B**.

Examples of commutative quadratic subspaces are easy to construct. Take any set $P_1, P_2, \ldots, P_k$ of idempotent matrices such that $P_i P_j = 0$ for all $i \neq j$. Let **B** be the vector space spanned by $P_1, P_2, \ldots, P_k$. Then **B** is a commutative quadratic subspace of **A**. Interestingly enough, every commutative quadratic subspace arises in this way.

P 13.2.5 Let **B** be a commutative quadratic subspace of **A**. Then there exists a basis $P_1, P_2, \ldots, P_k$ of **B** such that each P_i is idempotent and $P_i P_j = 0$ for all $i \neq j$.

PROOF. By Corollary 13.2.3, there exists a basis $B_1, B_2, \ldots, B_k$ of the commutative quadratic subspace **B** such that each B_i is idempotent. Consider the following matrices

$$B_1 B_2,\ B_1 - B_1 B_2,\ B_2 - B_1 B_2,\ B_3, \ldots, B_k.$$

It is clear that these $(k+1)$ idempotent matrices span the vector space **B**. (Why?) The first three matrices are very special. (What is special about them?) From these matrices, we can select a basis, $C_1, \ldots, C_k$

for the subspace **B** such that $C_1C_2 = 0$. (Why?) Consider the following matrices

$$C_1C_3,\ C_2C_3,\ C_1 - C_1C_3,\ C_2 - C_2C_3,\ C_3 - (C_1C_3 + C_2C_3),\ C_4, \ldots, C_k.$$

These matrices have the following properties: (1) Each one of them is idempotent. (2) The product of any two of the first five matrices is zero. (3) The matrices span **B**. We can then select a basis $D_1, D_2, \ldots, D_k$ for the subspace such that $D_1D_2 = D_1D_3 = D_2D_3 = 0$. This process can be continued until we get a basis of **B** possessing the properties stipulated by the theorem.

Complements

13.2.1 Let **A** be the collection of all 2×2 symmetric matrices. Are there quadratic subspaces of **A** which are not commutative?
13.2.2 Let **A** be the collection of all 3×3 symmetric matrices. Exhibit a quadratic subspace of **A** which is not commutative.

13.3. Commutators of Quadratic Subspaces

In all the examples presented so far, one particular feature can be identified. In each subspace there is one element Π such that it commutes with every member of the space. One can call Π as a commutator of the subspace. In this section, we will prove the existence and uniqueness of a commutator for any given quadratic subspace. We need some preliminary results which will help to achieve the objective.

P 13.3.1 If A and B are any two members of **B**, a quadratic subspace of **A**, then there exists $T \in \mathbf{B}$ such that $Sp(A) + Sp(B) = Sp(T)$.

PROOF. Recall that $Sp(A)$ is the space spanned by the column vectors of A. Let $A = \lambda_1P_1 + \lambda_2P_2 + \ldots + \lambda_rP_r$ be a sparse spectral decomposition of A. Let $P = P_1 + P_2 + \ldots + P_r$. It is clear that $P \in \mathbf{B}$ and $Sp(A) = Sp(P)$. Let

$$T = P + (I - P)B^2(I - P).$$

Note that $T \in \mathbf{B}$. (Why?) Further, $P[(I - P)B] = 0$. Consequently,

$$\begin{aligned} Sp(T) &= Sp(P + (I - P)B^2(I - P)) = Sp(P) \oplus Sp((I - P)B^2(I - P)) \\ &= Sp(P) \oplus Sp(B(I - P)) = Sp(A) + Sp(B). \end{aligned}$$

P 13.3.2 Let **B** be a quadratic subspace of **A**. Then there exists a matrix $\Pi \in \mathbf{B}$ such that $\Pi B = B$ for all $B \in \mathbf{B}$ with the following additional properties.

(1) Π is unique.
(2) $\Pi B = B\Pi$ for all $B \in \mathbf{B}$.
(3) $A \in \mathbf{B}$ and $\rho(A) = \rho(\Pi) \Rightarrow Sp(A) = Sp(\Pi)$.
(4) $A \in \mathbf{B}$ and $\rho(A) = \rho(\Pi) \Rightarrow \mathbf{B} = \{ABA : B \in \mathbf{B}\}$.

PROOF. Let the matrices $B_1, B_2, \ldots, B_k$ span the subspace **B**. By **P 13.3.1**, there exists $T_i \in \mathbf{B}$ such that

$$Sp(B_1) + Sp(B_2) + \ldots + Sp(B_i) = Sp(T_i)$$

for $i = 2, 3, \ldots, k$. Let $\Pi = T_m T_m^+$. By Corollary 13.2.3, $\Pi \in \mathbf{B}$. The matrix Π will do the trick.

13.4. Estimation of Variance Components

Consider a linear model $Y = X\beta + \varepsilon$, where:

(1) Y is a random vector of order $n \times 1$.
(2) X is a deterministic known matrix of order $n \times p$.
(3) β is an unknown parameter vector of order $p \times 1$ in the p-dimensional Euclidean space $\mathbf{R}^p$.
(4) ε is of order $n \times 1$ with $E\varepsilon = 0$ and $D(\varepsilon) = \Sigma\theta_i V_i$.
(5) $\theta = (\theta_1, \theta_2, \ldots, \theta_m)$ is an unknown vector belonging to the parameter space

$$\Omega = \{\theta' = (\theta_1, \theta_2, \ldots, \theta_m) : \theta_i V_i + \ldots + \theta_m V_m \text{ is p.d.}\}.$$

(6) The matrices $V_1, V_2, \ldots, V_m$ are known linearly independent symmetric matrices with $V_m = I_n$.
(7) The subset Ω of $\mathbf{R}^m$ has a non-empty interior.
(8) The random vector ε has a multivariate normal distribution.

We will present later an example of a situation in which such a model known as a variance component model arises.

If we can find a complete sufficient statistic for $\beta \in \mathbf{R}^p, \theta \in \Omega$ in the family $N_n(X\beta, \sum_{i=1}^{m} \theta_i V_i)$ of distributions, we can find good estimators of β and θ. The symbol N_n stands for n-variate normal distribution. Let $\mathbf{B}$ be the vector space spanned by the matrices $V_1, V_2, \ldots, V_m$. The following result is instrumental in paving a way for a good estimation of the variance components and the parameter vector β.

P 13.4.1 Let $N_n(X\beta, \sum_{i=1}^{m} \theta_i V_i)$, $\beta \in \mathbf{R}^p$, $\theta \in \Omega$ be a family of n-variate normal distributions with Ω having a non-empty interior. Suppose that the following two conditions are met:

(1) The space $\mathbf{B}$ spanned by the matrices $V_1, V_2, \ldots, V_m$ is a quadratic subspace of $\mathbf{A}$.

(2) $Sp(V_iX) \subset Sp(X)$ for each i. (Recall that $Sp(X)$ is the vector space spanned by the columns of X.)

Then the vector statistic $(X'Y, Y'V_1Y, Y'V_2Y, \ldots, Y'V_mY)$ is a complete sufficient statistic for the family.

PROOF. If $A \in \mathbf{B}$, we have already seen that its Moore-Penrose inverse A^+ also belongs to $\mathbf{B}$. Let $\theta' = (\theta_1, \theta_2, \ldots, \theta_m) \in \Omega$. Thus we have $(\theta_i V_i + \ldots + \theta_m V_m)^{-1} \in \mathbf{B}$. But we can write

$$(\sum_{i=1}^{m} \theta_i V_i)^{-1} = \sum_{i=1}^{m} \theta_i^*(\theta) V_i$$

for some functions $\theta_1^*(\cdot), \theta_2^*(\cdot), \ldots, \theta_m^*(\cdot)$ on Ω, since $V_1, V_2, \ldots, V_m$ form a basis for the vector space $\mathbf{B}$. Let us analyze the second condition of the theorem. The condition $Sp(V_iX) \subset Sp(X)$ implies that $V_iX = XC_i$ for some matrix C_i. Assume, without loss of generality, that $\text{Rank}(X) = p$, i.e., the rank of X is full. (Why?) Let $f(\cdot; \beta, \theta)$ be the joint density function of the components of the random vector Y. Then

$$f(y; \beta, \theta) = c(\theta) \exp\{-(1/2)(y - X\beta)'(\sum_{i=1}^{m} \theta_i V_i)^{-1}(y - X\beta)\}, \ y \in \mathbf{R}^n,$$

where $c(\theta)$ is the normalizing factor depending only on θ. The expression in the exponent after excluding the factor - (1/2) simplifies to

$$y'(\sum_{i=1}^{m} \theta_i^*(\theta) V_i) y - 2\beta' X'(\sum_{i=1}^{m} \theta_i^*(\theta) V_i) y + \beta' X'(\sum_{i=1}^{m} \theta_i^*(\theta) V_i) X\beta$$

which after some further simplification reduces to

$$y'(\sum_{i=1}^{m}\theta_i^*(\theta)V_i)y - 2\beta'X'(\sum_{i=1}^{m}\theta_i^*(\theta)V_i)X(X'X)^{-1}X'y$$
$$+ \beta'X'(\sum_{i=1}^{m}\theta_i^*(\theta)V_i)X\beta.$$

In all these simplifications only the middle term is worked upon. The density can now be written as

$$\begin{aligned}f(y;\beta,\theta) &= c^*(\beta,\theta)\exp\{-(1/2)[\sum_{i=1}^{m}\theta_i^*(\theta)y'V_iy \\ &\quad - 2\beta'X'(\sum_{i=1}^{m}\theta_i^*(\theta)V_i)X(X'X)^{-1}X'y]\} \\ &= c^*(\beta,\theta)\exp\{\sum_{i=1}^{m}\varphi_i(\theta)y'V_iy + A(\beta,\theta)X'y\},\end{aligned}$$

where $c^*(\beta,\theta) = c(\theta)\exp\{-(1/2)\beta'X'(\sum_{i=1}^{m}\theta_i^*(\theta)V_i)X\beta\}$, and

$$\varphi_i(\theta) = -(1/2)\theta_i^*(\theta),\ i = 1, 2, \ldots, m,$$
$$A(\beta,\theta) = \beta'X'(\sum_{i=1}^{m}\theta_i^*(\theta)V_i)X(X'X)^{-1}.$$

The family $f(\cdot\,;\beta,\theta)$, $\beta \in \mathbf{R}^p$, $\theta \in \Omega$ of densities constitute an exponential family and from the general theory of such families, it follows that statistic $(X'Y, Y'V_1Y, Y'V_2Y, \ldots, Y'V_mY)$ is complete and sufficient.

Once we identify a complete and sufficient statistic for a family of distributions, it becomes a simple task to identify Uniformly Minimum Variance Unbiased Estimators (UMVUE) of parameters of interest. In the following result, we follow this cue. The basic problem is to estimate linear functions of the components of β unbiasedly with minimum variance. This leads to the question as to what linear functions of the components of β can be estimated unbiasedly. This question has been tackled earlier. It suffices to estimate unbiasedly $X\beta$. (Why?) The next problem is to estimate linear functions of the variance components. Let $\lambda' = (\lambda_1, \lambda_2, \ldots, \lambda_m)$ be a given vector. Let $P = X(X'X)^{-1}X'$, and $S = \text{tr}(V_i(I_n - P)V_j)$. The matrix S is of order $m \times m$ with (i,j)-th element given by $\text{tr}(V_i(I_n - P)V_j)$. We have seen that there exists an unbiased quadratic estimator of $\lambda'\theta$ if $\lambda \in Sp(S)$, i.e., there exists a vector α such that $\lambda = S\alpha$. (See Sections 12.7 - 12.10.)

P 13.4.2 Let $Y = X\beta + \varepsilon$ be a variance components linear model satisfying all the conditions stipulated in **P 13.4.1**. Let $\lambda'\theta$ be an estimable linear function of the variance components, i.e., $\lambda = S\alpha$ for some vector $\alpha' = (\alpha_1, \alpha_2, \ldots, \alpha_m)$. Then:

(1) The UMVUE of $X\beta$ is $X(X'X)^{-1}X'Y$.

(2) The UMVUE of $\lambda'\theta$ is $\sum_{i=1}^{m} \alpha_i Y'(V_i - PV_iP)Y$.

PROOF. (1) It is enough to show that $X(X'X)^{-1}X'Y$ is an unbiased estimator of $X\beta$. (This is one of the benefits of complete sufficiency of the vector statistic of **P 13.4.1**.) Note that $E(X(X'X)^{-1}X'Y) = X\beta$.

(2) In a similar vein, it suffices to show that the estimator proposed in (2) is an unbiased estimator of $\lambda'\theta$. Before proceeding with the task, we do a simple computation to show that $PV_jP = V_jP$ for each j. Since $Sp(V_jX) \subset Sp(X)$, there exists a matrix C such that $V_jX = XC$. Hence

$$\begin{aligned} PV_jP &= X(X'X)^{-1}X'V_jX(X'X)^{-1}X' = X(X'X)^{-1}C'X'X(X'X)^{-1}X' \\ &= X(X'X)^{-1}C'X' = X(X'X)^{-1}X'V_j = PV_j. \end{aligned}$$

For the evaluation of the desired expectation, we need to recall the formula for the expectation of quadratic form of a random vector. Note that

$$\begin{aligned} &E\sum_{i=1}^{m} \alpha_i Y'(V_i - PV_iP)Y \\ &= \sum_{i=1}^{m} \alpha_i \,\text{tr}((V_i - PV_iP)\text{Disp}(Y)) + \sum_{i=1}^{m} \alpha_i \,\beta' X'(V_i - PV_iP)X\beta \\ &= \sum_{i=1}^{m} \alpha_i \,\text{tr}((V_i - V_iP)\sum_{j=1}^{m} \theta_j V_j) + \sum_{i=1}^{m} \alpha_i \,\beta'(X'V_iX - X'PV_iPX)\beta \end{aligned}$$

$$= \sum_{j=1}^{m} \theta_j \sum_{i=1}^{m} \alpha_i \operatorname{tr}((V_i - V_i P)V_j) + \sum_{i=1}^{m} \alpha_i \beta'(X'V_iX - X'V_iPX)\beta$$

$$= \sum_{j=1}^{m} \theta_j \sum_{i=1}^{m} \alpha_i \operatorname{tr}(V_i(I_n - P)V_j) + 0 = \sum_{j=1}^{m} \theta_j \lambda_j,$$

in view of the fact that $\lambda = S\alpha$. This completes the proof.

We would like to indicate the use of **P 13.4.2** in the context of design of experiments. We will focus specifically on Balanced Incomplete Block Designs (BIBD). Suppose we wish to compare the performance of some ν treatments, $T_1, T_2, \dots, T_\nu$. We select b blocks each containing k experimental units. The treatments are assigned to the experimental units subject to the following conditions:

(1) All treatments are replicated the same number of times r.
(2) Every pair of treatments occur the same number of times λ together in the blocks.

Such a design is called a BIBD. The word "incomplete" in BIBD stems from the fact that $k < \nu$. Consider one particular BIBD adopted for the purported problem of comparing the performance of the treatments. Let Y_{ij} be the response from the j-th unit of the i-th block and $i(j)$ the subscript of the treatment that has been assigned to the j-th unit of the i-th block, $j = 1, 2, \dots, k$ and $i = 1, 2, \dots, b$.

The following is a model with the parameters defined in a natural way that would explain the variation one observes in the responses:

$$Y_{ij} = \mu + \alpha_i + \beta_{i(j)} + \varepsilon_{ij}, \; i = 1, 2, \dots, b; \; j = 1, 2, \dots, k,$$

(1) μ is the general effect,
(2) α_i is the effect of the i-th block,
(3) β_j is the effect of the j-th treatment,
(4) ε_{ij} is the random error,
(5) μ and β_j's are deterministic unknown parameters,
(6) α_i's are independent identically distributed normal random variables with common mean zero and variance σ_α^2,
(7) ε_{ij}'s are independent identically distributed normal random variables with common mean zero and variance σ^2,
(8) α_i's and ε_{ij}'s are mutually independent.

The model is an example of a mixed linear model. The goal now is to put the mixed linear model in the format of $Y = X\beta + \varepsilon$. Let Y be the column vector of all the responses stacked block by block. It is of order $n \times 1$, where $n = bk$. Let $\beta' = (\mu, \beta_1, \beta_2, \dots, \beta_\nu)$, $\alpha' = (\alpha_1, \alpha_2, \dots, \alpha_b)$,

$$X = \begin{bmatrix} 1_k & X_1 \\ \cdot & \cdot \\ 1_k & X_b \end{bmatrix}, \text{ and } U = \begin{bmatrix} U_1 \\ \cdot \\ U_b \end{bmatrix},$$

where 1_k is a column vector of order $k \times 1$ in which each entry is unity. The matrix X is a block matrix of order $n \times (\nu + 1)$ and each X_i is of order $k \times \nu$. The first row of X_i has 1 in the $i(1)$-th column and zeros elsewhere. Remember that the first unit of the i-th block received the treatment $T_{i(1)}$ and $Y_{i1} = \mu + \alpha_i + \beta_{i(1)} + \varepsilon_{i1}$. The second row of X_i has 1 in the $i(2)$-th column and zeros elsewhere. Every other row of X_i has a similar description. The matrix U is a block matrix of order $n \times b$. Each U_i is of order $k \times b$. The i-th column of U_i is 1_k and the rest of the columns are all zero vectors. Finally, let η be the column vector of order $n \times 1$ consisting of ε_{ij}'s stacked block by block. All this effort is needed to write merely

$$Y = X\beta + U\alpha + I_n\eta.$$

The reader is strongly urged to get a grip on this particular form of writing the mixed linear model. One can identify the vector ε with $U\alpha + I_n\eta$. The dispersion matrix of ε works out to be $D(\varepsilon) = \sigma_\alpha^2 UU' + \sigma^2 I_n$. The matrix UU' is block diagonal $n \times n$ matrix with the i-th diagonal block as J_k, a matrix of order $k \times k$ in which every entry is equal to unity. Thus we are back in the mould of a variance components model with $m = 2$, $V_1 = UU'$, $V_2 = I_n$, $\theta_1 = \sigma_\alpha^2$, and $\theta_2 = \sigma^2$. Observe that $V_1^2 = kV_1$.

Note: A seminal paper on quadratic subspaces is by Seely (1971). For further nourishment, pursue Seely (1971), Seely and Zyskind (1977), and Rao and Kleffe (1988).

CHAPTER 14

INEQUALITIES WITH APPLICATIONS IN STATISTICS

In this chapter we provide a number of inequalities involving vectors and matrices, which have applications in statistics. Proofs are given for some propositions, while for others references to original sources, where proofs can be found, are given. Applications to statistical problems are indicated.

14.1. Some Results on nnd and pd Matrices

Let us recall that an $n \times n$ Hermitian matrix A (i.e., $A = A^*$) is said to be non-negative definite (nnd) if $x^*Ax \geq 0$ for every $x \in \mathbf{C}^n$ and positive definite (pd) if $x^*Ax > 0$ for every non-zero $x \in \mathbf{C}^n$. The following propositions characterize nnd and pd matrices.

P 14.1.1 A is nnd if and only if there exists a matrix U such that $A = U^*U$, where U is such that $\text{Rank}(A) = \text{Rank}(U)$. In fact, we can choose U such that $U = U^*$ and write $A = U^2 = UU$. A is pd if $\text{Rank}(U) = n$.

The proposition follows easily from the spectral decomposition of A. In fact U can be chosen to be nnd. In that event we say that U is a square root of A. An nnd matrix is also called Gramian. A square root of A is computed as follows. Write the spectral decomposition of A in the style

$$A = \lambda_1^2 P_1 P_1^* + \lambda_2^2 P_2 P_2^* + \ldots + \lambda_k^2 P_k P_k^*,$$

where $k = \text{Rank}(A)$, $\lambda_1^2, \lambda_2^2, \ldots, \lambda_k^2$ are positive eigenvalues of A and $P_1, P_2, \ldots, P_k$ the corresponding orthonormal eigenvectors of A. An nnd (also called a Gramian) square root U of A can be taken as

$$U = \lambda_1 P_1 P_1^* + \lambda_2 P_2 P_2^* + \ldots + \lambda_k P_k P_k^*. \tag{14.1.1}$$

P 14.1.2 Let Y be $n \times r$ matrix and denote by $Y^{\perp}$, the matrix whose columns are orthogonal to those of Y and such that $\text{Rank}(Y) + \text{Rank}(Y^{\perp}) = n$. Then

$$P_Y = Y(Y^*Y)^- Y^* \quad \text{and } I - P_Y, \tag{14.1.2}$$

are nnd, where $(Y^*Y)^-$ is any g-inverse.

PROOF. Let $a \in \mathbf{C}^n$ be arbitrary. Write $a = Ya_1 + Y^{\perp}a_2$ for some vectors a_1 and a_2. (How?) Note that

$$\begin{aligned} a^* P_Y a &= a^* Y(Y^*Y)^- Y^* a = a_1^* Y^* Y(Y^*Y)^- Y^* Y a_1 \\ &= a_1^*(Y^*Y)a_1 = (a_1^* Y^*)(Y a_1) \geq 0. \end{aligned}$$

Consequently, P_Y is nnd. In a similar manner,

$$a^*(I - P_Y)a = (a_2^*(Y^{\perp})^*)(Y^{\perp}a_2) \geq 0,$$

from which we conclude that $I - P_Y$ is nnd.

We have seen the matrix P_Y before. It is the orthogonal projector from $\mathbf{C}^n$ onto $Sp(Y)$. See Chapter 7.

P 14.1.3 Let A be nnd and consider a partition

$$A = \begin{bmatrix} A_{11} & A_{12} \\ A_{21} & A_{22} \end{bmatrix}, \tag{14.1.3}$$

where A_{11} and A_{22} are square matrices. Then $A_{11} - A_{12}A_{22}^- A_{21}$ is nnd and unique for any g-inverse A_{22}^- of A_{22}.

PROOF. By **P 14.1.1**, one can write $A = U^*U$ for some matrix U. Partition $U = (X|Y)$, where the number of columns of X is the same as the number of columns of A_{11}. Note that

$$A = U^*U = \begin{bmatrix} X^* \\ Y^* \end{bmatrix} [X|Y] = \begin{bmatrix} X^*X & X^*Y \\ Y^*X & Y^*Y \end{bmatrix} = \begin{bmatrix} A_{11} & A_{12} \\ A_{21} & A_{22} \end{bmatrix},$$

$$\begin{aligned} A_{11} - A_{12}A_{22}^- A_{21} &= X^*X - X^*Y(Y^*Y)^- Y^*X \\ &= X^*(I - P_Y)X. \end{aligned} \tag{14.1.4}$$

Since $I - P_Y$ is nnd, $A_{11} - A_{12}A_{22}^- A_{21}$ is nnd, and unique (Why?).

The matrix $A_{11} - A_{12}A_{22}^{-}A_{21}$ is called the Schur complement of A with respect to A_{22}.

P 14.1.4 Let A be pd and partition A^{-1} as

$$A^{-1} = \begin{bmatrix} A_{11} & A_{12} \\ A_{21} & A_{22} \end{bmatrix}^{-1} = \begin{bmatrix} A^{11} & A^{12} \\ A^{21} & A^{22} \end{bmatrix} \tag{14.1.5}$$

where A_{11} is a square matrix. Then

$$(1) \qquad A^{11} = (A_{11} - A_{12}A_{22}^{-1}A_{21})^{-1}, \tag{14.1.6}$$

$$(2) \qquad A^{11} - A_{11}^{-1} \quad \text{is nnd.} \tag{14.1.7}$$

PROOF. From (14.1.5), we have the equations

$$A^{11}A_{11} + A^{12}A_{21} = I,$$
$$A^{11}A_{12} + A^{12}A_{22} = 0.$$

Eliminating A^{12} we have an equation in A^{11} which gives (14.1.6). To prove (14.1.7), we note that $A_{11} - (A^{11})^{-1} = A_{12}A_{22}^{-1}A_{21}$, which is nnd. This shows that $A_{11} \geq (A^{11})^{-1}$ which implies $(A_{11})^{-1} \leq A^{11}$. In particular, from (14.1.7), we have that if $A = (a_{ij})$ and $A^{-1} = (a^{ij})$, then $a^{11} \geq 1/a_{11}$.

Complements

14.1.1 If $A = B+C$, B is pd and C is skew symmetric, then $|A| \geq |B|$.

14.1.2 If A and B are real pd and of order $n \times n$, show that

(1) $|\lambda A + (1-\lambda)B| \geq |A|^{\lambda}|B|^{1-\lambda}$ for $0 \leq \lambda \leq 1$;

(2) $|A+B|^{1/n} \geq |A|^{1/n} + |B|^{1/n}$ (Minkowski's inequality).

14.1.3 If B is pd and $A - B$ is nnd, then

(1) $|A - \lambda B| = 0$ has all its roots $\lambda \geq 1$;

(2) $|A| \geq |B|$;

(3) $(B^{-1} - A^{-1})$ is nnd; and

(4) if A_r is a principal minor of A of order r and B_r is the corresponding minor of B, then $A_r - B_r$ is nnd.

14.1.4 Let x be a column vector of order $k \times 1$ and A be a square symmetric matrix of order $k \times k$, with all its elements non-negative. Then for any integer n,

$$(x'A^n x) > (x'Ax)^n / x'x$$

with equality if and only if x is an eigenvector of A. This inequality occurs in genetical theory (Mulholland and Smith (1959)).

14.1.5 If $\|Ax\| \leq \|Ax + By\|$ for all x and y, show that $A^*B = 0$, where $\|\cdot\|$ is the Euclidean norm.

14.1.6 If A is nnd and $a_{ii} = 0$, show that $a_{ij} = 0$ for all j.

14.1.7 Let A and B be two pd matrices with eigenvalues contained in the interval $[m, M]$, where $M \geq m > 0$. Let $0 \leq \lambda \leq 1$. Show that

$$\lambda A^2 + (1-\lambda)B^2 - [\lambda A + (1-\lambda)B]^2 \leq \frac{1}{4}(M-m)^2 I.$$

This is the reverse of the following convex matrix inequality

$$0 \leq \lambda A^2 + (1-\lambda)B^2 - [\lambda A + (1-\lambda)B]^2.$$

14.1.8 Let $A = (a_{ij})$ be an $n \times n$ complex matrix and $U = (U_{ij})$ be an $n \times n$ unitary matrix. Denote the Hadamard-Schur product of A and U by $A \cdot U = (a_{ij}u_{ij})$ and the largest singular value by $\sigma_{\max}$. Show that

$$\min_U \sigma_{\max}(A \cdot U) \leq \frac{1}{\sqrt{n}}\left(\sum_{i,j=1}^{n} |a_{ij}|^2\right)^{1/2}.$$

14.1.9 For any square complex matrices X and Y of the same size, show that

$$|\det(X+Y)|^2 \leq \det(I + XX^*)\det(I + Y^*Y).$$

14.1.10 Show that an nnd matrix A has a unique decomposition, $A = TT'$, where T is a triangular matrix with diagonal elements non-negative.

14.1.11 Let $H = A + iB$ be a Hermitian pd, where A and B are real matrices. Show that $|A| > |B|$ (Robertson inequality) and $|H| \leq |A|$ (Taussky inequality). Further, if α_i's are the eigenvalues of A and β_i's of $-iB$, both in decreasing order, show that $\alpha_i > \beta_i$.

14.1.12 If A and B are nnd matrices, show that AB has only non-negative eigenvalues.

14.1.13 Schur's majorization theorem: If A is a symmetric $n \times n$ matrix, show that the eigenvalues of A majorizes its diagonal elements.

14.1.14 Let $A \in M_n$ and pd be partitioned as

$$A = \begin{bmatrix} A_{11} & A_{12} \\ A_{21} & A_{22} \end{bmatrix},$$

where A_{11} and A_{22} are square matrices. Show that

$$|A| \leq |A_{11}|\ |A_{22}|$$

which is Fischer's inequality.
14.1.15 If A is nnd partitioned as in complement 14.1.14, show that

$$\begin{bmatrix} A_{11} & A_{12} \\ A_{21} & A_{21}A_{11}^{-1}A_{12} \end{bmatrix}$$

is nnd.
14.1.16 Let A be $n \times n$ positive definite matrix partitioned as in complement 14.1.14. Show that

$$A \cdot A^{-1} \geq \begin{bmatrix} A_{11} \cdot A_{11}^{-1} & 0 \\ 0 & A^{22} \cdot (A^{22})^{-1} \end{bmatrix},$$

where A^{22} is the partition conformal to A_{22} in the reciprocal matrix A^{-1}. As a consequence of the above result, show that $\lambda_1(A \cdot B) \geq \lambda_1(AB)$ when A and B are $n \times n$ nnd matrices, where λ_1 is a generic symbol for the largest eigenvalue.
14.1.17 Let A and B be non-negative definite Hermitian matrices of order n. Show that

1. $\lambda_{\min}(A \cdot B) \geq \lambda_{\min}(AB'),$
2. $\lambda_{\min}(A \cdot B) \geq \lambda_{\min}(AB).$

14.1.18 Oppenheim's inequality: Let A and B be $n \times n$ nnd matrices. Show that $|A \cdot B| \geq |B|\ a_{11} \ldots a_{nn}$.
14.1.19 Fiedler's inequality: If A is pd, show that $A \cdot A^{-1} - I$ is nnd.
14.1.20 Let A and B be $n \times n$ positive definite matrices. Show that $|A|\ |B| \leq |A \cdot B| \leq (a_{11}a_{22} \cdots a_{nn})(b_{11}b_{22} \cdots b_{nn})$.
14.1.21 Let A and B be $n \times n$ symmetric matrices and let the elements of B be non-negative. Show that $(\lambda_1(A \cdot B), \ldots, \lambda_n(A \cdot B))'$ majorizes

$(\lambda_1(A), \dots, \lambda_n(A))'$. If A also has all non-negative entries, show that $\Pi_{i=1}^n \lambda_i(A \cdot B) \geq \Pi_{i=1}^n \lambda_i(A)$.

14.1.22 Consider the projection operator $P_Y = Y(Y^*Y)^- Y^*$ on the space $Sp(Y)$. Let H be a matrix whose columns form an orthonormal basis of $Sp(Y)$. Show that $P_Y = P_H = HH^*$. Get an explicit form of H by using the s.v.d. of Y.

14.1.23 Let $B = (B_1|B_2)$ be such that $B_1'B_2 = 0$. Show that $P_B - P_{B_1}$ is nnd.

14.1.24 (Ostrowski-Taussky Theorem). Let $A \in M_n$, and $H = 2^{-1}(A + A^*)$ be pd. Show that $|H(A)| \leq |A|$.

14.2. Cauchy-Schwartz and Related Inequalities

Cauchy-Schwartz inequality has numerous applications in statistics. Some well-known applications are in showing that the correlation coefficient lies between -1 and 1 and in deriving a lower bound to the variance of an unbiased estimator such as the Cramér-Rao lower bound.

P 14.2.1 Let x and y be two vectors of real elements. Then

$$(x'y)^2 \leq (x'x)(y'y) \tag{14.2.1}$$

with equality if and only if $y = ax$ for some scalar a or $x = ay$ for some scalar a.

PROOF. Let $x \neq 0$. A simple demonstration of (14.2.1) is as follows:

$$0 \leq \Sigma(y_i - \frac{\Sigma x_i y_i}{\Sigma x_i^2} x_i)^2 = \Sigma y_i^2 - \frac{(\Sigma x_i y_i)^2}{\Sigma x_i^2}. \tag{14.2.2}$$

For equality in (14.2.2), we must have $y_i = [(\Sigma x_i y_i)/\Sigma x_i^2] x_i$ for all i, i.e., $(\Sigma x_i y_i)/(\Sigma x_i^2)$ is constant, say a.

If $p_1, \dots, p_n$ are non-negative numbers, it follows from (14.2.1) that the weighted version

$$(\Sigma p_i x_i y_i)^2 \leq (\Sigma p_i x_i^2)(\Sigma p_i y_i^2) \tag{14.2.3}$$

is also true.

P 14.2.2 Let x and y be two real vectors and A be an nnd matrix. Then

(1) $(x'Ay)^2 \leq (x'Ax)(y'Ay)$, (14.2.4)

(2) $(x'y)^2 \le (x'Ax)(y'A^{-1}y)$, if A^{-1} exists, (14.2.5)
(3) $(x'x)^2 \le (x'Ax)(x'A^{-1}x)$, if A^{-1} exists. (14.2.6)

P 14.2.3 Let x and y be two complex vectors and A be a Hermitian nnd matrix. Then

(1) $|x^*y|^2 \le (x^*x)(y^*y)$, (14.2.7)
(2) $|x^*Ay|^2 \le (x^*Ax)(y^*Ay)$, (14.2.8)
(3) $|x^*y|^2 \le (x^*Ax)(y^*A^{-1}y)$, if A^{-1} exists, (14.2.9)

where $|\cdot|$ represents the absolute value.

P 14.2.4 Constrained Cauchy-Schwartz inequality. Let x and y be two vectors of order $n \times 1$ and B be a matrix of order $n \times n$. Then

(1) $(x'y)^2 \le (x'P_Bx)(y'y)$ if $y \in Sp(B)$, where P_B is the projection operator on $Sp(B)$,
(2) $(x'y)^2 \le (y'B^-y)(x'Bx)$ if B is nnd, $y \in Sp(B)$, and B^- is any g-inverse of B.

P 14.2.5 The integral version of the Cauchy-Schwartz inequality is

$$\Big(\int_A fg\, dv\Big)^2 = \int_A f^2\, dv \int_A g^2\, dv,$$

where f and g are real functions defined on a set A, and f and g are square integrable with respect to a measure v.

P 14.2.6 The Cauchy-Schwartz inequality for non-null vectors x and y can be stated in the form

$$x^*x - (x^*y)(y^*y)^{-1}(y^*x) \ge 0.$$

A more general version of the inequality is obtained when x and y are replaced by matrices X and Y, yielding that

$$X^*X - (X^*Y)(Y^*Y)^-(Y^*X) \ge_L 0, \tag{14.2.10}$$

i.e., the matrix in (14.2.10) is nnd. Using Complement 14.1.3, we also have the weaker version in terms of determinants that

$$|X^*X| \ge |X^*Y(Y^*Y)^-Y^*X|. \tag{14.2.11}$$

The Cauchy-Schwartz inequality gives an upper bound to $x'y$. We mention without proof some complementary inequalities which give a lower bound to $x'y = (x_1y_1 + \ldots + x_ny_n)$.

P 14.2.7 Let $0 < m_1 \le x_i \le M_1$, $0 < m_2 \le y_i \le M_2$, and $0 < m < \gamma_i \le M$ for every $i = 1, 2, \ldots, n$. Further, let ξ_i $(i = 1, \ldots, n)$ be real numbers. Then the following inequalities hold.

(1) $(\Sigma\gamma_i)(\Sigma\frac{1}{\gamma_i}) \le \frac{(M+m)^2}{4Mm}n^2$. (Schweitzer)

(2) $(\Sigma x_i^2)(\Sigma y_i^2) \le \frac{(M_1M_2+m_1m_2)^2}{4m_1m_2M_1M_2}(\Sigma x_iy_i)^2$. (Polya-Szegö)

(3) $(\Sigma\gamma_i\xi_i^2)(\Sigma\frac{\xi_i^2}{\gamma_i}) \le \frac{(M+m)^2}{4Mm}(\Sigma\xi_i^2)^2$. (Kantorovich)

(4) $(\Sigma x_i^2\xi_i^2)(\Sigma y_i^2\xi_i^2) \le \frac{(M_1M_2+m_1m_2)^2}{4m_1m_2M_1M_2}(\Sigma x_iy_i\xi_i^2)^2$.
(Greub-Rheinboldt)

The inequality (4) can written in a form to include the lower and upper bounds to the inner product. (See Diaz and Metcalf (1964).)

$$1 \le \frac{(\Sigma x_i^2\xi_i^2)(\Sigma y_i^2\xi_i^2)}{(\Sigma x_iy_i\xi_i^2)^2} \le \frac{(M_1M_2+m_1m_2)^2}{4M_1M_2m_1m_2}. \tag{14.2.12}$$

14.3. Hadamard Inequality

P 14.3.1 For a nonsingular real $n \times n$ matrix $B = (b_{ij})$

$$|B|^2 \le \Pi_{i=1}^n(\Sigma_{k=1}^n b_{ik}^2). \tag{14.3.1}$$

As special cases, we have

(1) $|B| \le 1$, if $\Sigma_{j=1}^n b_{ij}^2 = 1$, $i = 1, \ldots, n$, (14.3.2)

(2) $|B| \le M^n n^{n/2}$, if $|b_{ij}| \le M$ for all i and j. (14.3.3)

PROOF. To prove (14.3.1), we consider the pd matrix $A = BB'$ and apply the inequality $|A| \le a_{11} \ldots a_{nn}$, the product of the diagonal elements of A which is proved as follows. Consider the expansion

$$|A| = a_{11}\begin{vmatrix} a_{22} & \cdots & a_{2n} \\ \cdot & \cdots & \cdot \\ a_{n2} & \cdots & a_{nn} \end{vmatrix} + \begin{vmatrix} 0 & a_{12} & \cdots & a_{1n} \\ a_{21} & a_{22} & \cdots & a_{2n} \\ \cdot & \cdot & \cdots & \cdot \\ a_{n1} & a_{n2} & \cdots & a_{nn} \end{vmatrix}.$$

Since A is pd, the matrix $(a_{ij})_{2\leq i\leq n,\,2\leq j\leq n}$ is also pd, and the second term is negative (Why?). Hence

$$|A| \leq a_{11} \begin{vmatrix} a_{22} & \dots & a_{2n} \\ \cdot & \dots & \cdot \\ a_{n2} & \dots & a_{nn} \end{vmatrix}.$$

The result $|A| \leq a_{11}a_{21}\dots a_{nn}$ follows by induction. The result (14.3.1) follows by observing that $a_{ii} = b_{i1}^2 + \dots + b_{in}^2$.

14.4. Hölder's Inequality

P 14.4.1 Let $x_i, y_i \geq 0$, $i = 1, \dots, n$ and $(1/p) + (1/q) = 1$ with $p > 1$. Then (using Σ to indicate summation over $i = 1, \dots, n$),

$$\Sigma x_i y_i \leq (\Sigma x_i^p)^{1/p} (\Sigma y_i^q)^{1/q}, \tag{14.4.1}$$

with equality if and only if y_i's are proportional to x_i^{p-1}'s. The integral version of Hölder's inequality for functions $f, g \geq 0$ is given by

$$\int_R f g dv \leq (\int_R f^p dv)^{1/p} (\int g^q dv)^{1/q}. \tag{14.4.2}$$

PROOF. First, we establish by differentiation or otherwise

$$\min_{x\geq 0} \left\{ t(x) = \frac{x^p}{p} + \frac{x^{-q}}{q} \right\} = 1,$$

and the minimum is attained at $x = 1$. Substituting $x = u^{1/q}v^{-1/p}$, $u, v \geq 0$ in $t(x) \geq 1$, we find

$$uv \leq \frac{u^p}{p} + \frac{v^q}{q} \tag{14.4.3}$$

with equality when $v = u^{p-1}$. Now let

$$u_k = x_k/(\Sigma x_k^p)^{1/p}, v_k = y_k/(\Sigma y_k^q)^{1/q}.$$

Substituting in (14.4.3) and summing over k, we have

$$\Sigma u_k v_k \leq \frac{\Sigma u_k^p}{p} + \frac{\Sigma v_k^q}{q} = \frac{1}{p} + \frac{1}{q} = 1,$$

which gives the desired inequality. The integral version is proved in a similar manner.

P 14.4.2 Let $x_i^{(r)}$, $i = 1, 2, \ldots, n$ be a finite sequence of non-negative numbers for each $r = 1, 2, \ldots, m$. Let $p_1, p_2, \ldots, p_m$ be m numbers each > 1 such that

$$\frac{1}{p_1} + \frac{1}{p_2} + \ldots + \frac{1}{p_m} \leq 1.$$

Then

$$\sum_{i=1}^{n} \Pi_{j=1}^{m} x_i^{(j)} \leq \Pi_{j=1}^{m} [\sum_{i=1}^{n} (x_i^{(j)})^{p_j}]^{1/p_j}. \tag{14.4.4}$$

The result is easily established.

From Hölder's inequality, it is easy to deduce that if $x_i, y_i \geq 0$ and $p \geq 1$, then

$$[\Sigma(x_i + y_i)^p]^{1/p} \leq (\Sigma x_i^p)^{1/p} + (\Sigma y_i^p)^{1/p}, \tag{14.4.5}$$

which is Minkowski's inequality, where all summations are from 1 to n.

14.5. Inequalities in Information Theory

(1) Let Σa_i and Σb_i be convergent sequences of positive numbers such that $\Sigma a_i \geq \Sigma b_i$. Then

$$\Sigma a_i \log \frac{b_i}{a_i} \leq 0 \tag{14.5.1}$$

with equality being attained if and only if $a_i = b_i$. Further, if $a_i \leq 1$ and $b_i \leq 1$ for all i, then

$$2\Sigma a_i \log \frac{a_i}{b_i} \leq \Sigma a_i(a_i - b_i)^2. \tag{14.5.2}$$

To prove the inequalities, note that for $x > 0$, the expansion of $\log x$ at $x = 1$ yields

$$\log x = (x-1) - (x-1)^2(2y^2)^{-1} \quad \text{with } y \in (1, x). \tag{14.5.3}$$

Using the expansion (14.5.3) for each term in (14.5.1), we have

$$\Sigma a_i \log \frac{b_i}{a_i} = (\Sigma b_i - \Sigma a_i) - \Sigma a_i(b_i - a_i)^2(2a_i^2 y_i^2)^{-1} \leq 0 \tag{14.5.4}$$

thus proving (14.5.1). In (14.5.4), $a_i^2 y_i^2 \in (a_i^2, b_i^2)$, and if $a_i \leq 1, b_i \leq 1$, the maximum value of y_i^2 is not greater than unity. Hence

$$-\Sigma a_i(b_i - a_i)^2(2a_i^2 y_i^2)^{-1} \leq \frac{-\Sigma a_i(b_i - a_i)^2}{2}. \tag{14.5.5}$$

Combining (14.5.4) and (14.5.5), we obtain (14.5.2).

The results (14.5.1) and (14.5.2) are true if a_i and b_i are non-negative and the summations are extended over values of i for which $a_i > 0$. In other words we admit the possibility of some of the b_i being zero (but not the a_i).

(2) Let f and g be non-negative and integrable functions with respect to a measure μ and S be the region in which $f > 0$. Then

$$\int_S (f - g) \geq 0 \Rightarrow \int_S f \log \frac{f}{g}\, d\mu \geq 0 \tag{14.5.6}$$

with equality only when $f = g$ *a.e.*$[\mu]$.

The proof is the same as that of (14.5.1) with summations replaced by integrals. The reader may directly deduce the inequalities (14.5.1) and (14.5.6) by applying Jensen's inequality (14.6.6), given in the next section.

14.6. Convex Functions and Jensen's Inequality

A function $f(x)$ is said to be convex, if for $\alpha, \beta > 0, \alpha + \beta = 1$

$$f(\alpha x + \beta y) \leq \alpha f(x) + \beta f(y) \quad \text{for all } x, y. \tag{14.6.1}$$

For such a function, we shall first show that at any point x_0, $f'_+(x_0)$, $f'_-(x_0)$, the right and left derivatives, respectively, exist. Let $x_0 < x_1 < x_2$. Choosing $\alpha = (x_2 - x_1)/(x_2 - x_0)$, $\beta = (x_1 - x_0)/(x_2 - x_0), x = x_0$, and $y = x_2$ so that $\alpha x + \beta y = x_1$, and using (14.6.1), we see that

$$(x_2 - x_0)f(x_1) \leq (x_2 - x_1)f(x_0) + (x_1 - x_0)f(x_2), \tag{14.6.2}$$

after multiplication by $(x_2 - x_0)$. Adding $x_0 f(x_0)$ to both sides and rearranging the terms in (14.6.2), we have

$$\frac{f(x_1) - f(x_0)}{x_1 - x_0} < \frac{f(x_2) - f(x_0)}{x_2 - x_0}, \tag{14.6.3}$$

which shows that $[f(x) - f(x_0)]/(x - x_0)$ decreases as $x \to x_0$. By adding $x_1 f(x_1)$ to both sides of (14.6.2) and rearranging the terms, we have

$$\frac{f(x_0) - f(x_1)}{x_0 - x_1} \leq \frac{f(x_2) - f(x_1)}{x_2 - x_1}. \tag{14.6.4}$$

Equation (14.6.4) in terms of $x_{-1} < x_0 < x_1$ becomes

$$\frac{f(x_{-1}) - f(x_0)}{x_{-1} - x_0} \leq \frac{f(x_1) - f(x_0)}{x_1 - x_0},$$

which shows that $[f(x) - f(x_0)]/(x - x_0)$ is bounded from below, and since it decreases as $x \to x_0$, the right-hand derivative $f'_+(x_0)$ exists. Similarly, $f'_-(x_0)$ exists and obviously $f'_-(x_0) \leq f'_+(x_0)$. Let L be such that $f'_-(x_0) \leq L \leq f'_+(x_0)$. Then, for all x,

$$f(x) \geq f(x_0) + L(x - x_0), \tag{14.6.5}$$

for if $x_1 > x_0$,

$$\frac{f(x_1) - f(x_0)}{x_1 - x_0} \geq f'_+(x_0) \geq L,$$

and the reverse relation is true when $x_1 < x_0$. The inequality (14.6.5) has important applications; it leads to Jensen's inequality.

P 14.6.1 (Jensen's Inequality). If X is a random variable such that $E(X) = \mu$ and $f(\cdot)$ is a convex function, then

$$E[f(X)] \geq f[E(X)] \tag{14.6.6}$$

with equality if and only if X has a degenerate distribution at μ or f is a linear function.

PROOF. Consider the inequality (14.6.5) with μ for x_0,

$$f(X) \geq f(\mu) + L(X - \mu), \tag{14.6.7}$$

and take expectations of both sides. The expectation of the second term on the right-hand side of (14.6.7) is zero, yielding the inequality (14.6.6).

Let $f(x)$ be a convex function of a vector $x \in R^n$. Then a result analogous to (14.6.7) is

$$f(x) \geq f(\mu) + L'(x - \mu) \tag{14.6.8}$$

where $L \in R^n$. Using (14.6.8), we find that the result (14.6.6) is true for a convex function of a vector variable.

P 14.6.2 Let $x_1, \ldots, x_n$ be positive real numbers and $\mu_1 \ldots, \mu_n$ be arbitrary real numbers. Then

$$(\sum_{i=1}^{n} x_i^{-1})^{-1} \leq (\sum_{i=1}^{n} x_i \mu_i^{-2})(\sum_{i=1}^{n} \mu_i^{-1})^{-2}. \tag{14.6.9}$$

PROOF. By the Cauchy-Schwartz inequality (14.2.1)

$$(\sum_{i=1}^{n} x_i^{-1})(\sum_{i=1}^{n} x_i \mu_i^{-2}) \geq (\sum_{i=1}^{n} \mu_i^{-1})^2$$

from which (14.6.9) follows.

P 14.6.3 (Inequality on Harmonic Mean) Let $X_1, \ldots, X_n$ be positive random variables. Then the expected value of their harmonic mean is not greater than the harmonic mean of their expected values.

PROOF. Let $E(X_i) = \mu_i$, $i = 1, \ldots, n$. Considering the inequality (14.6.9) after multiplying both sides by n and taking expectations, we get

$$E\ [n(\sum_{i=1}^{n} x_i^{-1})^{-1}] \leq n\ (\sum_{i=1}^{n} \mu_i^{-1})^{-1}, \tag{14.6.10}$$

which is the desired result. See Rao (1996).

14.7. Inequalities Involving Moments

Let X be a random variable such that $E(X) = \mu$. Assume that the moments

$$E(X - \mu)^r = \mu_r,\ r = 1, \ldots, 2N$$

exist. The quadratic form $\sum_{i=1}^{N+1}\sum_{j=1}^{N+1} \mu_{i+j-2}\, y_i y_j$ in $y_1, \dots, y_{N+1}$

$$= E[y_1 + y_2(X-\mu) + \dots + y_{N+1}(X-\mu)^N]^2 \geq 0.$$

Hence the matrix whose (i,j)-th element is μ_{i+j-2} is nnd. This, however, is not a sufficient condition for μ_r to be a moment sequence of a random variable. The condition is necessary whether the moments are of a discrete or a continuous distribution, or calculated from a given set of observations. In particular, for $N = 2$,

$$\begin{bmatrix} 1 & 0 & \mu_2 \\ 0 & \mu_2 & \mu_3 \\ \mu_2 & \mu_3 & \mu_4 \end{bmatrix} \geq 0 \text{ or } \mu_2^3\left(\frac{\mu_4}{\mu_2^2} - \frac{\mu_3^2}{\mu_2^3} - 1\right) \geq 0 \tag{14.7.1}$$

i.e., $\beta_2 \geq 1 + \beta_1$, where $\beta_1 = \mu_3^2/\mu_2^3$ and $\beta_2 = \mu_4/\mu_2^2$ are measures of skewness and kurtosis of a distribution.

14.8. Kantorovich Inequality and Extensions

In this section, we focus on Kantorovich inequality and some of its variants. Application to some statistical problems is also broached.

P 14.8.1 If A is $n \times n$ Hermitian pd with eigenvalues $\lambda_1 \geq \dots \geq \lambda_n$, then

$$1 \leq \frac{x^*Ax}{x^*x}\frac{x^*A^{-1}x}{x^*x} \leq \frac{(\lambda_1+\lambda_n)^2}{4\lambda_1\lambda_n}. \tag{14.8.1}$$

for all nonzero vectors x.

PROOF. The left-hand side inequality follows from the Cauchy-Schwartz inequality (14.2.6). To prove the right-hand side, we first observe that A and A^{-1} have the spectral decomposition

$$A = P\Lambda P^*, \ A^{-1} = P\Lambda^{-1}P^*,$$

for some unitary matrix P and $\Lambda = \text{Diag}\{\lambda_1, \lambda_2, \dots, \lambda_n\}$, so that the middle term in (14.8.1) can be written as

$$\frac{y^*\Lambda y}{y^*y} \cdot \frac{y^*\Lambda^{-1}y}{y^*y} \tag{14.8.2}$$

with $y = P^*x$. Thus we need only to find the upper bound of (14.8.2) involving diagonal matrices Λ and Λ^{-1} instead of general matrices A and A^{-1}. Taking the logarithm of (14.8.2), finding the derivative with respect to y, and setting the derivative to zero, we have the equation

$$\xi_1 \Lambda y + \xi_2 \Lambda^{-1} y = 2y, \tag{14.8.3}$$

where $\xi_1 = y^*y/y^*\Lambda y$ and $\xi_2 = y^*y/y^*\Lambda^{-1}y$. The equation can be written as

$$\xi_1 \lambda_i y_i + \frac{\xi_2}{\lambda_i} y_i = 2y_i, \; i = 1, \dots, n. \tag{14.8.4}$$

The equations (14.8.4) are soluble if two y_i's corresponding to two different values of λ_j are nonzero. Let y_i and y_j be nonzero. Then the equations for ξ_1 and ξ_2 are

$$\xi_1 \lambda_i + \frac{\xi_2}{\lambda_i} \; = \; \xi_1 \lambda_j + \frac{\xi_2}{\lambda_j} \; = \; 2$$

giving

$$\xi_1^{-1} = \frac{\lambda_i + \lambda_j}{2}, \; \xi_2^{-1} = \frac{\lambda_i + \lambda_j}{2\lambda_i \lambda_j}$$

and the value of (14.8.2) is then

$$\frac{1}{\xi_1 \xi_2} = \frac{(\lambda_i + \lambda_j)^2}{4\lambda_i \lambda_j}$$

which attains the maximum value, when $\lambda_i = \lambda_1$ and $\lambda_j = \lambda_n$,

$$\frac{(\lambda_1 + \lambda_n)^2}{4\lambda_1 \lambda_n}. \tag{14.8.5}$$

The right-hand side of (14.8.1) is proved.

Application: Consider the statistical regression problem with one concomitant variable

$$y_i = \beta x_i + \epsilon_i, \; i = 1, \dots, n.$$

If the covariance matrix of ϵ_i's is $\sigma^2 A$, then the weighted least squares estimate of β has the variance $(x'A^{-1}x)^{-1}$, where $x' = (x_1, \dots, x_n)$. If we estimate β by the ordinary least squares method, then its variance is $x'Ax/(x'x)^2$. The inefficiency of the latter estimate is

$$\frac{(x'x)^2(x'A^{-1}x)^{-1}}{x'Ax} = \frac{(x'x)^2}{(x'Ax)(x'A^{-1}x)} \geq \frac{4\lambda_1\lambda_n}{(\lambda_1+\lambda_n)^2}$$

using the inequality (14.8.1). Thus the worst possible value for inefficiency is $4\lambda_1\lambda_n/(\lambda_1+\lambda_n)^2$.

If we consider the general linear model

$$Y = X\beta + \epsilon, \tag{14.8.6}$$

where X is an $n \times m$ matrix and $\text{cov}(\epsilon) = \sigma^2 A$, then the covariance matrix of the weighted least squares estimator of β is $(X'A^{-1}X)^{-1}$ and that of the ordinary least squares estimator is $(X'X)^{-1}(X'AX)(X'X)^{-1}$. We may consider various measures of inefficiency based on the roots of the determinantal equation

$$|(X'X)^{-1}X'AX(X'X)^{-1} - \theta(X'A^{-1}X)^{-1}| = 0. \tag{14.8.7}$$

One measure is the product of the roots

$$E_1 = \frac{|X'AX|\,|X'A^{-1}X|}{|X'X|^2} = \theta_1\theta_2\cdots\theta_m \tag{14.8.8}$$

which is similar to the middle expression in (14.8.1) with x replaced by a matrix X. Bloomfield and Watson (1975) and Knot (1975) showed that

$$1 \leq E_1 \leq \Pi_{i=1}^{s}\frac{(\lambda_i+\lambda_{n-i+1})^2}{4\lambda_i\lambda_{n-i+1}}, \tag{14.8.9}$$

where $s = \min\{m, n-m\}$. Another measure is $E_2 = \theta_1 + \ldots + \theta_m$. Khatri and Rao (1981, 1982) showed that

$$m \leq \theta_1 + \ldots + \theta_m \leq \sum_{i=1}^{s}\frac{(\lambda_i+\lambda_{n-i+1})^2}{4\lambda_i\lambda_{n-i+1}} + t, \tag{14.8.10}$$

where $t = 0$ if $s = m$ and $t = 2m - n$ if $s = n - m$.

Another measure of inefficiency is

$$E_3 = \text{tr}((X'X)^{-1}X'AX(X'X)^{-1} - (X'A^{-1}X)^{-1}).$$

Rao (1985c) showed that when $X'X = I$,

$$0 \leq E_3 \leq \sum_{i=1}^{s}(\sqrt{\lambda_i} - \sqrt{\lambda_{n-i+1}})^2, \tag{14.8.11}$$

where $s = \min(m, n-m)$. The inequality when $m = 1$ was earlier given by Styan (1983).

A fourth measure of inefficiency is

$$E_4 = \text{tr}(PA^2P - (PAP)(PAP)),$$

where $P = X(X'X)^{-1}X'$ is the projection operator. Bloomfield and Watson (1975) showed that

$$0 \leq E_4 \leq \frac{1}{4}\sum_{i=1}^{s}(\lambda_i - \lambda_{n-i+1})^2. \tag{14.8.12}$$

The expressions (14.8.9) - (14.8.12) are generalizations of the Kantorovich inequality (14.8.1). The proofs are given in the references cited.

Complements

14.8.1 (Strang (1960).) Let A be an $n \times n$ nonsingular matrix with singular values $\delta_1 \geq \delta_2 \geq \ldots \geq \delta_n > 0$ and define

$$\omega_i = (\delta_i + \delta_{n-i+1})^2/4\delta_i\delta_{n-i+1}.$$

Then

$$\frac{(x'Ay)(y'A^{-1}x)}{(x'x)(y'y)} \leq \omega_1.$$

14.8.2 (Khatri and Rao (1981).) Let A and ω_i be as in Complement 14.8.1. Consider

$$g(X,Y) = |X'AP_YA^{-1}X|/|X'X|$$
$$f(X,Y) = \text{tr}(P_XAP_YA^{-1}),$$

where X and Y are $n \times k$ and $n \times s$ matrices of ranks k and s, respectively, with $s \geq k$, $P_X = X(X'X)^{-1}X'$ and $P_Y = Y(Y'Y)^{-1}Y'$. Then

$$g(X,Y) \leq \prod_{i=1}^{\min(k,n-k)} \omega_i,$$

$$f(X,Y) \leq \sum_{i=1}^{k} \omega_i \quad \text{if } n \geq 2k,$$

$$\leq \sum_{i=1}^{n-k} \omega_i + 2k - n \quad \text{if } n < 2k.$$

14.8.3 (Greub and Rheinboldt (1959).) Let G and H be pd commuting matrices with eigenvalues $\lambda_1 \geq \ldots \geq \lambda_n > 0$ and $\mu_1 \geq \ldots \geq \mu_n > 0$ respectively. Then

$$\frac{(x'G^2x)(x'H^2x)}{(x'GHx)^2} \leq \frac{(\lambda_1\mu_1 + \lambda_n\mu_n)^2}{4\lambda_1\lambda_n\mu_1\mu_n}.$$

14.8.4 If $\lambda_1 \geq \lambda_2 \geq \ldots \geq \lambda_n > 0$, show that

$$\max_{i,j}[(\lambda_i + \lambda_j)^2/4\lambda_i\lambda_j] = (\lambda_1 + \lambda_n)^2/4\lambda_1\lambda_n.$$

14.8.5 (Khatri and Rao (1982)) A measure of inefficiency alternative to (**14.8.8**) is

$$\prod_{i=1}^{m}(1 - \frac{1}{\theta_i}) = |I - (X'A^{-1}X)^{-1}X'X(X'AX)^{-1}X'X|$$

$$\leq \prod_{j=1}^{\min(m,n-m)} \frac{(\lambda_j - \lambda_{n-m+j})^2}{(\lambda_j + \lambda_{n-m+j})^2}.$$

14.8.6 If $B(X,Y) = X'A^{-1}Y(Y'A^{-1}Y)^{-1}Y'A^{-1}X$ and $A(X) = X'A^{-1}X - X'X(X'AX)^{-1}X'X$, then

$$\sup_{Y} |B(X,Y)| = |A(X)|.$$

CHAPTER 15

NON-NEGATIVE MATRICES

In this chapter, we will examine some of the features of the world of non-negative matrices. non-negative matrices occur naturally in several areas of application. From the statistical side, non-negative matrices figure prominently in Markov Chains. Some models in Genetics are based on non-negative matrices. Leontief models in Economics derive sustenance from non-negative matrices. We will touch upon these applications. One of the most prominent results in the area of non-negative matrices is the Perron-Frobenius Theorem. In the next section, we will dwell upon this remarkable result.

15.1. Perron-Frobenius Theorem

We need to set some definitions in place to pave the way for an enunciation of the Perron-Frobenius Theorem. The concept of irreducible matrix is central to the development of this section.

DEFINITION 15.1.1. Let $A = (a_{ij}) \in \mathbf{M}_n$.

(1) The matrix A is said to be non-negative if $a_{ij} \geq 0$ for all i and j. (If A is non-negative, we use the symbol $A \geq_e 0$ or $0 \leq_e A$, the suffix e denotes entry wise.) In the general theory of matrices, the symbol $A \geq_L 0$, alternatively $A \geq 0$, is used to indicate that A is non-negative definite.

(2) The matrix A is said to be positive if $a_{ij} > 0$ for all i and j. (The symbol that is used in this context is $A >_e 0$.)

The concepts of nonnegativity and positivity perfectly make sense for matrices not necessarily square. If A and B are two matrices of the same order; we say that $A \geq_e B$ or $A - B \geq_e 0$ to mean that if $A = (a_{ij})$ and $B = (b_{ij})$, then $a_{ij} \geq b_{ij}$ for all i and j.

DEFINITION 15.1.2. A non-negative matrix $A \in \mathbf{M}_n (n \geq 2)$ is said to be reducible if there exists a permutation matrix $P \in \mathbf{M}_n$ such that PAP' is of the form

$$PAP' = \begin{bmatrix} B & 0 \\ C & D \end{bmatrix}, \tag{15.1.1}$$

where $B \in \mathbf{M}_r$ and $D \in \mathbf{M}_{n-r}$, and $0 \in \mathbf{M}_{r,n-r}$ is the null matrix, and $r > 1$.

What reducibility means is that if we can find some rows of A such that these rows and the corresponding columns of A are permuted, the resultant matrix has a structure stipulated in (15.1.1). The next question is how to identify reducible matrices. In the following proposition, we focus on this problem. We take $n \geq 2$ in all the propositions.

P 15.1.3 Let $A \in \mathbf{M}_n$ be a non-negative matrix. The matrix A is reducible if and only if there exists a nonempty proper subset I of $\{1, 2, \ldots, n\}$ such that $a_{ij} = 0$ for every $i \notin I$ and $j \neq I$.

PROOF. Sufficiency. Let $I = \{i_1 < i_2 < \ldots < i_k\}$ and $I^c = \{1, 2, \ldots, n\} - I = \{j_1 < j_2 < \ldots < j_{n-k}\}$. Let σ be the permutation map from $\{1, 2, \ldots, n\}$ to $\{1, 2, \ldots, n\}$ defined by $\sigma(t) = i_t$, for $t = 1, 2, \ldots, k$, and $\sigma(t) = j_{t-k}$, for $t = k+1, k+2, \ldots, n$. Let P be the permutation matrix associated with the permutation map σ. One can verify that PAP' is of the form (15.1.1). The necessity is clear.

One of the characteristic features of reducible matrices is the following. Suppose A is a matrix already in the reduced form (15.1.1). Then A^k, for any positive integer k, is also reducible. More generally, if A is reducible then A^k is reducible for any positive integer k. (Why?) The notion of reducibility can be defined for matrices not necessarily non-negative. We do not need the definition in generality. The negation of reducibility is one of the key concepts in this section.

DEFINITION 15.1.4. A non-negative matrix $A \in \mathbf{M}_n$, $n \geq 2$, is said to be irreducible if it is not reducible.

A trivial example of an irreducible matrix is any positive matrix. Another example is a 2×2 matrix with main diagonal elements zero and off diagonal elements unity. Matrices of order 1×1 are summarily excluded from discussion in this chapter. For us, n is always ≥ 2. The following is a characterization of irreducible matrices.

P 15.1.5 Let $A = (a_{ij}) \in \mathbf{M}_n$ be a non-negative matrix. The following statements are equivalent.

(1) A is irreducible.
(2) $(I + A)^{n-1}$ is positive.
(3) For any i and j with $1 \leq i \leq n$ and $i \leq j \leq n$, there exists a positive integer $k = k(i,j)$ such that $k \leq n$ and $a_{ij}^{(k)} > 0$, where $A^k = (a_{ij}^{(k)})$.

PROOF. (1) $\Rightarrow$ (2). Let $y \geq 0$ be a non-zero vector of order $n \times 1$ and $z = y + Ay = (I + A)y$. Let us compute how many non-zero elements z has. Since $Ay \geq 0$, z has at least as many non-zero elements as y. Could it be possible that y and z have exactly the same number of non-zero elements? Suppose it is possible. By rearranging the elements of y, if necessary, we can write $y' = (u', 0)$ with $u > 0$. Perforce, the vector z partitions as, $z' = (v', 0)$ with u and v being of the same order and $v > 0$. Partition A accordingly, i.e.,

$$z = \begin{bmatrix} v \\ 0 \end{bmatrix} = \begin{bmatrix} u \\ 0 \end{bmatrix} + \begin{bmatrix} A_{11} & A_{12} \\ A_{21} & A_{22} \end{bmatrix} \begin{bmatrix} u \\ 0 \end{bmatrix}.$$

This implies that

$$0 = 0 + A_{21}u.$$

Since $u > 0$, we have $A_{21} = 0$. Consequently, A is reducible. (Why?) This contradiction shows that z has more non-zero elements than y. Repeat this argument by taking z in the place of y. We find a vector s such that $s = (I + A)z = (I + A)^2 y$ has more nonzero entries than z. Repeating this argument at most $(n - 1)$ times, we find that $(I + A)^{n-1} >_e 0$. This is true for every non-zero vector $y \geq 0$. Hence $(I + A)^{n-1} >_e 0$. (Why?)

(2) $\Rightarrow$ (3). Since $(I + A)^{n-1} > 0$ and $A \geq_e 0$, we have

$$A(I + A)^{n-1} = \sum_{r=0}^{n-1} \binom{n-1}{r} A^{r+1} >_e 0.$$

Consequently, for any (i, j), the $(i, j)^{th}$ entry of $A, A^2, \dots,$ or A^n is positive.

(3) $\Rightarrow$ (1). Suppose A is reducible. There exists a permutation matrix P such that

$$PA^kP' = \begin{bmatrix} B_k & 0 \\ C_k & D_k \end{bmatrix},$$

for every $k \geq 1$. (Why?) Consequently, we can find $i \neq j$ such that the $(i,j)^{th}$-entry, $(PA^kP')_{ij} = 0$ for all $k \geq 1$. Let $P = (p_{ij})$. Thus we have

$$\sum_{r=1}^{n}\sum_{s=1}^{n} p_{ir}a_{rs}^{(k)}p_{js} = 0 \text{ for all } k \geq 1.$$

For some r and s, $p_{ir} = 1 = p_{js}$. Therefore, $a_{rs}^{(k)} = 0$ for all $k \geq 1$. This is a contradiction to (3). This completes the proof.

There is another concept closely related to irreducibility. For primitive matrices, one could obtain a stronger version of Perron-Frobenius Theorem.

DEFINITION 15.1.6. A non-negative matrix A is said to be primitive if A^k is positive for some positive integer k.

It is clear that every primitive matrix is irreducible. The converse is not true. Look at the case of 2×2 matrix with diagonal elements zero and off diagonal elements unity.

We now introduce the notion of the modulus of a matrix. Let $A = (a_{ij})$ be a matrix of any size. We define $m(A) = (|a_{ij}|)$. The following properties of the operation of modulus are easy to establish.

P 15.1.7

(1) If A and B are two matrices such that AB is defined, then $m(AB) \leq_e [m(A)][m(B)]$.
(2) If A is a square matrix, then $m(A^k) \leq_e [m(A)]^k$ for all positive integers k.
(3) If A and B are square matrices of the same order such that $m(A) \leq_e m(B)$, then $\|m(A)\|_F \leq \|m(B)\|_F$, where $\|\cdot\|_F$ stands for the Frobenius norm.

We now focus on the spectral radii of matrices. Recall that the spectral radius $\rho_s(A)$ of a square matrix A is the maximum of the absolute values of the eigenvalues of A.

P 15.1.8 If $A, B \in \mathbf{M}_n$ and $m(A) \leq B$, then $\rho_s(A) \leq \rho_s[m(A)] \leq \rho_s[(B)]$. (In other words, the spectral radius $\rho_s(\cdot)$ is monotonically increasing on the set of all non-negative matrices in $\mathbf{M}_n$.)

PROOF. (1) Note that for any positive integer $k, m(A^k) \leq_e [m(A)]^k \leq_e B^k$. It now follows that $(\|m(A^k)\|_F)^{1/k} \leq (\|m(A^k)\|)^{1/k} \leq (\|B^k\|)^{1/k}$. By taking the limit as $k \to \infty$ now, we note that $\rho_s(A) \leq \rho_s[m(A)] \leq \rho_s(B)$. See **P 11.2.17**.

P 15.1.9 Let $A = (a_{ij}) \in \mathbf{M}_n$ be a non-negative matrix, and B a principal submatrix of A. Then $\rho_s(B) \leq \rho_s(A)$. In particular, $\max_{1 \leq i \leq n} a_{ii} \leq \rho_s(A)$.

PROOF. Define a matrix $C \in \mathbf{M}_n$ as follows. Place the entries of B in C in exactly the same position wherever they come from A. The remaining entries of C are zeros. We note that $\rho_s(B) = \rho_s(C)$ (Why?) and $C \leq_e A$. [We use the result $0 \leq_e A_1 \leq_e A_2 \Rightarrow \rho_s(A_1) \leq \rho_s(A_2)$.]

P 15.1.10 Let $A = (a_{ij})$ be a non-negative matrix such that all the row sums of A are equal to the same number α. Then $\rho_s(A) = \alpha$. If all the column sums of A are equal to the same number β, then $\rho_s(A) = \beta$.

PROOF. Recall the form of the induced ℓ_∞-norm on $\mathbf{M}_n$. For $B = (b_{ij})$, $\|B\|_{\infty,in} = \max_{1 \leq i \leq n} \sum_{j=1}^{n} |b_{ij}|$. See **P 11.2.8**. This norm is a matrix norm. Further, recall that $\rho_s(B) \leq \|B\|$ for any matrix norm $\|\cdot\|$. Thus we observe that $\|A\|_{\infty,in} = \alpha \leq \rho_s(A)$. On the other hand, note that α is an eigenvalue of A with eigenvector $(1, 1, \ldots, 1)'$. Therefore, $\alpha \leq \rho_s(A)$. This proves the first part of the proposition. For the second part, use the matrix norm, $\|\cdot\|_{1,in}$. See **P 11.2.11**.

P 15.1.11 Let $A = (a_{ij}) \in \mathbf{M}_n$ be a non-negative matrix with row sums $r_1, r_2, \ldots, r_n$ and column sums $c_1, c_2, \ldots, c_n$. Then

$$(1) \qquad \min_{1 \leq i \leq n} r_i \leq \rho_s(A) \leq \max_{1 \leq i \leq n} r_i,$$

$$(2) \qquad \min_{1 \leq i \leq n} c_i \leq \rho_s(A) \leq \max_{1 \leq i \leq n} c_i.$$

PROOF. (1) Let $\alpha = \min_{1 \leq i \leq n} r_i$. We show that $\alpha \leq \rho_s(A)$. If $\alpha = 0$, this inequality is trivially true. Suppose $\alpha > 0$. Let $B = (b_{ij})$ be defined by $b_{ij} = \frac{\alpha a_{ij}}{r_i}$. Clearly, $0 \leq_e B \leq_e A$ and every row sum of B is equal to α. Consequently, $\alpha = \rho_s(B) \leq \rho_s(A)$. The other inequalities can be established in a similar vein.

P 15.1.12 Let $A = (a_{ij}) \in \mathbf{M}_n$ be a non-negative matrix. Let $x' = (x_1, x_2, \ldots, x_n)$ be positive. Then

$$(1) \qquad \min_{1\leq i\leq n} \frac{1}{x_i} \sum_{j=1}^{n} a_{ij}x_j \leq \rho_s(A) \leq \max_{1\leq i\leq n} \frac{1}{x_i} \sum_{j=1}^{n} a_{ij}x_j,$$

$$(2) \qquad \min_{1\leq j\leq n} x_j \sum_{i=1}^{n} \frac{a_{ij}}{x_i} \leq \rho_s(A) \leq \max_{1\leq j\leq n} x_j \sum_{i=1}^{n} \frac{a_{ij}}{x_i}.$$

Further, if $\alpha, \beta \geq 0$ and $\alpha x \leq_e Ax \leq_e \beta x$, then $\alpha \leq \rho_s(A) \leq \beta$. If $\alpha x < Ax$, then $\alpha < \rho_s(A)$; if $Ax <_e \beta x$, then $\rho_s(A) < \beta$.

PROOF. First, we realize that the matrices $S^{-1}AS$ and A have the same set of eigenvalues for any non-singular matrix S. This implies that $\rho_s(A) = \rho_s(S^{-1}AS)$. Let $S = \text{diag}\{x_1, x_2, \ldots, x_n\}$. It is clear that $S^{-1}AS \geq_e 0$. Identify the product $S^{-1}AS = (a_{ij}x_i^{-1}x_j)$. **P 15.1.11** as applied to the matrix $S^{-1}AS$ establishes the inequalities (1) and (2). For the second part, suppose $\alpha x \leq_e Ax$. This means that $\alpha x_i \leq \sum_{j=1}^{n} a_{ij}x_j$ for every i. Consequently, $\alpha \leq \min_{1\leq i\leq n} \frac{1}{x_i} \sum_{j=1}^{n} a_{ij}x_j \leq \rho_s(A)$. If $\alpha x < Ax$, we can find $\alpha_0 > \alpha$ such that $\alpha_0 x \leq_e Ax$. (Why?) This gives us $\alpha < \alpha_0 \leq \rho_s(A)$. The other inequalities are established analogously.

We need now some facts about simple roots of polynomials and spectral radii of matrices. Let $p(x)$ be a polynomial of degree n in x. Let λ be a root of the polynomial.

LEMMA 15.1.13. *The root λ is simple (of multiplicity one) if and only if* $p'(\lambda) = \frac{d}{dx}p(x)\big|_{x=\lambda} \neq 0$.

PROOF. It is left to the reader.

Let us discuss some aspects of the determinantal equation $p(x) = |xI - A| = 0$ for any given matrix A. Let A_{ii} be the principal submatrix of A obtained from A by deleting the i-th column and i-th row of A. One can verify that

$$\frac{d}{dx}p(x) = \sum_{i=1}^{n} |xI - A_{ii}|,$$

where I stands for the identity matrix of appropriate order in the above.

P 15.1.14 (The Perron-Frobenius Theorem for Positive Matrices.) Let $A = (a_{ij}) \in \mathbf{M}_n$ be a positive matrix. Then the following statements are valid.

(1) $\rho_s(A) > 0$.
(2) $\rho_s(A)$ is an eigenvalue of A.
(3) There exists a positive eigenvector of A corresponding to the eigenvalue $\rho_s(A)$.
(4) If μ is any other eigenvalue of A, then $|\mu| < \rho_s(A)$. (What this means, in particular, is that we cannot find an eigenvalue μ of A such that $|\mu| = \rho_s(A)$.)
(5) The multiplicity of the eigenvalue $\rho_s(A)$ is one, i.e., $\rho_s(A)$ is not a repeated eigenvalue.

PROOF. (1) Since A is positive, it follows that $\rho_s(A) > 0$, by **P 15.1.11.**

To prove (2) and (3), let μ be an eigenvalue of A such that $|\mu| = \rho_s(A)$. Let x be a corresponding eigenvector and $|x| = m(x)$. Then

$$\rho_s(A)|x| = |\mu||x| = |\mu x| = |Ax| \leq_e |A||x| = A|x|.$$

Let $y = A|x| - \rho_s(A)|x|$. It is clear that $y \geq_e 0$. If $y = 0$, then $\rho_s(A)$ is an eigenvalue of A with a corresponding eigenvector $|x|$. Since $|x| \geq_e 0, |x| \neq 0$, and A is positive, $A|x| >_e 0$. Further, $|x| = [\rho_s(A)]^{-1}A|x| >_e 0$. This establishes (2) and (3) in case $y = 0$. Suppose $y \neq 0$. Set $u = A|x|$ which is, obviously, positive. Note that, since A is positive,

$$0 <_e Ay = A(A|x| - \rho_s(A)|x|) = Az - \rho_s(A)z,$$

which means that $Az >_e \rho_s(A)z$. By **P 15.1.12**, $\rho_s(A) > \rho_s(A)$, which is not possible. Hence $y = 0$. Thus (2) and (3) are established.

(4) Let μ be an eigenvalue of A such that $\mu \neq \rho_s(A)$. By the very definition of spectral radius, $|\mu| \leq \rho_s(A)$. We claim that $|\mu| < \rho_s(A)$. Suppose $|\mu| = \rho_s(A)$. Let x be an eigenvector of A corresponding to the eigenvalue μ. Following the argument presented in the proof of (2) and (3) above, it follows that $|x| >_e 0$ and $|x|$ is an eigenvector of A corresponding to the eigenvalue $\rho_s(A)$. Let x_i be the i-th component of x. The equation $Ax = \mu x$ implies that

$$\rho_s(A)|x_i| = |\mu||x_i| = |\mu x_i| = |\sum_{j=1}^{n} a_{ij}x_i| \leq \sum_{j=1}^{n} a_{ij}|x_j| = \rho_s(A)|x_i|,$$

for each i. Thus equality must prevail throughout in the above. This means that the complex numbers $a_{ij}x_j$, $j = 1, 2, \ldots, n$ must lie on the same ray in the complex plane. Let θ be their common argument. Then $e^{-i\theta}a_{ij}x_j > 0$ for all j. Since $a_{ij} > 0$, we have $\omega = e^{-i\theta}x >_e 0$. The vector ω is also an eigenvector of A corresponding to the eigenvalue μ of A, i.e., $A\omega = \mu\omega$. Since $\omega >_e 0$, $\mu \geq 0$. (Why?) Trivially, $\mu\omega \leq_e A\omega \leq_e \mu\omega$. By **P 6.1.12**, $\mu \leq \rho_s(A) \leq \mu$. This contradiction establishes the claim.

(5) First, we establish the following result. If $0 <_e C \leq_e D$ and $\rho_s(C) = \rho_s(D)$, then $C = D$. We now know that there exist positive eigenvectors x and y such that $Cx = \rho_s(C)x$ and $Dy = \rho_s(D)y$. We would like to show that x is also an eigenvector of D corresponding to the eigenvalue $\rho_s(C)$ of D. Note that $0 <_e \rho_s(C)x = \rho_s(D)x = Cx \leq_e Dx$. Let $z = Dx - \rho_s(D)x$. It is clear that $z \geq_e 0$. We claim that $z = 0$. Suppose $z \neq 0$. Let $y = Dx$. Note that $0 <_e Dz = D(y - \rho_s(d)x) = Dy - \rho_s(D)y$ leading to the inequality $Dy >_e \rho_s(D)y$. By **P 15.1.12**, $\rho_s(D) < \rho_s(D)$. This contradiction establishes that $z = 0$. Thus the vector x is a common eigenvector of C and D corresponding to the same eigenvalue $\rho_s(C)$, i.e., $Cx = Dx$. Since $x >_e 0$ and $C \leq_e D$, we have $C = D$. (Why?)

Let $p(x) = |xI - A|$. Let A_{ii} be the principal submatrix of A obtained from A by deleting the i-th row and i-th column of A. Write

$$|xI - A_{ii}| = \Pi_{j=1}^{n-1}(x - \mu_i),$$

where $\mu_1, \mu_2, \ldots, \mu_{n-1}$ are the eigenvalues of A_{ii}. Since $\rho_s(A_{ii}) \leq \rho_s(A)$, $|\mu_i| \leq \rho_s(A)$ for every i. See **P 15.1.9**. We can say something stronger. Since A is positive, $\rho_s(A_{ii}) < \rho_s(A)$, by what we have seen at the beginning of the proof of (5). Equivalently, $|\mu_i| < \rho_s(A)$ for every i. Consequently, $|xI - A_{ii}| > 0$ if $x \geq \rho_s(A)$. In particular, $|\rho_s(A)I - A_{ii}| > 0$ for all i. Observe that

$$\frac{d}{dx}p(x)\Big|_{x=\rho_s(A)} = \sum_{i=1}^{n} |\rho_s(A)I - A_{ii}| > 0.$$

This shows that the eigenvalue $\rho_s(A)$ is simple. This completes the proof.

The spectral radius $\rho_s(A)$ of a positive matrix is called the Perron root of A. The associated positive eigenvector $x' = (x_1, x_2, \ldots, x_n)$, i.e., $Ax = [\rho_s(A)]x$, $x >_e 0$, with $\sum_{i=1}^{n} x_i = 1$ is called the right Perron vector of A. Note that A' is also positive. The spectral radius remains the same. The right Perron vector y of A' is called the left Perron vector of A.

P 15.1.14 is usually called Perron's theorem. A similar statement has been established by Frobenius in the environment of irreducible matrices. We now focus on irreducible matrices. The extension of **P 15.1.14** revolves around comparing the eigenvalues of A and those of $I + A$. Some results in this connection are worth noting.

P 15.1.15 Let $A \in \mathbf{M}_n$ with eigenvalues $\lambda_1, \lambda_2, \ldots, \lambda_n$. Then the eigenvalues of $I + A$ are $1 + \lambda_1, 1 + \lambda_2, \ldots, 1 + \lambda_n$. Further, $\rho_s(I + A) \leq 1 + \rho_s(A)$. If A is non-negative, then $\rho_s(I + A) = 1 + \rho_s(A)$.

PROOF. The first part of the result follows easily. Note that $\rho_s(I + A) = \max_{1 \leq i \leq n} |1 + \lambda_i| \leq 1 + \max_{1 \leq i \leq n} |\lambda_i| = 1 + \rho_s(A)$. If $A \geq_e 0$, then $1 + \rho_s(A)$ is an eigenvalue of $I + A$.

P 15.1.16 Let A be a non-negative matrix such that A^k is positive for some positive integer k, i.e., A is primitive. Then the assertions of **P 15.1.14** hold. [This is easy to establish.]

P 15.1.17 (Perron-Frobenius Theorem) Let $A \in \mathbf{M}_n$ be a non-negative irreducible matrix. Then:

(1) $\rho_s(A) > 0$.
(2) $\rho_s(A)$ is an eigenvalue of A.
(3) There exists a positive eigenvector x of A corresponding to the eigenvalue $\rho_s(A)$ of A.
(4) The eigenvalue $\rho_s(A)$ is simple.

PROOF. Since A is irreducible, $(I + A)^{n-1}$ is positive. **P 15.1.14** becomes operational for the matrix $(I + A)^{n-1}$. Now, (1), (2), (3), and (4) follow. Use **P 15.1.15** and **P 15.1.16**.

There is one crucial difference between **P 15.1.14** and **P 15.1.17**. If the matrix A is irreducible, it is possible that there is eigenvalue λ of A such that $|\lambda| = \rho_s(A)$.

Complements

15.1.1 Let $A = (a_{ij})$ be a matrix of order 3×3 with exactly one entry of A equal to zero. Characterize the position of this single zero in the matrix so that A becomes reducible.

15.1.2 Let $A = (a_{ij})$ be a matrix of order 3×3 with exactly two entries of A equal to zero. Characterize the position of these zeros in the matrix so that A becomes reducible.

15.1.3 Let $A = (a_{ij})$ be a non-negative matrix such that $a_{ii} > 0$ for all i. Show that A is primitive.

15.1.4 If A is irreducible, show that A' is irreducible.

15.1.5 Let $A >_e 0$. Suppose x and y are eigenvectors of A corresponding to the eigenvalue $\rho_s(A)$ of A. Show that $x = \alpha y$ for some number α. (The eigenvalue $\rho_s(A)$ is of geometric multiplicity one.)

15.1.6 Let

$$A = \begin{bmatrix} 1-\alpha & \beta \\ \alpha & 1-\beta \end{bmatrix}, \quad 0 < \alpha, \ \beta < 1.$$

Determine the Perron root and Perron right vector of A. Examine the asymptotic behavior of A^k as $k \to \infty$.

15.1.7 Let A be a positive matrix. Assess the asymptotic behavior of A^k as $k \to \infty$. More precisely, show that $[A/\rho_s(A)]^k$ converges as $k \to \infty$. Show that the limit matrix L is given by $L = xy'$, where $Ax = [\rho_s(A)]x$, $x >_e 0$, $y'A = [\rho_s(A)]y'$, $y >_e 0$, and $x'y = 1$. (The proof of **P 15.3.2** can be adapted in this case by taking L in the place of Q.) Show that exactly the same conclusion is valid for a primitive matrix A.

15.1.8 Let A be a positive matrix and x the right Perron vector of A. Show that $\rho_s(A) = \Sigma\Sigma a_{ij}x_j$, the summation being over $i, j = 1, \ldots, n$.

15.1.9 If A is a positive non-singular matrix, demonstrate that A^{-1} cannot be non-negative.

15.1.10 Establish a statement analogous to **P 15.1.14** for primitive matrices. Prove this statement using **P 15.1.14**.

15.1.11 Let $f_n, n \geq 1$ be the Fibonacci sequence, i.e., $f_1 = f_2 = 1$, $f_n = f_{n-1} + f_{n-2}, n \geq 3$. One can show that $\lim_{n\to\infty} \frac{f_n}{f_{n-1}} = \frac{1-\sqrt{5}}{2}$, the golden ratio. For any odd positive integer $n = 2m + 1$, define a matrix

$A_n = (a_{ij}^{(n)})$ of order $n \times n$ by

$$a_{ij}^{(n)} = \begin{cases} 1 & \text{if } |i-j| = 1, \\ 1 & \text{if } i = j = m+1, \\ 0 & \text{otherwise.} \end{cases}$$

Show that $\rho_s(A_n) \leq \sqrt{5}$. *Hint: Use* **P 15.1.12**.

15.1.12 Let A be a non-negative matrix. Show that $\rho_s(A)$ is an eigenvalue of A. Show that there exists a vector $x \geq 0$ such that $Ax = [\rho_s(A)]x$.

15.1.13 For the following matrices, examine which of the properties (1), (2), (3), (4), and (5) of Theorem 15.1.14 are violated:

$$\begin{bmatrix} 0 & 1 \\ 0 & 0 \end{bmatrix}; \begin{bmatrix} 1 & 0 \\ 0 & 1 \end{bmatrix}; \begin{bmatrix} 0 & 1 \\ 1 & 0 \end{bmatrix}.$$

15.2. Leontief Models in Economics

We begin with a description of Leontief's Model for an economic system involving inputs and outputs of the industries comprising the economy. Suppose an economy has n industries and each industry produces (output) only one commodity. Each industry requires commodities (inputs) from all the industries, including itself, of the economy for the production of its commodity. No input from outside the economy under focus is needed. This is an example of a closed economy. The problem is to determine suitable "prices" to be charged for these commodities so that to each industry total expenditure equals total income. Such a price structure represents equilibrium for the economy. Let us fix the notation and formulate the problem. Let

$$\begin{aligned} a_{ij} &= \text{the fraction of the total output of the } j\text{-th industry} \\ &\quad \text{purchased by the } i\text{-th industry, } i, j = 1, 2, \ldots, n. \end{aligned}$$

It is clear that $a_{ij} \geq 0$. Further, $a_{1j} + \ldots + a_{nj} = 1, j = 1, \ldots, n$. Let $A = (a_{ij})$. Thus A is a non-negative matrix and each column of A sums up to unity. Consequently, the spectral radius $\rho_s(A)$ of A is unity. (Why?) We assume that A is known. The matrix A is called the input-output matrix of the economy. Let p_i = price for the i-th industry for its total output, $i = 1, 2, \ldots, n$. The equilibrium condition can be stated as follows.

Total expenditure incurred by the i-th industry is equal to the total income of the i-th industry, i.e.,

$$a_{i1}p_1 + \ldots + a_{in}p_n = p_i, \; i = 1, \ldots, n.$$

These equations can be rewritten as $Ap = p$, where $p' = (p_1, p_2, \ldots, p_n)$. The objective is to determine the price vector p. We are back in the realm of non-negative matrices. Since $\rho_s(A) = 1$, we are looking for the eigenvector of A corresponding to the eigenvalue $\rho_s(A)$ of A. This problem falls into the realm of the Perron-Frobenius theorem. In practice, one looks for a positive solution (viable solution) of $Ap = p$. If A is irreducible, we know that p exists and is positive.

As an example, suppose an economy has four industries: a steel plant, an electricity generating plant, a coal mine, and an iron ore mining facility. Twenty percent of the output of the steel plant is used by itself, 30 percent of the output of the steel plant is used by the electricity generating plant, 15 percent by the coal mine, and 35 percent by the iron ore facility. Twenty percent of the electricity generated is used by the steel plant, 25 percent by itself, 25 percent by the coal mine, and 30 percent by the iron ore facility. Thirty percent of the coal produced by the coal mine is used by the steel plant, 30 percent by the electricity generating plant, 20 percent by itself, and 20 percent by the iron ore facility. Finally, 80 percent of iron ore produced by the iron ore mining facility is used by the steel plant, 10 percent by the electricity generating plant, 10 percent by the coal mine, and nothing for itself. The corresponding input-output matrix works out to be

$$A = \begin{bmatrix} 0.20 & 0.20 & 0.30 & 0.80 \\ 0.30 & 0.25 & 0.30 & 0.10 \\ 0.15 & 0.25 & 0.20 & 0.10 \\ 0.35 & 0.30 & 0.20 & 0.00 \end{bmatrix}. \tag{15.2.1}$$

The basic problem is to determine how much the total output of each industry is to be priced so that total expenditure equals total income for each industry. Note that the matrix A is irreducible. There exists a positive vector p satisfying $Ap = p$. As a matter of fact, any multiple of p constitutes a solution to the equilibrium equation.

Now we begin with a description of an open economy. Suppose there are n industries in an economy each producing only one type of commodity. Portions of these outputs are to be used in the industries within the economy but there is also some demand for these commodities outside the economy. The prices of units of these commodities are fixed and known. Let us introduce some notation. Let d_i = monetary value of the output of the i-th industry demanded by sources outside the economy, $i = 1, 2, \ldots, n$. Let $d' = (d_1, d_2, \ldots, d_n)$. Since the prices are known, the sources outside the economy can compute how much is the monetary value of the commodities they are seeking from the economy. The vector d is known. Denote by c_{ij}, the monetary value of the output of the i-th industry needed by the j-th industry in order to produce one unit of monetary value of its output, $i, j = 1, 2, \ldots, n$. Let $C = (c_{ij})$. Clearly, C is non-negative. The matrix C is called the consumption matrix of the economy. Normally, $\sum_{i=1}^{n} c_{ij} \leq 1$ for each i. If the sum is equal to 1, the industry is not profitable. Finally, let x_i be the monetary value of the total output of the i-th industry, $i = 1, 2, \ldots, n$. The objective is to determine the values of x_i's so that the needs of the industries within the economy and the demands from sources outside the economy are exactly met. Let $x' = (x_1, x_2, \ldots, x_n)$. Note that $\sum_{j=1}^{n} c_{ij}x_j$ is the monetary value of the output of the i-th industry needed by all the industries inside the economy. Consequently, $x - Cx$ represents monetary values of excess outputs of the industries. We set $x - Cx = d$ to match the excess output with the demand. The objective is to determine $x \geq 0$ so that

$$(I - C)x = d \tag{15.2.2}$$

is satisfied. If $(I - C)$ is nonsingular, there is a unique solution to the system (15.2.2) of equations. The solution may not be non-negative. If $(I - C)^{-1}$ is non-negative, we will then have a unique non-negative solution x satisfying (15.2.2) for any given demand vector $d \geq 0$. The following results throw some light on solutions to the system (15.2.2).

P 15.2.1 Let $C \in \mathbf{M}_n$ be a non-negative matrix. Then $(I - C)^{-1}$ exists and non-negative if and only if there exists a non-negative vector x such that $x > Cx$. (The condition $x > Cx$ means that there is some production schedule x such that each industry produces more than it consumes.)

PROOF. If $C = 0$, the result trivially holds. Assume that $C \neq 0$. Suppose $(I-C)^{-1}$ exists and non-negative. There exists a non-negative non-zero vector x such that $(I-C)^{-1}x = dx$, where $d = \rho_s[(I-C)^{-1}]$. See Complement 15.1.12. We will show that $d > 1$. The equation $(I-C)^{-1}x = dx$ implies that $x = \left(\frac{d}{d-1}\right)Cx$. Since $x \geq 0, x \neq 0, C \geq 0$, it follows that $d > 1$. Moreover, $[d/(d-1)] > 1$. Hence $x > Cx$. Suppose for some vector $x \geq 0, x > Cx$. It means that x better be positive. If this is the case, we can find $0 < \lambda < 1$ such that $Cx < \lambda x$. (Why?) This implies that $C^k x < \lambda^k x$ for all $k \geq 1$. Consequently, $\lim_{k\to\infty} C^k = 0$. Since $(I-C)(I+C+C^2+\ldots+C^m) = I - C^{m+1}$, which converges to 0 as $m \to \infty$, it follows that the series $I + C + C^2 + \ldots$ is convergent and is equal to $(I-C)^{-1}$. Thus $I - C$ is invertible. It is clear that $(I-C)^{-1} \geq 0$. This completes the proof.

The following results are consequences of **P 15.2.1**.

COROLLARY 15.2.2. Let C be a non-negative matrix such that each of its row sums is less than one. Then $(I-C)^{-1}$ exists and is non-negative.

COROLLARY 15.2.3. Let C be a non-negative matrix such that each of its column sums is less than one. Then $(I-C)^{-1}$ exists and is non-negative.

The essential difference between the closed model and open model are the following.

(1) In the closed model, the outputs of the industries are distributed among themselves. In the open model, an attempt is made to satisfy an outside demand for the outputs.

(2) In the closed model, the outputs are fixed and the problem is to determine a price structure for the outputs so that the total expenditure and total income for each industry are equal. In the open model, the prices are fixed and the problem is to determine a production schedule meeting the internal and external demands.

Complements

15.2.1 For the input-output matrix A of (15.2.1), examine equilibrium solutions and interpret them.

15.2.2 Three neighbors have backyard vegetable gardens. Neighbor A grows tomatoes, neighbor B grows corn, and neighbor C lettuce. They agree to divide their crops among themselves. Neighbor A keeps half of the tomatoes he produces, gives a quarter of his tomatoes to neighbor B, and a quarter to neighbor C. Neighbor B shares his crop equally among themselves. Neighbor C gives one-sixth of his crop to neighbor A, one-sixth to neighbor B, and the rest he keeps himself. What prices the neighbors should assign to their respective crops if the equilibrium condition for a closed economy is to be satisfied, and if the lowest-priced crop is to realize \$75?

15.2.3 A town has three main industries: a coal mine, an electricity generating plant, and a local railroad. To mine \$1 worth of coal, the mining operation needs 25 cents worth of electricity and makes use of 25 cents worth of transportation facilities. To produce \$1 worth of electricity, the generating plant requires 65 cents worth of coal, 5 cents worth of its own electricity, and 5 cents worth of transportation needs. To provide \$1 worth of transportation, the railroad requires 55 cents of coal for fuel and 10 cents worth of electricity. In a particular period of operation, the coal mine receives orders for \$50,000 of coal from outside and the generating plant receives orders for \$25,000 of electricity from outside. Determine how much each of these industries should produce in the period under focus so that internal and external demands are exactly satisfied.

(Source: C. Rorres and H. Anton (1984).)

15.3. Markov Chains

Let $X_0, X_1, \ldots$ be a stochastic process, i.e., a sequence of random variables. Assume that each random variable takes values in a finite set $\{1, 2, \ldots, k\}$. The set $\{1, 2, \ldots, k\}$ is called the state space of the process and members of the set are called the states of the process. In this case, it is easy to explain what a stochastic process means. For every $n \geq 0$ and $i_0, i_1, i_2, \ldots, i_n \in \{1, 2, \ldots, k\}$, the probabilities, $\Pr\{X_0 = i_0, X_1 = i_1, \ldots, X_n = i_n\}$, are spelled out. Let $p_i = \Pr\{X_0 = i\}$, $i = 1, 2, \ldots, k$. The vector $p' = (p_1, p_2, \ldots, p_k)$ is called the initial distribution of the process, i.e., p is the distribution of X_0.

A physical evolution of a stochastic process can be described as follows. Suppose a particle moves over the states at times $0, 1, 2, \ldots$ in a random manner. Let X_n be the state in which the particle is at time $n, n \geq 0$. The joint distribution of the process $X_n, n \geq 0$ describes the random movement of the particle over time. Let us introduce the notion of a Markov chain.

DEFINITION 15.3.1. The process $X_n, n \geq 0$ is called a Markov chain if the conditional probability

$$\begin{aligned}
&\Pr\{X_{n+1} = j | X_0 = i_0, X_1 = i_1, \ldots, X_{n-1} = i_{n-1}, X_n = i\} \\
&\quad = \frac{\Pr\{X_0 = i_0, X_1 = i_1, \ldots, X_{n-1} = i_{n-1}, X_n = i, X_{n+1} = j\}}{\Pr\{X_0 = i_0, X_1 = i_1, \ldots, X_{n-1} = i_{n-1}, X_n = i\}} \\
&\quad = \Pr\{X_{n+1} = j | X_n = i\} = \frac{\Pr\{X_n = i, X_{n+1} = j\}}{\Pr\{X_n = i\}} = p_{ij}, \text{ (say)}
\end{aligned}$$

for all $i_0, i_1, \ldots, i_{n-1}, i$, and $j \in \{1, 2, \ldots, k\}$ and $n \geq 0$.

The definition means several things: the conditional probability that $X_{n+1} = j$ given the past $\{X_0 = i_0, X_1 = i_1, \ldots, X_{n-1} = i_{n-1}, X_n = i\}$ depends only on the immediate past $\{X_n = i\}$; the conditional probability does not depend on n. The numbers p_{ij}'s are called one-step transition probabilities. The number p_{ij} is the conditional probability of moving from the state i to state j in one step from time n to time $(n+1)$.

Let $P = (p_{ij})$. The matrix P is called the transition probability matrix. It has the special property that $\sum_{j=1}^{k} p_{ij} = 1$ for every i. The matrix P is an example of what is called a stochastic matrix. In the case of a Markov chain, the knowledge of the initial distribution p and the transition matrix P is enough to determine the joint distribution of any finite subset of the random variables $X_n, n \geq 0$. For example, the distribution of X_n is $(p')P^n$. The joint distribution of X_2 and X_3 is given by

$$\begin{aligned}
\Pr\{X_2 = i, X_3 = j\} &= \Pr\{X_2 = i\} \Pr\{X_3 = j | X_2 = i\} \\
&= \Pr\{X_2 = i\} p_{ij},
\end{aligned}$$

for any i and j in the state space. The entry $\Pr\{X_2 = i\}$ is the i-th component of $(p')P^2$.

One of the basic problems in Markov chain is to assess the asymptotic behavior of the process. One determines $\lim_{n\to\infty} \Pr\{X_n = i\}$ for every state i. Assume that P is irreducible. Irreducibility in the context of Markov chains has a nice physical interpretation. With positive probability one can move from any state to any state in a finite number of steps, i.e., $p_{ij}^{(n)} > 0$ for some $n \geq 1$, where $P^n = (p_{ij}^{(n)})$. See **P 15.1.5**. Note that the spectral radius $\rho_s(P) = 1$. The conclusion of the main result of this section, i.e., **P 15.3.2** is not valid for irreducible transition matrices. Let us assume that P is primitive. Look up Complement 15.1.10. Let $q' = (q_1, q_2, \ldots, q_k)$ be the left Perron vector of P, i.e., $q > 0$, $q'P = q'$, and $\sum_{s=1}^{k} q_s = 1$. The right Perron vector of A is $x' = (1, 1, \ldots, 1)$, i.e., $Px = x$. We want to show that the limiting distribution of X_n is q'.

P 15.3.2 If the transition matrix P is primitive, then

$$\lim_{n\to\infty} P^n = Q,$$

where all the rows of Q are identical and equal to q'. Consequently, the limiting distribution of X_n is q', i.e., $\lim_{n\to\infty} p'P^n = q'$.

PROOF. Observe the following properties of the matrices P and Q.

(1) Q is idempotent.
(2) $P^mQ = QP^m = Q$ for all $m \geq 1$.
(3) $Q(P - Q) = 0$.
(4) $(P - Q)^m = P^m - Q$ for all $M \geq 1$.
(5) Every non-zero eigenvalue of $P - Q$ is also an eigenvalue of P. This can be proved as follows. Let λ be a non-zero eigenvalue of $P - Q$. Let ω be a non-zero vector such that $(P - Q)\omega = \lambda\omega$. Then $Q(P - Q)\omega = \lambda Q\omega = 0$. This implies that $Q\omega = 0$ and $P\omega = \lambda\omega$.
(6) $\rho_s(P) = 1$ is not an eigenvalue of $P - Q$, i.e., $I - (P - Q)$ is invertible. This can be proved as follows. Suppose 1 is an eigenvalues of $P - Q$. Then there exists a non-zero vector ω such that $(P - Q)\omega = \omega$. This implies that $Q\omega = 0$ and ω is an eigenvector of P corresponding to the eigenvalue 1 of A. Since the algebraic multiplicity of the eigenvalue 1 is one, $\omega = \alpha x$ for

some non-zero α, where x is the right Perron vector of P. Since $Q\omega = 0$, we must have $Qx = 0$. This is not possible.

(7) Let $\lambda_1\lambda_2, \ldots, \lambda_{k-1}, 1$ be the eigenvalues of P. Assume that $|\lambda_1| \leq |\lambda_2| \leq \ldots \leq |\lambda_{k-1}| < 1$. Then $\rho_s(P - Q) \leq |\lambda_{k-1}| < 1$. From Property 5 above, $\rho_s(P - Q) = |\lambda_s|$ for some s, or $= 1$, or $= 0$. From Property 6 above, we must have $\rho_s(P - Q) \leq |\lambda_{k-1}|$.

(8) $P^m = Q + (P - Q)^m$ for all $m \geq 1$. Since $\rho_s(P - Q) < 1$, $\lim_{m\to\infty} (P - Q)^m = 0$. Consequently, $\lim_{m\to\infty} P^m = Q$.

The last property proves the professed assertion.

P 15.3.2 is the fundamental theorem of Markov chains. It asserts that whatever the initial distribution p may be, the limiting distribution of X_n is q', the left Perron vector of P. For primitive stochastic matrices, obtaining the left Perron vector of P is tantamount to solving the equations $q'P = q'$ in unknown q.

Complements

15.3.1 Let $P = \begin{bmatrix} 0 & 1 \\ 1 & 0 \end{bmatrix}$. Show that for the transition matrix P, the limit of P^m as $m \to \infty$ does not exist.

15.3.2 If P is a stochastic matrix, show that P^m is also stochastic for any positive integer m.

15.3.3 An urn contains a black and b red balls. At time n, draw a ball at random from the urn, note its color, put the ball back into the urn, and add $c > 0$ balls of the same color to the urn. Let X_n be the color of the ball drawn at n-th time, $n \geq 1$. Obtain the joint distribution of X_1, X_2, and X_3. Evaluate the conditional probabilities,

$$\Pr\{X_3 = \text{black } |X_1\text{=black, } X_2\text{=black}\},$$
$$\Pr\{X_3 = \text{black } |X_1\text{=red, } X_2\text{=black}\},$$
$$\Pr\{X_3 = \text{black } |X_2\text{=black}\}.$$

Show that $X_n, n \geq 1$ is not a Markov chain.

15.3.4 Show that the transition matrix

$$\begin{bmatrix} 0.0 & 0.5 & 0.5 \\ 0.5 & 0.5 & 0.0 \\ 0.5 & 0.0 & 0.5 \end{bmatrix}$$

is primitive. Obtain the limiting distribution of the Markov chain driven by the above transition matrix.

15.3.5 A country is divided into three demographic regions. It is found that each year 5% of the residents of Region 1 move to Region 2 and 5% to Region 3. Of the residents of Region 2, 15% move to Region 2 and 10% to Region 3. Finally, of the residents of Region 3, 10% move to Region 1 and 5% to Region 2. Obtain the limiting distribution of the underlying Markov chain.
(Source: Rorres and Anton (1984).)

15.4. Genetic Models

Gregor Mendel is generally credited with the formulation of laws of inheritance of traits from parents to their offspring. One of the basic problems in genetics is to examine the propagation of traits over several generations of a population. Each inherited trait such as eye color, hair color, is usually governed by a set of two genes, designated by the generic symbols A and a. Plants and animals are composed of cells. Each cell has a collection of chromosomes. Chromosomes carry hereditary genes. Each human being has roughly 100,000 pairs of genes. Each individual in the population has one of the pairings AA, Aa, or aa. These pairings are called genotypes. If the genes correspond to color of eyes in human beings, the human being with genotype AA or Aa will have brown eyes, and the one with aa will have blue eyes. In such a case, we say that the gene A dominates a, or equivalently, the gene a is recessive.

In what is called autosomal inheritance, an offspring will receive one gene from each parent. If the father is of the genotype AA, the offspring will receive the gene A from the father. If the father is of the genotype Aa, the offspring will receive either A or a from the father with equal probability. Similar considerations do apply to mothers. If the father is of the genotype AA and the mother is of the genotype Aa, the offspring will receive the gene A from the father and either gene A or gene a from the mother with equal probability. Consequently the genotype of the offspring is either AA with probability $1/2$ or Aa with probability $1/2$. If the genes correspond to eye color, the offspring will have brown eyes with probability one. (Why?) If both father and mother are of the same genotype Aa, the offspring is of genotype AA, Aa, or aa with probabilities $1/4$, $1/2$, or $1/4$, respectively, If the genes correspond to eye color, the offspring will have brown eyes with probability 3/4 or blue eyes with probability 1/4. In the following table, we list the possible genotype of offspring along with their probabilities.

Genotypes of parents	Genotype of offspring AA	Aa	aa
AA & AA	1	0	0
AA & Aa	1/2	1/2	0
AA & aa	0	1	0
Aa & Aa	1/4	1/2	1/4
Aa & aa	0	1/2	1/2
aa & aa	0	0	1

No distinction is made of (father, mother) pairings of genotypes $(AA,\ Aa)$ and $(Aa,\ AA)$.

Let us look at a simple inbreeding model. Suppose in the 0-th generation, the population consists of a proportion of p_0 individuals of genotype AA, q_0 of genotype Aa, and r_0 of genotype aa. Clearly, $p_0 + q_0 + r_0 = 1$. Suppose mating takes place between individuals of the same genotype only. Let in the population,

p_n = proportion of genotype AA in the n-th generation,

q_n = proportion of genotype Aa in the n-th generation, and

r_n = proportion of genotype aa in the n-th generation.

We would like to determine the limiting behavior of these proportions. In the first generation, we will have $p_1 = p_0 + (1/4)q_0$, $q_1 = (1/2)q_0$, $r_1 = r_0 + (1/4)q_0$. These equations can be rewritten as

$$\begin{bmatrix} p_1 \\ q_1 \\ r_1 \end{bmatrix} = \begin{bmatrix} 1 & 1/4 & 0 \\ 0 & 1/2 & 0 \\ 0 & 1/4 & 1 \end{bmatrix} \begin{bmatrix} p_0 \\ q_0 \\ r_0 \end{bmatrix}.$$

Let $x_n' = (p_n, q_n, r_n), n \geq 0$ and A the 3×3 matrix that appears in the above linear equations. It is clear that the vectors x_n's are governed by the equation,

$$x_n = A^n x_0,\ n \geq 0.$$

Note that A is non-negative matrix. Let us determine the limiting behavior of the sequence A^n, $n \geq 0$. The eigenvalues of A are $\lambda_1 =$

$1, \lambda_2 = 1$, and $\lambda_3 = 1/2$ with corresponding eigenvectors chosen to be

$$\begin{bmatrix} 1 \\ 0 \\ 0 \end{bmatrix}, \begin{bmatrix} 0 \\ 0 \\ 1 \end{bmatrix}, \text{ and } \begin{bmatrix} 1 \\ -2 \\ 1 \end{bmatrix}.$$

Let

$$P = \begin{bmatrix} 1 & 0 & 1 \\ 0 & 0 & -2 \\ 0 & 1 & 1 \end{bmatrix}, \text{ and } P^{-1} = (1/2) \begin{bmatrix} 2 & 1 & 0 \\ 0 & 1 & 2 \\ 0 & -1 & 0 \end{bmatrix}.$$

Note that $A = P\Delta P^{-1}$, where $\Delta = \text{diag}\{1, 1, 1/2\}$. It now transpires that $A^n = P\Delta^n P^{-1}$ for all $n \geq 0$. Consequently, for any $n \geq 0$,

$$A^n = \begin{bmatrix} 1 & 1/2 - (1/2)^{n+1} & 0 \\ 0 & (1/2)^n & 0 \\ 0 & 1/2 - (1/2)^{n+1} & 1 \end{bmatrix}.$$

More explicitly,

$$p_n = p_0 + [\frac{1}{2} - (\frac{1}{2})^{n+1}]q_0, q_n = (\frac{1}{2})^n q_0, r_n = r_0 + [\frac{1}{2} - (\frac{1}{2})^{n+1}]q_0.$$

Consequently,

$$\lim_{n\to\infty} p_n = p_0 + (\frac{1}{2})q_0, \lim_{n\to\infty} q_n = 0, \lim_{n\to\infty} r_n = r_0 + (\frac{1}{2})q_0.$$

In the long run, individuals of genotype Aa will disappear! Only pure genotypes AA and aa will remain in the population.

Let us look at another model called selective breeding model. In the 0-th generation, a population has a proportion p_0 of individuals of genotype AA, a proportion q_0 of genotype Aa, and a proportion r_0 of genotype aa. In a special breeding program, all individuals are mated with individuals of genotype AA only. Let in the population

p_n = proportion of genotype AA in the n-th generation,

q_n = proportion of genotype Aa in the n-th generation, and

r_n = proportion of genotype aa in the n-th generation.

We would like to examine the limiting behavior of these proportions. In the first generation, we have

$$p_1 = p_0 + (\frac{1}{2})q_0, \; q_1 = r_0 + (\frac{1}{2})q_0, \; r_1 = 0.$$

These equations can be rewritten as

$$\begin{bmatrix} p_1 \\ q_1 \\ r_1 \end{bmatrix} = \begin{bmatrix} 1 & 1/2 & 0 \\ 0 & 1/2 & 1 \\ 0 & 0 & 0 \end{bmatrix} \begin{bmatrix} p_0 \\ q_0 \\ r_0 \end{bmatrix}.$$

Let $x_n' = (p_n, q_n, r_n), n \geq 0$ and A the 3×3 matrix that appears in the above linear equations. It is clear that the vectors x_n's are governed by the equation, $x_n = A^n x_0$, $n \geq 0$. Note that A is non-negative matrix. Let us determine the limiting behavior of the sequence A^n, $n \geq 0$. The eigenvalues of A are $\lambda_1 = 1, \lambda_2 = \frac{1}{2}$, and $\lambda_3 = 0$ with corresponding eigenvectors chosen to be

$$\begin{bmatrix} 1 \\ 0 \\ 0 \end{bmatrix}, \begin{bmatrix} 1 \\ -1 \\ 0 \end{bmatrix}, \text{ and } \begin{bmatrix} 1 \\ -2 \\ 1 \end{bmatrix}.$$

Let

$$P = \begin{bmatrix} 1 & 1 & 1 \\ 0 & -1 & -2 \\ 0 & 0 & 1 \end{bmatrix}, \text{ and } P^{-1} = (1/2) \begin{bmatrix} 1 & 1 & 1 \\ 0 & -1 & -2 \\ 0 & 0 & 1 \end{bmatrix},$$

Note that $A = P\Delta P^{-1}$, where $\Delta = \text{diag}\{1, 1/2, 0\}$. It now transpires that $A^n = P\Delta^n P^{-1}$ for all $n \geq 0$. Consequently, for any $n \geq 1$,

$$A^n = \begin{bmatrix} 1 & 1-(1/2)^n & 1-2^{1-n} \\ 0 & (1/2)^n & (1/2)^{n-1} \\ 0 & 0 & 0 \end{bmatrix}.$$

More explicitly,

$$p_n = 1 - (\frac{1}{2})^n q_0 - (\frac{1}{2})^{n-1} r_0, \; q_n = (\frac{1}{2})^n q_0 + (\frac{1}{2})^{n-1} r_0, \; r_n = 0.$$

Consequently, $\lim_{n\to\infty} p_n = 1$, $\lim_{n\to\infty} q_n = 0$, $\lim_{n\to\infty} r_n = 0$.

In the long run, individuals of genotype Aa and aa will disappear! Only the pure genotypes AA will remain in the population. As a matter of fact, individuals of genotype aa will disappear in the fist generation itself.

Let us look at a simple restrictive breeding model. In the 0-th generation, a proportion p_0 of individuals is of genotype AA, a proportion q_0 of genotype Aa, and a proportion r_0 of genotype aa. Suppose only (AA, AA) and (AA, Aa) matings are allowed. This model is feasible if one wishes to eliminate certain genetic diseases from the population. In many genetic diseases such as cystic fibrosis (predominant among Caucasians), sickle cell anemia (predominant among Blacks), and Tay-Sachs disease (predominant among East European Jews), the relevant normal gene A dominates the recessive gene a. If an individual is of genotype Aa, he or she will be normal but a carrier of the disease. If an individual is of genotype aa, he or she will have the disease and the offspring will be at least a carrier of the disease. One would like to see the effect of the policy of preventing the sufferers of the disease to mate. Let p_0 be the proportion of the population of genotype AA and q_0 the proportion of the population of genotype Aa. Since the mating is restricted, the population is taken to be those individuals of genotype AA or Aa. Consequently, $p_0 + q_0 = 1$. Let in the population

p_n = proportion of genotype AA in the n-th generation, and

q_n = proportion of genotype Aa in the n-th generation.

One can check that for every $n \geq 0$,

$$\begin{bmatrix} p_n \\ q_n \end{bmatrix} = A^n \begin{bmatrix} p_0 \\ q_0 \end{bmatrix}, \text{with } A = \begin{bmatrix} 1 & 1/2 \\ 0 & 1/2 \end{bmatrix}.$$

A direct computation of powers of A yields

$$p_n = 1 - (\frac{1}{2})^n p_0, \quad q_n = (\frac{1}{2})^n p_0.$$

Consequently, $\lim_{n\to\infty} p_n = 1$ and $\lim_{n\to\infty} q_n = 0$. In the long run, the carriers of the disease will disappear!

15.5. Population Growth Models

One of the important areas in Demography is a critical examination of how population grows over a period of interest. The so-called "Leslie Model" describes the growth of the female portion of a human or animal population. In this section, we will describe the mechanics of this model, in which a non-negative matrix appears. We will outline the limiting behavior of the powers of this matrix in order to shed light on the long-term growth of the population.

Suppose the maximum age attained by any female in the population is M years. We classify the female population into some k age classes of equal length. Say, the age classes are: $[0, M/k), [M/k, 2M/k), \ldots, [(k-1)M/k, M]$. When the study is initiated, note down

$$p_i^{(0)} = \text{number of females in the population falling into the age group } [(i-1)M/k, iM/k),\ i = 1, 2, \ldots, k.$$

The vector $p^{(0)\prime} = (p_1^{(0)}, p_2^{(0)}, \ldots, p_k^{(0)})$ is called the initial age distribution vector. The main objective is to examine how the age distribution changes over time. We would like to examine the age distribution of the female population at times $t_0 = 0,\ t_1 = M/k,\ t_2 = 2M/k$, and so on. As time progresses, the composition of the classes varies because of three biological processes: birth, death, and aging. These biological processes may be described by the following demographic parameters. Let

$$a_i = \text{the average number of daughters born to a single female during the time she is in the } i\text{-th age class},\ i = 1, 2, \ldots, k,$$

$$b_i = \text{the proportion of females in the } i\text{-th class expected to survive and pass into the next class},\ i = 1, \ldots, k.$$

It is clear that $a_i \geq 0$ for every i. Assume that $0 < b_i \leq 1$ for every $i = 1, 2, \ldots, k-1$. Assume that at least one a_i is positive. Let $p^{(m)\prime} = (p_1^{(m)}, p_2^{(m)}, \ldots, p_k^{(m)})$ be the age distribution of the females at time t_m, where $p_i^{(m)}$ is the number of females in the i-th age class, $i = 1, 2, \ldots, k$. The vectors $p^{(m)}$'s satisfy the following recurrent relation:

$$p_1^{(m)} = a_1 p_1^{(m-1)} + a_2 p_2^{(m-1)} + \ldots + a_k p_k^{(m-1)},$$
$$p_i^{(m)} = b_{i-1} p_{i-1}^{(m-1)},\ i = 2, 3, \ldots, k.$$

These equations can be put in a succinct form:

$$p^{(m)} = Lp^{(m-1)},\ m = 1, 2, 3, \ldots,$$

where

$$L = \begin{bmatrix} a_1 & a_2 & a_3 & \ldots & a_{k-1} & a_k \\ b_1 & 0 & 0 & \ldots & 0 & 0 \\ 0 & b_2 & 0 & \ldots & 0 & 0 \\ \cdot & \cdot & \cdot & \ldots & \cdot & \cdot \\ \cdot & \cdot & \cdot & \ldots & \cdot & \cdot \\ \cdot & \cdot & \cdot & \ldots & \cdot & \cdot \\ 0 & 0 & 0 & \ldots & b_{k-1} & 0 \end{bmatrix},$$

the so-called Leslie matrix. Clearly, L is a non-negative matrix. It now follows that $p^{(m)} = L^m p^{(0)}$. In order to facilitate the computation of powers of L, finding eigenvalues and eigenvectors of L is helpful. The eigenvalues of L are the roots of the determinantal equation

$$\begin{aligned} 0 = p(\lambda) &= |L - \lambda I| \\ &= \lambda^k - a_1\lambda^{k-1} - a_2 b_1 \lambda^{k-2} - a_3 b_1 b_2 \lambda^{k-3} - \ldots - a_k b_1 \ldots b_{k-1}. \end{aligned}$$

Since at least one a_i is positive, there is at least one non-zero root of the polynomial equation. Consequently, the spectral radius $\rho_s(L) > 0$. We record, without proof, some facts concerning the Leslie matrix. For some details, the reader may refer to Rorres and Anton (1984).

(1) The eigenvalue $\rho_s(L)$ is simple.

(2) The vector x, given below, is positive and is an eigenvector corresponding to the eigenvalue $\rho_s(L)$:

$$x' = (1, b_1/\rho_s(L), b_1 b_2/[\rho_s(L)]^2, \ldots, b_1 b_2 \ldots b_{k-1}/[\rho_s(L)]^{k-1}).$$

(3) If two successive entries a_i and a_{i+1} are positive, then $|\lambda| < \rho_s(L)$ for any eigenvalue λ of L different from $\rho_s(L)$.

Assume that a_i and a_{i+1} are positive for some i. It now follows that $[\rho_s(L)]^{-m} L^m p^{(0)}$ converges to a constant vector q which depends only on $p^{(0)}$. If m is large, $p^{(m)} \cong [\rho_s(L)]^m q \cong [\rho_s(L)] p^{(m-1)}$. What this means is that the age distribution is a scalar multiple of the preceding age distribution.

Complements

15.5.1 Comment on the eigenvalues of the following Leslie matrix:

$$\begin{bmatrix} 0 & 4 & 2 \\ 1/4 & 0 & 0 \\ 0 & 1/8 & 0 \end{bmatrix}.$$

Let $M = 15$ and $k = 3$. If the initial age distribution is given by $p^{(0)} = (1000,\ 900,\ 800)$, examine the age distribution after 15 years.

15.5.2 Comment on the eigenvalues of the following Leslie matrix:

$$\begin{bmatrix} 0 & 0 & 2 \\ 1/2 & 0 & 0 \\ 0 & 1/3 & 0 \end{bmatrix}.$$

15.5.3 (Fibonacci numbers) In 1202 Leonardo of Pisa, also called Fibonacci, posed the following problem. A pair of rabbits do not produce any offspring during their first month of life. However, starting with the second month, each pair of rabbits produces one pair of offsprings per month. Suppose we start with one pair of rabbits and none of the rabbits produced from this pair die. How many pairs of rabbits will be there at the beginning of each month. Let u_n be the pair of rabbits at the beginning of the n-th month. Establish the recurrence relation $u_n = u_{n-1} + u_{n-2}$ and show that

$$u_n = 5^{-1/2}\ (\lambda_1^{n+1} - \lambda_2^{n+1})$$

where λ_1 and λ_2 are eigenvalues of the matrix

$$A = \begin{bmatrix} 1 & 1 \\ 1 & 0 \end{bmatrix}.$$

Notes: The books by Horn and Johnson (1985, 1991) provide a comprehensive treatment on matrices. Some sections of this chapter are inspired by these works. The book by Rorres and Anton (1984) gives a good account of many applications of matrices to real world problems. Their influence is discernible in some of the examples presented here.

CHAPTER 16

MISCELLANEOUS COMPLEMENTS

Some topics, not covered under the different themes of the earlier chapters, which have applications in statistics and econometrics are assembled in this chapter. The proofs are omitted in most cases, but references to original papers and books where proofs and other details can be found are given.

16.1. Simultaneous Decomposition of Matrices

In Section 5.5 we have given a number of results on simultaneous diagonalization of matrices. We consider more general results in this section.

DEFINITION 16.1.1. Two matrices $A, B \in \mathbf{M}_n$ are said to be simultaneously diagonalizable if there exists a nonsingular transformation T such that T^*AT and T^*BT are both diagonal.

DEFINITION 16.1.2. Two matrices $A, B \in \mathbf{M}_n$ are said to be diagonalizable by contragredient transformations if there exists a nonsingular transformation T such that T^*BT and $T^{-1}A(T^{-1})^*$ are diagonal.

A typical example where both the definitions hold is the case of two Hermitian commuting matrices which are simultaneously diagonalizable by a unitary transformation.

We quote here a number of theorems on simultaneous diagonalizability of two matrices under various conditions given in Rao and Mitra (1971b).

P 16.1.3 Let $A \in \mathbf{M}_n$ be a Hermitian matrix and $B \in \mathbf{M}_n$ be an nnd matrix with rank $r \leq n$ and $N \in \mathbf{M}_{n,n-r}$ with $\rho(N) = n - r$ be such that $N^*B = 0$. Then the following hold.

(1) There exists a matrix $L \in \mathbf{M}_{n,r}$ such that $L^*BL = I(\in \mathbf{M}_r)$ and $L^*AL = \Delta(\in \mathbf{M}_r)$ diagonal.

(2) A necessary and sufficient condition that there exists a nonsingular transformation T such that T^*AT and T^*BT are both diagonal is (using the notation $\rho(B)$ = rank of B)

$$\rho(N^*A) = \rho(N^*AN).$$

(3) A necessary and sufficient condition that there exists a nonsingular transformation such that T^*BT and $T^{-1}A(T^{-1})^*$ are diagonal (i.e., A and B are reducible by contragredient transformations) is

$$\rho(BA) = \rho(BAB).$$

(4) If in addition A is nnd, then there exists a nonsingular transformation T such that T^*AT and T^*BT are both diagonal.

(5) If in addition A is nnd, then there exists a nonsingular transformation T such that T^*BT and $T^{-1}A(T^{-1})^*$ are diagonal.

P 16.1.4 Let A and B be Hermitian and B nonsingular. Then, there exists a nonsingular transformation T such that T^*AT and T^*BT are both diagonal if and only if there exists a matrix L such that $LAB^{-1}L^{-1}$ is diagonal with real diagonal elements (i.e., AB^{-1} is semisimple or similar to a diagonal matrix).

For details regarding the above theorems and further results, the reader is referred to Chapter 6 in Rao and Mitra (1971b).

16.2. More on Inequalities

In Chapters 10 and 14 we have discussed a number of inequalities which are useful in solving optimization problems and in establishing bounds to certain functions. We quote here some results from a recent thesis by Liu (1995).

P 16.2.1 (Matrix-trace versions of Cauchy-Schwartz inequality) Let $B \in \mathbf{M}_n$ be nnd, $A \in \mathbf{M}_{n,m}$ be such that $Sp(A) \subset Sp(B)$, $Z \in \mathbf{M}_{n,m}$ be arbitrary and B^+ be the Moore-Penrose inverse of B. Then

(1) $(\text{tr}Z'A)^2 \leq (\text{tr}Z'BZ)(\text{tr}A'B^+A)$
with equality if and only if BZ and A are proportional.

(2) $\text{tr}[(Z'A)^2] \le \text{tr}[Z'BZA'B^+A]$
with equality if and only if BZA' is symmetric

(3) $A^+B^+(A^+)' \ge (A'BA)^+$ (in Löwner sense)
where A^+ and B^+ are Moore-Penrose inverses of A and B respectively.

As special cases of (1) we have

$$(\text{vec}A)(\text{vec}A)' \le [(\text{vec}A)'(I \otimes B^+)\text{vec}A](I \otimes B)$$
$$cc' \le (c'B^-c)B \quad \text{for } c \in Sp(B).$$

For an application of the last result in statistics and econometrics see Toutenburg (1992, pp.286-287).

P 16.2.2 Let $B \in \mathbf{M}_n$ be nnd, $A \in \mathbf{M}_n$ symmetric and $T \in \mathbf{M}_{n,k}$ such that $\rho(T) = k$, $Sp(T) \subset Sp(B)$ and $T'BT = I_k$. Further let B^+ be the Moore-Penrose inverse of B and $\lambda_1 \ge \ldots \ge \lambda_n$ be the eigenvalues of B^+A. Then the following maxima or minima with respect to T hold.

(1) $\max(\text{tr}T'AT) = \lambda_1 + \ldots + \lambda_k$

(2) $\min(\text{tr}T'AT) = \lambda_{n-k+1} + \ldots + \lambda_n$

(3) $\max[\text{tr}(T'AT)^2] = \lambda_1^2 + \ldots + \lambda_k^2$

(4) $\min[\text{tr}(T'AT)^2] = \lambda_{n-k+1}^2 + \ldots + \lambda_n^2$

(5) $\max[\text{tr}(T'AT)^{-1}] = \lambda_{r-k+1}^{-1} + \ldots + \lambda_r^{-1}$ for $A > 0$, $r = \rho(B)$

(6) $\min[\text{tr}(T'AT)^{-1}] = \lambda_1^{-1} + \ldots + \lambda_k^{-1}$ for $A > 0$.

The optimum values are reached when $T = (t_1|\cdots|t_k)$, where $B^{1/2}t_i$ are orthonormal eigenvectors of $(B^+)^{1/2}A(B^+)^{1/2}$ associated with the eigenvalues $\lambda_i(B^+A)$, $i = 1, \ldots, k$.

P 16.2.3 For $A \in \mathbf{M}_{m,n}$, $\mathbf{B}_1 \in \mathbf{M}_m$ and $\mathbf{B}_2 \in \mathbf{M}_n$ are nnd, and $T \in \mathbf{M}_{m,k}$ such that $Sp(T) \subset Sp(B_1)$ and $T'B_1T = I_k$ and $W \in \mathbf{M}_{n,k}$ such that $Sp(W) \subset Sp(B_2)$ and $W'B_2W = I_k$, we have

$$\max_{T,W} \text{tr}(T'AW) = \lambda_1 + \ldots + \lambda_k$$
$$\max_{T,W} \text{tr}(T'AW)^2 = \lambda_1^2 + \ldots + \lambda_k^2$$

where λ_i^2 are the eigenvalues of $B_1^+ A B_2^+ A'$.

This theorem is useful in the study of canonical correlations in multivariate analysis. (See Lin (1990) and Yanai and Takane (1992).)

P 16.2.4 **(Matrix Kantorovich-type inequalities)** Let $B \in \mathbf{M}_n$ and nnd with $\rho(B) = b$ and $A \in \mathbf{M}_{n,r}$ such that $Sp(A) \subset Sp(B)$ and $\rho(A) = a \leq \min(b, r)$. Further let $\lambda_1 \geq \ldots \geq \lambda_b > 0$ be the eigenvalues of B. Then:

(1) $A^+B^+(A^+)' \leq \frac{(\lambda_1+\lambda_b)^2}{4\lambda_1\lambda_b}(A'BA)^+$
with equality if and only if $A = 0$ or
$A'BA = \frac{\lambda_1+\lambda_b}{2} A'A$ and $A'B^+A = \frac{\lambda_1+\lambda_b}{2\lambda_1\lambda_b} A'A$.

(2) $A^+B(A^+)' - (A'B^+A)^+ \leq (\sqrt{\lambda_1} - \sqrt{\lambda_b})^2 (A'A)^+$
with equality if and only if $A = 0$ or $\lambda_1 = \lambda_b$ or
$A'BA = (\lambda_1 + \lambda_b - \sqrt{\lambda_1\lambda_b}) A'A$ and $A'B^+A = (\lambda_1\lambda_b)^{-1/2} A'A$.

(3) $A'B^2A \leq \frac{(\lambda_1+\lambda_b)^2}{4\lambda_1\lambda_b} A'BAA^+BA$
with equality if and only if $A = 0$ or
$A'BA = \frac{2\lambda_1\lambda_b}{\lambda_1+\lambda_b} A'A$ and $A'B^2A = \lambda_1\lambda_b A'A$.

(4) $A'B^2A - A'BAA^+BA \leq \frac{1}{4}(\lambda_1 - \lambda_b)^2 A'A$
with equality if and only if $A = 0$ or $\lambda_1 = \lambda_b$ or
$A'BA = \frac{\lambda_1+\lambda_b}{2} A'A$ and $A'B^2A = \frac{\lambda_1^2+\lambda_b^2}{2} A'A$.

Liu (1995) established the following inequalities from the above inequalities, where $C \cdot D$ is the Hadamard-Schur product.

P 16.2.5 Let λ_1 and λ_b the maximum and minimum of eigenvalues of $C \otimes D$ where C and D are pd matrices. Then

(1) $(C \cdot D)^{-1} \leq C^{-1} \cdot D^{-1} \leq \frac{(\lambda_1+\lambda_b)^2}{4\lambda_1\lambda_b}(C \cdot D)^{-1}$

(2) $C \cdot D - (C^{-1} \cdot D^{-1})^{-1} \leq (\sqrt{\lambda_1} - \sqrt{\lambda_b})^2 I$

(3) $(C \cdot D)^2 \leq C^2 \cdot D^2 \leq \frac{(\lambda_1+\lambda_b)^2}{4\lambda_1\lambda_b}(C \cdot D)^2$

(4) $(C \cdot D)^2 - C^2 \cdot D^2 \leq \frac{1}{4}(\lambda_1 - \lambda_b)^2 I$

(5) $(C \cdot D) \leq (C^2 \cdot D^2)^{1/2} \leq \frac{\lambda_1+\lambda_b}{2\sqrt{\lambda_1\lambda_b}} C \cdot D$

(6) $(C^2 \cdot D^2)^{1/2} - C \cdot D \leq \frac{(\lambda_1-\lambda_b)}{4(\lambda_1+\lambda_b)} I$

P 16.2.6 (Schöpf (1960)) Let $A \in \mathbf{M}_n$ be pd with eigenvalues $\lambda_1 \geq \ldots \geq \lambda_n$ and let for $y \neq 0$, $\mu_t = y^* A^t y$. Then

$$1 \leq \frac{\mu_{t+1}\mu_{t-1}}{\mu_t^2} \leq \frac{(\lambda_1 + \lambda_n)^2}{4\lambda_1\lambda_n}.$$

16.3. Miscellaneous Results on Matrices

P 16.3.1 **(Toeplitz-Hausdorff Theorem)** The set

$$W(A) = \{< x, Ax >: \|x\| = 1\}$$

is a closed convex set.

Note that $W(UAU^*) = W(A)$ for any unitary U and $W(aA + bI) = aW(A) + bW(I)$ for all $a, b \in \mathbf{C}$. Also $\lambda(A) \in W(A)$, for any $\lambda(A)$.

P 16.3.2 For any matrix A, the series

$$\exp A = I + A + \frac{1}{2!}A^2 + \ldots + \frac{1}{n!}A^2 + \ldots + \frac{1}{n!}A^n + \ldots$$

converges. This is called the exponential of A. Then $\exp A$ is invertible and $(\exp A)^{-1} = \exp(-A)$. Conversely every invertible matrix can be expressed as the exponential of a matrix.

COROLLARY 16.3.3. Every unitary matrix can be expressed as the exponential of a skew-symmetric matrix.

P 16.3.4 Let $w(A) = \sup |< x, Ax >|$ over $\|x\| = 1$. Then

(1) $w(A)$ defines a matrix norm,
(2) $w(UAU^*) = w(A)$ for all unitary U, and
(3) $w(A) \leq \|A\| \leq 2w(A)$ for all A, where $\|A\| = \sup \|Ax\|$.

P 16.3.5 **(Weyl's Majorant Theorem)** Let $A \in \mathbf{M}_n$ with singular value $\sigma_1 \geq \ldots \geq \sigma_n$ and eigenvalues $\lambda_1, \ldots, \lambda_n$ arranged in such a way that $|\lambda_1| \geq \ldots \geq |\lambda_n|$. Then for every function $\phi : \mathbf{R}_+ \to \mathbf{R}_+$ such that $\phi(e^t)$ is convex and monotonic increasing in t, we have

$$(\phi(|\lambda_1|), \ldots, \phi(|\lambda_n|)) \ll_w (\phi(\sigma_1), \ldots, \phi(\sigma_n)).$$

In particular

$$(|\lambda_1|^p, \ldots, |\lambda_n|^p) \ll_w (\sigma_1^p, \ldots, \sigma_n^p)$$

for all $p \geq 0$.

P 16.3.6 **(Converse of Weyl's Majorant Theorem)** If $\lambda_1, \ldots, \lambda_n$ are complex numbers and $\sigma_1, \ldots, \sigma_n$ are positive real numbers ordered as $|\lambda_1| \geq \ldots \geq |\lambda_n|$ and $\sigma_1 \geq \ldots \geq \sigma_n$ and if

$$|\lambda_1| \ldots |\lambda_k| \leq s_1 \ldots s_k \quad \text{for } 1 \leq k \leq n$$
$$|\lambda_1| \ldots |\lambda_n| = s_1 \ldots s_n$$

then there exists a matrix $A \in \mathbf{M}_n$ with eigenvalues $\lambda_1 \ldots \lambda_n$ and singular values $\sigma_1, \ldots, \sigma_n$.

P 16.3.7 **(Fischer's Inequality)** Let $P_1, \ldots, P_r$ be a family of mutually orthogonal projectors in $\mathbf{C}^n$ such that $P_1 \oplus \ldots \oplus P_r = I_n$. Then for $A \geq 0$

$$\det A \leq \det(P_1 A P_1 + \ldots + P_r A P_r).$$

[The matrix $P_1AP_1 + \ldots + P_rAP_r$ is called the **pinching** of A.]

P 16.3.8 **(Aronszaju's Inequality)** Let $A \in \mathbf{M}_n$ be Hermitian matrix partitioned as

$$A = \begin{pmatrix} B & X \\ X^* & C \end{pmatrix}$$

where $B \in \mathbf{M}_k$. Let the eigenvalues of B, C and A be $\beta_1 \geq \ldots \geq \beta_k$; $\gamma_1 \geq \ldots \geq \gamma_{n-k}$; and $\alpha_1 \geq \ldots \geq \alpha_n$ respectively. Then

$$\alpha_{i+j-1} + \alpha_n \leq \beta_i + \gamma_j \text{ for all } i, j \text{ with } i + j - 1 \leq n.$$

P 16.3.9 Let A and B be pd matrices. Then the following hold.

(1) $\|A^s B^s\| \leq \|AB\|^s$ for $0 \leq s \leq 1$.

(2) $\|AB\|^t \leq \|A^t B^t\|$ for $t \geq 1$.

(3) $\lambda_1(A^s B^s) \leq \lambda_1^s(AB)$ for $0 \leq s \leq 1$.

(4) $[\lambda_1(AB)]^t \leq \lambda_1(A^t B^t)$ for $t \geq 1$.

(5) Let $\lambda^{1/t}(\cdot) = ((\lambda_1(\cdot))^{1/t}, \ldots, (\lambda_n(\cdot))^{1/t})'$. Then

$$\lambda^{1/t}(A^t B^t) \ll_w \lambda^{1/u}(A^u B^u) \text{ for all } 0 < t \leq u < \infty.$$

(6) **(Araki-Lieb-Thirring Inequality)**

$$\text{tr}[(B^{1/2} A B^{1/2})]^{st} \leq \text{tr}[(B^{t/2} A^t B^{t/2})^s]$$

where $A \geq 0$, $B \geq 0$ and s and t are positive real numbers with $t \geq 1$.

(7) **(Lieb-Thirring Inequality)** Let A and B be nnd matrices and m, k be positive integers with $m \geq k$. Then

$$\text{tr}[(A^k B^k)^m] \leq \text{tr}[(A^m B^m)^k].$$

In particular

$$\text{tr}[(AB)^m] \leq \text{tr}(A^m B^m).$$

P 16.3.10 (**n-dimensional Pythagorean Theorem**) Let $x_1, \ldots, x_n$ be orthogonal vectors in $\mathbf{R}^n$, and 0 denote the origin. Let the volume of the simplex $(x_1, \ldots, x_n)$ be V and that of the simplex $(0, x_1, \ldots, x_i, x_{i+1}, \ldots, x_n)$ be V_i. Then

$$V^2 = V_1^2 + \ldots + V_n^2.$$

A formal proof is given by S. Ramanan and K.R. Parthasarathy.

For proofs of the propositions in this section, and further results, reference may be made to Bhatia (1991).

P 16.3.11 Let $A_1, \ldots, A_n$ be pairwise commuting and pd matrices. Then

$$\mathcal{A}(A) \geq \mathcal{G}(A) \geq \mathcal{H}(A)$$

where $\mathcal{A}(A) = (A_1 + \ldots + A_n)/n$, $\mathcal{G}(A) = (A_1 \ldots A_n)^{1/n}$ and $\mathcal{H}(A) = n(A_1^{-1} + \ldots + A_n^{-1})^{-1}$. This is a generalization of the classical inequality connecting arithmetic, geometric and harmonic means. (Maher (1994).)

P 16.3.12 **(Geršgorin Theorem)** Let $A \in \mathbf{M}_n$ and $a_{ij} \in \mathbf{C}$ be the elements of A, $i, j = 1, \ldots, n$ and

$$\xi_i = \left(\sum_{j=1}^{n} |a_{ij}| \right) - |a_{ij}|.$$

Then every eigenvalue of A lies in at least one of the disks

$$\{z : |z - a_{ii}| \leq \xi_i\}, \ i = 1, \ldots, n$$

in the complex z-plane.

Furthermore, a set of m disks having no point in common with the remaining $(n - m)$ disks, contains m and only m eigenvalues of A.

P 16.3.13 Let $A \in \mathbf{M}_n$ as in **P 16.3.12** and

$$\nu = \max_{1 \leq j \leq n} (|a_{j1}| + \ldots + |a_{jn}|), \ \zeta = \max_{1 \leq k \leq n} (|a_{1k}| + \ldots + |a_{nk}|).$$

Then the eigenvalues of A lie in the disk

$$\{z \in \mathbf{C} : |z| \leq \min(\nu, \zeta)\}.$$

Furthermore $|\det A| \leq \min(\eta^n, \zeta^n)$.

P 16.3.14 Let $A \in \mathbf{M}_n$ as in **P 16.3.12**, $d_j = |a_{jj}| - \xi_j$, $j = 1, \ldots, n$ and $d = \min\{d_1, \ldots, d_n\} > 0$. Then $|\lambda_i| \geq d$, $i = 1, \ldots, n$, where $\lambda_1, \ldots, \lambda_n$ are the eigenvalues of A. Hence $|\det A| \geq d^n$.

P 16.3.15 **(Schur Theorem)** Let $A \in \mathbf{M}_n$, $\|\cdot\|$ denote the Euclidean matrix norm and $\lambda_1, \ldots, \lambda_n$, the eigenvalues of A. Then

$$|\lambda_1|^2 + \ldots + |\lambda_n|^2 \leq \|A\|^2$$
$$(\text{Re}\lambda_1)^2 + \ldots + (\text{Re}\lambda_n)^2 \leq \|B\|^2$$
$$(\text{Im}\lambda_1)^2 + \ldots + (\text{Im}\lambda_n)^2 \leq \|C\|^2$$

where Re and Im are real and imaginary parts, $B = (A + A^*)/2$ and $C = (A - A^*)/2$. Equality in any one of the above three relations implies equality in all three and occurs if and only if A is normal.

16.4. Toeplitz Matrices

Matrices whose entries are constant along each diagonal arise in many applications and are called Toeplitz matrices. Formally $T \in \mathbf{M}_n$ is Toeplitz if there exist scalars $c_{-p+1}, \ldots, c_0, \ldots, c_{p-1}$ such that t_{ij}, the (i,j)-th element of T is c_{j-i}. Thus

$$T = \begin{bmatrix} c_0 & c_1 & c_2 & \cdots & c_{p-1} \\ c_{-1} & c_0 & c_1 & \cdots & c_{p-2} \\ c_{-2} & c_{-1} & c_0 & \cdots & c_{p-3} \\ \cdot & \cdot & \cdot & \cdots & \cdot \\ c_{-p+1} & c_{-p+2} & c_{-p+3} & \cdots & c_0 \end{bmatrix} \tag{16.4.1}$$

is Toeplitz.

The special case of (16.4.1) when $c_{-i} = c_i$ and the matrix is positive definite arises in a linear prediction problem using an autoregressive model in time series,

$$x_t + a_1 x_{t-1} + \ldots + a_p x_{t-p} = \epsilon_t, \tag{16.4.2}$$

where, considering (16.4.2) as a stationary process, we have

$$E(\epsilon_t) = 0,\ E(x_t) = 0,\ E(x_t\epsilon_t) = 0$$
$$E(x_t x_{t-j}) = E(x_t x_{t+j}) = c_j.$$

Multiplying both sides of (16.4.2) by x_{t-j} and taking expectations, we have

$$c_j + a_1 c_{|j-1|} + \ldots + a_p c_{|j-p|} = 0,\ j = 1, \ldots, p$$

which can be written using a special case of (16.4.1),

$$\begin{bmatrix} c_0 & c_1 & \cdots & c_{p-1} \\ c_1 & c_0 & \cdots & c_{p-2} \\ \cdot & \cdot & \cdots & \cdot \\ c_{p-1} & c_{p-2} & \cdots & c_0 \end{bmatrix} \begin{bmatrix} a_1 \\ a_2 \\ \cdot \\ a_p \end{bmatrix} = - \begin{bmatrix} c_1 \\ c_2 \\ \cdot \\ c_p \end{bmatrix}. \tag{16.4.3}$$

If c_i are known, we can estimate a_i by solving equation (16.4.3) and use the estimates $\hat{a}_1, \ldots, \hat{a}_j$ in predicting x_t given $x_{t-1}, \ldots, x_{t-p}$ by the formula

$$\hat{x}_t = -(\hat{a}_1 x_{t-1} + \ldots + \hat{a}_p x_{t-p}). \tag{16.4.4}$$

If c_i's are not known, we can estimate them from the available observations in the time series, $x_1, \ldots, x_n$, by the formula

$$\hat{c}_i = \frac{1}{n_r} \sum_{a=1}^{n_r} x_a x_{a+r} \tag{16.4.5}$$

where n_r in (16.4.5) $= n - r$, the number of terms in the summation. The estimates of c_i are used in equation (16.4.3) to solve for a_i.

Because of the special structure of the Toeplitz matrix in (16.4.3), the computations involved in estimating a_i are somewhat simple. Efficient algorithms have been developed by Durbin (1960) for solving (16.4.3), by Levinson (1947) when the vector on the right in (16.4.3) is arbitrary and by Trench (1964) to find the inverse of the Toeplitz matrix on the left-hand side of (16.4.3). Excellent accounts of these algorithms can be found in Golub and Loan (1989).

We illustrate only the Durbin's algorithm to show how the Toeplitz form of the matrix of equations (16.4.3) provides a simple approach. First we introduce what are called persymmetric matrices.

DEFINITION 16.4.1. A matrix $B = (b_{ij}) \in \mathbf{M}_n$ is called persymmetric if $b_{ij} = b_{n-j+1,n-i+1}$ for all i and j. [What this concept means is that the matrix is symmetric around the diagonal running from the southwest corner to the northeast corner.]

An example of a 3×3 persymmetric matrix is as follows.

$$B = \begin{bmatrix} a & b & c \\ d & e & b \\ f & d & a \end{bmatrix}.$$

Let $E_n \in \mathbf{M}_n$ be defined by

$$E_n = \begin{bmatrix} 0 & 0 & \ldots & 0 & 1 \\ 0 & 0 & \ldots & 1 & 0 \\ \cdot & \cdot & \ldots & \cdot & \cdot \\ 1 & 0 & \ldots & 0 & 0 \end{bmatrix}$$

which is, indeed, a permutation matrix with the special name of exchange matrix. If $x' = (x_1, \ldots, x_n)$, note that $(E_n x)' = (x_n, \ldots, x_1)$. Further $E_n^{-1} = E_n$. The following results are easy to establish.

P 16.4.2 Let $B \in \mathbf{M}_n$. Then:

(1) B is persymmetric if and only if $B = E_n B' E_n$.
(2) If B is persymmetric and nonsingular, then B^{-1} is persymmetric.
(3) If $T \in \mathbf{M}_n$ is Toeplitz (of the form (16.4.1)), then T is persymmetric. The converse is not true.

We revert to the problem of solving equation (16.4.3). First by dividing each row by c_0, we can rewrite equation (16.4.3) as

$$T_p \alpha_p = \rho_p \tag{16.4.6}$$

where

$$T_p = \begin{bmatrix} 1 & r_1 & \dots & r_{p-1} \\ r_1 & 1 & \dots & r_{p-2} \\ \cdot & \cdot & \dots & \cdot \\ r_{p-1} & r_{p-2} & \dots & 1 \end{bmatrix}$$

$$r_i = c_i/c_0, \ \alpha'_p = (a_1, \dots, a_p), \ \rho'_p = (r_1, \dots, r_p).$$

Suppose that we are able to obtain a solution of

$$T_m y_m = \rho_m \tag{16.4.7}$$

in y_m for a given $m \in [1, p]$. Then, we can obtain a solution of

$$T_{m+1} z_{m+1} = \rho_{m+1} \tag{16.4.8}$$

provided that $m + 1 \leq p$. Rewrite (16.4.8) as

$$\begin{bmatrix} T_m & E_m \rho_m \\ (E_m \rho_m)' & 1 \end{bmatrix} \begin{bmatrix} u \\ v \end{bmatrix} = \begin{bmatrix} \rho_m \\ r_{m+1} \end{bmatrix} \tag{16.4.9}$$

where $z'_{m+1} = (u', v)$ and u is an m-vector. These equations give

$$T_m u + E_m \rho_m v = \rho_m$$
$$\rho'_m E_m u + v = r_{m+1}$$

from which we have

$$\begin{aligned} u &= T_m^{-1}(\rho_m - E_m \rho_m v) \\ &= y_m - T_m^{-1} E_m \rho_m v. \end{aligned} \tag{16.4.10}$$

Since T_m is Toeplitz, $T_m = E_m T_m E_m$, we have $T_m^{-1} E_m = E_m T_m^{-1}$ and

$$u = y_m - v E_m y_m.$$

Once we have y_m from (16.4.7), the computation of u from (16.4.10) is simple provided that we know v. Note that

$$\begin{aligned} v &= r_{m+1} - \rho_m' E_m u \\ &= r_{m+1} - \rho_m' E_m (y_m - v E_m y) \\ &= r_{m+1} - \rho_m' E_m y_m - v \rho_m' E_m y_m. \end{aligned}$$

Consequently

$$\begin{aligned} v(1 + \rho_m' y_m) &= r_{m+1} - \rho_m' E_m y_m \\ v &= \frac{r_{m+1} - \rho_m' E_m y_m}{1 + \rho_m' y_m}. \end{aligned}$$

Thus, $z_{m+1}' = (u', v)$ depends only on y_m. We continue the recursive process starting with any m, until we reach the value p. This is called Durbin's algorithm.

Let us look at the eigenvalue decomposition of a Toeplitz matrix. The eigenvalues of Toeplitz matrices do not have an explicit form. However, some asymptotic results on the behavior of eigenvalues are available. We look at a special form of Toeplitz matrices. Let $a_{-m}, a_{-(m-1)}, \ldots, a_{-1}, a_0, a_1, \ldots, a_{m-1}, a_m$ be a set of $2m+1$ numbers. Define $a_t = 0$ for all $|t| > m$. For each $n \geq 1$, let $T_n = (t_{ij}^{(n)})$ be a matrix of order $n \times n$ with $t_{ij}^{(n)} = a_{j-i}$. For example, if $m = 2$ and $n = 5$,

$$T_5 = \begin{bmatrix} a_0 & a_1 & a_2 & 0 & 0 \\ a_{-1} & a_0 & a_1 & a_2 & 0 \\ a_{-2} & a_{-1} & a_0 & a_1 & a_2 \\ 0 & a_{-2} & a_{-1} & a_0 & a_1 \\ 0 & 0 & a_{-2} & a_{-1} & a_0 \end{bmatrix}.$$

If n is large compared to m, a substantial number of entries in the upper right-hand corner and lower right-hand corner of the matrix T_n are zeros. Let $\alpha_{n,1}, \alpha_{n,2}, \ldots, \alpha_{n,n}$ be the eigenvalues of T_n. Let $f(\cdot)$ be the Fourier transform of the sequence $a_{-m}, a_{-(m-1)}, \ldots, a_{-1}, a_0, a_1, \ldots, a_{m-1}, a_m$, i.e.,

$$f(\lambda) = \sum_{k=-m}^{m} a_k e^{ik\lambda}, \; 0 \le \lambda \le 2\pi.$$

It can be shown that for any possible integer s,

$$\lim_{n\to\infty} \frac{1}{n} \sum_{k=1}^{n} \alpha_{n,k}^{s} = \frac{1}{2\pi} \int_0^{2\pi} [f(\lambda)]^s \, d\lambda.$$

If T_n is Hermitian for each n, then for any continuous $F(\cdot)$ on the appropriate interval,

$$\lim_{n\to\infty} \frac{1}{n} \sum_{k=1}^{n} F(\alpha_{n,k}) = \frac{1}{2\pi} \int_0^{2\pi} F[f(\lambda)] \, d\lambda.$$

The appropriate interval for F is $[\min_\lambda \; f(\lambda), \; \max_\lambda \; f(\lambda)]$. In this case, one can show that the eigenvalues of T_n lie in this appropriate interval. For further exposition in this connection, the reader can refer to Gray (1972) and Grenander and Szegö (1958).

Complements

16.4.1 Let $\mathbf{V}$ be the collection of all $n \times n$ Toeplitz matrices. Show that $\mathbf{V}$ is a vector space and its dimension is $2n-1$. Obtain a basis of $\mathbf{V}$.

16.4.2 Determine the inverse of the following Toeplitz matrix of order $n \times n$.

$$T = \begin{bmatrix} 1 & -1 & 0 & \ldots & 0 & 0 \\ 0 & 1 & -1 & \ldots & 0 & 0 \\ 0 & 0 & 1 & \ldots & 0 & 0 \\ \cdot & \cdot & \cdot & \ldots & \cdot & \cdot \\ 0 & 0 & 0 & \ldots & 1 & -1 \\ 0 & 0 & 0 & \ldots & 0 & 1 \end{bmatrix}.$$

For this matrix, $a_0 = 1, \; a_1 = -1$, and all other $a_i = 0$.

16.4.3 Determine the inverse of the following Toeplitz matrix of order $n \times n$.

$$T = \begin{bmatrix} 1 & 0 & 0 & \dots & 0 & 0 & 0 \\ -2 & 1 & 0 & \dots & 0 & 0 & 0 \\ 1 & -2 & 1 & \dots & 0 & 0 & 0 \\ \cdot & \cdot & \cdot & \dots & \cdot & \cdot & \cdot \\ 0 & 0 & 0 & \dots & -2 & 1 & 0 \\ 0 & 0 & 0 & \dots & 1 & -2 & 1 \end{bmatrix}.$$

In this example, $a_0 = 1$, $a_{-1} = -2$, $a_{-2} = 1$, and all other $a_i = 0$.

Hint: T^{-1} is lower triangular and Toeplitz.

16.4.4 Determine the inverse of the following Toeplitz matrix T of order $n \times n$.

$$T = \begin{bmatrix} 2 & -1 & 0 & \dots & 0 & 0 & 0 \\ -1 & 2 & -1 & \dots & 0 & 0 & 0 \\ 0 & -1 & 2 & \dots & 0 & 0 & 0 \\ \cdot & \cdot & \cdot & \dots & \cdot & \cdot & \cdot \\ 0 & 0 & 0 & \dots & -1 & 2 & -1 \\ 0 & 0 & 0 & \dots & 0 & -1 & 2 \end{bmatrix}.$$

In this example, $a_0 = 2$, $a_1 = a_{-1} = -1$, and all other $a_i = 0$. The matrix T is an example of a tri-diagonal matrix.

Hint: Study the pattern of inverses for $n = 1, 2, 3$, and 4.

16.5. Restricted Eigenvalue Problem

In statistical applications, it is sometimes necessary to find the optimum values of a quadratic form $x'Ax$ subject to the conditions $x'Bx = 1$, where B is pd, and $C'x = t$ (see Rao (1964b)). A simple solution exists when $t = 0$.

P 16.5.1 The stationary values of $x'Ax$ when x is restricted to $x'Bx = 1$ and $C'x = 0$ is attained at the eigenvectors of $(I - P)A$ with respect to B, where P is the projection operator

$$P = C(C'B^{-1}C)^{-}C'B^{-1}.$$

PROOF. Introducing Lagrangian multipliers λ and μ, we consider the expression

$$x'Ax - \lambda(x'Bx - 1) - 2x'C\mu$$

and equate its derivative to zero

$$\begin{aligned} Ax - \lambda Bx - C\mu &= 0 \\ C'x &= 0 \\ x'Bx &= 1. \end{aligned} \tag{16.5.1}$$

Multiplying the first equation in (16.5.1) by $I - P$, we have the equation

$$(I - P)Ax = \lambda Bx \tag{16.5.2}$$

which is required to be proved.

In the special case $A = aa'$ where a is a vector, $x'Ax$ has the maximum value when $x \propto B^{-1}(I - P)a$ which is an important result in problems of genetic selection. Another problem of interest in this connection is to find the maximum of $x'Ax$ when x is restricted by the inequality condition $C'x \geq_e 0$ in addition to $x'Bx = 1$. This leads to a quadratic programming problem as shown in Rao (1964a).

In the general case of the condition $C'x = t$, a solution is given by Gander, Golub and Matt (1989).

A more general eigenvalue problem which occurs in statistical problems is to find the stationary values of

$$\phi(x) = (x - b)^* A(x - b)$$

subject to the condition $x^*x = 1$, where A is a Hermitian matrix and b is a given vector. For computational aspects of this problem reference may be made to Forsyth and Golub (1965).

16.6. Product of Two Raleigh Quotients

We consider the problem of finding the stationary values of

$$\frac{x'Cx}{(x'Ax)^{1/2}(x'Bx)^{1/2}} \tag{16.6.1}$$

where A and B are pd matrices and C is a symmetric matrix, all of the same order. The square of (16.6.1) is the product of two Raleigh coefficients $(x'Cx/x'Ax)$ and $(x'Cx/x'Bx)$. The special case of (16.6.1) with $C = I$ and A is of order 2×2 originally arose in attempts to design control systems with minimum norm feedback matrices (Kouvaritakis and Cameron (1980) and Cameron and Kouvaritakis (1980)) and also in the study of the stability of multivariate nonlinear feedback systems (Cameron (1983)). The general case of (16.6.1) occurs in the analysis of familial data when multiple homologous measurements are available on father and son (say), and the objective is to determine a linear combination of measurements which has the maximum parent-offspring correlation (Rao and Rao (1987)).

P 16.6.1 The stationary values of (16.6.1) are

$$\lambda_i^2 \nu_i, \ i = 1, 2, \ldots$$

where λ_i and ν_i are solutions of the equations

$$\begin{aligned} 2Cx &= \lambda(A + \nu B)x, \\ \nu &= x'Ax/x'Bx. \end{aligned} \tag{16.6.2}$$

PROOF. Differentiating (16.6.1) with respect to x, we obtain

$$\frac{x'Cx}{x'Ax}Ax + \frac{x'Cx}{x'Bx}Bx = 2Cx. \tag{16.6.3}$$

Writing $\lambda = x'Cx/x'Ax$ and $\lambda\nu = x'Cx/x'Bx$, we get the desired equations (16.6.2).

A computational algorithm for solving equations (16.6.2) and some worked out examples are given in Rao and Rao (1987). An investigation into an efficient algorithm for solving (16.6.3) will be a useful contribution.

16.7. Matrix Orderings and Projection

Let us consider the real vector space $\mathbf{R}^n$ endowed with the ordinary inner product $< x, y > = x'y$, where $x, y \in \mathbf{R}^n$. A matrix $P \in \mathbf{M}_n$ is called a projector if $P^2 = P$ (i.e., P is idempotent). Let us introduce the following conventions of ordering two matrices A and B. (See Trenkler (1994) for more detailed discussion.)

Löwner ordering (Löwner (1934)). Let $A, B \in \mathbf{M}_n$. $A \leq_L B$ if and only if $B - A = CC'$ for some matrix C (i.e., $B - A$ is nnd matrix).

Star ordering (Drazen (1978)). Let $A, B \in \mathbf{M}_{m,n}$. $A \leq_* B$ if and only if $A'A = A'B$ and $AA' = BA'$.

Rank subtractive ordering (Hartwig (1980)). Let $A, B \in \mathbf{M}_{m,n}$. $A \leq_{rs} B$ if and only if $\rho(B - A) = \rho(B) - \rho(A)$.

The following proposition characterizes projection operators in terms of the above matrix orderings.

P 16.7.1 (Hartwig and Styan (1987)). P is a projector if and only if one of the following conditions is satisfied.

(1) $P \leq_{rs} I$
(2) $P \leq_* Q$ for some projector Q
(3) $P \leq_{rs} Q$ for some projector Q
(4) $P \leq_L Q$ for some projector Q such that $PQ = QP$ and all eigenvalues of P are 0 and 1.

P 16.7.2 (Baksalary and Mitra (1991)). P is an orthogonal projector (i.e., P is idempotent and symmetric) if and only if one of the following conditions is satisfied.

(1) $P \leq_* I$
(2) $P \leq_{rs} P'P$
(3) $P \leq_* Q$ for some orthogonal projector Q
(4) $P \leq_L Q$ for some orthogonal projector Q and all eigenvalues of P are 0 and 1
(5) $0 \leq_L P \leq_L Q$ and $(Q - P)^2 = (Q - P)$ for some orthogonal projector Q.

16.8. Soft Majorization

In Chapter 9, we discussed majorization of one real vector by another. We introduce here some alternative definitions applicable to real or complex vectors and mention some inequalities on eigenvalues.

DEFINITION 16.8.1. For real or complex vectors $v, w (\in \mathbf{C}^n)$, w is said to majorize v, denoted by $v \ll w$, if and only if v is a convex combination of the permutations w_π of w (w_π denotes a vector obtained from w by permuting its components according to the permutation π of $\{1, \ldots, n\}$), i.e., $v = \Sigma a_\pi w_\pi$, $\Sigma a_\pi = 1$, with a_π as real numbers. [Note that $v \ll w$ defined as above implies $v_1 + \ldots + v_n = w_1 + \ldots + w_n$, where $v' = (v_1, \ldots, v_n)$ and $w' = (w_1, \ldots, w_n)$.]

DEFINITION 16.8.2. With $v, w(\in \mathbf{C}^n)$ and w_π as in Definition 16.8.1, w is said to softly majorize v, denoted by $v \ll_s w$, if and only if $v = \Sigma z_\pi w_\pi$ (finite sum), where z_π are complex numbers such that $\Sigma|z_\pi| = 1$.

We use the notation Eig A to denote the n-vector of eigenvalues of $A \in \mathbf{M}_n$, including multiplicities and ordered arbitrarily. A matrix norm is called unitarily invariant if $\|U^*AV\| = \|A\|$ for every $A \in \mathbf{M}_{m,n}$ and unitary $U \in \mathbf{M}_m$ and $V \in \mathbf{M}_n$, in which case we denote the norm of A by $\|A\|_{ui}$. A matrix norm is called weakly unitarily invariant norm if $\|U^*AU\| = \|A\|$ for every $A \in \mathbf{M}_m$ and unitary $U \in \mathbf{M}_m$, in which case we denote the norm of A by $\|A\|_{wui}$.

P 16.8.3 **(Lidskii Theorem)** Let A and B be Hermitian matrices with $\alpha' = (\alpha_1, \dots, \alpha_n)$ and $\beta' = (\beta_1, \dots, \beta_n)$ as corresponding vectors of eigenvalues, where the components are arranged in decreasing order of magnitude. Then

$$\alpha - \beta \ll \text{Eig}(A - B)$$

which implies

$$\|\text{diag}\{\alpha_i\} - \text{diag}\{\beta_i\}\|_{ui} \le \|A - B\|_{ui}.$$

P 16.8.4 (Bhatia and Holbrook (1989)) For any $A, B \in \mathbf{M}_n$ and normal such that $A - B$ is also normal, there is an ordering of Eig A and Eig B such that

$$\text{Eig } A - \text{Eig } B \ll \text{Eig}(A - B).$$

P 16.8.5 (Sunder (1982)) If A, B and $(A - B)$ are normal as in **P 16.8.4**, then for each $\|\cdot\|_{wui}$, there is an ordering of Eig A and Eig B, which may depend on the norm, such that

$$\|A - B\|_{wui} \le \|\text{diag}\{\text{Eig } A\} - \text{diag } \{\text{Eig } B\}\|_{wui}.$$

P 16.8.6 (Bhatia and Holbrook (1989)) For A and $B \in \mathbf{M}_n$ and normal, the following statements are equivalent.

(1) $\|A\|_{wui} \le \|B\|_{wui}$ for every such norm.
(2) Eig $A \ll_s$ Eig (B).

16.9. Circulants

Let $C_0, C_1, C_2, \ldots, C_{n-1}$ be n numbers. The circulant C of order $n \times n$ based on these numbers is defined by

$$C = \begin{bmatrix} C_0 & C_1 & C_2 & \ldots & C_{n-2} & C_{n-1} \\ C_{n-1} & C_0 & C_1 & \ldots & C_{n-3} & C_{n-2} \\ C_{n-2} & C_{n-1} & C_0 & \ldots & C_{n-4} & C_{n-3} \\ \cdot & \cdot & \cdot & \ldots & \cdot & \cdot \\ C_2 & C_3 & C_4 & \ldots & C_0 & C_1 \\ C_1 & C_2 & C_3 & \ldots & C_{n-1} & C_0 \end{bmatrix}.$$

The elements of each row (column) of C are identical to those of the previous row (column), but are moved one position to the right (down) and wrapped around.

One can recognize from the structure of C that C is a Toeplitz matrix. The entries of C can be described using the modulo operation. Let $C = (C_{ij})$ be a matrix of order $n \times n$. The matrix C is a circulant if $C_{ij} = C_{j-i(\text{mod}\, n)}$ for all $1 \leq i, j \leq n$ for some numbers $C_0, C_1, \ldots, C_{n-1}$. The circulant C based on the numbers $C_0, C_1, \ldots, C_{n-1}$ can be denoted symbolically as $C(C_0, C_1, \ldots, C_{n-1})$. There is another way to look at circulants. Let

$$\Pi = \begin{bmatrix} 0 & 1 & 0 & \ldots & 0 & 0 \\ 0 & 0 & 1 & \ldots & 0 & 0 \\ 0 & 0 & 0 & \ldots & 0 & 0 \\ \cdot & \cdot & \cdot & \ldots & \cdot & \cdot \\ 0 & 0 & 0 & \ldots & 0 & 1 \\ 1 & 0 & 0 & \ldots & 0 & 0 \end{bmatrix}.$$

The matrix Π is a permutation matrix of order $n \times n$ and can be identified as the circulant $C(0, 1, 0, \ldots, 0, 0)$. The matrix Π is called the forward-shift permutation matrix. This is the permutation matrix associated with the permutation map $\sigma : \{0, 1, 2, \ldots, n-1\} \to \{0, 1, 2, \ldots, n-1\}$ given by $\sigma(0) = 1$, $\sigma(1) = 2, \ldots, \sigma(n-2) = n-1$, and $\sigma(n-1) = 0$. One can show that a matrix C of order $n \times n$ is a circulant if and only if $C\Pi = \Pi C$.

Circulants occur as natural models of dispersion matrices in a certain statistical context. The variances and covariances of say 8 measurements can be modeled in the following way.

(1) $\text{var}(X_i) = \sigma_0$, $i = 1, 2, \ldots, 8$.

(2) $\text{cov}(X_i, X_j) = \sigma_{|i-j|},\ i \neq j$.

The dispersion matrix Σ of the measurements has the following structure

$$\Sigma = \begin{bmatrix} \sigma_0 & \sigma_1 & \sigma_2 & \sigma_3 & \sigma_4 & \sigma_5 & \sigma_6 & \sigma_7 \\ \sigma_1 & \sigma_0 & \sigma_1 & \sigma_2 & \sigma_3 & \sigma_4 & \sigma_5 & \sigma_6 \\ \sigma_2 & \sigma_1 & \sigma_0 & \sigma_1 & \sigma_2 & \sigma_3 & \sigma_4 & \sigma_5 \\ \sigma_3 & \sigma_2 & \sigma_1 & \sigma_0 & \sigma_1 & \sigma_2 & \sigma_3 & \sigma_4 \\ \sigma_4 & \sigma_3 & \sigma_2 & \sigma_1 & \sigma_0 & \sigma_1 & \sigma_2 & \sigma_3 \\ \sigma_5 & \sigma_4 & \sigma_3 & \sigma_2 & \sigma_1 & \sigma_0 & \sigma_1 & \sigma_2 \\ \sigma_6 & \sigma_5 & \sigma_4 & \sigma_3 & \sigma_2 & \sigma_1 & \sigma_0 & \sigma_1 \\ \sigma_7 & \sigma_6 & \sigma_5 & \sigma_4 & \sigma_3 & \sigma_2 & \sigma_1 & \sigma_0 \end{bmatrix}.$$

Note that Σ is a circulant.

We now proceed to obtain spectral decomposition of a circulant. Ideas from group theory play a prominent role in the decomposition. Let $G = \{0, 1, 2, \dots, n-1\}$. The set G is a group under the binary operation of addition modulo n. Let $C_0, C_1, C_2, \dots, C_{n-1}$ be a set of numbers. These numbers can be viewed as a function defined on the group G. The Fourier transform $\hat{C}_{(\cdot)}$ of the function $C_{(\cdot)}$ is given by

$$\hat{C}_j = \sum_{k=0}^{n-1} e^{2\pi ijk/n} C_k,$$

where $i = \sqrt{-1}$. The function $C_{(\cdot)}$ can be recovered from its Fourier transform $\hat{C}_{(\cdot)}$. More precisely,

$$C_j = \frac{1}{n} \sum_{k=0}^{n-1} e^{-2\pi ijk/n} \hat{C}_k.$$

Define the Fourier matrix

$$F = (f_{ik}),\ f_{jk} = \frac{1}{n}\ e^{2\pi ijk/n}.$$

One can verify that F is a unitary matrix. Let C be the circulant based on $C_0, C_1, \dots, C_{n-1}$. It also follows that

$$C = F^*\ \text{Diag}\{\hat{C}_0, \hat{C}_{-1}, \hat{C}_{-2}, \dots, \hat{C}_{-(n-1)}\} F.$$

This is a spectral decomposition of C. The eigenvalues of C are given by $\hat{C}_0, \hat{C}_{-1}, \hat{C}_{-2}, \ldots, \hat{C}_{-(n-1)}$.

We will close this section with a description of cyclic designs. Suppose we want to compare the performance of some seven treatments on some experimental units. The experimental units are arranged in blocks of size 3. The units in blocks are homogeneous but there may be differences in blocks. The problem is to allocate the treatments to units in such a way that each treatment is replicated exactly the same number of times. The following is one such arrangement.

Block	Treatments
1	1 2 4
2	2 3 5
3	3 4 6
4	4 5 7
5	5 6 1
6	6 7 2
7	7 1 3

The starting block has treatments 1,2, and 4. The treatments in the following block are taken to be those by repeatedly adding 1 (mod 7) to the treatments in the preceding block. This is an example of a cyclic design. Note that this is indeed a Balanced Incomplete Block Design. Every treatment is replicated three times. Every pair of treatments appears exactly in one block. Let $N = (n_{ij})$ be the incidence matrix of the design, i.e.,

$$n_{ij} = \begin{cases} 1 & \text{if treatment } j \text{ appears in block } i, \\ 0 & \text{otherwise.} \end{cases}$$

Note that

$$N = \begin{bmatrix} 1 & 1 & 0 & 1 & 0 & 0 & 0 \\ 0 & 1 & 1 & 0 & 1 & 0 & 0 \\ 0 & 0 & 1 & 1 & 0 & 1 & 0 \\ 0 & 0 & 0 & 1 & 1 & 0 & 1 \\ 1 & 0 & 0 & 0 & 1 & 1 & 0 \\ 0 & 1 & 0 & 0 & 0 & 1 & 1 \\ 1 & 0 & 1 & 0 & 0 & 0 & 1 \end{bmatrix}.$$

Observe that N is a circulant. An excellent source for material on circulants is the book by Davis (1979).

Complements

16.9.1 Show that the collection of all circulants of order $n \times n$ is a vector space.
16.9.2 Show that only two circulants commute.
16.9.3 If C is a non-singular circulant, show that C^{-1} is a circulant. Using the spectral composition of C, identify the inverse of C.
16.9.4 Show that all circulants are simultaneously diagonalizable.
16.9.5 Explore the possibility of using a circulant for developing a magic square in which all rows sums, column sums, principal diagonal sums are identical.
16.9.6 Let $y_i,\ i = 0, \pm 1, \pm 2, \ldots$ be a doubly infinite sequence of numbers. Let

$$z_i = \frac{y_{i-1} + y_i + y_{i+1}}{3}, \quad i = 0, \pm 1, \pm 2, \ldots .$$

Identify the infinite matrix which transforms the sequence $\{y_i\}$ to $\{z_i\}$. Show that C is a circulant.

16.10. Hadamard Matrices

A matrix of order $n \times n$ whose entries are $+1$ or -1 is called a Hadamard matrix of order n if $HH' = nI_n$, which implies $H'H = nI_n$ and H is an orthogonal matrix. The following are some examples of Hadamard matrices:

$$\begin{bmatrix} 1 & 1 \\ 1 & -1 \end{bmatrix} \quad \text{and} \quad \begin{bmatrix} 1 & 1 & 1 & 1 \\ 1 & 1 & -1 & -1 \\ 1 & -1 & 1 & -1 \\ 1 & -1 & -1 & 1 \end{bmatrix}.$$

If H is a Hadamard matrix of order $n \geq 4$, then necessarily $n = 4t$ for some positive integer t. It has been conjectured that a Hadamard matrix of order $4t$ exists for every $t \geq 1$. The conjecture is not yet resolved. However, a number of results on the existence of Hadamard matrices of specific orders are available in the literature. For example, there exists a Hadamard matrix of order 2^k for every $k \geq 1$. For the use of Hadamard matrices in design of experiments, a good reference is Hedayat and Wallis (1976).

16.11. Miscellaneous Exercises

16.11.1 Verify that det$A$, where $A \in \mathbf{M}_n$ is the Fibonacci matrix,

$$A = \begin{bmatrix} 1 & 1 & 0 & \dots & 0 \\ -1 & 1 & 1 & \dots & 0 \\ 0 & -1 & 1 & \dots & 0 \\ \cdot & \cdot & \cdot & \dots & \cdot \\ 0 & 0 & 0 & \dots & 1 \end{bmatrix}$$

is exactly the n-th term of the Fibonacci sequence

$$1, 2, 3, 5, 8, 13, \dots$$

where $a_n = a_{n-1} + a_{n-2}$ $(n \geq 3)$.

16.11.2 Let J_n denote the tridiagonal Jacobi matrix

$$J_n = \begin{bmatrix} a_1 & b_1 & 0 & \dots & 0 \\ c_1 & a_2 & b_2 & \dots & 0 \\ 0 & c_2 & a_3 & \dots & 0 \\ \cdot & \cdot & \cdot & \dots & \cdot \\ 0 & 0 & 0 & \dots & a_n \end{bmatrix}$$

Show that $|J_n| = a_n|J_{n-1}| - b_{n-1}c_{n-1}|J_{n-1}|$, $n \geq 3$.

16.11.3 **(Fredholm Theorem)** The equation $Ax = b$, $(A \in \mathbf{M}_{m,n}$, with elements in $\mathbf{C}$, $b \in C^m)$ is solvable if and only if b is orthogonal to all solutions of the homogeneous equation $A^*y = 0$.

16.11.4 Let $A \in \mathbf{M}_n$ and Hermitian with eigenvalues $\lambda_1 \geq \dots \geq \lambda_n$. Let $\mu_1 \geq \dots \geq \mu_{n-r}$ be the eigenvalues of A subject to r linear constraints as in Section 16.5. Then show that

$$\lambda_i \geq \mu_i \geq \lambda_{i+r}, \; i = 1, \dots, n-r.$$

In particular, if there is only one constraint, then

$$\lambda_1 \geq \mu_1 \geq \lambda_2 \geq \dots \geq \mu_{n-1} \geq \lambda_n.$$

16.11.5 Define

$$a = \max_{r,s}|a_{rs}|, \; b = \max_{r,s}|b_{rs}|, \; c = \max_{r,s}|c_{rs}|$$

where $A = (a_{ij}) \in \mathbf{M}_n$, $B = (b_{ij}) = 2^{-1}(A+A)^*$, $C = (c_{ij}) = 2^{-1}(A-A^*)$. Then, if λ is any eigenvalue of A, show that

$$|\lambda| \le na, \ |\mathrm{Re}\,\lambda| \le nb, \ |\mathrm{Im}\,\lambda| \le nc.$$

This is a corollary to the Schur theorem (**P 16.3.15**).

16.11.6 Prove that a matrix G is the Moore-Penrose inverse of A if and only if

$$GAA^* = A^* \text{ and } G = BA^*$$

for some matrix B.

16.11.7 Let $A \in \mathbf{M}_{m,r}$ and $B \in \mathbf{M}_{r,n}$ each of rank $r \le \{\min(m,n)\}$. Show that

$$(AB)^+ = B^+A^+.$$

16.11.8 Let $A \in \mathbf{M}_n$ and $A = HU = U_1H_1$ be polar decomposition of A. Show that $A^+ = U^*H^+ = H_1^+U_1^*$ are those of A^+.

16.11.9 (Bahadur's expansion of a probability density function). The basic idea behind Bahadur's work is the following.

Let $f(\cdot)$ be a function defined on a finite set of elements

$$\mathbf{A} = \{\alpha_1, \dots, \alpha_k\} \tag{16.11.1}$$

and denote the vector

$$f' = (f(\alpha_1), \dots, f(\alpha_k)) \tag{16.11.2}$$

where $f(\alpha_i)$ is the value of f at α_i, $i = 1, \dots, k$. Further let $f_1, \dots, f_k$ be k functions defined on $\mathbf{A}$ such that they are not linearly dependent. Then, it is clear from our knowledge of vector spaces that any function f defined on $\mathbf{A}$ belongs to the vector space $\mathbf{V}_k = Sp(f_1, \dots, f_k)$, i.e., f can be written as

$$f = b_1 f_1 + \dots + b_k f_k \tag{16.11.3}$$

for some scalars $b_1, \dots, b_k$. Let us introduce an inner product $< \cdot, \cdot >$ in $\mathbf{V}_k$ and choose $f_1, \dots, f_k$ such that they are orthonormal with respect to the inner product. In such a case, the coefficients in (16.11.3) can be obtained in a simple way as

$$b_i = < f, f_i >, \ i = 1, \dots, k \ . \tag{16.11.4}$$

Bahadur (1961) used this idea in obtaining an expansion for the joint density function of m random variables

$$p(x_1, \ldots, x_m) = Pr.\ (X_1 = x_1, \ldots, X_m = x_m)$$

where each X_i can take the value 0 or 1. There are 2^m combinations $(x_1, \ldots, x_m)$ which are the elements of the set $\mathbf{A}$ as in (16.11.1).

Define

$$p(x_i) = Pr.\ (X_i = x_i), \tag{16.11.5}$$

$$E(X_i) = \pi_i, \tag{16.11.6}$$

$$\mathbf{V}(X_i) = E(X_i - \pi_i)^2 = \sigma_i^2, \tag{16.11.7}$$

$$E[(x_{i_1} - \pi_{i_1})(x_{i_2} - \pi_{i_2}) \ldots (x_{i_s} - \pi_{i_s})] = \left(\prod_{j=1}^{s} \sigma_{i_j}\right) \rho_{i_1 \ldots i_s}. \tag{16.11.8}$$

The numbers $\rho_{ij}, \rho_{ijk}, \ldots$ represent correlations of various orders.

Let us consider the functions

$$\begin{aligned} p(x_1, \ldots, x_m) &= 1 \text{ for all } (x_1, \ldots, x_m) \\ &= \frac{x_i - \pi_i}{\sigma_i}, \; i = 1, \ldots, m, \\ &= \left(\frac{x_i - \pi_i}{\sigma_i}\right)\left(\frac{x_j - \pi_j}{\sigma_j}\right), \; i \neq j = 1, \ldots, m, \\ &\qquad \ldots \qquad \ldots \\ &= \left(\frac{x_1 - \pi_1}{\sigma_1}\right) \ldots \left(\frac{x_m - \pi_m}{\sigma_m}\right), \end{aligned} \tag{16.11.9}$$

which are

$$1 + \binom{m}{1} + \binom{m}{2} + \ldots + 1 = 2^m$$

in number. Define the inner product of two functions p_r and p_s as

$$\sum_{(x_1, \ldots, x_m) \in \mathbf{A}} p_r(x_1, \ldots, x_m) p_s(x_1, \ldots, x_m) \prod_{i=1}^{m} p(x_i) \tag{16.11.10}$$

where $p(x_i)$, $i = 1, \ldots, m$ are fixed marginal densities arising out of $p(x_1, \ldots, x_m)$ as in (16.11.5).

It can be easily checked that with respect to the inner product (16.11.10), the functions (16.11.9) are orthonormal. Then, we have the expansion

$$\frac{p(x_1, \ldots, x_m)}{p(x_1)\ldots p(x_m)}$$
$$= b_0 1 + \sum_{i=1}^{m} b_i \left(\frac{x_i - \pi_i}{\sigma_i}\right) + \sum\sum_{i \neq j}^{m} b_{ij} \left(\frac{x_i - \pi_i}{\sigma_i}\right)\left(\frac{x_j - \pi_j}{\sigma_j}\right)$$
$$+ \ldots + b_{12\ldots k}\left(\frac{x_1 - \pi_1}{\sigma_1}\right)\ldots\left(\frac{x_m - \pi_m}{\sigma_m}\right) \qquad (16.11.11)$$

Computing the coefficients using the formula (16.11.4) we find

$$b_0 = 1, \ b_1 = \ldots = b_n = 0$$
$$b_{ij} = \rho_{ij}, \ldots, \ b_{12\ldots k} = \rho_{12\ldots k} \qquad (16.11.12)$$

where $\rho_{ij}, \rho_{ijk}, \ldots$ are as defined in (16.11.8). Equation (16.11.11) provides an expansion of $p(x_1, \ldots, x_m)$.

In statistical applications, we can estimate π_i, σ_i and $\rho_{ij}, \rho_{ijk}, \ldots$ from the sample observations on $(X_1, \ldots, X_m)$. In practical applications higher order correlations may be negligible in which case $p(x_1, \ldots, x_m)$ can be approximated by the first few terms in (16.11.11).

Note: Some references to material covered in this Chapter, besides those mentioned in the text, are Lancaster and Tismenetsky (1985), Bhatia (1991) and Bapat (1993).

Books by Dhrymes (1978), Muirhead (1982) and Amemiya (1985) contain numerous examples of applications of matrix algebra to econometrics and statistics.

REFERENCES

Amemiya, T. (1985). *Advanced Econometrics*, Basil Blackwell, Oxford.

Ammann, L.P. and J.W. Van Ness (1988). A routine for converting regression algorithm into corresponding orthogonal regression algorithms, *ACM Trans. Math. Software*, **14**, 76-87.

Ammann, L.P. and J.W. Van Ness (1989). Standard and robust orthogonal regression, *Communication in Statistics Simulation Comput.*, **18**, 145-162.

Anderson, T.W. (1976). Estimation of linear functional relationships: Approximate distributions and connections with simultaneous equations in econometrics (with discussion), *J. Roy. Statist. Soc. B*, **38**, 1-36.

Anderson, T.W. (1980). Estimation of linear statistical relationships, *Ann. Statist.*, **12**, 1-45.

Anderson, T.W. and Sawa Takamitsu (1982). Exact and approximate distributions of the maximum likelihood estimator of a slope coefficient, *J. Roy. Statist. Soc. B*, **44**, 52-62.

Anderson, T.W. (1984). *An Introduction to Multivariate Statistical Analysis*, Second edition, Wiley, New York.

Ando, T. (1982). Majorization, doubly stochastic matrices and comparison of eigenvalues, *Research Institute of Applied Electricity*, Hokkaido University, Sapporo, Japan.

Bahadur, R.R. (1961). On classification based on responses to n dichotomous items, In *Studies in Item Analysis* (ed. H. Solomon), Stanford Univ. Press, Stanford, 158-168.

Baksalary, J.K. and S.K. Mitra (1991). Left-star and right-star partial orderings, *Linear Algebra and its Applications*, **149**, 73-98.

Baksalary, J.K. and S. Puntanen (1991). Generalized matrix versions of the Cauchy-Schwarz and Kantorovich inequalities, *Aequationes Math.*, **41**, 103-110.

Bapat, R.B. (1993). *Linear Algebra and Linear Models*, Hindustan Book Agency, New Delhi.

Barnett, S. (1990). *Matrix, Methods and Applications*, Clarendon Press, New York.

Bao, P.C. and J. Rokne (1987). Closest normal matrix bound, *BII*, **27**, 585-598.

Bhatia, Rajendra (1991). *Matrix Analysis*, Springer-Verlag, New York.

Bhatia, R. and A.R. Holbrook (1989). A softer, stronger Lidskii theorem, *Proc. Indian Acad Sci. (Math. Sci.)*, **99**, 75-83.

Bloomfield, P. and G.S. Watson (1975). The inefficiency of least squares, *Biometrika*, **62**, 121-128.

Bose, R.C., S. Chowla and C.R. Rao (1944). On the integral order mod p of quadratics $x^2 + ax + b$ with applications to the construction of minimum functions for GF(p^2) and to some number theory results, *Bull. Cal. Math. Soc.*, **15**, 153-174.

Bose, R.C., S. Chowla and C.R. Rao (1945a). On the roots of a well-known congruence, *Proc. Nat. Acad. Sci.*, **14**, 193.

Bose, R.C., S. Chowla and C.R. Rao (1945b). Minimum functions in Galois fields, *Proc. Nat. Acad. Sci.*, **14**, 191.

Bose, R.C., S.S. Shrikhande and E.T. Parker (1960). Further results on the construction of mutually orthogonal latin squares and the falsity of Euler's conjecture, *Canad. J. Math.*, **12**, 189-203.

Cameron, R. (1983). Minimizing the product of two Raleigh quotients, *Linear and Multivariate Algebra*, **13**, 177-178.

Cameron, R. and B. Kouvaritakis (1980). Minimizing the norm of output feedback controllers used in pole placement: a dyadic approach, *Int. J. Control*, **32**, 759-770.

Datta Biswa Nath (1995). *Numerical Linear Algebra and Applications*, Brooks/Cole.

Davis, P.J. (1979). *Circulant Matrices*, Wiley, New York.

Diaz, J.B. and F.T. Metcalf (1964). Inequalities complementary to Cauchy's inequality for sums of real numbers, *J. Math. Anal.*, **9**, 59-74.

Dhrymes, P.J. (1978). *Introductory Econometrics*, Springer-Verlag, New York.

Drazen, M.B. (1978). Natural structures on semigroups with involution, *Bull. Amer. Math. Soc.*, **84**, 139-141.

Durbin, J. (1960). The fitting of time series models, *Rev Inst. Int. Stat.*, **28**, 233-243.

Eckart, C. and G. Young (1936). The approximation of one matrix by another of lower rank, *Psychometrika*, **1**, 211-218.

Edelman, A. (1997). The probability that a random real Gaussian matrix has k real eigenvalues, related distributions and the circular law, *J. Multivariate Analysis*, **60**, 203-232.

Forsythe, G.E. and G.H. Golub (1965). On the stationary values of a second degree polynomial on the unit sphere, *SIAM J. Appl. Math. Soc.*, **94**, 1-23.

Fuller, W.A. (1987). *Measurement Error Models*, Wiley, New York.

Gander, W., G.H. Golub and V. von Matt (1989). A constrained eigenvalue problem, *Linear Algebra and its Applications*, **114/115**, 815-839.

Gleser, L.J. (1981). Estimation in a multivariate "errors in variables" regression model: Large sample results, *Ann. Statist.*, **9**, 24-44.

Golub, G.H. and C.F. Van Loan (1983). *Matrix Computations*, North Oxford Academic, Oxford.

Graham, A. (1981). *Kronecker Products and Matrix Calculus with Applications*, Ellis Horwood Limited, Chichester, England.

Gray, R.M. (1972). On the asymptotic eigenvalue distribution of Toeplitz matrices, *IEEE Trans. Information Theory*, **18**, 725-730.

Grenander, Ulf and Gabor Szegö (1958). *Toeplitz Forms and Their Applications*, Univ. of California Press.

Greub, W. and W. Rheinboldt (1959). On a generalization of an inequality of L.V. Kantorovich, *Proc. Amer. Math. Soc.*, **10**, 407-415.

Halmos, P.R. (1958). *Finite-Dimensional Vector Spaces*, Van Nostrand, New York.

Halmos, P.R. (1972). Positive approximations of operators, *Indiana Univ. Math. J.*, **21**, 951-960.

Hartley, H.O., J.N.K. Rao and G. Kiefer (1969). Variance estimation with one unit per stratum, *J. Am. Statist. Assoc.*, **68**, 189-192.

Hartwig, R.E. (1980). How to partially order regular elements, *Mathematica Japonica*, **25**, 1-13.

Hartwig, R.E. and G.P.H. Styan (1987). Partially ordered idempotent matrices. In *Proc. Second Int. Tampere Conference in Statistics*, (eds. T. Pukkila and S. Puntanen), 361-383.

Hedayat, A. and W.D. Wallis (1978). Hadamard matrices and their applications, *Ann. Statist.*, **6**, 1184-1238.

Horn, R. and C.A. Johnson (1985). *Matrix Analysis*, Cambridge University Press, Cambridge, UK.

Horn, R. and C.A. Johnson (1991). *Topics in Matrix Analysis*, Cambridge University Press, Cambridge, UK.

Ikobe, Y., T. Inagaki and S. Miyamoto (1987). The monotonicity theorem, Cauchy's interlace theorem, and the Courant-Fischer theorem, *Amer. Math. Monthly*, **94**, 352-354.

Khatri, C.G. and C.R. Rao (1981). Some extensions of the Kantorovich inequality and statistical applications, *J. Multivariate Analysis*, **11**, 498-505.

Khatri, C.G. and C.R. Rao (1982). Some generalizations of Kantorovich inequality, *Sankhyā A*, **44**, 91-102.

Knott, M. (1975). On the minimum efficiency of least squares, *Biometrika* **62**, 129-132.

Kouvaritakis, B. and R. Cameron (1980). Pole placement with minimized norm controllers, *Proc. IEEE*, **127**, 32-36.

Ky Fan (1951). Maximum properties and inequalities for eigenvalues of completely continuous operators, *Proc. Nat. Acad. Sci.*, **37**, 760-766.

Ky Fan and A.J. Hoffman (1955). Some matrix inequalities in the space of matrices, *Proc. Amer. Math. Soc.*, **6**, 111-116.

Levinson, N. (1947). The Weiner RMS error criterion in filter design and prediction, *J. Math. Phys.*, **25**, 261-278.

Lin, C.T. (1990). Extremes of determinants and optimality of canonical variables, *Commun. Statist. Simul. Comp.*, **19**, 1415-1430.

Liu, S. (1995). *Contributions to Matrix Calculus and Applications in Econometrics*, Book No.106 of the Tinbergen Institute Research Series, Ph.D. Thesis.

Löwner, K. (1934). Über monotone Matrixfunktionen, *Mathematishe Zeitschrift*, **38**, 177-216.

Magnus, J.R. and H. Neudecker (1991). *Matrix Differential Calculus with Applications in Statistics and Econometrics*, Wiley, Chichester.

Maher, P.J. (1994). Means for matrices, *Int. J. Math. Educ. Sci. Technol.*, **25**, 591-623.

Mann, H.B. (1949). *Analysis and Design of Experiments*, New York, Dover Publication.

Marsaglia, G. (1967). Bounds on the ranks of sums of matrices. Trans. Fourth Prague Conference on Information Theory, Statistical Decision Functions and Random Processes, *Czech Acad, Sciences*, 455-462.

Marsaglia, G. and G.P.H. Styan (1974). Equalities and inequalities for ranks of matrices, *Linear and Multilinear Algebra*, **2**, 269-292.

Marshall, A.W. and I. Olkin (1979). *Inequalities: Theory of Majorization and its Applications*, Academic Press, New York.

Marshall, A.W. and I. Olkin (1990). Matrix versions of the Cauchy and Kantorovich inequalities, *Aequationes Math.*, **40**, 89-93.

Mirsky, L. (1960). Symmetric gauge functions and unitarily invariant norms, *Quarterly J. of Mathematics*, Oxford, Second series, **11**, 50-59.

Mirsky, L. (1990). *An Introduction to Linear Algebra*, Dover Publication.

Muirhead, R.J. (1982). *Aspects of Multivariate Statistical Theory*, Wiley, New York.

Mullholand, H.P. and C.A.B. Smith (1959). An inequality arising in genetical theory, *Amer. Math. Monthly*, **66**, 673-683.

Pearson, K. (1901). On lines and planes of closest fit to systems of points in space, *Philosophical Magazine*, **2** (sixth series), 559-572.

Raghavarao, D. (1971). *Constructions and Combinatorial Problems in Design of Experiments*, John Wiley, New York.

Rao, C.R. (1945a). Generalization of Markoff's theorem and tests of linear hypotheses, *Sankhyā*, **7**, 9-16.

Rao, C.R. (1945b). Markoff's theorem with linear restrictions on parameters, *Sankhyā*, **7**, 16-19.

Rao, C.R. (1945c). Finite geometries and certain derived results in theory of numbers, *Proc. Nat. Inst. Sci.*, **11**, 136-149.

Rao, C.R. (1946a). Difference sets and combinatorial arrangements derivable from finite geometrics, *Proc. Nat. Inst. Sci.*, **12**, 123-135.

Rao, C.R. (1946b). On the linear combination of observations and the general theory of least squares, *Sankhyā*, **7**, 237-256.

Rao, C.R. (1947). Factorial experiments derivable from combinatorial arrangements of arrays, *J. Roy. Statist. Soc.*, **9**, 128-140.

Rao, C.R. (1949). On a class of arrangements, *Edin. Math. Proc.*, **8**, 199-225.

Rao, C.R. (1951). A theorem in least squares, *Sankhyā*, **11**, 9-12.

Rao, C.R. (1955). Analysis of dispersion for multiply classified data with unequal numbers of cells, *Sankhyā*, **15**, 253-280.

Rao, C.R. (1959). Some problems involving linear hypotheses in multivariate analysis, *Biometrika*, **46**, 49-58.

Rao, C.R. (1962). A note of generalized inverse of a matrix with applications to problems in mathematical statistics, *J. Roy. Statist. Soc. B*, **24**, 152-158.

Rao, C.R. (1964a). Problems of selection involving programming techniques. In *Proc. IBM Scientific Computing Symposium on Statistics*, 29-51.

Rao, C.R. (1964b). The use and interpretation of principal component analysis in applied research, *Sankhyā A*, **26**, 329-358.

Rao, C.R. (1965). The theory of least squares when the parameters are stochastic and its applications to the analysis of growth curves, *Biometrika*, **52**, 447-458.

Rao, C.R. (1967). Calculus of generalized inverses of matrices: Part I - general theory, *Sankhyā A*, **29**, 317-350.

Rao, C.R. (1968). A note on a previous lemma in the theory of least squares and some further results, *Sankhyā A*, **30**, 259-266.

Rao, C.R. and S.K. Mitra (1968a). Simultaneous reduction of a pair of quadratic forms, *Sankhyā A*, **30**, 313-322.

Rao, C.R. and S.K. Mitra (1968b). Some results in estimation and tests of linear hypotheses under the Gauss-Markov model, *Sankhyā A*, **30**, 281-290.

Rao, C.R. and S.K. Mitra (1969). Conditions for optimality and validity of least squares theory, *Ann. Math. Statist.*, **40**, 1716-1724.

Rao, C.R. (1971). Unified theory of linear estimation. *Sankhyā A*, **33**, 371-394.

Rao, C.R. and S.K. Mitra (1971a). Further contributions to the theory of generalized inverse of matrices and its applications, *Sankhyā A*, **33**, 289-300.

Rao, C.R. and S.K. Mitra (1971b). *Generalized Inverse of Matrices and its Applications*, Wiley, New York (Japanese Translation, Tokyo, 1973).

Rao, C.R. (1972a). Estimation of variance components in linear models. *J. Am. Statist. Assoc.*, **67**, 112-115.

Rao, C.R. (1972b). A note on IPM method in the unified theory of linear estimation, *Sankhyā A*, **34**, 285-288.

Rao, C.R., P. Bhimasankaram and S.K. Mitra (1972). Determination of a matrix by its subclasses of g-inverse, *Sankhyā A*, **24**, 5-8.

Rao, C.R. (1973a). Unified theory of least squares, *Communications in Statistics*, **1**, 1-8.

Rao, C.R. (1973b). Representation of best unbiased estimators in the Gauss-Markov model with a singular dispersion matrix, *J. Multivariate Analysis*, **3**, 276-292.

Rao, C.R. (1973c). *Linear Statistical Inference and its Applications*, Wiley, New York.

Rao, C.R. and S.K. Mitra (1973). Theory and application of constrained inverse of matrices, *SIAM J. Appl. Math.*, **24**, 473-488.

Rao, C.R. (1974). Projectors, generalized inverses and the BLUE's, *J. Roy. Statist. Soc.*, **35**, 442-448.

Rao, C.R. and S.K. Mitra (1974). Projections under semi-norms and generalized inverse of matrices, *Linear Algebra and Appl.*, **9**, 155-167.

Rao, C.R. (1975). Theory of estimation of parameters in the general Gauss-Markov model, In a *Survey of Statistical Design and Linear Models*, (ed. J.N. Srivastava), North Holland, 475-487.

Rao, C.R. and S.K. Mitra (1975). Extension of a duality theorem concerning g-inverse of matrices. *Sankhyā A*, **37**, 439-445.

Rao, C.R. (1976a). Estimation of parameters in a linear model - Wald Lecture 1, *Ann. Statist.*, **4**, 1023-1037, with a correction in Vol. 7, 696.

Rao, C.R. (1976b). On a unified theory of linear estimation in linear models - A review of recent results, In *Perspectives in Probability*, Papers in honor of M.S. Bartlett, (ed. J. Gani), Academic Press, New York, 89-104.

Rao, C.R. (1978a). Least squares theory for possibly singular models, *The Canadian J. Statist.*, **6**, 19-23.

Rao, C.R. (1978b). A note on the unified theory of least squares, *Commun. Statist. Theor. Meth. A*, **7(5)**, 409-411.

Rao, C.R. (1978c). Choice of best linear estimators in the Gauss-Markov model with a singular dispersion matrix, *Commun. Statist. Theor. Meth. A*, **7(13)**, 1199-1208.

Rao, C.R. and H. Yanai (1979). General definition and decomposition of projectors and some applications to statistical problems, *J. Statist. Planning and Inference*, **3**, North-Holland, 1-17.

Rao, C.R. (1979a). Estimation of parameters in the singular Gauss-Markov model, *Commun. Statist. Theor. Meth. A*, **8(14)**, 1353-1358.

Rao, C.R. (1979b). Separation theorems for singular values of matrices and their applications in multivariate analysis, *J. Multivariate Analysis*, **9**, 362-377.

Rao, C.R. (1980). Matrix approximations and reduction of dimensionality in multivariate statistical analysis, In *Multivariate Analysis V*, (ed. P.R. Krishnaiah), North-Holland, 3-22.

Rao, C.R. and Jurgen Kleffe (1980). Estimation of variance components, *Handbook of Statistics*, (ed. P.R. Krishnaiah), North-Holland, **1**, 1-40.

Rao, C.R. (1981). A lemma on g-inverse of a matrix and a computation of correlation coefficients in the singular case, *Commun. Statist. Theor. Meth. A*, **10**, 1-10.

Rao, C.R. (1984). Optimization of functions of matrices with applications to statistical problems, In *W.G. Cochran's Impact on Statistics*, (ed. Poduri, S.R.S. Rao), John Wiley, New York, 191-202.

Rao, C.R. (1985a). Matrix derivatives: Applications in statistics, In *Encyclopedia of Statistical Sciences* Vol.5, (eds. Kotz-Johnson), John Wiley, New York, 320-325.

Rao, C.R. (1985b). A unified approach to inference from linear models, In *Proc. First International Tampere Seminar on Linear Statistical Models and Their Applications*, (eds.T. Pukkila and S. Puntanen), University of Tampere, Finland, 9-36.

Rao, C.R. (1985c). The inefficiency of least squares: Extension of Kantorovich inequality, *Linear Algebra and its Applications*, **70**, 249-255.

Rao, C.R. (1985d). Tests for dimensionality and interactions of mean vectors under general and reducible covariance structures, *J. Multivariate Analysis*, **16**, 173-184.

Rao, C.R. and H. Yanai (1985a). Generalized inverse of linear transformations: A geometric approach, *Linear Algebra and its Applications*, **66**, 87-98.

Rao, C.R. and H. Yanai (1985b). Generalized inverses of partitioned matrices useful in statistical applications, *Linear Algebra and its Applications*, **70**, 105-113.

Rao, C.R. (1987). Estimation in linear models with mixed effects: A unified theory, In *Proc. Second International Tampere Conference in*

Statistics, (eds. T. Pukkila and S. Puntanen), University of Tampere, Finland, 73-98.

Rao, C.R. and C.V. Rao (1987). Stationary values of the product of two Raleigh coefficients: homologous canonical variates, *Sankhyā B*, **49**, 113-125.

Rao, C.R. and J. Kleffe (1988). *Estimation of Variance Components and Applications*, North-Holland.

Rao, C.R. (1989). A lemma on optimization of a matrix function and a review of the unified theory of linear estimation, In *Statistical Data Analysis and Inference*, (ed. Y. Dodge), Elsevier Science Publishers B.V., 397-418.

Rao, C.R. (1995). A review of canonical coordinates and an alternative to correspondence analysis using Hellinger distance, *Qüestiio*, **19**, 15-63.

Rao, C.R. and H. Toutenburg (1995). *Linear Models: Least Squares and Alternative Methods*, Springer-Verlag.

Rao, C.R. (1996). Seven inequalities in statistical estimation theory, *Student*, **1**, 149-158.

Rorres, C. and H. Anton (1984). *Applications of Linear Algebra*, 3rd edition, John Wiley, New York.

Schopf, A.H. (1960). On the Kantorovich inequality, *Numer. Math.*, **2**, 344-346.

Seely, J. (1971). Quadratic subspaces and completeness, *Ann. Math. Stat.*, **42**, 710-721.

Seely, J. and G. Zyskind (1971). Linear spaces and minimum variance unbiased estimation, *Ann. Math. Stat.*, **42**, 691-703.

Srivastava, M.S. and C.G. Khatri (1979). *An Introduction to Multivariate Statistics*, North-Holland, New York.

Strang, W.G. (1960). On the Kantorovich inequality, *Proc. Amer. Math. Soc.*, **11**, 468.

Styan, G.P.H. (1983). On some inequalities associated with ordinary least squares and the Kantorovich inequality. In *Festschrift for Eino Haikala on his Seventieth Birthday*, Univ. of Tampere, 158-166.

Sunder, V.S. (1982). On permutations, convex hulls and normal operators, *Linear Algebra and Appl.*, **48**, 403-411.

Trenkler, G. (1994). Characterizations of oblique and orthogonal projectors, *Proc. Int. Conference on Linear Statistical Inference, LINSTAT'93* (eds. T. Calinski and R. Kala), 255-270.

Toutenburg, H. (1992). *Lineare Modelle*, Physica-Verlag, Heidelberg.

Trench, W.F. (1964). An algorithm for the inversion of finite Toeplitz matrices, *J. ACM*, **16**, 592-601.

Van Huffel, S. and J. Vandewalle (1991). *The Total Least Squares Problem: Computational Aspects and Analysis*, Frontiers in Applied Mathematics Ser., Vol.9, SIAM, Philadelphia.

Van Huffel, S. and H. Zha (1993). The total least squares. In *Handbook of Statistics 9, Computational Statistics*, Elsevier Science Publishers B.V., 377-408.

Von Neumann, J. (1937), Some matrix inequalities and metrization of metric spaces, *Tomsk Univ. Rev.*, **1**, 286-299.

Yanai, H. and Y. Takane (1992). Canonical correlation analysis with linear constraints, *Linear Algebra Appl.*, **176**, 75-89.

INDEX

[Items are generally listed under broad headings like *inequalities, factorization of matrix, matrix types, matrix reduction, generalized inverses* and so on.]